4th International Conference on Electrical, Electronics, Informatics, and Vocational Education (ICE-ELINVO 2021)

Journal of Physics: Conference Series Volume 2111

Yogyakarta, Indonesia
5 October 2021

ISBN: 978-1-7138-5082-3
ISSN: 1742-6588

Printed from e-media with permission by:

Curran Associates, Inc.
57 Morehouse Lane
Red Hook, NY 12571

Some format issues inherent in the e-media version may also appear in this print version.

Institute of Physics
Dirac House, Temple Back
Bristol BS1 6BE UK

Phone: 44 1 17 929 7481
Fax: 44 1 17 920 0979

techtracking@iop.org

Additional copies of this publication are available from:

Curran Associates, Inc.
57 Morehouse Lane
Red Hook, NY 12571 USA
Phone: 845-758-0400
Fax: 845-758-2633
Email: curran@proceedings.com
Web: www.proceedings.com

TABLE OF CONTENTS

Preface

Peer review declaration

Electric Ducted Fan (EDF) Rocket Attitude Telemetry Using 2.4 GHz Radio Frequency 1
M F Alfatih, W Hancoyo

IoT-Based 3D Printer Development for Student Competence Improvement 9
Eko Prianto, Herlambang Sigit Pramono, Yuchofif

Design Smart Grid Hybrid in Faculty of Engineering Universitas Negeri Yogyakarta.................................... 20
Muhamad Ali, Bt Djoko Laras, Muhfizaturrahmah, P S Deny

LQR state feedback controller with precompensator for magnetic levitation system 28
A Winursito, G N P Pratama

A Simple Brushless Motor and Propeller Test Stand for Experiment from Home.. 36
Erwan Eko Prasetiyo

Application of LQR Full-State Feedback Controller for Rotational Inverted Pendulum 43
N Setiawan, G N P Pratama

The Application of Innovation Visual Animation Media (VAM) at Electrical in Automotive Vocational Education 51
Muhammad Nurtanto, Farid Mutohhari, Fadli Rozaq, Suyitno Suyitno, Alias Masek

Development of Hydraulic Cylinder Excavator Learning Media Based on Augmented Reality with Shapr 3D.. 61
A Setiyawan, L C Manggalasari, T A Prasetya, Towip Towip, W Noviansyah

Design and Performance Analysis of 600 W On-Grid Solar Power Plant System based on Maximum Power Point Tracking (MPPT) 70
Nurhening Yuniarti, Alex Sandria Jaya Wardhana, Bagas Woro Saputra

Development Mobile Application Android Based for Sensor and Transducer 76
Y I Hatmojo, S M Azis

Development and Effectiveness of Drone as a Learning Media in Islamic Boarding School 84
I W Djatmiko, S Yatmono, A C Nugraha

Effect of Mesh Topologies to Predict Cooling Effect on Evaporative Cooling Duct.. 94
A Jamaldi, Sarjito, A S Nurrohkayati, N T Atmoko

Fuzzy Time Series for Forecasting Railway Passengers in Indonesia.. 101
Muhammad Fatih Rizqon, Handaru Jati

Internet of Things-based Hydroponic: Literature Review 109
Alan Hadinata, Mashoedah

Stress Distribution Analysis of the Rectangular and Star-Blade for Plastic Crusher Machine Using Finite Element Analysis 117
Nalendro Mataram, Sigiet Haryo Pranoto, Rizqi Ilmal Yaqin, Anis Siti Nurrohkayati, Noer Aden Bahry, Nuruddin Wahyu Eko Saputro, Sabaruddin Syach

Oxygen Availability Control and Monitoring System in Hospitals using IoT 127
Nikko Aji Bayu Nugraha, Mashoedah

Information System for Disaster Mitigation Using Google Data Traffic 133
Handaru Jati, Muhammad Izzuddin Mahali, Satriyo Agung Dewanto

Experimental Analysis of Heat Transfer on the Tubular Tube Heater with a Variable of Twisted Tape Inserts and Wire Coil to Prevent the Icing Contaminated Tailplane Stall 144
G Marausna, F S Reza, F Jayadi

The Design of Mini CNC Assembly Simulator Based on Virtual Reality Toolkit (VRTK) 150
C A J Malik, S Hadi

Innovation Media Learning: Online Project-Based Learning (O-PBL) on Drawing Competence in Automotive Engineering Using Video on YouTube 157
Haryanto, Wahyu Mustafa Kusuma, Farid Mutohhari, Muhammad Nurtanto, Suyitno Suyitno

Utilization of Electronic Waste for Energy-saving Lamp Circuits 166
Zamtinah, I Mustaqim, H S Pramono

Analysis of the Impact of Distance and Temperature in Determining Transfer Power Using Laser Applications 175
I A Rahardjo, S Muslim, F Wadjdi, M Rifan, A Sunawar, M Subekti, Sakirah

Economic Analysis Development and Provider Elective Decision Model on Auto-ID Technology Investment 179
Anis Siti Nurrohkayati, Iwan Vanany

Design and Development of Water Distribution Monitoring System in Regional Drinking Water Companies (PDAM) Based On Internet Of Things 188
Efrizon, M Irmansyah, Era Madona, N Anggara, Yultrisna

Developing an engaging whiteboard animation video for vitamins 196
S Purwaningsih, A A Anggraeni

Microcontroller Based Portable Incubator Monitoring Tool with Short Message Service (SMS) notifications 204
Muhammad Irmansyah, Efrizon, Anggara Nasution, Era Madona

Development of Training Kit Basic Aerial Robotics Using RC Plane Robot for Robotics Course 213
F N Aziz, Mashoedah

Improving online learning interactivity with 3D virtual classroom models 221
R Wardani, H Jati, Nurkhamid, Y Indrihapsari, P Setialana, A M Budiyanto, M N Ardiansyah

Development of Mobile Augmented Reality Based Media for An Electrical Measurement Instrument 231
Y M Nursita, S Hadi

Utilization Extract, Transform, Load For Developing Data Warehouse In Education Using Pentaho Data Integration 239
A D Barahama, R Wardani

Evaluating UI in Digital Skills Training: Human Centered-Design approach from UI Sanbercode Bootcamp 247
Fiducia, Suprapto

Need Analysis of Interactive Multimedia Development for Virtual Learning in Pandemic 255
J F Lempas, S Soenarto, A D Barahama

Item Selection on The Moodle-Based Computerized Adaptive Test 262
Haryanto, Y Neng-Shu, S Hadi, M Ali, A F Husna, A S J Wardhana

Development of IoT-Based Smart Home Electrical Protection System 268
R Y Renaldo, Edy Supriyadi

IoT Enabled Ventilator Monitoring System for Covid-19 Patients 279
Mashoedah, Umi Rochayati, Indra Hidayatulloh, Arya Sony, Ferda Ernawan, Fardiansyah, Apri Nuryanto

Performance of Frequency Regulation on Power in the Development of Variable Speed Drive Training Kit as a Learning Media for Motor Control Practices 286
T H T Maryadi, A S J Wardhana, E S Damarwan, T Sukisno

Development of Android-based Mobile Application "Cyber Gen" for Genetic Counselling Implementation among Thalassemia Patients 294
Henri Setiawan, Doni Setiawan, Suhanda, Acep Hidayatul Mustopa

Redesign The Electricity System of PT Barata Indonesia To Reduce Annual Power Loss 303
Muhamad Ali, Zaenal Arifin

Mobile Augmented Reality Application for Component Identification 311
Y I Hatmojo, T H T Maryadi, R Badarudin, B Indrawati

Performance of Trainers kits for Industrial Automation Based on Programmable Logic Controllers 319
Sukir, Sa'Adilah, Hartoyo, Sunaryo, B P Saifullizam

Electric Machines Virtual Laboratory: Testing of DC Motor 329
R Badarudin, D Hariyanto, M Ali, S Meunmany

Development of Electrical Circuit Learning Media Using Virtual Simulation 339
E S Damarwan, M L Hakim, A S J Wardhana, N Kholis

The Design of Sensor and Transducer Learning Media based on Augmented Reality 348
D Hariyanto, Zamtinah, T Triasih, M S Wati, G K Kassymova

The Effects of Rain Intensity and Water Elevation with Energy Productivity of Kracak Hydro Power Plant, Bogor Regency – West Java 358
I A Rahardjo, Suyitno, M Subekti, W Gunawan, R P Wijayanto, I Mahir, Sakirah

Web-Based Recruitment Selection Application Development with Analytical Hierarchy Process (AHP) Method 364
G J T Maruanaya, R F Maruanaya, D Hariyanto

Face Tracking for Flying Robot Quadcopter based on Haar Cascade Classifier and PID Controller 374
A S Priambodo, F Arifin, A Nasuha, A Winursito

Smart Air Filter Based on Activated Charcoal and Aloe Vera Using Fuzzy Logic Algorithm 383
F Arifin, W R Gusti, A Nasuha, A S Priambodo, A Winursito

Design of Robust Heart Abnormality Detection System based on Wavelet Denoising Algorithm 391
A Winursito, F Arifin, A Nasuha, A S Priambodo, Muslikhin

Prototype Development Of Distance Detection System Based On The Internet Of Things Using Esp 8266 Wifi Nodemcu Module 398
Deny Budi Hertanto, Rustam Asnawi, Faranita Surwi, Nurman Setiawan

Utilization of Used Electric Vehicle Batteries for The Design of Mini-Generating Systems 403
Khairunnisa, Muhammad Afif Amalul Arifidin, Hartoyo

Development of Unmanned Aerial Vehicle (UAV) Fixed-Wing for Monitoring, Mapping and Dropping applications on agricultural land 413
M L Hakim, H Pratiwi, A C Nugraha, S Yatmono, A S J Wardhana, E S Damarwan, T Agustianto, S Noperi

Analysis of Energy Consumption and Energy Efficiency in Government Office Buildings of Bengkulu, Indonesia 420
I A Rahardjo, M Djaohar, M Subekti, Parjiman, I Zakir, I Mahir, Sakirah, A Lubis

The Effect of Inclination Angle of The Solar Panel on The Resulting Output Voltage 426
A S J Wardhana, E S Damarwan, Ml Hakim

Performance efficiency of virtual laboratory based on Unity 3D and Blender during the Covid-19 pandemic 432
M A Hamid, S A Rahman, I A Darmawan, M Fatkhurrokhman, M Nurtanto

Development of Ball Detection System with YOLOv3 in a Humanoid Soccer Robot 442
A C Nugraha, M L Hakim, S Yatmono, M Khairudin

Smart Infusion and Web Based Monitoring Infusion Fluids in Isolation Room Based on Fuzzy Logic 447
M Yamin, S A Habir, W O S Nur Alam, L Surimi

Author Index

Preface of the 4th International Conference on Electrical, Electronics, Informatics, and Vocational Education (ICE-ELINVO) 2021

D Hariyanto[1]

[1] Conference Chair of ICE-ELINVO 2021
[1] Department of Electrical Engineering Education, Faculty of Engineering, Universitas Negeri Yogyakarta, Indonesia

E-mail: didik_hr@uny.ac.id

ICE-ELINVO 2021 is the fourth international conference on Electrical, Electronics, Informatics, and Vocational Education organized by the Department of Electrical Engineering Education in cooperation with the Department of Electronics and Informatics Education, Faculty of Engineering, Universitas Negeri Yogyakarta. The 4th ICE-ELINVO 2021 was held virtually on October 5th, 2021 due to the Covid-10 pandemic and some of its restrictions issues. The main purpose of this conference is to bring together the professional practices, industrial experts, researchers, vocational teachers, academia, and decision-makers including the government to a unified communication forum for sharing their recent research and developmental activities in Electrical, Electronics, Informatics, Mechatronics, and Vocational Education.

This year's conference theme is the "INNOVATION ON RESEARCH IN ELECTRICAL, ELECTRONICS, INFORMATICS, MECHATRONICS, AND VOCATIONAL EDUCATION IN SUPPORTING SUSTAINABLE SOCIETIES". In the new normal era, developments and innovations in technology are more advanced with a variety of sophistication that can facilitate almost all of our work. Industry 4.0 technologies such as mobile internet, automation of knowledge work, the internet of things, cloud technology, advanced robotics, artificial intelligence, embedded systems, and autonomous vehicles, provide new opportunities for humans to support intelligent societies. This conference is initiated to contribute cutting-edge ideas to technology growth.

The conference model is a joint conference of four international conferences in the Faculty of Engineering, Universitas Negeri Yogyakarta: ICE-ELINVO, ICTVT, ICOVEMAT, and ICSI. This conference is divided into two sessions, including a plenary session and a parallel session. The plenary session features four inspiring keynote speeches delivered by Prof. Yu-Ren Wu, Ph.D. from National Central University, Taiwan (R.O.C.), Associate Professor Dr. Nirmal Nair from the University of Auckland, New Zealand, Hyungwook Choi from Korean National Agency for Administrative City Construction, Republic of Korea (R.O.K.), and Prof. Slamet PH, MA, MEd, MA, MLHR, Ph.D. from Universitas Negeri Yogyakarta, Indonesia.

There were more than 100 participants attended the conference. The committee received 84 papers of which 68 papers were eligibly sent to a peer-review stage. Just after a thorough reviewing process, there were 56 papers presented in the parallel session. All accepted papers have been then submitted to the Journal of Physics: Conference Series for publication. The participants came from Indonesia, Malaysia, Lao PDR, Japan, Kazakhstan, and Germany.

The success of the ICE-ELINVO 2021 is due to the support of many parties. For that, we would like to express our utmost gratitude to the invited speakers, presenters, authors, and co-authors for contributing excellent papers. We are also grateful to all the technical program committee members involved in evaluating submitted papers, considering the merit through detailed reviews, and selecting papers of the highest quality for oral presentation. We would like to thank also to the organizing committee members and all other committees for their endless dedication. We hope it will have a fruitful and great value to everyone.

Peer review declaration

All papers published in this volume of **Journal of Physics: Conference Series** have been peer reviewed through processes administered by the Editors. Reviews were conducted by expert referees to the professional and scientific standards expected of a proceedings journal published by IOP Publishing.

- **Type of peer review: Single-blind / Double-blind / Triple-blind / Open / Other**
  In our conference, we do a single-blind peer review where the author does not know who the reviewers are. Submitting papers, editor decisions, review processes were conducted through our system http://ice-elinvo.uny.ac.id/. Reviews were conducted by expert referees to the professional and scientific standards.

- **Describe criteria used by Reviewers when accepting/declining papers. Was there the opportunity to resubmit articles after revisions?**
  Six criteria must be met by the papers that are included in the ICE-ELINVO 2021 international conference. All reviewers must refer to these six criteria.
  1. Is the paper relevant to the conference theme?
  2. Does it contribute to the theoretical or practical domains?
  3. Does it state the problem statement?
  4. Does it comprise an appropriate research methodology or proposed method?
  5. Does it describe indicative results?
  6. Does it fulfill the requirements of the paper template?

  Each criterion has a value from the range 1-4. The reviewer will evaluate each criterion, and then the system will add up all the scores for each criterion. The maximum total value is 24.
  The next stage is that the reviewer will send back the paper reviewed to the author through the system. Finally, authors are given two weeks to correct and resubmit.

- **Conference submission management system:**
  http://ice-elinvo.uny.ac.id/
  (our University has our own management system for reviewing articles)

- **Number of submissions received:**
  84 papers

- **Number of submissions sent for review:**
  68 papers

- **Number of submissions accepted:**
  56 papers

- **Acceptance Rate (Number of Submissions Accepted / Number of Submissions Received X 100):**
  66.67%

Journal of Physics: Conference Series **2111** (2021) 011002 doi:10.1088/1742-6596/2111/1/011002

- **Average number of reviews per paper:**
  2

- **Total number of reviewers involved:**
  48

- **Any additional info on review process (ie plagiarism check system):**
  Each manuscript submitted to the ICE-ELINVO 2021 international conference was checked for plagiarism using Turnitin.

- **Contact person for queries:**
  Muhammad Luthfi Hakim, Department of Electrical Engineering Education, Faculty of Engineering, Universitas Negeri Yogyakarta, Indonesia.
  Email: luthfihakim93@uny.ac.id

Electric Ducted Fan (EDF) Rocket Attitude Telemetry Using 2.4 GHz Radio Frequency

M F Alfatih[1] and W Hancoyo[2]

[1]Aerospace Engineering, Sekolah Tinggi Teknologi Kedirgantaraan Yogyakarta
[2] Department of Electrical Engineering, Diponegoro University

E-mail: faiz.alfatih@sttkd.ac.id

Abstract. On a rocket, control requires monitoring to determine its position and attitude. Rocket stance includes roll, pitch, and yaw angles. Monitoring can be done manually without the help of tools while the object's position is still within reach. It is difficult to carry out monitoring when the object is moving far and beyond the reach of the operator. This observation requires a monitoring system to monitor the object further. This research applies a telemetry system to a target tracking rocket. The measurement results from the IMU and GPS sensors on the rocket body will be processed by the ATMega 2560 microcontroller and sent via a 2.4 GHz RF signal. to Ground Control Station. The rocket telemetry data that has been sent to the GCS can be viewed by the operator through the GUI system in a computer program using the C# language in Visual Studio. Tests on latitude and longitude are carried out by tracking the trajectory of objects while testing for roll, pitch, and yaw angles is carried out by placing objects according to the reference angle. The results of the position obtained the values of altitude, latitude, and longitude of 0.5364 m, 0.000012°, and 0.000023°, respectively. An attitude tests for roll position, pitch, and compass heading have values of 0.35°, 0.07, and 2.90°, respectively. The telemetry data transmission distance test is still well-received at a distance of 200 with a test speed of 70 KM/hour.

Keywords: *Telemetry, GPS, IMU, Attitude, and Position*

1. Introduction

Moving objects such as control rockets require monitoring to determine the position and *attitude of the* object. *Attitude* includes *roll, pitch,* and *yaw* angles. Monitoring can be done manually without the aid of tools when the object's position is still within reach. It is difficult to carry out monitoring when objects are moving far away and are out of reach of the operator. This monitoring requires a monitoring system to further monitor the object [1].

One method that can be used to determine the position and *attitude of an* object is by combining *Global Positioning System* (GPS) technology and inertial sensors which are incorporated in the Inertial Navigation System. GPS is a radio navigation and positioning system [2] using navigation satellites owned and managed by the United States Department of Defence. An inertia sensor is an electronic device that can measure and detect angular levels using a combination of an accelerometer, gyroscope [3], and magnetometer or a compass sensor. Accelerometer, gyroscope, and magnetometer data can be combined so that the resulting angle has a better value than the angle on each sensor. Combination of sensors can be done using complementary filters [3].

Research on *seekers* and rocket control technology that has been carried out by Ahmad Dahlan University in collaboration with LAPAN [4] and BATAN [5]. This *seeker* system uses a PbS infrared detector which can capture radiation at a wavelength of 1.8 - 2.6 μm. This research has not used image processing technology [4].

Research conducted by Muhammad Hanifudin Al Fadli from Diponegoro University, used the IMU sensor for the control rocket's attitude telemetry which can measure the angle of the control rocket in the range of ± 15 ° with an average error of 0.108° for the x-axis and 0.137° for the axis. y. In this study, the results of telemetry-sensing rocket attitude were not discussed specifically because it was more specialized in research on the stability of the rocket control using the PID control method [5]. Research by Meita Sukma Listiyana from Diponergoro University. This study applies a telemetry system for UAV object attitude data acquisition using radio signals with a frequency of 433 MHz. In this study, the results of the maximum transmission range between the antenna *transmitter* and *receiver* as far as 125 meters [6].

Based on these problems and some research references that have been carried out, we conducted research on the radio frequency control rocket telemetry system. This rocket is also equipped with a long-range measurement system with a frequency of 2.4 GHz which allows the operator to monitor the rocket's attitude and position. The operator monitors the operation of the rocket on the GUI (*Graphical User Interface*) on the GCS (*Ground Control Station*).

2. System Overview and Design

2.1. System Flow Chart

The design of the software at the reception of telemetry data is done by designing a Graphical User Interface. The purpose of making GUI in this study is to monitor or monitor the position and attitude of objects in a place by the operator within a distance that cannot be reached. The GUI is created using the Visual Studio application using the C# programming language. The flow chart of the GUI display for the control rocket parameter data acquisition system is shown in Figure 1.

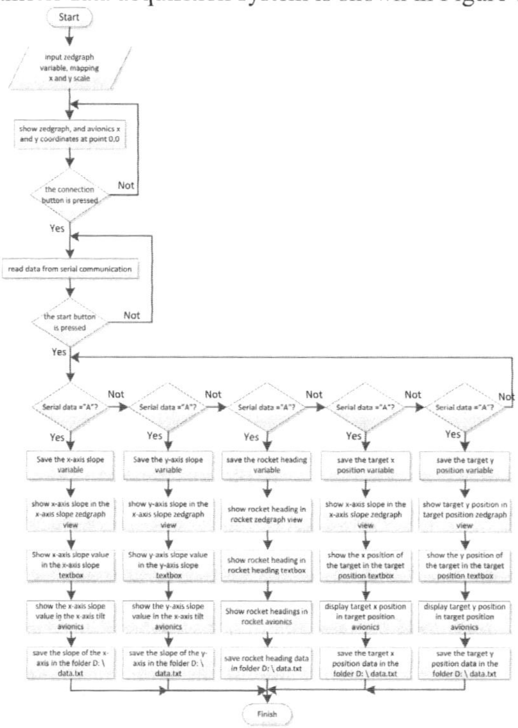

Figure 1. Flow Chart of the Telemetry Program

 2111 (2021) 012001 doi:10.1088/1742-6596/2111/1/012001

2.2. Hardware Design

In telemetry communication, the Atmega2560 microcontroller is designed to receive serial data from the Raspberry Pi, read data from the IMU GY-91 and GPS sensors with target coordinate data in the form of x, y from the frame, roll angle to the x axis, pitch angle to the y axis, yaw axis about the z axis.

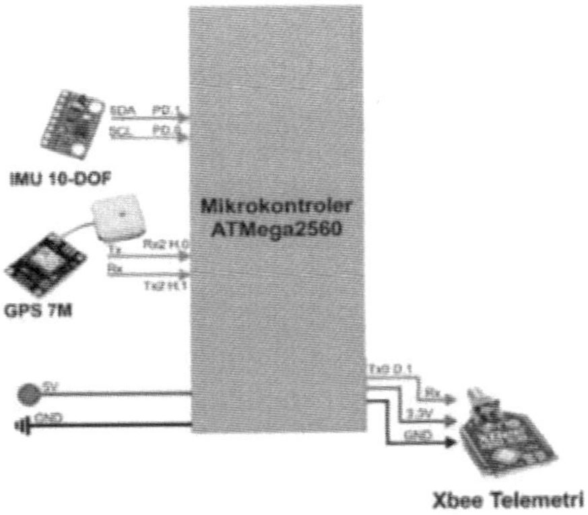

Figure 2. Telemetry data transmission device.

The schematic of the telemetry data transmission hardware circuit can be seen in Figure 2. Data from the GPS sent in the form of latitude coordinate values (latitude) and longitude (longitude).

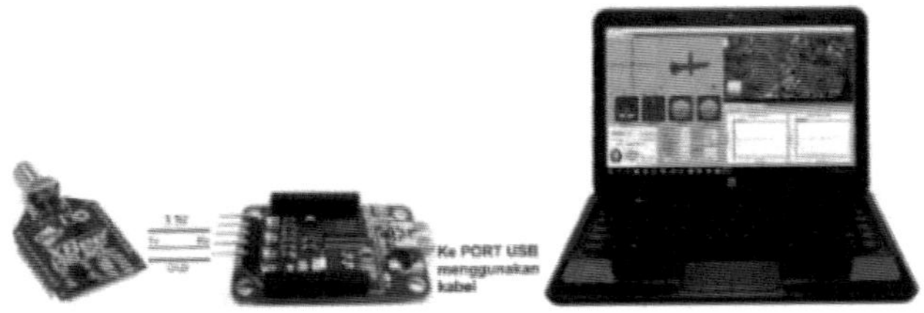

Figure 3. Telemetry data receiving device.

The schematic of the telemetry data reception hardware circuit can be seen in Figure 3. The control rocket data acquisition software is used to view the data sent by the rocket control which will then be displayed in a real time GUI display. The data is displayed in the form of a 2-dimensional avionics system and in the form of graphic data.

3. Result and Discussion

3.1. Altitude Testing

Altitude test results are carried out statically. Table 1 is a table of the results of the altitude test at three test points at the Diponegoro University stadium.

Table 1. Altitude Test Result

No.	Altitude Reff (m)	Measured Altitude (m)	Error (m)
1	200	200,516	0.516
2	203.86	204.44	0.58
3	205.66	206,346	0.686
4	209.26	209,698	0.438
5	214.06	214,522	0.462
Average error			0.5364

Altitude testing by static means to get an average error value of 0.5364.

3.2. Testing latitude and longitude.

3.2.1. Static Testing

Static testing was carried out at three Geodetic Benchmark points in the Diponegoro University campus area. Bench mark I is located at the coordinates 7.049572S, 110.439143E. The test at point BM I was carried out by placing the object right on the BM GD-17 peg. Static GPS testing at Bench mark I point gets an error value at latitude of 0.000012° and longitude of 0.000023°. Every 1° implements a distance of 111,322 Km according to the length of the earth's equatorial line of 40,070 KM. The latitude error value is 1.335864 meters, and the longitude error is 2.560406 meters.

The benchmark II point with the code BM UNDIP GD 15 is located at coordinates 7.050281S, 110.438812E. Static GPS testing at the bench*mark* II point gets an error value at latitude of 0.000016° and longitude *of* 0.000032°. Every 1° implements a distance of 111,322 Km according to the length of the earth's equatorial line of 40,070 KM. The latitude error value is 1.781152 meters, and the longitude error is 3.56234 meters.

The benchmark III point with the code BM UNDIP GD 01 is located at coordinates 7.050281S, 110.438812E. Static GPS testing at the benchmark III point gets an error value at latitude of 0.000005° and longitude of 0.000009°. Every 1° implements a distance of 111,322 Km according to the length of the earth's equatorial line of 40,070 KM. The latitude error value is 0.55661 meters, and the longitude error is 1.001898 meters.

3.2.2. Dynamic Testing

Dynamic GPS testing is carried out by tracking the coordinates of the object's displacement route from three Geodetic Benchmark points in the Diponegoro University campus area. Tracking starts from point BM III, namely with the code BM UNDIP GD-01. The path that is traversed starts from the coordinates -7,053634, 110,439628 to the coordinates 7.050281S, 110.438812E. The results of dynamic GPS testing with tracking can be seen in Figure 4.

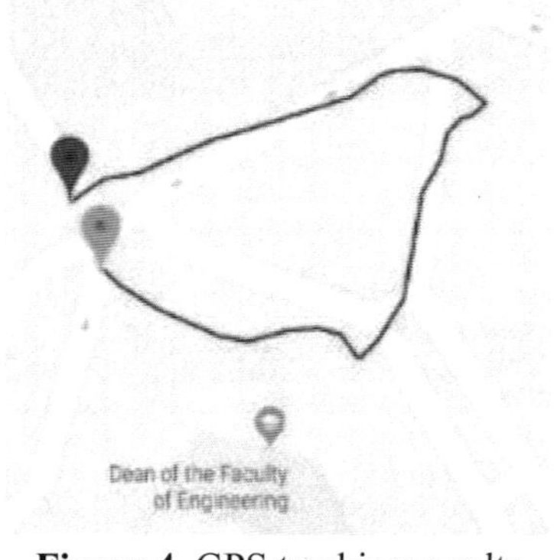

Figure 4. GPS tracking results

In Figure 8, the tracking for testing latitude and longitude data is in accordance with the path traversed by the object, but there are still differences in the results of the tracking test. Based on the test data in Appendix A, the average error obtained is 0.000012° and the average error at longitude is 0.000014°. The error in the distance is obtained the latitude 1.335864 meters error and the longitude error 1.558508.

3.3. Testing at Roll Angle (X).

The roll angle test is carried out by varying the angle on the x-axis. The error value of the test results is obtained by absolute reduction of the reference point value with the object angle. The roll angle test results in an unstable graph, so that the value taken is the angle with the value farthest from the reference angle. The test results can be seen in Table 2.

Table 2. Testing The Roll Angle on The X-Axis

No.	Reference Angle (°)	Rocket Angle (°)	Error (°)
1	-90	-90.12	0.12
2	-60	-60.68	0.68
3	-45	-42,732	0.78
4	-30	-27,824	0.17
5	0	0.216	0.216
6	30	29,342	0.27
7	45	43,644	0.22
8	60	57,796	0.17
9	90	87,676	0.85
Average error			0.35 °

From table 2 it can be seen that the 90° roll angle test has the farthest error value of 0.85°, the 0° roll angle test has the smallest error value of 0.06° and the average roll angle test error is 0.35°.

3.4. Testing at The Pitch Angle (Y)

Pitch angle testing is done by varying the angle on the y-axis. The error value of the test results is obtained by absolute reduction of the reference angle value with the object angle. The pitch angle test results in an unstable graph, so that the value taken is the angle with the farthest value from the reference angle. The test results can be seen in table 3.

Table 3. Testing The Pitch Angle on The Y-Axis

No.	Reference Angle (°)	Rocket Angle (°)	Error (°)
1	-90	-89.58	0.42
2	-60	-58.75	1.25
3	-45	-45.51	0.51
4	-30	-31.32	1.32
5	0	0.07	0.07
6	30	30.11	0.11
7	45	45.69	0.69
8	60	60.20	0.20
9	90	90.64	0.64
Average error			0.61

From table 3 the pitch angle test of -30° has the farthest error value of 1.32°, the 0° pitch angle test has the smallest error value of 0.07° and the average error value of the pitch angle test is 0.07°.

3.5. Testing at The Compass Heading Angle (Z)

Testing the compass heading angle on the rocket is done by aligning the object with a smartphone equipped with a compass application. The test is carried out with a variety of angles ranging from 0° to 6 times with a multiple test angle of 60°. The results of the object angle test against the compass can be seen in Table 4.

Table 4. Testing the Heading Angle on The Z-Axis of the North Magnetic Pole Of The Earth

No.	Reference Angle (°)	Rocket Angle (°)	Error (°)
1	0	1.30	1.30
2	60	59.40	0.60
3	120	121.50	1.50
4	180	182.52	2.52
5	240	234.65	5.35
6	300	306.09	6.09
Average error			2.90

From table 4 in the heading test for the compass angle of 300°, the farthest error value is 6.09°, the heading test for the compass angle of 60° has the smallest error value of 0.60° and the average error for the heading test for the compass angle is 2.90°.

3.6. 2.4 GHz Radio Frequency Data Transfer Test

3.6.1. Static Testing

Static testing is done by placing the object at a test distance. The results of the data transfer test results on the position and attitude of the object using a 2.4 GHz radio frequency. The test results show that the position and *attitude* data of the object can only be received well up to a distance of 200 meters. Position and attitude data of objects are not well received at distances exceeding 250 meters

3.6.2. Dynamic Testing

The dynamic test is carried out by carrying the object using a motorcycle at the test speed. This test is also carried out using a test point for the distance between the transmitter and receiver. The test results of the transfer of data on the position and attitude of objects using a 2.4 GHz radio frequency dynamically are shown in Tables 1 to 4.

Table 5. Testing Receiving Data with Variations in Object Speed

No	Motorcycle Speed	Receiving Data
1	20 Km/h	good
2	40 Km/h	good
3	50 Km/h	good
4	60 Km/h	good
5	70 Km/h	good

The test results on the object show that. The data sent by the rocket in the form of attitude and position can be well received at the tested speed variations.

3.7. GUI System Testing

The test was carried out in the area of the Diponegoro University Multipurpose Building. The test data is done by observing the coordinates of the location, 2-dimensional visuals and graphs of roll, pitch, and yaw. Figure 14 shows the position and attitude data of the object in that place.

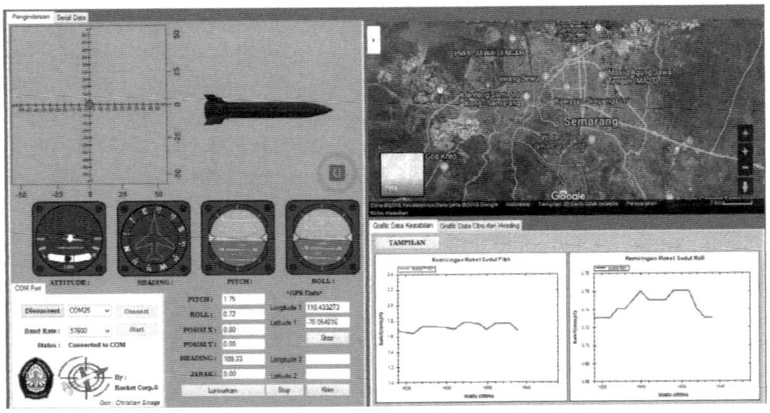

Figure 5. GUI rocket position and attitude.

Based on the examiner who shows the rank and attitude of the condition object in the multipurpose building, Diponegoro University, it can be monitored properly. The data obtained from this height are as follows:

1) The latitude and longitude values are -7.054016, 110.433273
2) The value of the roll angle obtained is 0.72 ° against the x-axis.
3) The pitch angle value obtained is 1.75 ° to the y-axis.
4) The yaw angle value obtained is 109.33 ° to the compass angle.

4. Conclusion

Based on the tests and analyses carried out in the analysis, it can be concluded that the altitude test results statically get an average error value of 0.5364 meters. Static latitude and longitude test results data received from GPS have an average error value at latitude 0.000012° and longitude 0.000023°. Data from the dynamic latitude and longitude test results received from GPS have an average error value at latitude 0.000011° and longitude 0.000021. The attitude test for roll position has an average error value of 0.35°. The attitude test for pitch attitude has an average error value of 0.07°. The attitude test for the compass heading position has an average error value of 2.90°. The transmission range between the transmitter and receiver antenna is as far as 200 meters. Testing the transmission range of the telemetry antenna shows that until the object speed reaches 70 km/h or equivalent to 19.4 m/s, data can be received well at a distance of 200 meters. The object position and attitude data acquisition system has been successfully designed and the sensor reading value is still within the designed tolerance.

5. References

[1] AA Farghani, R. Sumiharto and SB Wibowo, "Prototype of Ground Control Station for Observation and Control of Fixed Winged Unmanned Aerial Vehicles," Indonesian Journal of Electronics and Instrumentation Systems (IJEIS), vol. 3, no. 1, pp. 1-10, 2013.

[2] A. Budiwati, "Sticks Istiwa ', Global Positioning System (GPS) and Google Earth to Determine Earth's Coordinating Points and Its Application in Determining the Direction of the Qibla," Al-Ahkam, vol. 26, no. 1, pp. 65-92, 2016.

[3] MA Akbar, Wahyudi and A. Hidayatno, "Design of Angle and Position Detection Software using Atmega 32 Microcontroller," Transmission, vol. 16, no. 1, pp. 49-53, 2014.

[4] Dani T, Priyatikanto R, Supriadi S, Rachman, Q Abdul, Amrullah A." Pengaruh Aktivitas Matahari Terhadap Kerapatan Atmosfer Atas Berdasarkan Data Posisi Gps Satelit Lapan-a2 (Solar Actvity Influence on the Upper Atmospheric Density Derived From Gps Data on Lapan-a2 Satellite)" 1412-808X, 2018.

[5] Widodo S, Abimanyu, Adi A, Rosisk, "Development of drone mounted aerial gamma monitoring system for environmental radionuclide surveillance in BATAN" 1742-6588, 2020.
[4] DHS Gunawan S. Prabowo, "Initial Prototype Design of IR Seeker in Missile Systems: Characterization, Modeling, Early Prototyping," in *Insinas* , 2012, pp. 118–122.
[5] Al Fadli, Muhammad Hanifudin, Munawar Agus Riyadi and Budi Setiyono. "Design of Infrared Guided Full Rocket System Using Simulated Image Processing Method in a Wind Tunnel." *Transient* 7.1 (2018): 152-159.
[6] Listiyana, Meita Suka, Sudjadi and Wahyudi. "Designing a Data Acquisition System for the Position and Attitude of Objects" *Transient* 7.1 (2018). 118–122.
[7] A. Beckett, "Airborne Infrared Target Tracking with the Nintendo Wii Remote Sensor," Texas A&M University, 2012.
[8] DJ Frarr, "Control of Missile Airframes," 1979.
[9] M. Rif'an, W. Djuriatno, N. Sulistiyanto, P. Siwindarto, M. Aswin and V. Nurdinawati, "Utilization of the 3 axis Gyroscope L3G4200D for rocket payload angle measurement," EECCIS, vol. 6, no. 2, pp. 177-182, 2012.
[10] YA Pramana, "Implementation of Microcontroller-Based Accelerometer, Gyroscope and Magnetometer to Display Object Position Using Inertial Navigation System (INS)," UNIKOM ELektro Technique, pp. 110, 2013.
[11] T. Ozyagcilar, "Implementing A Tilt-Compensated eCompass Using Accelerometer and Magnetometer Sensors," Freescale Semiconductor Application Note, vol. Document Number: AN4248, no. Rev. 4, pp. 1-22, 2015.
[12] R. Nurfansyah, Wahyudi and B. Setiyono, "Estimation of Object Orientation Angle Using 6 DOF IMU Sensor and 3 Axis Magnetometer Sensor," Transient, vol. 2, no. 3, pp. 770-775, 2013.
[13] DD Quoc, J. Sun, VN Le and NN Tan, "Sensor Fusion based on Complementary Algorithms using MEMS IMU," International Journal of Signal Processing, Image Processing and Pattern Recognition, vol. 8, no. 2, pp. 313-324, 2015

Journal of Physics: Conference Series **2111** (2021) 012002 doi:10.1088/1742-6596/2111/1/012002

IoT-Based 3D Printer Development for Student Competence Improvement

Eko Prianto[1], Herlambang Sigit Pramono[2], Yuchofif[3]

[1, 2, 3] Electrical Engineering Education Department, Universitas Negeri Yogyakarta

Email : eko_prianto@uny.ac.id

Abstract. This study aims to develop an IoT-based 3D Printer as a CNC Maintenance and Repair Learning Media, identify competencies that can be supported by an IoT-based 3D Printer Learning Media, and find out the implementation of 3D Printers in learning to improve student competence in the Mechatronics Engineering Education Study Program. The method used in this research is Research and Development using the ADDIE model, namely Analyze, Design, Development, Implementation, and Evaluation. The results of this study obtained an IoT-based 3D Printer machine that has been tested both in terms of function and feasibility which can be used in learning to improve the competence of students of the Mechatronics Engineering Education study program and the competencies supported by the learning media.

1. Introduction

Various challenges are faced by the world is facing challenges in the era of the fourth Industrial Revolution (Industry 4.0). New technologies and approaches that combine the physical and digital worlds will fundamentally change human civilization today. Humans are on the threshold of a technological revolution that will fundamentally change the way of life, work, and relate to one another. In addition, entering the technological era as well as the fourth industrial revolution led to changes in various sectors, including in the world of work and industry.

Educational institutions have the role and responsibility to act as quickly as possible to design action plans to prepare future people to face new challenges. Educational institutions need planning for action in different directions and different fields, such as laboratories, teachers, and curriculum [1].

Education has the task of organizing an educational process to prepare experts for the Industry 4.0 paradigm. The Industry 4.0 paradigm means creating an industry with digital informative technology and functions automatically. To be able to work in the industry, education must prepare a supportive competency profile. The scientific content of educational programs must contain the nature of interoperability for all technologies used together in digital production [2].

Graduates must have the appropriate competencies in meeting the needs of the current Industrial Revolution 4.0 Era. Industry 4.0 is known as a term for digitizing the production process of goods and services where decisions are made without involving humans by making computers connected and communicating with each other. The application of industry 4.0 is currently carried out in the fields of: Internet of Things (IoT), robotics, smart factories, 3D printing, routers to innovations in goods and service production systems that are integrated with information and computer technology (ICT).

In adapting to the era of the industrial revolution 4.0, the world of education requires major changes. Curriculum, learning, environmental, social, and cultural changes in the school environment need to be made in adapting to this era. To support educational attainment in the era of industrial revolution 4.0, facilities and infrastructure are needed in learning, one of which is practicum equipment. One of the factors that have a strong influence on the output in education is the fulfillment of the needs of school facilities used in learning such as practicum equipment.

Practical learning must be reproduced to achieve appropriate graduate competencies in the Industrial Revolution 4.0 era. Several supporting factors must be met to achieve success in practical learning in the Mechatronics Engineering Education study program. One of these factors is the facilities and infrastructure, especially in the field of learning media or practical equipment. The practicum equipment used in learning must support and adapt to the needs in the era of industrial revolution 4.0 with the characteristics of working automatically, 3D printing, working based on the internet of things (IoT), and Data of Things. According to Shuning Li [3], IoT and 3D printers are two important new technologies, these technologies are increasingly impacting many fields of industry and daily life. Therefore, students need to be introduced to this technology and get ready for future career opportunities.

One of the practical pieces of equipment that works with the principle as mentioned above is a 3D Printer machine. According to Assante et al [4], currently, the use of 3D printers is increasingly widespread as a supporting tool in education, so it is necessary to have special skills and competencies in using this technology. Günther [5], stated that the obstacle in implementing the practical use of digital technology, in this case, 3D printers, for students is the lack of the necessary equipment and technical knowledge in many universities. The 3D Printer design that will be developed has been equipped with modern computing technology. This IoT-based 3D printer learning media aims to make students not only skilled in operating a tool but can design and program to make an object.

1.1. 3D Printer Technology

3D Printer is a medium that forms a component using layers of material to create objects in 3 dimensions. 3D Printers can create physical objects from geometric representations by sequentially adding materials.

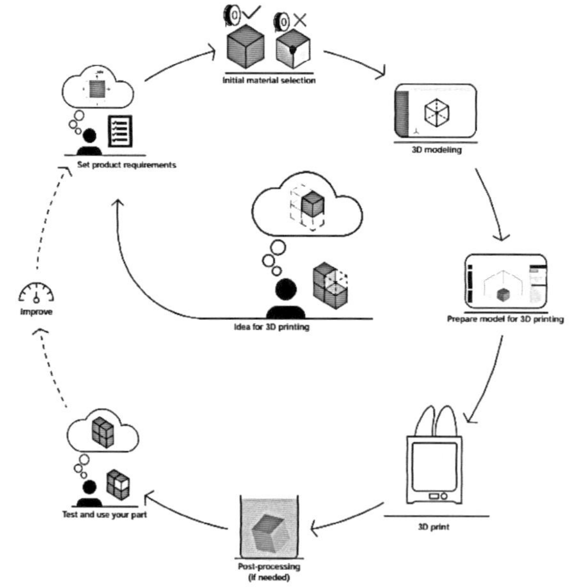

Figure 1. The work process in 3D Print technology [6]

3D Printer technology is derived from the technology of layer-by-layer structural formation of objects in three dimensions (3D) directly from computer-aided design (CAD) drawings. Conventional thermoplastics, ceramics, graphene-based materials, and metals are materials that can now be printed using 3D Print technology. 3D Print technology has the potential to revolutionize the industry and transform production lines. Production speed will increase and costs will be reduced further with the adoption of 3D printing technology. So that further consumer demand will have a greater influence on production. The process in 3D Print can be seen in Figure 1.

At present, 3D printing technology is increasingly being used for mass customization, production of all kinds of open-source designs in the field of agriculture, health care, automotive industry, locomotive industry and aviation industries [20].

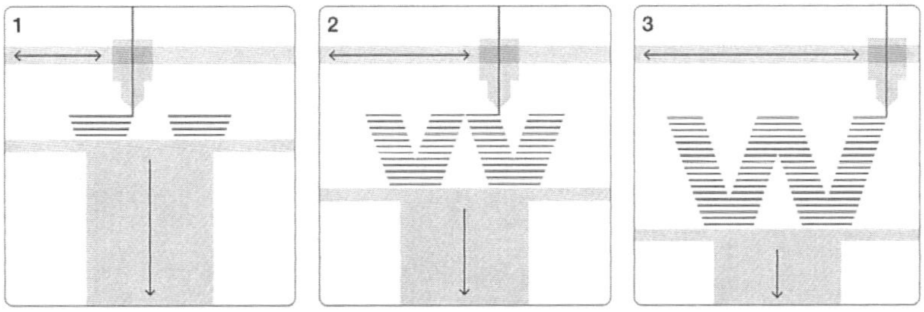

Figure 2. Formation of workpieces from layer to layer [7]

Shahrubudin [20] state that according to ASTM Standard F2792, ASTM categorizes 3D printing technology into seven groups, namely binding jetting, directed energy deposition, extrusion material, jetting material, powder bed fusion, sheet lamination, and vat photopolymerization. Currently, 3D printing technology is no longer limited to the use of prototypes but is increasingly being used to make various products. 3D Print technology is capable of producing components in a variety of materials, namely ceramics, metals, polymers, and their combinations in the form of hybrids, composites, or functionally graded materials (FGMs).

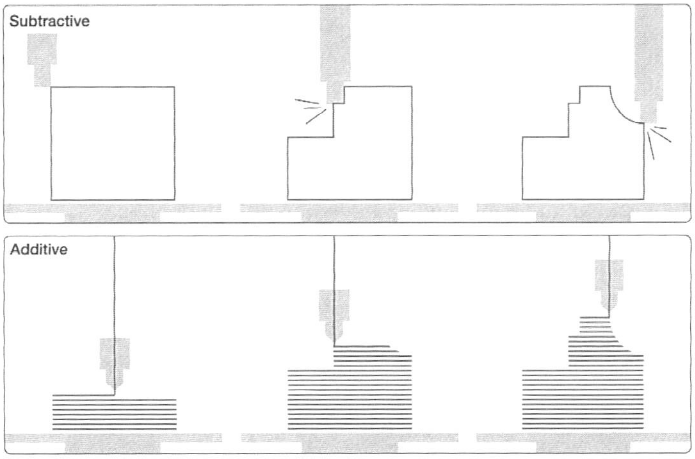

Figure 3. Difference between subtractive (CNC) and additive (3D Printer) technology [8]

A 3D printer machine consists of components, namely Motherboard, Frame, Motion Controller, Power Supply Unit, Print Material, Feeder System, Extruder, Print Bed, Connectivity, Interface. The motherboard serves as the brain of the 3D printer which directs the motion components according to the instructions sent from the computer and at the same time, interprets the signals from the sensors. The frame serves to unite all the components of the 3D printer and maintain stability in its operation. Motion controller/motion controller will receive instructions from the motherboard about the movement to be done, this component consists of a belt, stepper motors, threaded rods, and end stops.

The Power Supply Unit serves to provide power for the smooth operation of the 3D printer components.

Print material in the form of filament is used for FDM 3D printers. Filaments are available in coils. The filament is heated to a certain temperature and melted for printing on the print bed. There are two feeder systems in the feeder system which are most commonly used in 3D printer components, namely the Bowden system and the direct system. An extruder or also known as a print head extrudes the filament and prints it on the print bed. The print bed is the place where the product/design will be made. In terms of 3D printer connectivity, some 3D Printers only provide an ethernet or USB port for connection. However, nowadays there are many 3D printers available with Wi-Fi settings. Most 3D printers come with an LCD user interface.

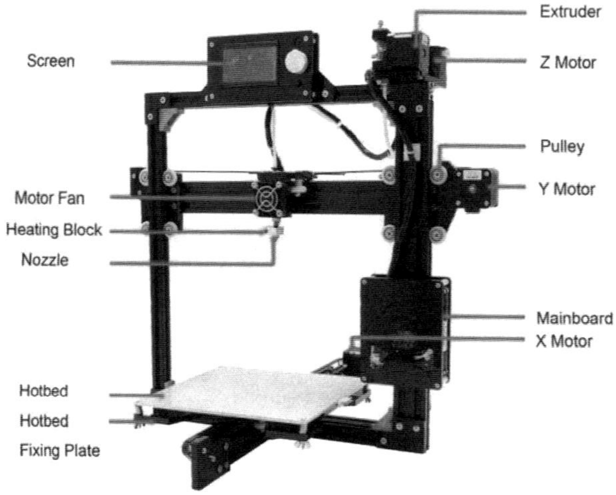

Figure 4. 3D Printer Parts [9]

1.2. Internet of Thing

The era of the industrial revolution 4.0 has used Cyber-Physical Systems, in other words, the industry has used the internet and big data for all processes in the industry. With the Revolutionary Era 4.0, the industry produces a smart factory where everyone uses the Internet of Things (IoT). IoT is a concept such as industrial machines, electric generators, vehicles, household appliances, to wearable devices that are interconnected through networks to exchange in real-time [10]. There are four principles in Industry 4.0, namely: (1) the ability of machines, devices, sensors, and humans to connect and communicate with one another through IoT; (2) the ability of information systems to create a physical world into a virtual world; (3) the ability of information systems to assist humans by gathering and visualizing comprehensive information and performing a range of tasks that are too strenuous or unsafe for humans; and (4) the ability of cyber systems to make their own decisions and perform tasks independently [11].

According to Handayani [12], the Internet of Things is the next evolution of the Internet where it takes a big leap in its ability to collect, analyze, and distribute data that turns into information, knowledge, and wisdom. The purpose of the Internet of Things is to connect devices to one another via the internet in the hope that the system can help people in doing a task or work [13].

Internet of Things also allows users to manage and optimize electronic and electrical equipment connected to the Internet [12]. Meanwhile, according to Hidayatulloh [14], the Internet of Things is a concept that aims to expand the benefits of continuously connected internet connectivity. This means that the Internet of Thing utilizes a programming argument in which every command of an argument produces an interaction and communication between machines that are connected automatically through the internet.

2. Research Methodology

A design developed by Mohamad Hasan Bin Tasneem and Gamal Talal Amer [15]. 3D Printers use BLDC motors because the accuracy and efficiency are much higher than the current use of stepper motors. In addition, it uses a ball screw instead of the main screw as it provides smoother, quieter, and more efficient movement.

Based on research conducted by Ishtiaq Ahmed, et al [16], 3D Printer is an additive manufacturing technique in which 3D objects are printed with the help of CAD (Computer-Aided Design) software. The process adopted is FDM technology, in which different materials such as PLA (polylactic acid), ABS (acrylonitrile butadiene styrene), HIPS (high impact polystyrene), etc. By heating one of the filament materials to its melting point and carried out layer by layer.

The results of research conducted by Jun Zhao [17], new technology in the design of 3D Printers use nozzles of different diameters in different stratification heights respectively on the printing pressure surface, internal support, and auxiliary support of the workpiece, and effectively improves printing speed and precision molding of the workpiece surface, with different wire materials, at a suitable temperature range to ensure the extrusion mechanism on the stability of the filament, and improve the shaping precision of the workpiece.

The research was conducted using the Research and Development method. This research was carried out based on the ADDIE model, namely Analysis, Design, Development, Implementation, and Evaluation. At the Analysis stage, the identification of what competencies are needed by students in terms of the design and operation of IoT-based 3D Printers. This is intended to see the extent of the specifications of the 3D Printer that will be made to support these competencies. The Design stage is carried out by designing an IoT-based 3D Printer according to the needs at the analysis stage. This design is carried out in the form of hardware and software. The results of the 3D Printer design that will be used are expected to be close to industry standards.

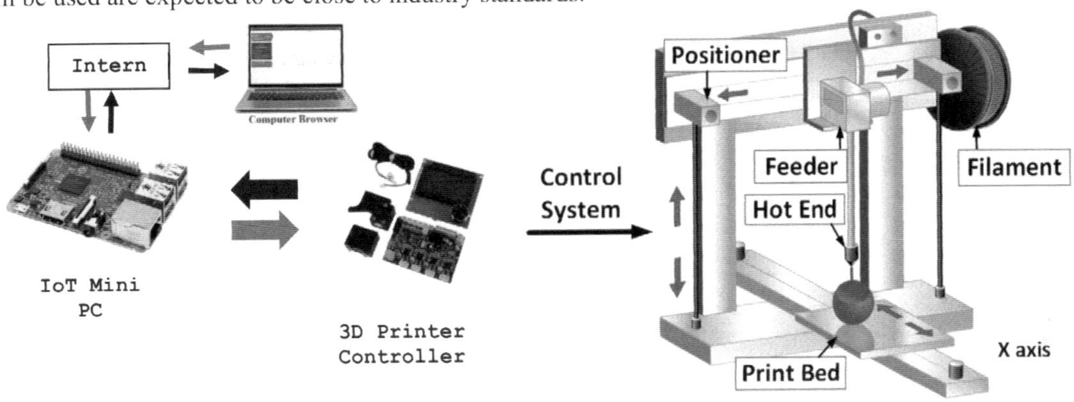

Figure 5. Initial design of an IoT-based 3D Printer system

The development stage is carried out by making the 3D Printer starting from assembling the hardware, then installing and connecting the software to the assembled hardware. Settings are also made on the software so that it can communicate with the hardware that has been prepared and can be controlled via the internet network. At this stage, validation of the media that has been made is carried out. The validation is carried out by experts and lecturers who teach courses that are following the objectives of this research. At the implementation stage, the implementation of the CNC maintenance and repair practice course is carried out using a 3D Printer. The selection of classes as respondents, step-by-step implementation of learning, and getting feedback from students and lecturers are carried out in this process. The evaluation stage is carried out to obtain improvements from the implementation that has been carried out. Improvements in hardware, software, lesson plans, and user manuals are carried out in this stage.

This study has the ultimate goal of designing a learning media in the form of an IoT-based 3D Printer to improve student competence in the Mechatronics Engineering Education study program. It is expected that the results of this study can be used by the study program and develop competencies of students, especially in terms of design, installation, programming, and control of a 3D Printer.

Data collection was carried out in two ways, namely literature studies related to the competencies needed by students in terms of design, installation, programming, and control of IoT-based 3D Printers and questionnaires to validate the results of the designs that had been implemented. Performance testing of the tools that have been made is also carried out in the data collection process by taking tool parameters that will produce specifications and tool manuals.

3. Discussion on the Results of the Redesign

The analysis stage is carried out by identifying what competencies are needed by students of the Mechatronics Engineering Education study program related to the operation and maintenance of 3D printers as one of the parts studied in the CNC maintenance and repair course. This analysis phase is carried out by studying literature obtained from various sources related to the operation of 3D printers for learning. The curriculum compiled by Erasmus+ supported by the European Commission [18] presents learning outcomes and a sequence of activities designed to learn 3D Printers which consist of; 1) Technical Basic, 2) Introduction to 3D Printing, 3) 3D Printing in education, 4) Design and/or print 3D objects, 5) Overview of CAM Processes, and 5) Set up a 3D Printer. In the curriculum of the UNY mechatronic engineering education study program, it has been studied regarding the introduction of 2D and 3D designs in the Engineering Drawing and CAD Practice course so that the operation of 3D printers in the CNC maintenance and repair course prioritizes the CAM process and prepares 3D Printers to print objects. In the discussion related to the CAM process, we will briefly discuss the CAM process, CAD and CAM processes, and 3D Printer materials. Materials related to preparing a 3D Printer to print objects consist of an introduction to the parts of an IoT-based 3D Printer, how to build an IoT-based 3D Printer, Setting up an IoT-based 3D Printer, and Maintenance and troubleshooting related to an IoT-based 3D Printer. The hardware and software produced at the design stage are presented in Figure 6.

Table 1. Topics and explanations of CNC maintenance and repair course related to 3D Printers.

Meetings and Topics	Learning Materials
Meeting 1: CAM Process	Overview of CAM processes, CAD and CAM processes, and 3D Printer materials.
Meeting 2: Setting up the 3D Printer	Introduction of IoT-based 3D Printer parts, building an IoT-based 3D Printer machine.
Meeting 3: The process of printing objects	IoT-based 3D Printer setup and Maintenance and troubleshooting related to IoT-based 3D Printers.

The 3D printer performs the Additive Process from the Gcode 3D design that has been made using Filament material as the construction medium. Raspberry Pi is used as a 3D printing bridge with users over the internet. As well as controlling and retrieving data needed for IoT from 3D printing such as working temperature data, camera monitoring, percentage of running processes, as well as cloud storage so that Gcode design data can be stored and used at any time. Octoprint is an agent providing 3D printing connectivity services to be able to connect to the internet. Gujar [19] explains, IoT-based Octoprint provides a web interface to control 3D printers, enabling users to start printing jobs by sending Gcode to a 3D printer connected via a USB port. Octoprint is free, open-source software that allows to remotely perform and monitor all aspects of a 3D printer using a Raspberry Pi.

This Octoprint has its own Operating System called Octopi which is installed into the Raspberry Pi so that IoT can work. In addition, Octoprint is also tasked with retrieving the required data from 3D printing via a USB cable. The internet router/modem serves to connect the Raspberry Pi device to the internet so that the IoT system can work properly. Local devices consisting of Smartphones, PCs, or laptops connected to a local network, the function is to access Octoprint so that it can function as a controller of the IoT 3D printing system. A local network is a local internet network (LAN) emitted by a router/modem connected to the IoT system on Octoprint. Cloud or the Internet serves to store data so that it can be used in the IoT 3D printing system, usually containing the Gcode of the design to be printed. The Spaghetti Detective (TSD) is third-party software that can be integrated with Octoprint so that users can access the IoT 3D printing system from anywhere without having to connect to a local network. External devices are devices used by users who are outside the local network so that users can access IoT 3D printing anytime and anywhere as long as the device they have is connected to the TSD.

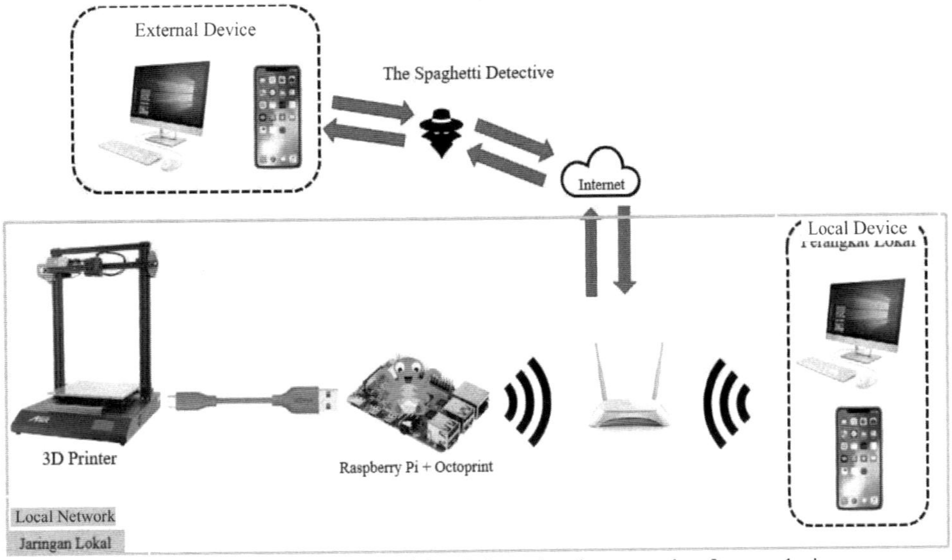

Figure 6. IoT-based 3D Printer hardware and software design

When the user is outside the local network and will operate the 3D printing machine, the user can use TSD as an intermediary. The TSD will later operate Octoprint through the existing internet network. Users simply enter the Gcode design that will be printed to the cloud that has been provided by TSD. Then after that, the user can start printing the design by pressing the start printing button and then selecting the Gcode file to be printed. At that time, TSD will send the Gcode file into the Octoprint system. In the Octoprint system, the Gcode file will be translated and later will be entered into the 3D machine via a USB cable. 3D printing that has received the Gcode file from Octoprint will then work on the file into a 3D form that is following what was designed by the user. Due to the model of accessing via the internet, the quality of the internet network connected to both local and external devices greatly affects the 3D print response in executing commands.

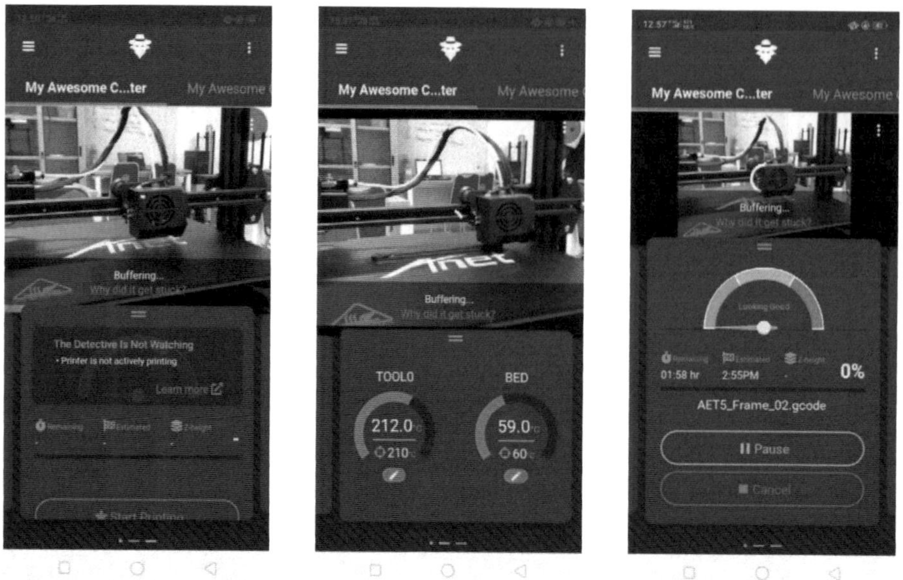

(a) (b) (c)

Figure 7. TSD display on the Smart phone screen

Figure 7a is an initial view of The Spaghetti Detective which is accessed using a smartphone. In this initial view, several parts are seen, namely the monitoring section and the control section. The monitoring section shows images in real-time that are connected to the camera on 3D Printing. While the control section has a start printing button to start the printing process.

In Figure 7b, showing the process when it has started the print process, here it is possible to monitor the temperature of the Hot End (left) and Hot Bed (Right) where this monitoring is based on the temperature setting that is set when carrying out the slicing design process. The temperature monitoring display is divided into two parts, namely monitoring the running temperature on the top side, and monitoring the target temperature on the bottom side.

As for the main display, monitoring of the 3D printing process can be carried out, such as estimated time, estimated completion, and the 3D printing process is declared good, as shown in Figure 7c. In addition, the Pause and cancellation process can also be carried out if needed. While the name of the file that is printed will be displayed on this menu to minimize errors in selecting the file to be processed.

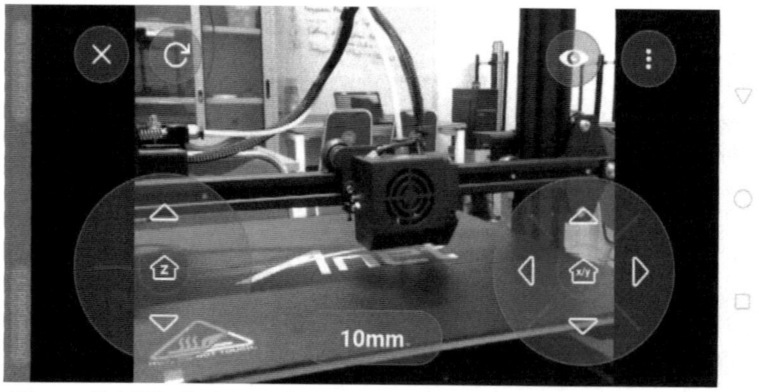

Figure 8. TSD display on Smartphone screen for 3D Printer axis control

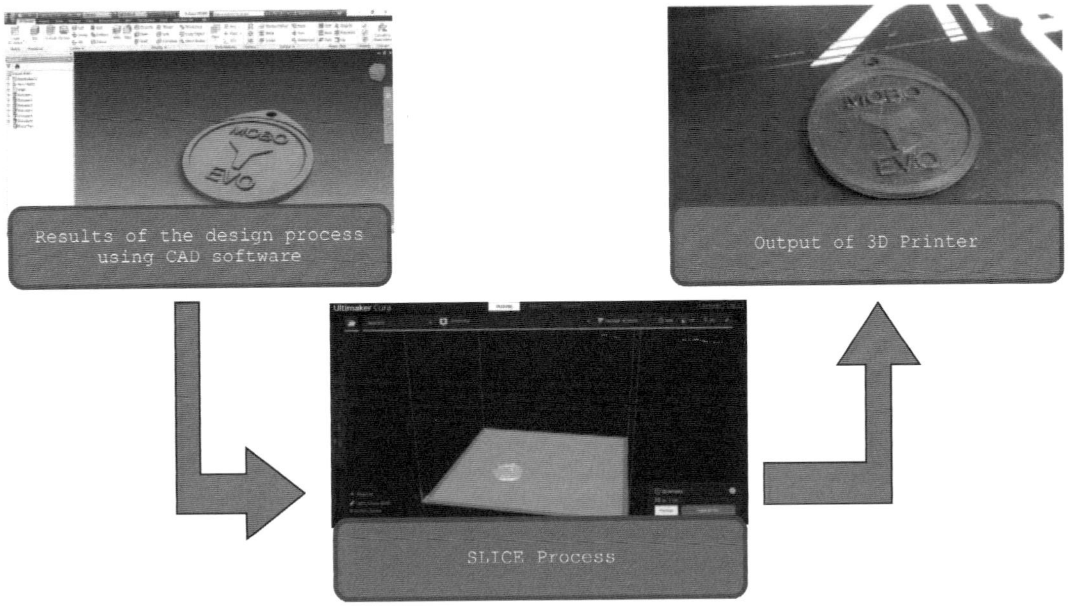

Figure 9. Process of operating the 3D Printer

Figure 8 shows the 3D printing axis control display where this menu can be accessed by double-tapping on the monitoring section of the home menu. In this view, you can control or adjust the position of the nozzle on the X, Y, and Z axes in 3D printing by clicking the arrows on the buttons. If you want to adjust the X, Y-axis then use the right-side button, then if you want to adjust the Z-axis use the left side button. If you want to do home coordinates or do machine zero points, then you can click the button with the symbol of the house and the axis in it. The button that says 10mm is a one-click movement distance setting, so when you click the arrow button once, the axis will move 10mm in the selected direction. Then to clean the button so that it doesn't interfere while monitoring, you can do it by clicking the eye button on the top right side of the screen. The process of operating the 3D printer carried out in the test is shown in figure 9.

An IoT-based 3D Printer machine has been tested for accuracy by providing input commands via gadgets (SmartPhone, Laptop) then measuring and comparing the command and the results of the machine movement. The measured movement is the movement in the direction of the X, Y, and Z axes with the command of 1 click movement or multiples thereof which results in a shift of the nozzle towards the selected axis. The results of the movement are then measured using a ruler, then compared between the measurement results (output) and the command (input).

Table 2. Data from the 3D Printer movement accuracy test results

X-Axis Direction in (mm)		**Y-Axis Direction in (mm)**		**Z-Axis Direction in (mm)**	
Input	**Measured Value**	**Input**	**Measured Value**	**Input**	**Measured Value**
0-10	10	0-10	10	0-10	10
10-20	10	10-20	10	10-20	10
20-30	10	20-30	10	20-30	10
30-40	10	30-40	10	30-40	10
40-50	10	40-50	10	40-50	10
50-60	10	50-60	10	50-60	10
60-70	10	60-70	10	60-70	10
70-80	10	70-80	10	70-80	10
80-90	10	80-90	10	80-90	10
90-10	10	90-10	10	90-10	10

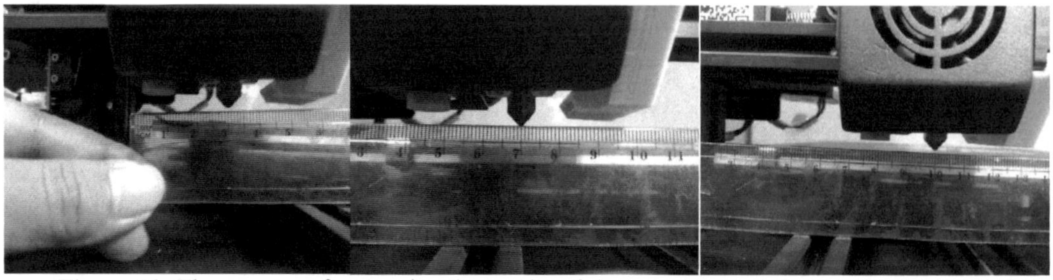
Figure 10. The process of measuring the results of the movement on the 3D Printer nozzle

From the results of testing the accuracy of the movement of the 3D Printer in table 2, it can be seen that there is no significant difference between the input value and the measured value on the movement of the nozzle towards the X-axis, Y-axis, and Z-axis. Testing the accuracy of tool movement is carried out for the setting value of 1 click (10 mm) to 10 clicks (100 mm) for the X, Y, and Z axes. The amount of this input value can be set via the device using the TSD application. Based on the measurement results, it can be concluded that the IoT-based 3D Printer that has been developed can be controlled and serve design processing via the internet with printer results that match the design input. The materials used in the implementation trials were filaments made from Polylactic Acid (PLA).

4. Conclusion
Based on the description above, IoT-based 3D Printers can support learning CNC maintenance and repair courses with relevant topics related to the CAM process, setting up a 3D Printer and the process of printing objects using a 3D Printer. This tool can be used by being operated or controlled via the internet so that the learning process can be done online.

The stages of research and development, namely analysis, design, development, implementation, and evaluation, have been carried out and resulted in IoT-based 3D Printer learning media, identification of student competencies in terms of the design and operation of 3D Printers, and learning modules that can be applied in learning. The design and manufacture of IoT-based 3D Printers as Learning Media consists of designing hardware and software where the hardware used consists of 3D Printers as processors. Raspberry Pi is used as a 3D printing bridge with users over the internet. Octoprint as software installed on the Raspberry Pi, is an agent providing 3D printing connectivity services so that you can connect to the internet. Local devices consisting of Smartphones, PCs, and laptops connected to a local network, which function to access Octoprint so that they can function as controllers of the IoT 3D printing system. The Spaghetti Detective (TSD) is third-party software that can be integrated with Octoprint so that users can access the IoT 3D printing system from anywhere without having to connect to a local network.

The performance of IoT-based 3D printers can be seen from the results of the movement accuracy test where there is no significant difference between the input value and the measured value from the test results of the tool and can carry out the process of forming objects using PLA materials by being controlled via the internet so that it is suitable for use as a practical learning medium.

References
[1] Huleihil, M. 2017. *3D printing technology as innovative tool for math and geometry teaching applications*. IOP Conf. Series: Materials Science and Engineering 164 (2017) 012023 doi:10.1088/1757-899X/164/1/012023.

[2] Zakoldaev, D A et al. (2019). *Ecosystem Of Education Competence Center Of The Industry 4.0*. Journal of Physics: Conference Series 1399 (2019) 044015 doi:10.1088/1742-6596/1399/4/044015.

[3] Shuning Li, Elizabeth Freije and Paul Yearling. 2017. Monitoring 3D Printer Performance using Internet of Things (IoT) Application. American Society for Engineering Education, Paper ID #19431.

[4] D. Assante, G. M. Cennamo and L. Placidi, "3D Printing in Education: an European perspective," *2020 IEEE Global Engineering Education Conference (EDUCON)*, 2020, pp. 1133-1138, doi: 10.1109/EDUCON45650.2020.9125311.

[5] J. Günther, L. Brehm, H. Günzel and A. Humpe, "Teaching 3D Printing Technology Hands-on," 2020 IEEE Global Engineering Education Conference (EDUCON), 2020, pp. 953-957, doi: 10.1109/EDUCON45650.2020.9125302.

[6] Ultimaker-How to design for FFF 3D printing. https://ultimaker.com/3d-printers

[7] Based on 3D HUBS, "Introduction to FDM 3D printing". https://www.3dhubs.com/knowledge-base/introduction-fdm-3d-printing

[8] "3D Printing vs. CNC machining". https://www.3dhubs.com/knowledge-base/3d-printing-vs-cnc-machining

[9] https://imall.com/product/Anet-A8-A2-3D-Printer-Kit-i3-DIY-Printing-Machine-Self-Assembly-w-Filament-Card-Tools/aliexpress.com/32828346821/144-27701050/en?&ViewedIn=other

[10] Furinto, Asnan. 2017. *Menelusuri Inovasi Esensi Berbagai Studi untuk Inspirasi Inovasi*. PT. Gramedia Pustaka Utama. Jakarta.

[11] M. Hermann, T. Pentek and B. Otto, "Design Principles for Industrie 4.0 Scenarios," *2016 49th Hawaii International Conference on System Sciences (HICSS)*, 2016, pp. 3928-3937, doi: 10.1109/HICSS.2016.488.

[12] Handayani, F. 2019. Tren Masif Internet of Things (IoT) di Perpustakaan. Jurnal Ilmu Perpustakaan dan Informasi: Vol 4. 194-209

[13] Hasiholan, C. Primananda, R & Amron, K. 2018. Impelementasi Konsep Internet of Things pada Sistem Monitoring Banjir Menggunakan Protokol MQTT. Jurnal Pengembangan Teknologi Informasi dan Ilmu Komputer: Vol 2. 6128-6135.

[14] Hidayatulloh, S. 2016. Internet of Things Bandung Smart City. Informatika: vol 3. 164-175

[15] Mohamad Hasan Bin Tasneem and Gamal Talal Amer. 2019. Design, Fabrication and Testing of a 3D Printer. *Proceedings of the International Conference on Industrial Engineering and Operations Management,* Pilsen, Czech Republic, July 23-26, 2019

[16] Ishtiaq Ahmed, dkk. 2017. *Design And Fabrication Of Portable 3d Printer*. Visvesvaraya Technological University, Belagavi.

[17] Jun Zhao. 2017. Design of 3D Metal FDM Printing Nozzle Based on Melt Forming. *Chemical Engineering Transactions*. Vol. 59, 2017 DOI: 10.3303/CET1759013

[18] Erasmus+. (2020). *3D Printing in VET (2019-1-EL01-KA202-062909) : Curriculum on 3D Printing*. Accessed from the internet on August 1, 2021 from URL; https://epale.ec.europa.eu/is/node/162121

[19] Gujar. M. P, Yogita Shinde, Ruksar Madaki, and Smina Nadaf. 2019. IoT based 3D Printer. International Research Journal of Engineering and Technology (IRJET) Volume: 06 Issue: 03, Maret 2019 e-ISSN: 2395-0056.

[20] N. Shahrubudin, T.C. Lee, R. Ramlan. 2019. An Overview on 3D Printing Technology: Technological, Materials, and Applications. *Procedia Manufacturing 35 (2019)* 1286–1296. Published by Elsevier B.V.

Design Smart Grid Hybrid in Faculty of Engineering Universitas Negeri Yogyakarta

Muhamad Ali[1], Djoko Laras, BT[2], Muhfizaturrahmah[3], Deny PS[4]

[1,2,3,4] Electrical Engineering Education Department Faculty of Engineering, Universitas Negeri Yogyakarta, Indonesia

Email: muhal@uny.ac.id, djoko_laras@uny.ac.id, muhfizaturrahmah@uny.ac.id

Abstract. The Faculty of Engineering, Universitas Negeri Yogyakarta (UNY), as one of the educational institutions in Engineering, still uses electrical energy from PT PLN, mainly generated from steam and gas power plants. Dependence on fossil energy can be reduced by utilizing renewable power plants, both solar and wind. For this reason, it is necessary to study the use of a Smart Grid system that can regulate electricity needs by optimizing renewable power plants. The Smart Grid components consist of solar power plants, wind power plants, batteries, inverters, and grid power sources from PLN integrated into the Smart grid system. We have designed the Smart Grid system through field observations and data processing with the HOMER Pro software to obtain an optimal hybrid power generation system and wind turbine. The study results indicate that the Faculty of Engineering, UNY has excellent potential to develop smart grids. The potential for solar energy is 418.393 kWh/year, and wind energy is 2.78 kWh/year. The Smart grid system is sufficient to meet the electricity consumption of only 205.5 kWh/year.

Keywords: Smart Grid, design, hybrid, UNY

1. Introduction

Electricity is an energy that has a vital role in modern human life. It has many advantages such as being practical, clean, easy to generate and distribute, reliable, safe, and environmentally friendly. As secondary energy, electricity is obtained through converting primary fuels such as oil, gas, coal, solar, geothermal, nuclear, and other direct energy [1], [2]. Power plants in Indonesia mainly use fossil fuels, gas, and coal, and some power plants use geothermal energy, water, photovoltaic, wind turbine, and waste. The need for electrical energy increases from year to year, prompting the Government to increase power generation. However, unfortunately, the policy of adding power plants uses coal-fired power plants. This condition creates potential energy crises, global warming, climate change, ecosystem damage, and acid rain [2], [3].

The Ministry of Energy and Mineral Resources has planned to increase the use of new and renewable energy in the future. Figure 1. shows the electrical energy mix plan until 2050. The use of new and renewable energy in 2016 was still deficient at less than 10%. Based on the long-term goal, the Government will increase renewable energy use to 23% in 2025 and 32% in 2050. This energy policy is very good in developing the electricity system in Indonesia, so it needs the support of all parties in the development of new and renewable energy [1], [3], [4]

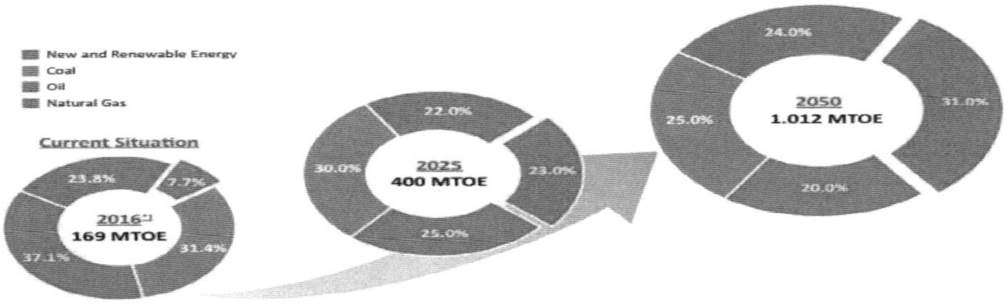

	2016	2025	2050
NRE Mix	7.7 %	23%	31%
Energy Supply	169 MTOE	400 MTOE	1.012 MTOE
Generation Capacity	59 GW (NRE 8 GW)	136 GW (NRE > 45 GW)	443 GW (NRE > 167 GW)
Electricity/Capita	956 kWh	2.500 kWh	7.000 kWh
Electrification Rate	91%	~100%	~100%

Figure 1. Research Method Design Smart Grid System

Several renewable energy power plants continue to be developed, but the high investment costs and low economic value make renewable energy development less attractive [3]. The use of fossil energy sources, especially coal, is influenced by the abundance of coal reserves in Indonesia and the low price compared to other power plants, especially renewable power plants. The development of new renewable energy at the Faculty of Engineering, UNY, was initiated in 2010. Several studies related to renewable power plants have been carried out at the Faculty of Engineering, UNY. Some of the activities include the development of street lighting with solar panels and supercapacitors, micro-hydropower plants, electric power plants, wind, and renewable energy power plants integrated with the grid [1]. However, all of them are limited to small units used for learning in the laboratory.

Energy sustainability and environmental conservation are important issues of global concern due to climate change and the increasing energy demand. A technologically advanced area inevitably requires greater electrical energy consumption and continues to rise at a level that may no longer be manageable if left unattended. Smart grids offer one of the more sustainable technology shift solutions such as distributed generation and microgrids. General public awareness and adequate attention from potential researchers and policymakers are essential [5], [6].

One way to meet people's electrical energy needs is to maintain fossil energy reserves and prevent global warming. The Government has launched the development of New Renewable Energy power plants. One of the problems of developing new renewable energy power plants is the economic value and intermittent nature. Concerning the monetary value of new renewable energy, the Government has issued a subsidy policy for the public and businesses that use renewable energy. Meanwhile, it is necessary to modify the conventional electric power system into a Smart Grid to solve the intermittent nature of energy. The Smart Grid system is one of the technologies that can effectively and efficiently address the integration of New and Renewable Energy into the electricity system on a large scale [8], [9], [10].

Conventional electric power systems based on fossil generators need to be changed with a Smart grid system with renewable power plants [9], [12]. The Conventional electric power system is built based on the location of the energy source, then very long transmission and distribution lines are constructed to reach consumers. The weakness of the conventional electric power system is that the energy flow is only in one direction from the generator to the consumer. The traditional electrical system is very prone to disturbances that cause blackouts. Along with the development of technology and consumer needs, the electric power system must meet the needs of consumers who, at the same time, can function as producers. The power plant is supplied from the center, but each element can also produce electricity sent to the grid.

For this reason, this article will try to provide a review of hybrid Smart Grid planning by utilizing the potential of solar energy, wind energy, electrical energy storage, and conventional generation in a Smart Grid system. This research is expected to make a real contribution to the development and implementation of renewable energy. The impact of implementing renewable energy can reduce dependence on fossil energy and reduce global warming.

2. Method

The stages of designing the Smart Grid system of the Faculty of Engineering UNY can be explained as shown in figure 2.

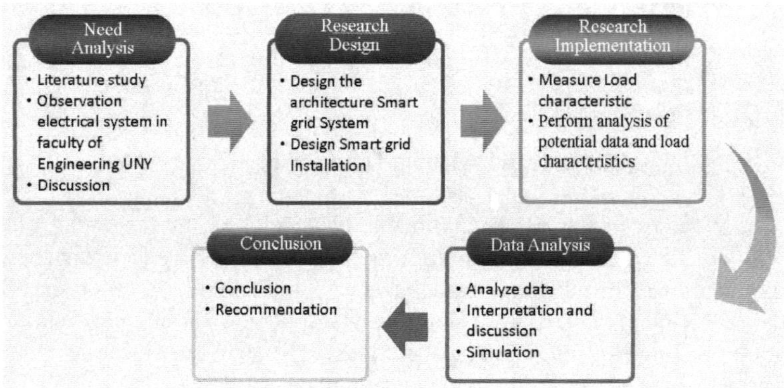

Figure 2. Research Method Design Smart Grid System

The first step is to analyze the need for developing a Smart Grid Hybrid at the Faculty of Engineering, UNY, through literature studies, observation, and discussions with related parties. The next stage is to design the architecture and installation of the Smart grid system in the Faculty of Engineering UNY. The third step is conducting research which consists of measuring load characteristics, identifying renewable energy, and measuring renewable energy potential at the Faculty of Engineering, UNY. The fourth step is to analyze the measurement data and conduct simulations and discussions. The final step is to draw conclusions and make recommendations to the parties concerned.

3. Results and Discussion

3.1. Smart Grid System Design

The Faculty of Engineering UNY is located at Karangmalang Campus, Special Region of Yogyakarta. The Faculty of Engineering has several buildings, including the Electrical and Electronic Engineering Education Department block, the Mechanical Engineering Education Department block, the Civil Engineering Education Department block, the culinary and fashion education department block, and the KPLT block. Planning for the development of a Smart Grid system at the Faculty of Engineering, UNY by optimizing New and Renewable Energy according to Figure 1 is as follows:

a. Solar Power Plant

The solar panels used in the Smart Grid system are Canadian Solar MaxPower CS6U-340M with a capacity of 0.340 kW at US$279.50 (Solaris Shop, 2021) and maintenance costs of US$2,7950 (includes online ordering fees, shipping costs), and installation costs).

b. Wind Power Plant

Wind turbines are also a significant component in this hybrid power generation system. The type of wind turbine that will be used for this simulation is AWS HC 3.3 kW with a purchase and replacement cost of US$25,000, operating and repair costs of US$250.

c. Export and Import KWH meter system from PT. PLN
The EXIM (Export-Import) kWh meter device is a tool for calculating electrical energy consumption that is installed in PLN customer homes and has a function to record the amount of production (export) and consumption (import) of customers.

d. Converter
The converter is a valuable component for converting DC electricity to AC. The converter type used in this simulation is REFUsol 46K-MV. The converter has an inverter and a rectifier section with an efficiency of 90% each. The converter fee is US$19,034.00.

e. Battery
The battery used is a Trojan SSIG 12255 with a specification of 12V 259 Ah. The cost of all battery kits is US$425.20, and maintenance costs are US$4.2520.

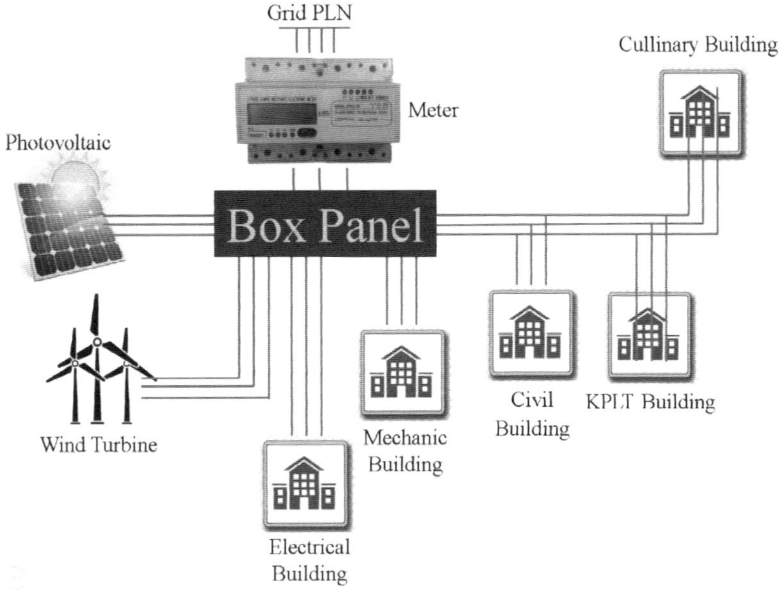

Figure 3. Design of Smart grid System at Faculty of Engineering UNY

Several renewable energy sources, such as water/hydro, photovoltaics, biomass, and wind/wind, need a study implemented into a power plant in the Faculty of Engineering Building.

Table 1. Renewable energy potential in Faculty of Engineering UNY

Energy sources	Location	Characteristic	Information
Water/Micro Hydro	Power House Faculty of Engineering Universitas Negeri Yogyakarta	No potential	The location is far from heavy water flows such as rivers, lakes, waterfalls, and other sources of water flow
Solar		Potential	The location is constantly exposed to sunlight throughout the day, and no buildings are blocking it so that it can be used for solar power generation
Biomass		No potential	The location is quite far from plantations, forests, and

		livestock, so the biomass potential is not optimal.
Wind	Potential	Wind at the Faculty of Engineering, UNY, has characteristics that change due to climate and seasons, so a more in-depth study of the utilization of wind power plants is needed.

The Smart grid design of the Faculty of Engineering UNY consists of two types of energy sources, namely fossil energy and new and renewable energy. Fossil energy consists of a grid component and a generator. Meanwhile, new and renewable energy consists of solar panels (PV) and wind turbines. Other components included in the hybrid power generation system are converters, battery storage, AC and DC buses. The following is the architectural design of the Faculty of Engineering UNY hybrid power plant system.

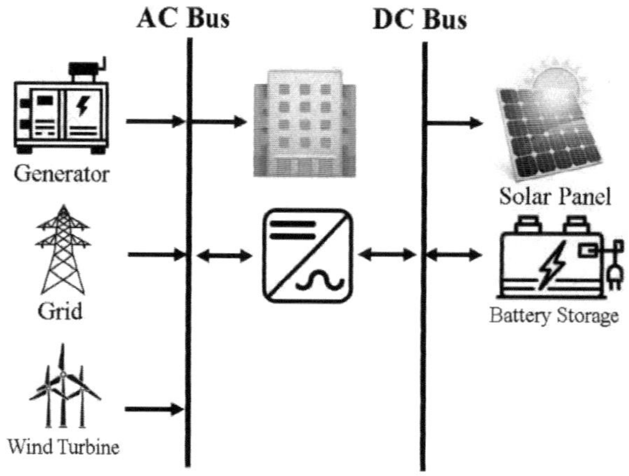

Figure 4. Hybrid Power Generation System Design

The architectural drawing of the Smart Grid system has two buses, namely the AC bus and the DC bus connected to the inverter. The DC bus is connected to the photovoltaic and battery storage system. The inverter functions to convert from DC electricity to AC and vice versa, as well as AC chargers and system performance regulators to regulate network voltage and frequency. For AC buses connected to the load or electrical load supplied, the grid, generator, and wind turbine.

4. Photovoltaic and Wind Turbin Energy Potential

Potential renewable energy sources consist of solar energy and wind energy. This data can be obtained from the NASA website database at https://eosweb.larc.nasa.gov/cgi-bin/sse by entering geographic information on the latitude and longitude of the FT UNY Building. Energy potential data in the form of radiation, sun clearness index, and wind speed for a year in tabular form. The average annual solar radiation obtained from the NASA database is 4.80 kWh/m^2/s, and the yearly average wind speed is 4.2 m/s. The distribution of solar energy potential data can be seen in Figure 4, while the wind energy potential is in Figure 5.

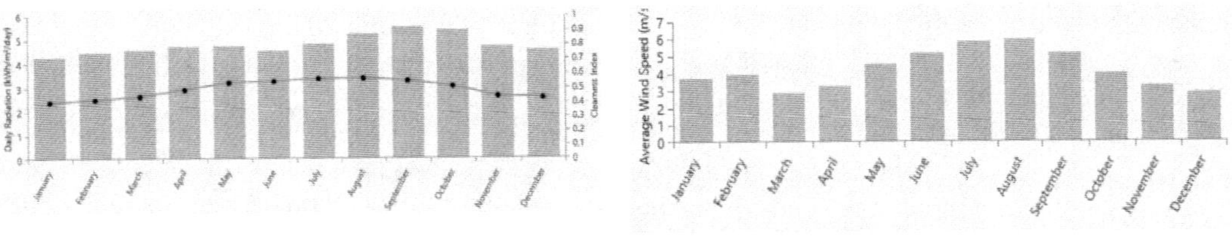

Figure 5. Photovoltaic and wind turbine Energy Potential

The daily load energy consumption per hour of the Faculty of Engineering UNY Building can be seen in Figure 6.

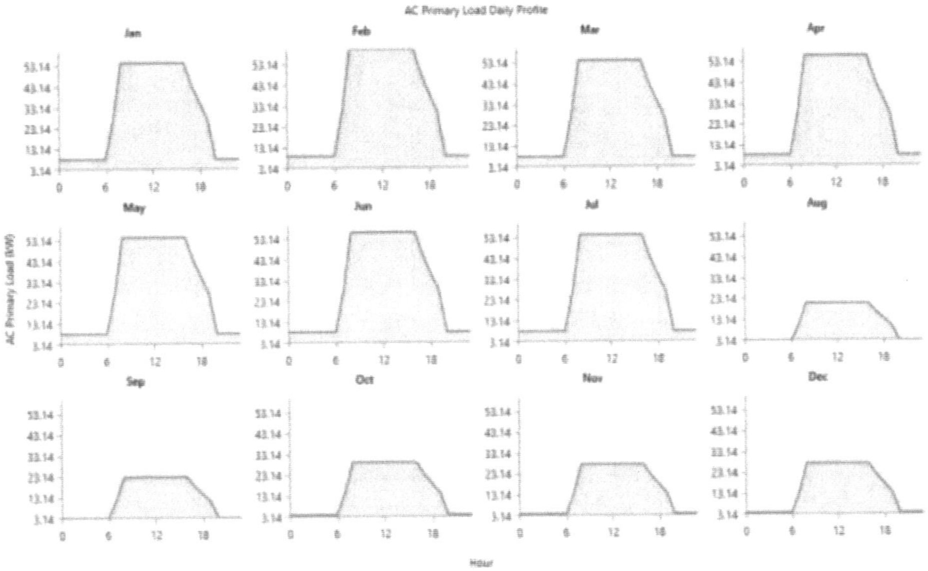

Figure 6. Annual Energy Profile

Figure 7 is the metric load of the UNY Faculty of Engineering Building with an average daily energy load of 562.98 kWh/d and average daily power of 23.46 kW. Peak power is 59.98 with a power factor of 0.79.

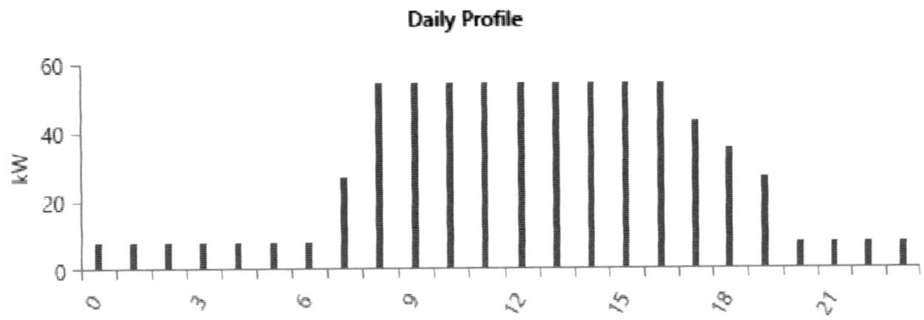

Figure 7. Daily Profile Electrical Load Faculty of Engineering UNY

From the sensitivity results of the Homer Pro simulation above, the results of the sensitivity case with wind speed (average) for wind turbines are 4.20 m/s. For optimization results with system, configurations are solar panels, wind turbines, generator sets, batteries, grids [11]. And inverters. The results of the system configuration that the Homer Pro software has processed are 278 kWh solar panels, 1 unit of a wind turbine, 440 kW generator set, 500 units of battery, inverter with a capacity of 96.6 kW, and grid or PLN network.

Table 2. Potential of Renewable Energy Production

Component	Production (kWh)	Percentage (%)
Photovoltaic	403.12	96,30
Wind Turbine	2.78	0,66
Grid	12.50	2,99
Total	**418.39**	**100,00**

Table 3. Energy Consumption per year

Energy Consumption	Energy Per Year (kWh/yr)	Percentage (%)
AC Primary Load	205,489	100
DC Primary Load	0	0
Total	**205,489**	**100**

The total production of electrical energy from solar panels is greater than that of wind turbines because sunlight is terrific throughout the year. In contrast, wind speed is not stable due to changes in direction and speed influenced by air pressure, climate, seasons, and other influences. The Smart Grid configuration system can meet the electrical power needs of the Faculty of Engineering and has an excess of 186,379 per year. This condition shows that the Smart Grid system can be developed at the Faculty of Engineering, UNY.

Smart grid implementation at the Faculty of Engineering, UNY, is possible by optimizing solar power plants and wind turbines. The potential solar and wind power sources at the location can meet the Faculty of Engineering, UNY's electricity needs for an entire year. The Smart grid system requires a battery to overcome the problem of intermittent electrical energy so that the electrical power supply can be fulfilled continuously, safely, and reliably.

5. Conclusion

As an engineering higher education institution, the Faculty of Engineering of UNY needs to develop a Smart Grid on campus. The study results show that the Faculty of Engineering, UNY, has excellent potential to create smart grids. The potential for solar energy is 418.393 kWh/year, and wind energy is 2.78 kWh/year. The Smart grid system is sufficient to meet the electricity consumption of only 205.5 kWh/year. Implementing the Smart Grid Hybrid will improve the quality of the electric power system and reduce dependence on fossil fuels.

References

[1] M. Ali, Alex, Eko, SD, Muhfizaturrohmah, Nurhening, Bagas, 2020, Design and Implementation of Trainer Kit for Hybrid On-Grid Solar Power Generation System, Journal of Physics: Conference Series, ICE-ELINVO 2020, oi:10.1088/1742-6596/1737/1/012002

[2] Anon 2015 Indonesia Energy Policy, Laws and Regulation Handbook Volume 1 Strategic Information and Basic Laws (Washington DC: International Business Publications)

[3] IRENA, 2017, Renewable Energy Prospects: Indonesia, a REmap analysis, International Renewable Energy Agency (IRENA), Abu Dhabi, www.irena.org/remap

[4] Agus Sugiyono, Anindhita, Laode M.A. Wahid, A. (2016). Pengembangan Energi untuk Mendukung Industri Hijau. In Outlook Energi Indonesia.

[5] Rashid Al-Badawi, Mohammad Abusara & Tapas Mallick (2015) A Review of Hybrid Solar PV and Wind Energy System, Smart Science, 3:3, 127-138, DOI: 10.1080/23080477.2015.11665647

[6] R. Ramakumar, 2012. "Integrated Renewable Energy System – Micro Grid (IRES-MG) For Sustainable Development," ICSUNIDO, Trieste, Italy,

[7] S. Kurtz, "Opportunities and challenges for the development of a mature concentrating photovoltaic power industry," Technical Report NREL/TP-520-43208, Revised November (2009)

[8] Maria LorenaTuballaaMichael LochinvarAbundo, 2016, A review of the development of Smart Grid technologies, Renewable and Sustainable Energy Reviews, Volume 59, June 2016, Pages 710-725, Renewable and Sustainable Energy Reviews, https://doi.org/10.1016/j.rser.2016.01.011

[9] Abdelaziz Mohamed, M., & Eltamaly, A. M. (2018). Modeling and Simulation of Smart Grid Integrated with Hybrid Renewable Energy Systems. 121(October 2017), 10–21. https://doi.org/10.1007/978-3-319-64795-1

[10] Hatziargyriou, N. (2014). Microgrid: Architecture and Control. In John Wiley and Sons Ltd (Vol. 1). https://doi.org/10.1038/ki.2013.249

[11] HOMER Energy LLC. (2019). HOMER Pro user manual. (August). Retrieved from https://www.homerenergy.com/products/pro/docs/3.13/index.html

[12] Iqbal, F., & Siddiqui, A. S. (2017). Optimal configuration analysis for a campus microgrid—a case study. Protection and Control of Modern Power Systems, 2(1). https://doi.org/10.1186/s41601-017-0055-z

LQR state feedback controller with precompensator for magnetic levitation system

A Winursito[1], G N P Pratama[2]

[1, 2]Department of Electronics and Informatics Engineering Education, Universitas Negeri Yogyakarta, Indonesia.

E-mail: anggunwinursito@uny.ac.id[1], gilang.n.p.pratama@uny.ac.id[2]

Abstract. Magnetic levitation system (MLS) is a nonlinear system that attracts the attention of many researchers, especially control engineers. It has wide range of application such as robotics, high-speed transportation, and many more. Unfortunately, it is not a simple task to control it. Here, we utilize state feedback controller with Linear-Quadratic Regulator (LQR) to regulate the position of a steel-ball in MLS. In addition, we also introduce the precompensator to nullify the steady-state errors. The linearized model, controller, and precompensator are simulated using MATLAB. The results and simulation verify that the state feedback controller and precompensator succeed to stabilize the position of steel-ball at the equilibrium for 0.1766 seconds and no steady-state errors.

1. Introduction

Technically, magnetic levitation system or simply MLS, is a realization of electromechanical coupling that can provide contactless motion and levitation of an object using a magnetic field. This is possible when the magnetic force is greater than the gravitational force. The magnetic field is generated from the electromagnetic coil and the current in the coil, which controls the magnitude of the magnetic force.

MLS has been widely applied in various fields, such as high-speed magnetic levitation trains [1], bearingless motor [2], robotics [3], and wind turbine [4]. It provides many advantages such as less frictional movement, high speed, environmental isolation, as well as relatively lower maintenance costs. Unfortunately, modeling and controlling this system is not easy. It is due to the nonlinear characteristics of MLS, which causes the system to become unstable. Many researchers, especially in the field of control engineering, have tried to overcome this problem by applying the feedback-linearization [5], PID controller [6], fuzzy logic controller [7], and sliding mode controller [8].

There are several components that necessary for MLS, such as controller, driver, electromagnet, and sensor [5]. In this paper, the model of MLS with 1-Degree-of-Freedom (1-DoF) is considered as depicted in Figure 1.

We present nonlinear dynamic model of MLS which is linearized to the operating point. The model of MLS is developed by investigating both electrical and mechanical parts. Furthermore, we utilize a precompensator to the reference of MLS. It is necessary, since the precompensator is expected to nullify the steady-state errors.

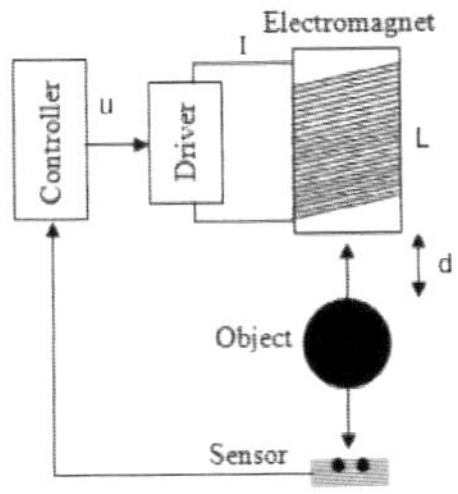

Figure 1. Model of a magnetic levitation system.

After introducing MLS, let us elaborate the organization of this paper. The model of MLS is briefly explained in Section 2. Later, in Section 3, the design of the controller and the precompensator are discussed. The results and analysis of this research can be found in Section 4. Last but not the least, the conclusion of this paper is explained in Section 5.

2. Model of Magnetic Levitation System

There are two main topics that will be explained in this section. First, in Subsection 2.1, we discuss the nonlinear model of magnetic levitation system. Later, the linearization of magnetic levitation system is presented in Subsection 2.2.

2.1. Nonlinear Equations

Let us consider the equation of motion of the steel-ball such as

$$\begin{aligned} m\ddot{x} &= F_g - F_m, \\ &= mg - \frac{K_c I^2}{x^2}, \end{aligned} \tag{1}$$

where respectively m, x, g, F_g, F_m, I, and K_c denote the mass of the steel-ball, its position, gravity constant, gravitational force, magnetic force, applied driving current, and force constant of electromagnet respectively. Based on equation (1), we can define a nonlinear function as

$$f(x, I) = g - \frac{K_c I^2}{mx^2}. \tag{2}$$

Meanwhile, the dynamics of applied driving current can be denoted as

$$\dot{I} = \frac{V}{L} - \frac{IR}{L}, \tag{3}$$

where consecutively R, L, and V are the resistance, inductance, and operating voltage.

Usually, the nonlinear system is expected to operate at an equilibrium point. In our case, we expect that $\ddot{x} = 0$, $\dot{x} = 0$, and $\dot{I} = 0$. Since $\ddot{x}$ and $\dot{I}$ are zeros, it indicates that the magnetic and gravitational force are equal. Hence, the steel-ball can be levitated at an equilibrium position.

Take a look back at equation (1), since $F_g = F_m$, we can calculate the applied driving current at the equilibrium as

$$I_e = x_e\sqrt{(mgK_c{}^{-1})}, \tag{4}$$

where respectively I_e and x_e are the applied driving current and position of the steel-ball at the equilibrium point. Furthermore, we can obtain the voltage at equilibrium point, V_e, as

$$\begin{aligned} V_e &= I_e R, \\ &= x_e\sqrt{(mgK_c{}^{-1})}\; R. \end{aligned} \tag{5}$$

2.2. Linearization

We can obtain the linearized model of MLS by applying Taylor Series Expansion as

$$\ddot{x} = c_1 (x - x_e) + c_2 (I - I_e), \tag{6}$$

where

$$\begin{aligned} c_1 &= \left.\frac{\partial f(x,I)}{\partial x}\right|_{x=x_e, I=I_e}, \\ &= \frac{2K_c I_e^2}{m x_e^3}, \end{aligned} \tag{7}$$

and

$$\begin{aligned} c_2 &= \left.\frac{\partial f(x,I)}{\partial I}\right|_{x=x_e, I=I_e}, \\ &= -\frac{2K_c I_e}{m x_e^2}. \end{aligned} \tag{8}$$

Let the input u be the operating voltage V_e, then we can define the state space as

$$\begin{aligned} \dot{\mathbf{x}} &= \mathbf{A}\mathbf{x} + \mathbf{B}u, \\ &= \begin{bmatrix} 0 & 1 & 0 \\ c_1 & 0 & c_2 \\ 0 & 0 & -\frac{R}{L} \end{bmatrix} \mathbf{x} + \begin{bmatrix} 0 \\ 0 \\ \frac{1}{L} \end{bmatrix} u, \end{aligned} \tag{9}$$

where the state vector be

$$\mathbf{x} = \begin{bmatrix} x_1 & x_2 & x_3 \end{bmatrix}^{\mathsf{T}}. \tag{10}$$

The state variables are x_1, x_2, and x_3 that respectively denote the position of the steel-ball, its velocity, and drive current. Meanwhile, the output can be defined as

$$\begin{aligned} y &= \mathbf{C}\mathbf{x}, \\ &= \begin{bmatrix} 1 & 0 & 0 \end{bmatrix} \mathbf{x}, \\ &= x_1. \end{aligned} \tag{11}$$

3. Controller Design and Precompensator

In this part, we will discuss about the design of control law and precompensator for magnetic levitation system. It exists a problem of parameterization to find a stable controller [9, 10, 11]. Here, we utilize LQR to obtain a stable controller. A brief explanation of LQR is presented in Subsection 3.1. Meanwhile, the precompensator is elaborated in Subsection 3.2.

3.1. Linear-Quadratic Regulator

Linear-Quadratic Regulator (LQR) is an optimal control that can be implemented either for Single-Input Single-Output or Multi-Input Multi-Output system. It offers not only a good stability, but it can guarantee the stability margin of the system [12]. It is worth to be mentioned that LQR can give us a better optimal energy consumption compared againsts PID and fuzzy controller [13]. We can see the block diagram of a full-state feedback controller using LQR as depicted in Figure 2.

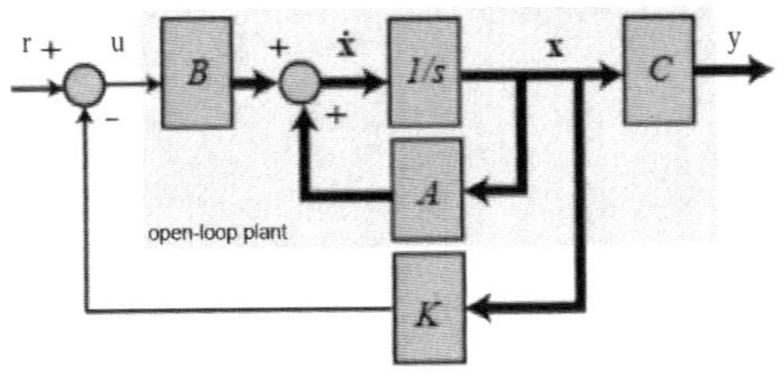

Figure 2. Block diagram of a LQR full-state feedback controller.

In general, LQR has an objective function such as

$$J(\mathbf{u}) = \int_0^\infty (\mathbf{x}^\mathsf{T}\mathbf{Q}\mathbf{x} + \mathbf{u}^\mathsf{T}\mathbf{R}\mathbf{u})dt, \tag{12}$$

that must be minimized. In this research, MLS has three states and one input, therefore $\mathbf{Q} \in \mathbb{R}^{3\times 3}$, a 3-by-3 matrix. On the other hand, we have $\mathbf{R} \in \mathbb{R}$, as a scalar. Both parameters, $\mathbf{Q}$ and $\mathbf{R}$, affect differently towards gain feedback $\mathbf{K}$. If we expect a larger gain feedback $\mathbf{K}$, we can increase $\mathbf{Q}$. On the contrary, if we have a larger $\mathbf{R}$, then we can obtain a smaller gain feedback $\mathbf{K}$ [14]. Next, we need to calculate the auxiliary matrix $\mathbf{S}$ by solving this Algebraic Ricatti Equation (ARE) where

$$\mathbf{A}^\mathsf{T}\mathbf{S} + \mathbf{S}\mathbf{A} - \mathbf{S}\mathbf{B}\mathbf{R}^{-1}\mathbf{B}^\mathsf{T}\mathbf{S} + \mathbf{Q} = 0. \tag{13}$$

It is necessary to determine the gain feedback $\mathbf{K}$ such as

$$\mathbf{K} = \mathbf{R}^{-1}(\mathbf{B}^\mathsf{T}\mathbf{S}). \tag{14}$$

3.2. Precompensator

The precompensator $\bar{N}$ is needed to handle the steady-state error. Here, the steady-state errors are expected to be zeros, where we can utilize a new value as the reference. In practice, the reference r is multiplied by $\bar{N}$ to obtain the new reference. The whole block diagram of state feedback controller with precompensator can be seen in Figure 3.

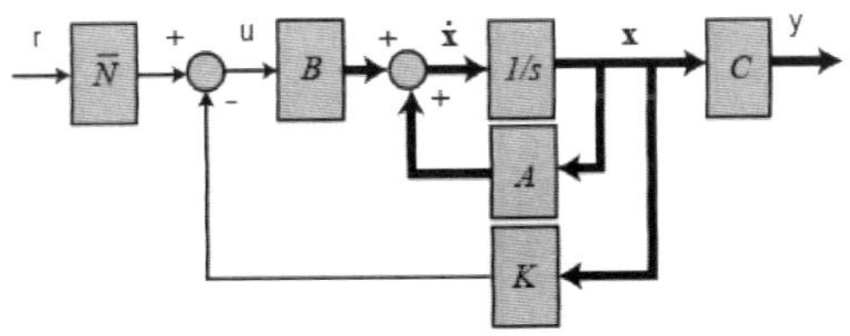

Figure 3. Block diagram of a state feedback controller with precompensator.

There are some steps in order to calculate the precompensator. First, we need to find $\mathbf{N}$ as

$$\mathbf{N} = \left[\begin{array}{c|c} \mathbf{A} & \mathbf{B} \\ \hline \mathbf{C} & 0 \end{array}\right]^{-1} \mathbf{Z}, \tag{15}$$

where

$$\mathbf{Z} = \begin{bmatrix} 0 & 0 & 0 & 1 \end{bmatrix}^\mathsf{T}. \tag{16}$$

After obtaining **N**, we can partitioned it as

$$\mathbf{N} = \begin{bmatrix} \alpha \\ \beta \end{bmatrix}, \tag{17}$$

where $\alpha \in \mathbb{R}^{3\times 1}$ is a column vector, meanwhile $\beta \in \mathbb{R}$ is a scalar. Eventually, the precompensator can be calculated as

$$\bar{N} = \mathbf{K}\alpha + \beta. \tag{18}$$

4. Results and Discussion

Here we present the results of this research, including the necessary parameters. In Subsection 4.1, the parameters of magnetic levitation system is discussed. Later, in Subsection 4.2, we will explain the system response with state feedback controller and precompensator.

4.1. System Parameters

Before stepping any further, it is necessary to define the parameters of MLS that used in this research. Those parameters can be seen in Table 1.

Table 1. Parameters of magnetic levitation system.

Parameters	Values
m	$0.008\,\mathrm{kg}$
K_c	$6.8823\times10^{-5}\,\mathrm{Nm^2A^{-2}}$
R	$2.5\,\mathrm{ohm}$
L	$0.01\,\mathrm{H}$

In the meantime, we also decide the equilibrium point for the position of the steel-ball $x_e = 0.01\,\mathrm{m}$. Now, we can determine the equlibrium point for the driving current by substituting those values into equation (4), which is $I_e = 0.3375\,\mathrm{A}$. Moreover, the operating voltage at equlibrium can be obtained using equation (5) as $V_e = 0.8538\,\mathrm{V}$.

4.2. Simulation and Analysis

After defining the parameters, we can investigate the stability of the open-loop system. First, let us substitute those parameters on Table 1 and with the gravity constant $g = 9.8\,\mathrm{m\,s^{-2}}$ into state matrix **A** and input vector **B**, then we have

$$\mathbf{A} = \begin{bmatrix} 0 & 1 & 0 \\ 1960 & 0 & -58.0717 \\ 0 & 0 & -250 \end{bmatrix}, \tag{19}$$

$$\mathbf{B} = \begin{bmatrix} 0 \\ 0 \\ 100 \end{bmatrix}. \tag{20}$$

The stability of open-loop system can be known by calculating the eigenvalues of **A**. Here, we have the eigenvector $\mathrm{eig}(\mathbf{A}) = \begin{bmatrix} 44.2719 & -44.2719 & -250 \end{bmatrix}^{\mathsf{T}}$. Clearly, the open-loop system is unstable, since it has a real positive eigenvalue.

Next, we can investigate either the system is controllable or not. It can be done by calculating the rank of controllability matrix. Here, the controllability matrix **Co** can be defined as

$$\mathbf{Co} = \begin{bmatrix} \mathbf{B} & \mathbf{AB} & \mathbf{A}^2\mathbf{B} \end{bmatrix}, \tag{21}$$

where **Co** is a 3-by-3 matrix, which corresponds to the number of state variables. Hence, by utilizing MATLAB, we can determine the rank of controllability matrix, which is 3. Since its rank is equal to the number of state variables, we can conclude that the system is controllable.

Now we can design the controller using LQR, where we need to determine the parameters, which are

$$\mathbf{Q} = \begin{bmatrix} 100 & 0 & 0 \\ 0 & 100 & 0 \\ 0 & 0 & 100 \end{bmatrix}, \tag{22}$$

and

$$\mathbf{R} = 1. \tag{23}$$

Those values are chosen to get a quick response of the closed-loop system [14].

Since the parameters of **C**, **B**, **Q**, and **R**, have been determined, we can solve the Algebraic Ricatti Equation in equation (13) using MATLAB. Therefore, we obtain the auxiliary matrix as

$$\mathbf{S} = \begin{bmatrix} 6964 & 136.25 & -7.3089 \\ 136.25 & 3.644 & -0.193 \\ -7.3089 & -0.193 & 0.0884 \end{bmatrix}. \tag{24}$$

Eventually, the equation (14) can be solved and the feedback gain be

$$\mathbf{K} = \begin{bmatrix} -730.8898 & -19.3003 & 8.8431 \end{bmatrix}. \tag{25}$$

Since the gain feedback has been obtained, we can determine the stability of the closed-loop system in the same way as the open-loop system. The closed-loop system has eigenvalues as $\text{eig}(\mathbf{A}-\mathbf{BK}) = \begin{bmatrix} -1029.2 & -24.3122 & -80.7729 \end{bmatrix}^\mathsf{T}$. It is apparent that all the real elements of eigenvector are negatives, hence we can declare that the closed-loop system is stable. Another thing we need to do is calculate the precompensator. Consider the equations on Subsection 3.2 and by substituting the parameters, then we have $\bar{N} = -348.0447$.

After elaborating the necessaries, we can simulate the response of the closed-loop system. Let the initial states be $\mathbf{x}_0 = \begin{bmatrix} 0 & 0 & 0 \end{bmatrix}^\mathsf{T}$. It appears that at first the steel-ball is stand still lying on the plate and there is no driving current flows. Meanwhile, the desired states be $\mathbf{x}_d = \begin{bmatrix} x_e & \dot{x}_e & I_e \end{bmatrix}^\mathsf{T}$, where $x_e = 0.01\,\text{m}$, $\dot{x}_e = 0\,\text{m}\,\text{s}^{-1}$, and $I_e = 0.3375\,\text{A}$. Here, the response of MLS with state feedback controller and precompensator can be seen in Figure 4.

The response is quickly converge to the desired position, which can be verified by computing the rise time and settling time. It has a rise time of 0.097 seconds, and the settling time is 0.1766 seconds, without overshoot. Another thing worth to be noted that it provides no steady-state errors. Furthermore, we can analyze the dynamics of the states as depicted in Figure 5.

Based on Figure 5, we have x_1, x_2, and x_3 that consecutively correspond to x_e, $\dot{x}_e$, and I_e. Since x_1 is already explained, then we will analyze the rests. Initially the steel-ball is stand still, then after the operating voltage is being applied, it will start levitating. We can see that x_2 start increasing until $x_1 \approx x_e$. Hence, in order to keep $x_1 = x_e$, then x_2 must be zero. In the meantime, after the operating voltage being applied, the driving current x_3 is flowing. We can see that x_3 is rising until it reaches the equilibrium I_e. It also indicates that the equilibrium for x_e has been met, since the driving current is proportional to the position of the steel-ball.

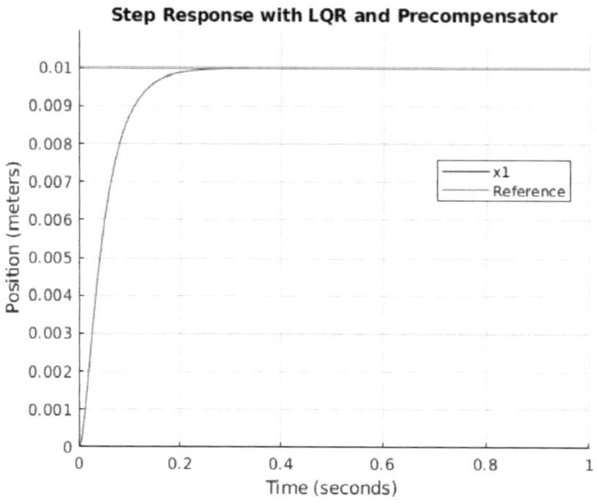

Figure 4. Step response of x_1.

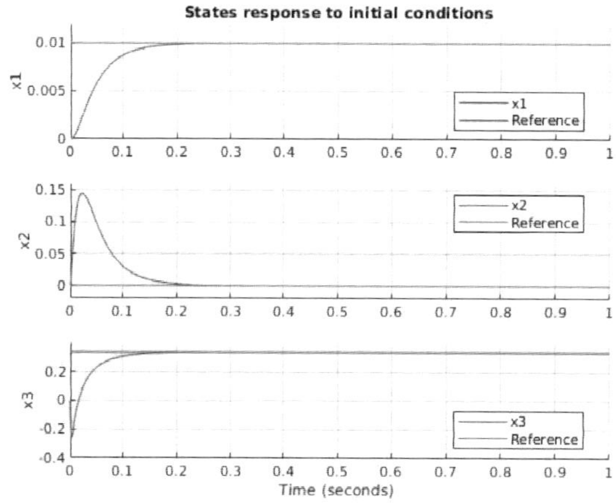

Figure 5. States response to initial conditions.

5. Conclusion

The modeling of magnetic levitation system and the design of state feedback controller using LQR are elaborated in this paper. The model magnetic levitation system is linearized at equilibrium points, which is necessary for designing the controller. We also implement a precompensator to nullify the steady-state errors. The model, controller, and precompensator are simulated using MATLAB. It has 0.097 seconds of rise time and without overshoot. Meanwhile, the settling time is 0.1766 seconds. Additionally, the precompensator prevails to guarantee zero steady-state errors. The results and simulation verify that the state feedback controller with precompensator can give us a quick response without steady-state errors neither overshoot.

References

[1] H.-W. Lee, K.-C. Kim, and J. Lee. Review of maglev train technologies. *IEEE Transactions on Magnetics*, 42(7):1917–1925, 2006.

[2] J. Asama, Y. Hamasaki, T. Oiwa, and A. Chiba. Proposal and analysis of a novel single-drive bearingless motor. *IEEE Transactions on Industrial Electronics*, 60(1):129–138, 2013.

[3] M. Hagiwara, T. Kawahara, T. Iijima, and F. Arai. High-speed magnetic microrobot actuation in a microfluidic chip by a fine v-groove surface. *IEEE Transactions on Robotics*, 29(2):363–372, 2013.

[4] C. V. Aravind, R. Rajparthiban, R. Rajprasad, and Y. V. Wong. A novel magnetic levitation assisted vertical axis wind turbine design procedure and analysis. In *2012 IEEE 8th International Colloquium on Signal Processing and its Applications*, pages 93–98, 2012.

[5] R. Uswarman, A. I. Cahyadi, and O. Wahyunggoro. Control of a magnetic levitation system using feedback linearization. In *2013 International Conference on Computer, Control, Informatics and Its Applications (IC3INA)*, pages 95–98, 2013.

[6] S. Hlangnamthip, C. Thammarat, and D. Puangdownreong. Cs-based pid controller design for maglev vehicle suspension system. In *2019 7th International Electrical Engineering Congress (iEECON)*, pages 1–4, 2019.

[7] N. Sahoo, A. Tripathy, and P. Sharma. Single axis control of ball position in magnetic levitation system using fuzzy logic control. In *IOP Conference Series: Materials Science and Engineering*, volume 323, 2018.

[8] R. Uswarman, A. I. Cahyadi, and O. Wahyunggoro. Modified sliding mode control with uncertainties behavior of a magnetic levitation system. In *2013 International Conference on Robotics, Biomimetics, Intelligent Computational Systems*, pages 194–199, 2013.

[9] T. Sakanushi, K. Yamada, Y. Ando, T. M. Nguyen, and S. Matsuura. A design method for simple multi-period repetitive controllers for multiple-input/multiple-output plants. *ECTI Transactions on Electrical Engineering, Electronics, and Communications*, 9(1):202–211, 2011.

[10] T. Sakanushi, K. Yamada, I. Murakami, J. Hu, and S. Matsuura. A design method for robust stabilizing simple multi-period repetitive controllersfor multiple-input/multiple-output plants. *ECTI Transactions on Electrical Engineering, Electronics, and Communications*, 10(1):1–13, 2012.

[11] T. Sakanushi, K. Yamada, Y. Ando, S. Matsuura, and J. Hu. A design method for simple repetitive controllers for multiple-input/multiple-output time-delay plants. *ECTI Transactions on Electrical Engineering, Electronics, and Communications*, 10(1):24–34, 2012.

[12] Priyatmadi, A. P. Sandiwan, H. Wijaya, and A. Cahyadi. Application of spsa lqr tuning on quadrotor. In *2016 6th International Annual Engineering Seminar (InAES)*, pages 32–36, 2016.

[13] H. Maghfiroh, A. Ataka, O. Wahyunggoro, and A. I. Cahyadi. Optimal energy control of dc motor speed control: Comparative study. In *2013 International Conference on Computer, Control, Informatics and Its Applications (IC3INA)*, pages 89–93, 2013.

[14] O. A. Dhewa, A. Dharmawan, and T. K. Priyambodo. Model of linear quadratic regulator (lqr) control method in hovering state of quadrotor. *Journal of Telecommunication, Electronic and Computer Engineering*, 9(3):135–143, 2017.

A Simple Brushless Motor and Propeller Test Stand for Experiment from Home

Erwan Eko Prasetiyo

Sekolah Tinggi Teknologi Kedirgantaraan, Jl. Parangtritis Km 4,5 Sewon, Bantul, D.I. Yogyakarta

Email: erwan.eko@sttkd.ac.id

Abstract. A brushless motor and propeller test stand is used to test brushless motors and propellers. This testing instrument is still only available in research laboratories. Students and researchers are unable to use laboratory facilities because of the Covid-19 epidemic, thus students must be able to do tests independently from home. Purchasing this testing instrument would be too expensive for students. It is essential to construct a brushless motor and propeller testing instrument at home using simple components that are easy to get on the marketplace. The design concept reads force data using a loadcell sensor and an HX711 driver, and current and voltage data with an INA 219 sensor. The brushless motor's rotational speed is controlled by a potentiometer. Force, current, voltage, and power are all examples of test results data. A 16x2 LCD is used to show data immediately. Data is also transmitted via a USB connection to a computer device for storage or additional analysis. This study proposes a simple brushless motor and propeller test stand that can measure forces from 0 gf to 1000 gf with an error rate of 0.72 %. The power that can be read ranges from 0 mW to 18960 mW, with a 0.59 % error rate.

1. Introduction

An unmanned vehicle requires a propeller driven by a motor. A Brushless motor is an electric technology that has a very high rotation, low energy consumption, and is not easy to heat. Unmanned vehicles use a brushless motor to rotate the propeller to produce thrust. Unmanned vehicles are designed by choosing an energy-efficient brushless motor [1]. Motor efficiency is related to the flying durability of an aircraft, considering that the battery resources used are limited. In unmanned vehicles with energy sources from batteries, the efficiency of using battery energy is very important to consider. Therefore, it is necessary to choose the right brushless motor and propeller.

The power from the battery is one of the factors that influence how well an unmanned aircraft performs in flight [2]. To extend the flight length of an unmanned aircraft, one of the attempts is to select a brushless motor with the lowest power consumption. This allows the aircraft to fly longer. Experiments and testing must be carried out to determine the brushless motor's power output. When the power consumption of a brushless motor is known, it may be used to determine an unmanned aircraft's flying performance.

Brushless motor and propeller testing can be done using a brushless motor and propeller test stand. The existence of this testing tool is still limited to research laboratories. The existence of the Covid-19 pandemic condition provides limitations for students and researchers in using laboratory facilities so that students must be able to do testing independently from home. However, this testing tool is still quite expensive for students if they have to buy it. Therefore, it is necessary to make a brushless motor and propeller testing tool independently.

Research and development on brushless motor and propeller testing tools have been carried out [3], [4]. A test bench for comparison of given motors and propellers is introduced [5]. A propeller-motor test bench is designed to design an efficient quadrotor. The Thrust Benchmarking System (TBS) is a low-cost, easy-to-manufacture sensor for measuring the performance of electric UAV propulsion systems [6]. Its applications include calculating a propeller's static thrust using basic empirical characterization modeling. The development of a laboratory tester for evaluating the effectiveness of propeller-engine groups (PEG) of electric unmanned aerial vehicles (UAV) [7]. A system has been developed that enables the measurement of motor amperage, the generated thrust of PEG, battery voltage, and motor temperature. Research to test the performance of the propeller is carried out by comparing the calculated data with experimental data [8]. The theoretical performance estimations for the motors and propellers are calculated and compared to experimental findings from static thrust testing [9]. The goal is to check the models' accuracy and make choosing the right brushless DC motor and propeller for flight applications easier.

This paper proposes a simple brushless motor and propeller test stand which can be used for experimental testing of brushless motors and propellers on a small scale, especially during home experiments. The design concept uses a load cell sensor and HX711 driver to read force data and an INA 219 sensor to read current and voltage data. The rotational speed of the brushless motor is regulated using a potentiometer. The test results data are in the form of force, current, voltage, and power. Data is displayed directly using a 16x2 LCD. Data can also be sent to a computer via a USB port for storage or further analysis.

2. Design A Simple Brushless Motor and Propeller Test Stand

A simple brushless motor and propeller test stand consists of 4 main parts, namely the input unit, the controller, the output unit, and the brushless motor mounting. The block diagram is shown in figure 1.

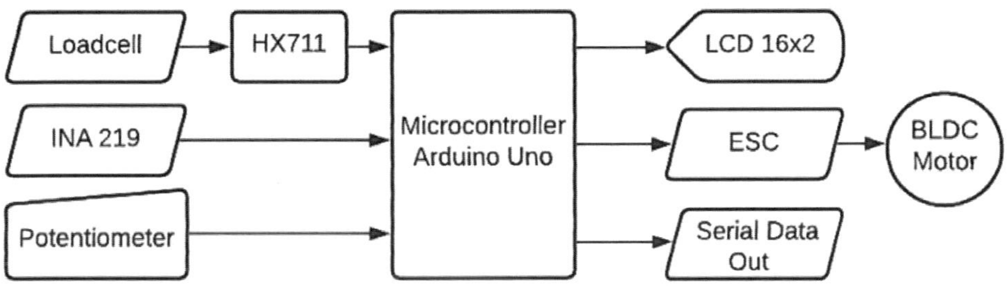

Figure 1. Block diagram of general system architecture.

2.1. The Input Unit

In the input unit, there are two sensors. Loadcell sensor to read force data with a capacity of 1Kg and INA219 sensor to read voltage and current data. The loadcell sensor is connected to the HX711 type module via pins A+, A-, E-, and E+. The HX711 module is connected to the Arduino controller via the I2C pin.

2.2. The Controller

The controller uses an Arduino Uno with an ATmega 328 processor. The controller is used to process data from the sensor, then sends data to a 16x2 LCD (Liquid Crystal Display) and sends data serially to a computer via a USB port.

2.3. The Output Unit

The output unit there is a display using a 16x2 LCD to display force, current, voltage, and power data. The reading data is also sent to the computer serially via the USB port so that the data can be displayed on the serial monitor and can also be stored for further analysis purposes.

2.4. The Brushless Motor Mounting

The brushless motor mounting serves to put the brushless motor to be tested. In addition, this brushless motor mounting also functions to transmit the force generated by the brushless motor to the loadcell sensor. Brushless motor mounting made of an acrylic board with a thickness of 3mm. All electronic components are placed on a wooden board measuring 26cm x 15cm x 1 cm. In addition to electronic components, this board is also a brushless motor mounting mount that unites all components into one integrated testing system. The overall configuration of the wiring diagram is shown in figure 2.

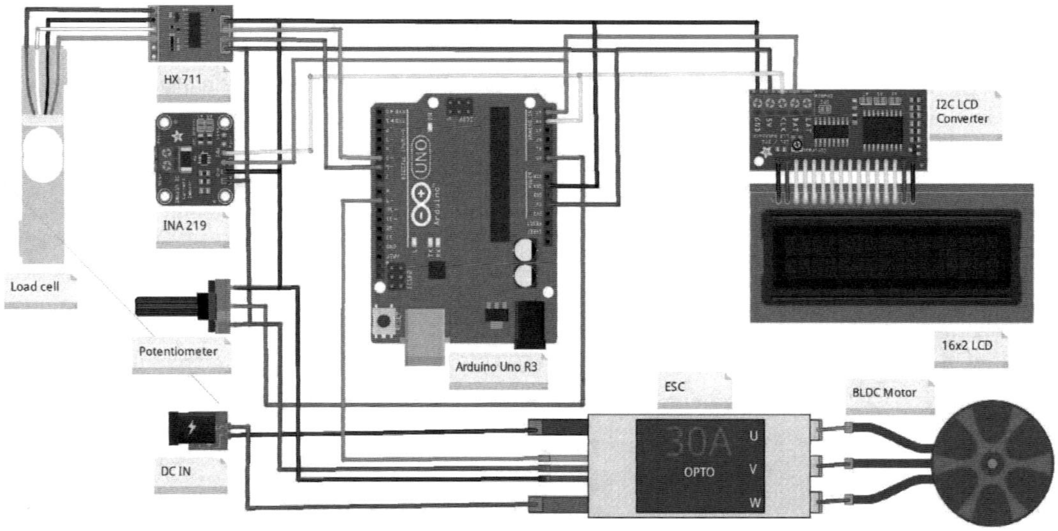

Figure 2. Wiring diagram.

The components needed to make a simple brushless motor and propeller test stand consist of two parts, namely electronic components and mechanical components. The electronic components are loadcell sensor, INA 219 sensor, Arduino Uno, 16x2 LCD, ESC, and battery. The mechanical components are the board and the brushless motor mounting. The list of components and estimated costs are presented in table 1.

Table 1. The Component list and cost estimation.

No	Name of Component	Quantity	Price per Item (IDR)	Total Price (IDR)
1	Loadcell 1 Kg and Modul HX711	1	45.000	45.000
2	INA219 sensor	1	35.000	35.000
3	Arduino Uno R3	1	55.000	55.000
4	LCD 16x2 Green Background	1	17.000	17.000
5	I2C LCD Converter	1	10.000	10.000
6	Hobbywing Skywalker 30A ESC	1	145.000	145.000
7	Potentiometer 50K	1	3.500	3.500
8	Battery Lipo Turnigy 1300mAh 3S 30C	1	265.000	320.000
9	Brushless Motor Mounting	1	20.000	265.000
10	Board 26cm x 15cm x 1 cm	1	50.000	50.000
	Total	10		945.500

3. Implementation and Testing

The implementation phase is divided into 2 (two) stages, namely hardware and software implementation. Hardware implementation starts from assembling the brushless motor mounting with the loadcell sensor and then installing it on the circuit board. The next stage is to install all the components on the circuit board according to the layout then connect the cables from one part to another according to the wiring diagram. If the wiring stage is complete, then proceed with checking to avoid installation errors, especially in the power supply section. After the hardware implementation is complete, then proceed with programming using the Arduino IDE. Programming is done in stages to ensure each part can function properly. The results of the implementation of A Simple Brushless Motor and Propeller Test Stand are shown in figure 3.

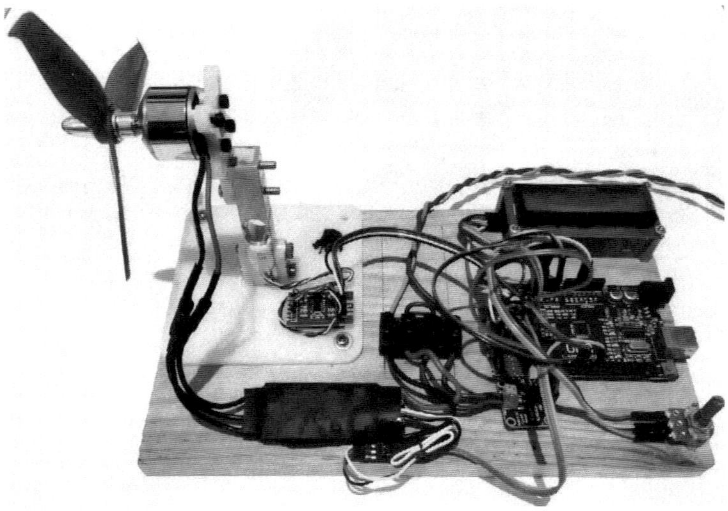

Figure 3. The implementation.

The performance testing of the Brushless Motor and Propeller Test Stand was carried out using 3 types of brushless motors and 1 type of propeller. The type of brushless motor used is type A2212(6T)-2200 KV, MR2205-2300 KV, and MT2204-2300 KV. The type of propeller used is type 5055 with 3 blades.

3.1. Sensor Function Testing

This test is to determine the accuracy and precision of reading data. The test is done by comparing the data from the sensor readings with the readings of the measuring instrument. The results of reading the current, voltage, and power data on the sensor are compared to the data readings on the measuring instrument. Each test was carried out 3 times and then the average value was taken. The results of the sensor function test are shown in table 2.

Table 2. The Results of Sensor Readings.

Instrument				Sensor				%Error			
V (V)	I (mA)	P (mW)	T (gf)	V (V)	I (mA)	P (mW)	T (gf)	V (V)	I (mA)	P (mW)	T (gf)
0	0,00	0	0	0,00	0,00	0	0	0,00	0,00	0,00	0,00
1	124,80	124,8	50	1,00	124,00	124	51	0,00	0,64	0,64	2,00
2	580,00	1160	100	2,00	582,00	1164	99	0,00	0,34	0,34	1,00

3	710,00	2130	150	2,99	712,00	2128,88	152	0,33	0,28	0,05	1,33
4	830,00	3320	200	3,94	830,00	3270,2	201	1,50	0,00	1,50	0,50
5	950,00	4750	300	4,94	945,00	4668,3	298	1,20	0,53	1,72	0,67
6	1040,00	6240	400	5,93	1044,00	6190,92	403	1,17	0,38	0,79	0,75
7	1140,00	7980	500	6,92	1140,00	7888,8	500	1,14	0,00	1,14	0,00
8	1230,00	9840	600	7,92	1234,00	9773,28	604	1,00	0,33	0,68	0,67
9	1320,00	11880	700	8,99	1322,00	11884,8	706	0,11	0,15	0,04	0,86
10	1410,00	14100	800	9,98	1408,00	14051,8	805	0,20	0,14	0,34	0,63
11	1500,00	16500	900	10,97	1499,00	16444	906	0,27	0,07	0,34	0,67
12	1580,00	18960	1000	11,98	1583,00	18964,3	1003	0,17	0,19	0,02	0,30
			Average					0,55	0,23	0,59	0,72

According to the test results, the INA219 sensor can read voltage values with an error percentage of 0.55 %, current values with an error percentage of 0.23 %, and power values with an error percentage of 0.59 %. The loadcell sensor can function properly to read the force value with an error percentage of 0.72 %. The results in this study are different when compared to the results of the INA219 sensor testing in previous studies. Sensor testing in the previous study was carried out at a voltage of 3V-5V with an error percentage of 0.2% and testing at a current of 2V-3V with an error percentage of 0.17% [10].

3.2. Power and Force Reading Performance Testing

Tests were carried out on three types of brushless motors with one type of propeller. The test is done by adjusting the speed of the brushless motor from minimum rotation to maximum rotation. The power supply voltage used is from a power supply with a capacity of 15VDC 50A. The observed data are power data and force. The test is carried out by adjusting the force from 0-200 gf in multiples of 10 gf and then observing the amount of power consumed by the brushless motor. The 0 – 200 gf interval was chosen because it refers to the maximum capability produced when tested with a propeller at a power supply voltage of 15VDC 50A.

Each test was taken three times, then the data that had been obtained was taken the average value. This is done to reduce the risk of errors during data collection. Data from the test results of power and force readings are presented in a graph as shown in figure 4.

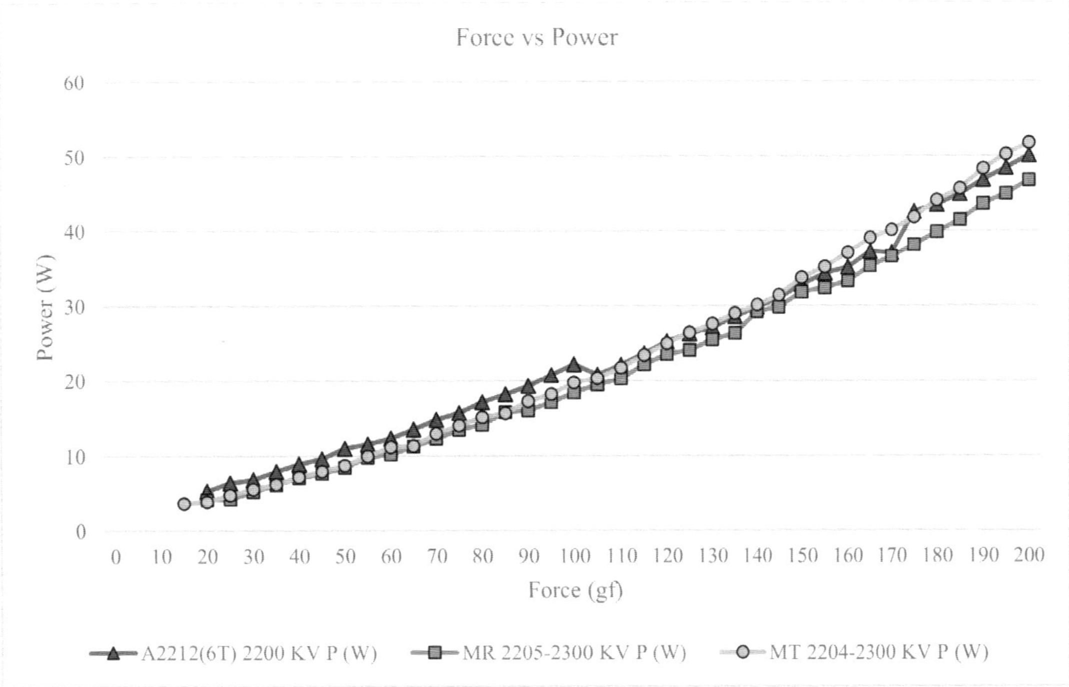

Figure 4. The reading test results in force and power with 0–200 gf interval of force.

Based on the test data, it can be seen that the Brushless Motor and Propeller Test Stand can test the three types of brushless motors with 3 blades propeller type 5055. The graph in figure 4 shows that the performance of the Brushless Motor and Propeller Test Stand can read data stably. When testing using a brushless motor, the A2212(6T)-2200 KV type and the MR2205-2300 KV type were able to read forces from 20 gf to 200 gf, while the MT2204-2300 KV type was able to read forces from 15 gf to 200 gf. The power that is read on the brushless motor on the type A2212(6T)-2200 KV is between 5.29 W and 50.06 W, on the type MR2205-2300 KV between 4.16 W and 46.76 W while on the MT2204-2300 KV type it is between 3.62 W and 51.74 W. This is enough to test a small capacity brushless motor.

4. Conclusion

The simple brushless motor and propeller test stand makes it simple to test a variety of brushless motors and propellers. The power and force generated by brushless motors and propellers may be tested using this instrument. This equipment may be used to perform experiments and test different types of brushless motors and propellers at home. This study proposes a simple brushless motor and propeller test stand that can measure forces from 0 gf to 1000 gf with an error rate of 0.72 %. The power that can be read ranges from 0 mW to 18960 mW, with a 0.59 % error rate.

References

[1] T. Túró, "Monitoring of military vehicles batteries," *ICMT 2015 - International Conference on Military Technologies 2015*, 2015, doi: 10.1109/MILTECHS.2015.7153716.
[2] N. Williard, W. He, M. Osterman, and M. Pecht, "Predicting Remaining Capacity of Batteries for UAVs and Electric Vehicle Applications," *Polymer*, 2010.
[3] F. Cunha, A. Fonseca, and L. Sousa, "Development of a Test Bench for UAV Propulsion Systems Characterisation," 2017.

[4] A. J. Verbin, T. T. Takahashi, and D. B. White, “Detailed Design of a Pulsed Plasma Thrust Stand,” 2017. doi: 10.2514/6.2017-4523.
[5] G. Dındış, B. Keskin, and A. Karamancioglu, “A Test Bench for Electrical Motor and Propeller Components of a Lightweight Quadcopter,” 2018, no. April.
[6] R. Virginio *et al.*, “Design and implementation of low cost thrust benchmarking system (TBS) in application for small scale electric UAV propeller characterization,” *Journal of Physics: Conference Series*, vol. 1130, no. 1, 2018, doi: 10.1088/1742-6596/1130/1/012022.
[7] V. A. Morozov, A. M. Yahin, A. E. Zhirova, and A. v. Parshin, “Hardware-software system for studying the effectiveness of propeller-engine groups for unmanned aerial vehicles,” *Journal of Physics: Conference Series*, vol. 1582, no. 1, 2020, doi: 10.1088/1742-6596/1582/1/012063.
[8] M. Scanavino, A. Vilardi, and G. Guglieri, “An Experimental Analysis on Propeller Performance in a Climate-controlled Facility,” *Journal of Intelligent and Robotic Systems: Theory and Applications*, vol. 100, no. 2, pp. 505–517, 2020, doi: 10.1007/s10846-019-01132-9.
[9] D. Muzar and E. Lanteigne, “Experimental Characterization of Brushless DC Motors and Propellers for Flight Application,” in *Proceedings of The Canadian Society for Mechanical Engineering International Congress 2016*, 2016, pp. 1–11. [Online]. Available: http://hdl.handle.net/10393/36011
[10] M. Y. Darmawan, M. S. Anrokhi, and A. Komarudin, “Rancang Bangun Sistem Pemantauan Kinerja Panel Surya Tipe Mono-Crystalline Silicon Berbasis IoT,” *Electrician*, vol. 13, no. 3, pp. 81–83, 2019, doi: 10.23960/elc.v13n3.2127.

Application of LQR Full-State Feedback Controller for Rotational Inverted Pendulum

N Setiawan[1] and G N P Pratama[2]

[1]Department of Electrical Engineering Education, Universitas Negeri Yogyakarta, Indonesia.
[2]Department of Electronics and Informatics Engineering Education, Universitas Negeri Yogyakarta, Indonesia.

E-mail: nurman.setiawan@uny.ac.id[1], gilang.n.p.pratama@gmail.com[2]

Abstract. The rotational inverted pendulum is an interesting subject for some researchers, especially control engineers. Its nonlinear and underactuated characteristic make it quite challenging to stabilize it. Hence, a proper control law is a must to make it stable. Here, in this paper, we present a control law using LQR (Linear-Quadratic Regulator) to stabilize the rotational inverted pendulum. The experiments are carried out by linearizing the model and simulate the response in MATLAB. The results show that the controller succeeds to stabilize the states of rotational inverted pendulum to their respective equilibrium points. Even more, it provides zero settling errors.

1. Introduction

Over the years, the inverted pendulum has been an interesting topic for many researchers, especially in control engineer. It is due to its nonlinearity and unstable behavior. Therefore, in order to stabilize it, we need to design a proper controller [1]. There are some approaches to design the controller, either using linear or nonlinear methods [2]. Codrons *et al.* [3] and Morimoto *et al.* [4], have designed a nonlinear controller for an inverted pendulum. Meanwhile, Sugie *et al.* [5] and Renou *et al.* [6] are applying linearization method to design the controller.

Unlike the linear controller, the nonlinear one is quite difficult to design. Nevertheless, we need to linearize the model before designing the controller. Here, we utilize Linear-Quadratic Regulator (LQR), a linear controller, to stabilize the rotational inverted pendulum with some advantages. First, LQR is such an optimal controller that can provide not only a good stability, but also the stability margin of a system is guaranteed [7]. It provides a better optimal energy compared againts PID and fuzzy controller [8]. Next, we do not need to do the loop-shaping to obtain the gain parameters. Lastly, it can work on Single-Input Multi-Output (SIMO) systems, which necessary for our rotational inverted pendulum.

Practically, the rotational inverted pendulum as depicted in Figure 1, is an underactuated system. It means, the number of degree-of-freedom (DoF) that needs to be controlled, is more than the number of actuators. Meanwhile, if those numbers are equal, thus the system is fully actuated. Despite the difficulties to stabilize it, the underactuated systems have some advantages over the other one. Their main advantages are the control efforts and energy consumption which are relatively lower than the fully actuated systems.

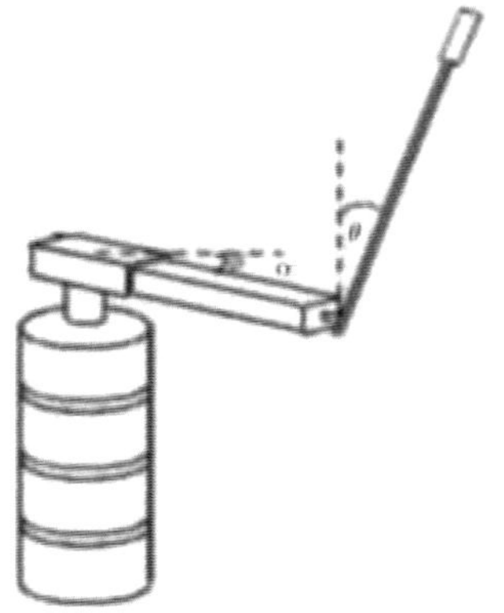

Figure 1. Rotational inverted pendulum.

Technically speaking, there are three main parts of rotational inverted pendulum, they are motor, rotational arm, and the pendulum [9]. The rotational arm can be fully rotated on a horizontal plane since it is connected to the motor shaft. Meanwhile, the pendulum is connected to the other end of the arm. It makes the pendulum can be rotated vertically.

It is necessary to mention that this research is conducted by modeling the rotational inverted pendulum, then designing the controller using Matlab. Furthermore, the organization of this paper is as follows. It has mentioned before, that the model of rotational inverted pendulum is necessary, and it is elaborated in Section 2. Later, in Section 3, the design of the controller is presented. The results and analysis of the controller can be seen in Section 4. Lastly, the conclusion of this paper can be found in Section 5.

2. Model of Rotational Inverted Pendulum

In this section, there are two main topics that will be presented. First, the nonlinear equations of the model can be found in Subsection 2.1. Next, we take linearization into account, it is briefly explained in Subsection 2.2.

2.1. Nonlinear Equations

We can use Euler-Lagrange equation to obtain the nonlinear model of the system as

$$\dot{\theta} = \frac{d\theta}{dt}, \tag{1}$$

$$\ddot{\theta} = (I + m(R^2 + l^2)\sin^2(\theta))^{-1}[-2ml\sin(\theta)\cos(\theta)\,\dot{\alpha}\dot{\theta}$$
$$-mlR\sin(\theta)\cos^2(\theta)\,\dot{\alpha}^2 - mRg\sin(\theta)\cos(\theta) + mRl\sin(\theta)\dot{\theta} + u], \tag{2}$$

$$\dot{\alpha} = \frac{d\alpha}{dt}, \tag{3}$$

$$\ddot{\alpha} = (I + m(R^2 + l^2)\sin^2(\theta))^{-1}\left[\frac{(I+mR^2+ml^2\sin^2(\theta))\beta}{R\cos(\theta)} - \gamma - u\right], \tag{4}$$

$$\beta = (l\sin(\theta)\cos(\theta)\dot{\alpha}^2 + g\sin(\theta)), \tag{5}$$

$$\gamma = 2ml^2\sin(\theta)\cos(\theta)\dot{\alpha}\theta \; - mRl\sin(\theta)\dot{\theta}^2, \tag{6}$$

where respectively θ and α are the pendulum and rotating arm angle. Meanwhile, I, m, l, R, u, and g that consecutively denote moment inertia of the DC motor, mass of the pendulum, length of the pendulum, length of the arm, input torque, and gravity constant.

Those equations can be represented into a nonlinear function such as

$$\dot{\boldsymbol{x}} = f(\boldsymbol{x}, u) \tag{7}$$

where the states be

$$\boldsymbol{x} = [\theta \quad \dot{\theta} \quad \alpha \quad \dot{\alpha}] \tag{8}$$

2.2. *Linearization*

In order to design a linear controller, we need to linearize those equations. We decide the equilibrium points such as the state vector **x** = [0 0 0 0] and the input torque $u = 0$, then it can be linearized as

$$\dot{\boldsymbol{x}} = \frac{\delta f(x,u)}{dx}\Big|_{x=0} + \frac{\delta f(x,u)}{du}\Big|_{u=0} \tag{9}$$

$$= \boldsymbol{Ax} + \boldsymbol{Bu} \tag{10}$$

Respectively the state matrix and input vector can be denoted as

$$\boldsymbol{A} = \begin{bmatrix} 0 & 1 & 0 & 0 \\ \frac{g(I+mR^2)}{Il} & 0 & 0 & \frac{Rb}{Il} \\ 0 & 0 & 0 & 1 \\ -\frac{mRg}{I} & 0 & 0 & -\frac{b}{I} \end{bmatrix} \tag{11}$$

$$\boldsymbol{B} = \begin{bmatrix} 0 \\ -\frac{R}{Il} \\ 0 \\ \frac{1}{I} \end{bmatrix} \tag{12}$$

where b is viscous coefficient. As for the output we have

$$y = \boldsymbol{Cx}$$
$$= [\theta \quad \alpha]^T \tag{13}$$

where the output matrix be

$$\boldsymbol{C} = \begin{bmatrix} 1 & 0 & 0 & 0 \\ 0 & 0 & 1 & 0 \end{bmatrix} \tag{14}$$

3. Controller Design

LQR is an optimal control that offers the robustness of a system with small steady state errors [10]. The block diagram of a full-state feedback controller can be seen in Figure 2.

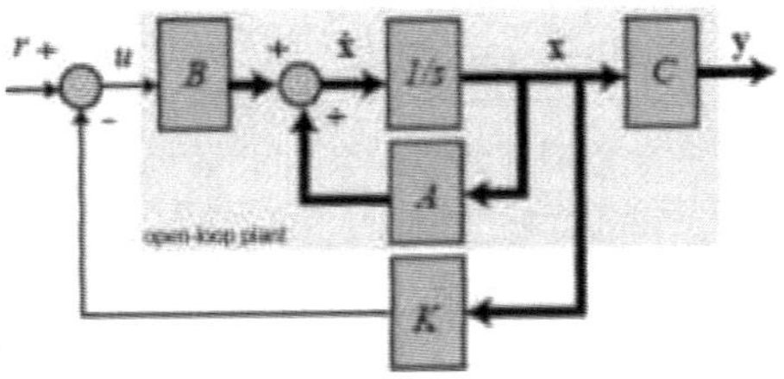

Figure 2. Block diagram of a LQR full-state feedback controller.

As an optimal control, it comes with a cost function such as

$$J(\boldsymbol{u}) = \int_0^\infty \left(\boldsymbol{x}^T \boldsymbol{Q} \boldsymbol{x} + \boldsymbol{u}^T \boldsymbol{R} \boldsymbol{u}\right) dt \tag{15}$$

that must be minimized. In our case, we have four states and one input, therefore **Q** is a 4-by-4 matrix, meanwhile **R** is a scalar. Both **Q** and **R** affect differently towards gain feedback **K**. We can obtain a larger gain feedback **K** by increasing **Q**. On the other hand, by assigning a larger value of **R**, we can obtain a smaller gain feedback **K** [10]. Next, we need to find an auxiliary matrix **S** by solving this Algebraic Ricatti Equation (ARE) where

$$\mathbf{A}^T\mathbf{S} + \mathbf{SA} - \mathbf{SBR}^{-1}\mathbf{B}^T\mathbf{S} + \mathbf{Q} = 0. \tag{16}$$

The auxiliary matrix **S** is needed to determine the gain feedback **K** such as

$$\mathbf{K} = \mathbf{R}^{-1}(\mathbf{B}^T\mathbf{S}). \tag{17}$$

4. Results and Discussion

This section covers the results of this paper, including the parameters and its analysis. The parameters of rotational inverted pendulum used in this paper is presented in Subsection 4.1. Then, in Subsection 4.2 we will elaborate system response with fully-state feedback controller.

4.1. *System Parameters*

Before any further, we need to define the parameters used in this simulation.

Table 1. Parameters of rotational inverted pendulum.

Parameters	Values
m	0.05kg
l	0.48m
R	0.57m
I	0.03264kgm^2
b	0.00351kgm^{2}s^{-1}

Consecutively, we have m, l, R, I, and b that denote mass of the pendulum, length of the pendulum, length of the arm, and viscous coefficient. Those are based on the previous research conducted by Cahyadi *et al.* [1], meanwhile as for the gravity constant we have $g = 9.8\text{ms}^{-2}$.

4.2. *Simulation and Analysis*

First we can analyze the stability of the open-loop system. Let us substitute those values on Table 1 and the gravity constant into state matrix **A** and input vector **B**, hence we have

$$\boldsymbol{A} = \begin{bmatrix} 0 & 1 & 0 & 0 \\ 7.052 & 0 & 0 & 0.1277 \\ 0 & 0 & 0 & 1 \\ -8.557 & 0 & 0 & -0.1075 \end{bmatrix} \tag{18}$$

$$\boldsymbol{B} = \begin{bmatrix} 0 \\ -8.3824 \\ 0 \\ 30.6373 \end{bmatrix} \tag{19}$$

We can get the information either the open-loop system is stable or not by computing the eigenvalues of **A**. Eventually, we have the eigenvalues as $eig(\boldsymbol{A}) = [0\ -2.7316\ 2.5764\ 0.0476]^T$. It is apparent that the open-loop system is unstable since there are some elements with real positive values. Furthermore, we need to investigate either the system is controllable or not. It is necessary since the controller can only be applied if the system is controllable. It can be achieved by calculating the rank of controllability matrix **Co**. The controllability matrix can be calculated as follows

$$\boldsymbol{Co} = [\boldsymbol{B}\ \boldsymbol{AB}\ \boldsymbol{A}^2\boldsymbol{B}\ \boldsymbol{A}^3\boldsymbol{B}]$$

$$\boldsymbol{Co} = \begin{bmatrix} 0 & -8.3824 & 3.9124 & -59.5333 \\ -8.3824 & 3.9124 & -59.5333 & 36.7950 \\ 0 & 30.6373 & -3.2935 & 72.0822 \\ 30.6373 & -3.2935 & 72.0822 & -41.2271 \end{bmatrix} \tag{20}$$

Indeed, **Co** is a 4-by-4 matrix, that corresponds to the number of state variables. Here, we have the rank of **Co** = 4, it means that the system is controllable.

After investigating the controllability of the system, we can move further to design a proper control law to make it stable. In order to design a control law using LQR, first we have to decide both **Q** and **R**. Here, we have

$$\boldsymbol{Q} = \boldsymbol{C}^T\boldsymbol{C},$$

$$\boldsymbol{Q} = \begin{bmatrix} 1 & 0 & 0 & 0 \\ 0 & 0 & 0 & 0 \\ 0 & 0 & 1 & 0 \\ 0 & 0 & 0 & 0 \end{bmatrix} \tag{21}$$

and

$$R = 1. \tag{22}$$

We chose those values to obtain an appropriately small feedback gain **K** that leads to a lower energy drive. Even though it will take longer settling time, the sparseness and small values of **Q** can give us zero settling error [10].

After obtaining the parameters of **C**, **B**, **Q**, and **R**, then we can solve the Algebraic Ricatti Equation using Matlab. Hence, we have the auxiliary matrix as

$$\mathbf{S} = \begin{bmatrix} 53.1138 & 24.4687 & 6.9955 & 6.1891 \\ 24.4687 & 11.3251 & 3.233 & 2.8702 \\ 6.9955 & 3.233 & 1.1882 & 0.8519 \\ 6.1891 & 2.8702 & 0.8519 & 0.736 \end{bmatrix}$$

Lastly, we can solve equation (17) to obtain the feedback gain as

$$\mathbf{K} = [\,-15.4872\ \ -6.9955\ \ -1\ \ -1.5095\,]\,. \tag{23}$$

Since the feedback gain has been obtained, we can define the control law for the full-state feedback system in Figure 2 as

$$u = r - \mathbf{Kx}. \tag{24}$$

As the control law is being implemented, we can investigate the stability of the closed-loop system in the same fashion as the open-loop system. The eigenvalues of the closed-loop are $eig(\boldsymbol{A} - \boldsymbol{BK}) = [-4.1307 + 3.7968i \quad -4.1307 - 3.7968i \quad -2.1193 + 0.2947i \quad -2.1193 - 0.2947i]^T$. Since all the real elements of eigenvector are negatives, then the closed-loop system is stable.

Now, we will simulate the response of the closed-loop system with initial states $\boldsymbol{x_0} = [\,0.5\ \ 0\ \ 0.4\ \ 0\,]^T$ and the desired states $\boldsymbol{x_d} = [\,0\ \ 0\ \ 0\ \ 0\,]^T$. First, we have the response of pendulum angle as depicted in Figure 3.

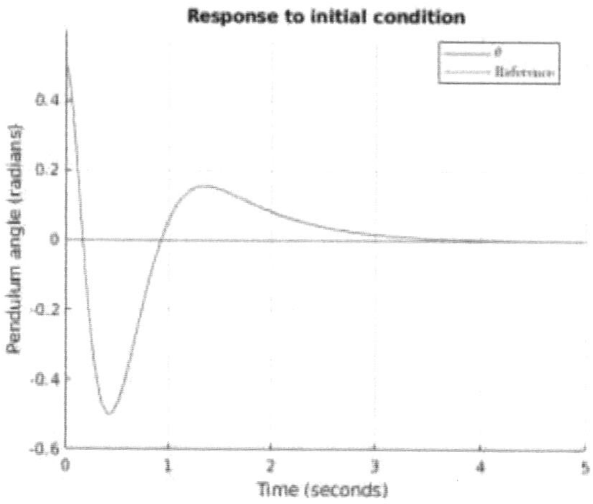

Figure 3. Response of pendulum angle θ.

It takes almost 4 seconds for the controller to regulate the pendulum angle from the initial condition at $\theta_0 = 0.5$rad to the equilibrium at $\theta_d = 0$rad. Nevertheless, it succeeds to reach the equilibrium without settling error. In corresponds to the pendulum angle, we have the angle rate $\dot{\theta}$ as shown in Figure 4.

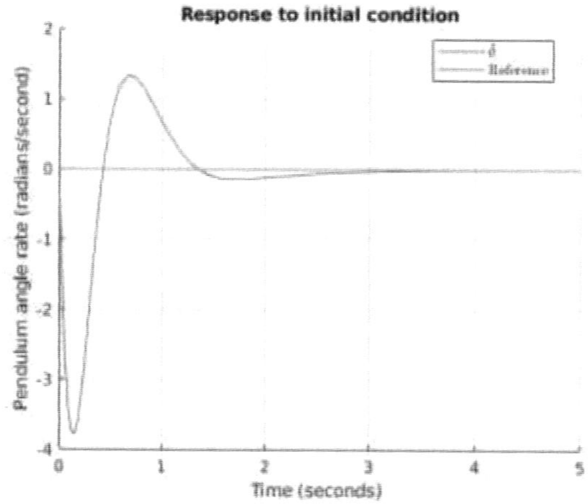

Figure 4. Response of pendulum angle rate $\dot{\theta}$.

Initially, we have angle rate $\dot{\theta}_0 = 0\ rad/s$. We can see that if the angle rate $\dot{\theta}$ is turning to negative values, then θ will follow, and vice versa. It is apparent that $\dot{\theta}$ will change over the time until the pendulum angle settling at $\theta_d = 0$rad. It is also worth to be mentioned that the angle rate indicates the direction of rotation, either it is clockwise or counterclockwise.

In our case, if the angle rate has negative values, it is supposed to rotate clockwise. On the contrary, it will rotate counterclockwise when the angle rate is positive. Just like the pendulum angle, the arm angle also affected by the changes on its rate. Here we have the response of arm angle α and its rate $\dot{\alpha}$ as illustrated in Figure 5 and Figure 6.

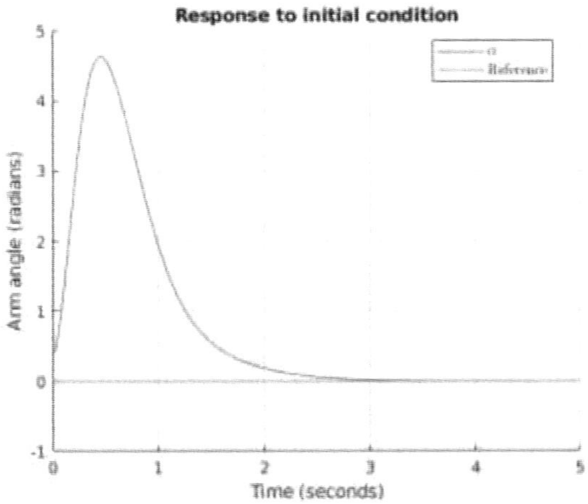

Figure 5. Response of arm angle α.

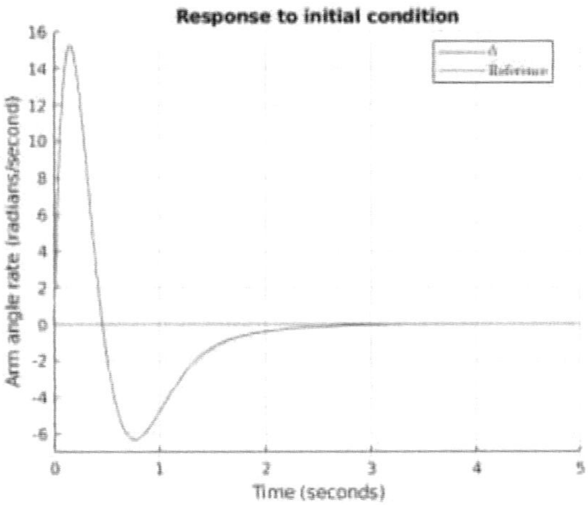

Figure 6. Response of arm angle rate $\dot{\alpha}$.

At first, we have the initial arm angle at α = 0.4rad, meanwhile $\dot{\alpha}$ = 0rad/s. It indicates that at the beginning the arm angle α is stand still, does not move. To regulate the arm angle α to its equilibrium at α_d = 0rad, then it needs to rotate. Therefore, as long as α has not settle the equilibrium, then the arm angle rate $\dot{\alpha}$ is not supposed to be 0rad/s. Eventually, both of them prevail to settle at their equilibrium points in 3 seconds.

Inspite that the responses are quite slow, it can still be tolerated. The settling times for both θ and α are not our main concern here. It is due to the fact that the controller is designed to stabilize the states to their equilibrium points with zero settling error.

5. Conclusion

The rotational inverted pendulum is an underactuated system that cannot be made stable by its mechanical structure. Therefore, we need a proper control law to make it stable. In this paper we present the design of LQR controller for rotational inverted pendulum by considering the sparseness and small value of **Q**. Although the response is supposed to be slower, but it can give us zero settling error. The experiments are conducted by employing Matlab to simulate the response of the system with our proposed controller. The results verify that the proposed controller can be used to stabilize the pendulum and arm angle to their respective equilibrium. Respectively, it takes around 3 and 4 seconds to stabilize the arm and pendulum angle without settling error. Indeed, our focus is to stabilize the arm angle and pendulum angle to their equilibrium points, therefore a sluggish response is still acceptable.

References

[1] A. I. Cahyadi, D. Isarakorn, T. Benjanarasuth, J. Ngamwiwit, and N. Komine. Application of coefficient diagram method for rotational inverted pendulum control. In *ICARCV 2004 8th Control, Automation, Robotics and Vision Conference, 2004.*, volume 3, pages 1769–1773, 2004.

[2] M. F. Hamza, H. J. Yap, I. A. Choudhury, A. I. Isa, A. Y. Zimit, and T. Kumbasar. Current development on using rotary inverted pendulum as a benchmark for testing linear and nonlinear control algorithms. *Mechanical Systems and Signal Processing*, 116:347–369, 2019.

[3] B. Codrons, F. De Bruyne, M. De Wan, and M. Gevers. Iterative feedback tuning of a nonlinear controller for an inverted pendulum with a flexible transmission. In *Proceedings of the 1998 IEEE International Conference on Control Applications (Cat. No.98CH36104)*, volume 2, pages 1281–1285, 1998.

[4] H. Morimoto and S. Kawamoto. Nonlinear control based on equilibrium point analysis for inverted pendulum. In *Proceedings of the 41st SICE Annual Conference. SICE 2002.*, volume 1, pages 238–242, 2002.

[5] T. Sugie and K. Fujimoto. Control of inverted pendulum systems based on approximate linearization: design and experiment. In *Proceedings of 1994 33rd IEEE Conference on Decision and Control*, volume 2, pages 1647–1648, 1994.

[6] S. Renou and L. Saydy. Real time control of an inverted pendulum based on approximate linearization. In *Proceedings of 1996 Canadian Conference on Electrical and Computer Engineering*, volume 2, pages 502–504, 1996.

[7] Priyatmadi, A. P. Sandiwan, H. Wijaya, and A. Cahyadi. Application of spsa lqr tuning on quadrotor. In *2016 6th International Annual Engineering Seminar (InAES)*, pages 32–36, 2016.

[8] H. Maghfiroh, A. Ataka, O. Wahyunggoro, and A. I. Cahyadi. Optimal energy control of dc motor speed control: Comparative study. In *2013 International Conference on Computer, Control, Informatics and Its Applications (IC3INA)*, pages 89–93, 2013.

[9] M. Antonio-Cruz, R. Silva-Ortigoza, C. A. Merlo-Zapata, M. G. Villarreal-Cervantes, D. Munoz-Carrillo, and V. M. Hernandez-Guzman. Modeling and construction of a furuta pendulum prototype. In *2014 International Conference on Mechatronics, Electronics and Automotive Engineering*, pages 98–103, 2014.

[10] O. A. Dhewa, A. Dharmawan, and T. K. Priyambodo. Model of linear quadratic regulator (lqr) control method in hovering state of quadrotor. *Journal of Telecommunication, Electronic and Computer Engineering*, 9(3):135–143, 2017.

The Application of Innovation Visual Animation Media (VAM) at Electrical in Automotive Vocational Education

Muhammad Nurtanto[1], Farid Mutohhari[2], Fadli Rozaq[3], Suyitno Suyitno[4], Alias Masek[5]

[1,2]Universitas Sultan Ageng Tirtayasa
[3]Politeknik Perkeretaapian Mediun Indonesia
[4]Universitas Muhammadiyah Purworejo
[5] Universiti Tun Hussein Onn Malaysia

E-mail: mnurtanto23@untirta.ac.id

Abstract. The purpose of this study is to apply the concept of learning to electricity in the automotive sector based on visual animation media (VAM) as a creative learning product and to assess learning products. One shoot case study was chosen as the research design. A total of 25 participants were involved in the VAM evaluation. Data collection uses product assessment sheets based on creativity, appearance, and product content validity. The results showed that it consisted of three important aspects, namely "mastery" of learning materials, "creative", and "artistic". Product creativity as an output is categorized as creative enough, the product display is quite good, and 64% of the products meet the content validity requirements, while 36% do not meet the exact criteria.

Keywords: Visual animation media, automotive vocational, innovation learning

1. Introduction

Nowadays, advancement of information and communication technology (ICT) has changed the way of knowledge transfer, particularly in the field of education [1]. Knowledge transfer has become easier with the integration of the three essential elements of the technology, particularly in the implementation into the learning process: namely, the technology-based learning, the integration of technology into learning, and environment [2]. Incorporation of these elements and the use of ICT could facilitate students' learning [3], while computers and its software can help teachers to do their tasks creatively, to support student learning [4].

Students who can utilize technology in learning can balance between the demands of work and learning [5]. It certainly strengthens the students' competence. Competence is defined as the knowledge, skills and abilities related to performance in carrying out the work [6]. Competence is also an underlying characteristic of a person associated with superior performance in the job [7]. Competencies required by the students during the learning process using the technology according to are the ability to manage e-learning by integrating a variety of software applications to create products and interaction with the learning content through the presentation of information in various formats (video, audio, animation, etc.).

Many teachers in vocational education institutions have been less concern on the attractiveness aspect of learning materials. The attractive design of the learning materials is possible to achieve, especially in the development of computer technology, using computer-based media, [8]. The quality of teaching, the way teachers teach, teacher education, and the performance of teachers during the teaching are important factors that affect students' achievement [9], because teaching is a collaborative learning process from various learning experiences, open-minded, open – hearted, teachers do not always become the centre of the learning process, and allows students play an active role, [10], [11].

Nurturing good teachers who are competent and able to provide interesting learning process, teacher education must be able to provide professional competence and capable of carrying out an interesting learning session with students. Teacher education must facilitate technology as an important tool for learning process. Consequently, teacher education program has the most influence on the implementation of the technology integration in education. With the benefits of the use of technology in the learning process, then the technology should be included in the learning environment [12], which emphasizes on problem solving [13]. The approach and strategy of technology integration in teacher education are very important in the practice of learning. Teacher education institution must implement digital learning context for learning more effectively and students get interactive learning experience [14].

Teaching quality determines the quality of education. Quality education starts from high quality teachers. The quality of teachers, attitude, dedication to the teaching profession, educational qualification, and professional training play an important role in modern education. Therefore, the teacher's role has paramount importance in generating enthusiasm and inspire students to learn, develop a person's intelligence and wisdom. Professional competence refers to the knowledge, skills, attitudes, capacity, and confidence to succeed in the profession and work. Teachers' professional competence includes a wide range of competencies in different fields such as pedagogical, cultural, communication, personality, and professional needed for effective learning.

Automotive Engineering Vocational Education Program is a program which prepares candidates for vocational education teachers in the field of Automotive Engineering. Based on the study, one of the problems encountered in automotive electrical system learning process is that students have difficulty in understanding automotive electrical circuits and how the system works. One of the solutions is the learning process of automotive electrical system learning material through creative learning activities by transforming the textual material into creative visual products in the form of animation. In addition, while learning textual material, students are required to transform a textual concept into the visual one and put it in the animation on the material being studied. In this study, the creativity of students is required to transform textual information into visual information which is more easily understood. Consequently, this requires creative thinking.

In this article, the invention was directed at developing ideas based on textual knowledge on automotive electrical system books. The ideas were then applied into visual concepts and manifested in a simple animation product. Animation was aimed at facilitating an understanding of the abstract concept of automotive electrical system into easily understood information through easily observed visualization or animation. According to Lowe and Schnotz [15], animation has become an important feature in technology-based learning. The rapid development of animation application in education is progressive with the advancement of information and communication technology.

Animation is the fast display of a static image sequence creating the illusion of motion to mimic the real motion. The practical feasibility of the animation use for learning process depends on the formulation and use of technology [16]. Therefore, the animation develops rapidly in the field of education. Creating a simple animation can be conducted in several stages, namely: making personal notes, character recognition, gaining inspiration, developing ideas, writing scripts, production planning, research/analysis, designing, creating storyboards, and creating. Animation is widely applied to increase interest and motivation, direct attention, illustrate the procedure, and explain how things work [15]. This study was aimed at (a) applying the concept of learning by creating by transforming textual learning materials into visual animation as a creative learning product, and (b) assessing product of the learning process based on the creativity, appearance of products, and the content validity of the product.

2. Research Methodology

A case study was implemented, involved 25 participants who underwent treatment, and further were assessed based on the products produced during learning process. The learning process was conducted using the concept of learning by creating through transforming the textual learning materials into visual material in the form of animation, as a creative learning product. The product was assessed based on the creativity, appearance, and the content validity.

The learning process was carried out with reference to the following guidelines: (a) the use of the technology during learning process to manage e-learning by integrating a variety of software applications to create a product and interaction with the learning material content through the presentation of information in various formats, in this case the animation; (b) the generative learning process of transforming textual learning material into something new in graphic form [17]; (c) the learning process was carried out by the productive process [18] through creative learning process to encourage students to find new ideas and products, be initiative, find relevant information allocating sufficient time (2 weeks) to transform the textual learning material (automotive electrical system) into a visual concept and put it in a creative visual product in the form of animation; (d) the creation of an animated visual learning media with the principle of fast static image display sequentially which creates the illusion of motion to mimic the real motion; and (e) the animation creation process was conducted in stages.

Participants in this study were the fifth semester students of Automotive Engineering Vocational Education program in the Universitas Negeri Semarang, who had completed the automotive electrical system course and currently enrol in the educational media course. The total number of participants was 25 students between the ages of 19-22 years. In this study, the learning process was conducted in the classroom for 3 weeks. The class activity in 1st week was: (a) researcher (as the lecturer) provided learning materials about animation and its example based on the Harrison and Hummell's concept, namely the creation of visual animation instructional media with the principle of fast static image display sequentially, which creates the illusion of motion to mimic the real motion; (b) students were asked to choose the topic of automotive electrical system learning material from the book to be transformed into a creative product; (c) students were given time a week to get ideas, initiatives, and to create scripts to transform the textual information into visual information; and (d) the students then designed the animation using the software such as MS Powerpoint, Flash, or other software. The activities in the 2nd week consisted of: (a) students presented a script and the design of their animation; (b) other students were given the opportunity to comment and provide input on the draft products; (c) the lecturer provided the input; (d) the presenting students were given time to respond to comments the inputs; and (e) the students were given an opportunity to improve their draft for the final product. The activities in the 3rd week consisted of: (a) each student presented the final product of visual media in the form of animation; (b) the final products were collected; and (c) the final products were assessed by the researcher with the instruments that have been prepared.

Data were collected using the product assessment sheet, comprises of several criteria, including the creativity, the appearance, and the content validity of the products. The creativity of the products was assessed from novelty (the product is original, surprising, and germinal), resolution (the product is valuable, logical, useful, and understandable), and the elaboration and synthesis (the product is organic, elegant, complex, and well-crafted) [19]. The assessment of the products was assessed in terms of attractiveness, layout, and smoothness of motion. The content validity of the product was assessed in terms of validity, the motion validity, and the validity of the overall system. The products were assessed using a Likert scale ranging from 1 to 5 points. The data were analysed using descriptive statistics in the form of average and percentage to determine the level of creativity, appearance, and the validity of the products.

The classification of score were made into the level which have evaluative meaning based on Azwar [50], if the score is $(M + 1{,}5s) < X$, then it is categorized as a very high; if the score is $(M + 0{,}5s) < X \leq (M + 1{,}5s)$, then it is categorized as high; if the score is $(M - 0{,}5s) < X \leq (M + 0{,}5s)$, then it is categorized as sufficient; if the score is $(M - 1{,}5s) < X \leq (M - 0{,}5s)$, then it is categorized as mediocre;

and if the score is X ≤ (M11,5s), then it is categorized as bad. M is the average results of the assessment, s is the standard deviation of the results of the assessment, and X is a score assessment.

3. Result and Discussion

The results of the assessment of learning by creating learning products were presented in the following table 1.

Table 1. The Results of The Product Creativity

No	Indikator	Assessment Aspect	The Average Score (1-5)
1	Novelty	Original	3.64
		Surprising	3.64
		Germinal	3.52
2	Resolution	Valuable	3.60
		Logical	3.68
		Useful	3.72
		Understandable	3.80
3	Elaboration and synthesis	Organic	3.48
		Elegant	3.52
		Complex	3.60
		Well-crafted	3.44
		Average score	**3.60**

The assessment condition was written as follows: if the score was 4.9 <X, then it was categorized as highly creative product; if the score was 4 <x ≤ 4.9, then it was categorized as creative; if the score was 3.2 <X ≤ 4, then it was categorized as quite creative; if the score was 2.3 <x ≤ 3.2, then it was categorized as uncreative; and if the score was X ≤ 2.3, then it was categorized as very uncreative. Table 1 indicates that the average score of the product is 3.60. Therefore, the product is generally categorized as quite creative. The distribution of the product creativity of the participants is shown in Figure 1.

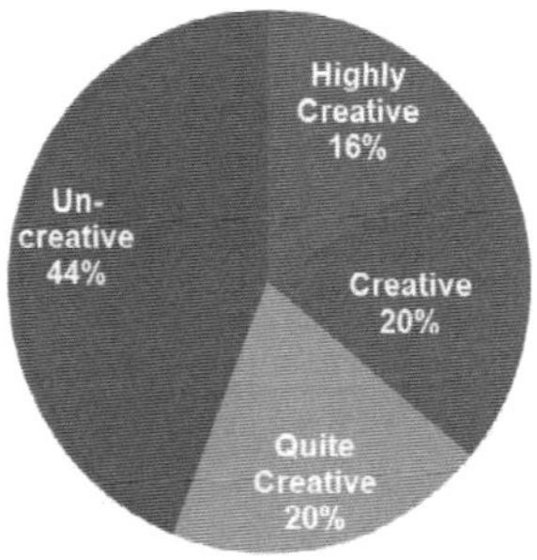

Figure 1. Distribution of Product Creativity

Pie chart indicates a large portion of products (44%) do not meet the creative criteria as expected. Some of the reasons were: the products did not contain an interesting idea, illogical visualization, also, the products were worthless and could not be used to facilitate learning, the message or information contained on the product was elusive, missing an element of motion that characterized the animation, and were not made carefully. Based on the appearance and the validity of the product, the results of the assessment as shown in table 2.

Table2. The Results of The Appearance and The Content Validity of The Product Assessment

No	Indicator	Assessment Aspect	The average score (1 – 5)
1	The appearance of the product	Attractiveness	3.44
		Layout	3.52
		the smoothness of the motion	3.20
		Average score	**3.39**
2	The product validity	The content validity	3.76
		The motion validity	3.68
		The validity of the overall system	3.76
		Average score	**3.73**

The assessment condition was written as follows: if the score was 4.6 <X, then it was categorized as very attractive product; if the score was 3.8 <x ≤ 4.6, then it was categorized as attractive; if the score was 3.0 < X ≤ 3.8, then it was categorized as quite attractive; if the score was 2.1 < X ≤ 3.0, then it was categorized as less attractive; and if the score was X ≤ 2.1, then it was categorized as very unattractive. From table 2, the average score of the product is 3.39. Therefore, the products were generally categorized as quite attractive. The distribution of the product appearance data is shown in Figure 2(a).

(a) (b)

Figure 2. Distribution of The Product Appearance (a), and The Validity of The Product (b)

The number of products which was categorized as bad was only 12 %. Some things that are not good in terms of the appearance that the resulting visualization was not beautiful and not attractive to look at; a placement of images, text, and moving elements were imbalanced; and the resulting motion was very rough. Animation as a tool to convey information to the learner must contain elements of validity. Therefore, the validity of the products must also be judged from the validity aspect. The validity product in this study was assessed in terms of the content validity, motion validity in the animation, and the validity of the overall system from the beginning to the product. The assessment result of the level of the product validity is shown in Figure 2 (b).

The validity of the products was divided into two categories: namely, the valid product and the invalid product. There were 36% of the products did not meet the elements of validity, in terms of content validity, motion of the animation validity, and the overall contents validity of the electrical system that were visualized. If there was one thing that contained incorrect information, then the product was categorized as not meeting the elements of validity. Common mistakes made by the students were the lack of learning material understanding. Information from the books being studied was partially or completely not understood. Therefore, the concept of product development was invalid. Consequently, it resulted in the invalid visualization. Overview of products produced in the learning in terms of the creative side, the view, and the truth of the contents can be seen on Figure 3.

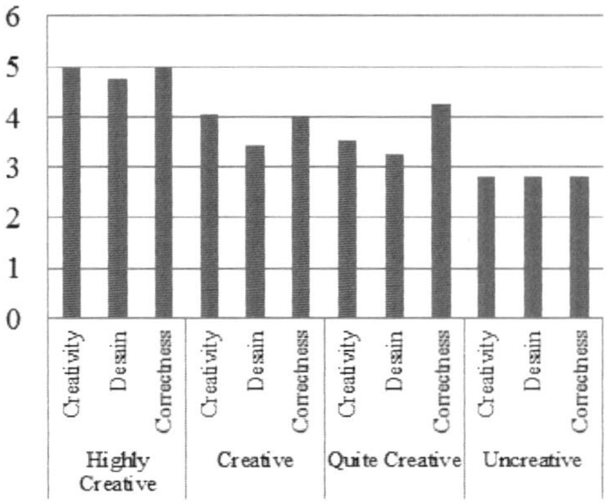

Figure 3. The Overview of Condition of Creativity, Appearance, And the Validity of The Product

Based on Figure 3, the creative products are the reflection of the students who have the creativity in making the products. Creative products were followed with a great visual appearance of the products and the validity of the content as well. The uncreative products were also followed by an unattractive appearance and the incorrect information about the products. Based on the above results, it can be described that students who were categorized as creative had good results with good content validity. Here is an example of screen shots form two visual products in the form of animation with automotive electrical system theme.

Another example of products in this study was the visualization of how the starting system works on the vehicle (see figure 4). When the system is working, the electrical current flow is visualized by the flow of the green line to all parts of the starter system at work. The idea of this product was to slowly describe the flow of electrical current. Therefore, the phenomenon of how the starter system works can be observable. Based on figure 4, the students can easily understand the working principles than reading the working principle of the starting system in the form of text in a textbook.

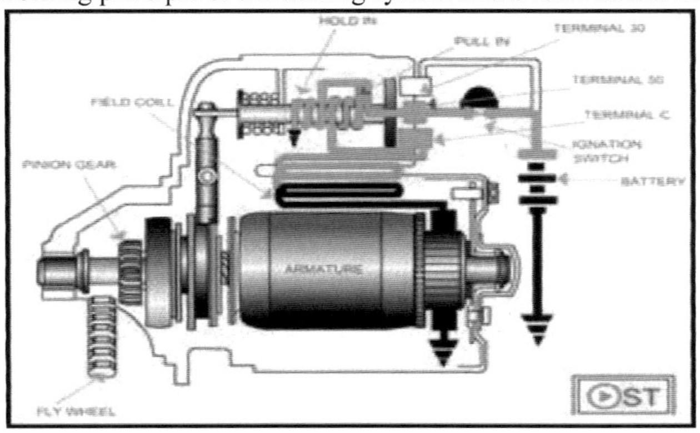

Figure 4. The Visualization of The Flow of Electrical Current to The Starter System at Work

Two examples of the above products were the result of creative learning activities that students found new ideas and generated product, be initiative, found relevant information to transform the textual material into a visual concept and put it in a creative visual product in the form of an animated visual media. This product is clearly beneficial for students who wanted to study automotive electrical systems for the first time. The application of the concept of learning by creating through by transforming the textual learning materials into visual animation as creative learning products in this study lead to the discovery of creative ideas of students. Creative ideas need to be put forth in a visual concept. The ability of students to create ideas should be supported also by the spirit of art and understanding of the content of the learning material. Without the sufficient knowledge and artistry, these ideas were difficult to be realized in the form of products which also should have the valid content and an attractive appearance. Creative and artistic element in the manufacture of the product supported the design and produced a good appearance to meet user requirements [19]. The elements of the learning material mastery supported the validity of the information presented in the product. Consequently, important aspects in learning by creating in this article consisted of three important things that the students need to have during the learning process, the "mastery" of the learning material, "creative", and artistic.

The mastery of the learning material was required as a fundament of the validity of the products. Students who were creative, although they did not understand the material, made an effort in a creative way to produce a good and valid product by studying the required learning materials, identifying difficulties and finding a solution. This was supported by the data (Figure 3) which indicates that the product made by the creative students, the validity of the content can be trusted. On the other hand, the products made by uncreative students, the validity of the contents was also unreliable (the average score of the validity was less than 60%). Products which were categorized as uncreative were 44% and were not suitable for the students' learning tools. Some possible causes of this impropriety where the students did not understand the learning material about electricity that was presented and did not try to understand it by learning. Creativity is a process to be sensitive to the problem and a lack of knowledge. Therefore, the impropriety of the students' products in this study can be caused by a lack of sensitivity to the problems and the low mastery of the learning material. Although, they had been given sufficient time (2 weeks), the time did not seem to put to good use by students who were not creative to produce a good product.

The uncreative products produced by the less creative students with low imaginative activity resulted in improper products. The ability to think creatively was what generated ideas for innovative products [20]. For students like this, they had to be encouraged by planting logic for creative learning [21]. Stimulating the creativity of students whose products were less creative could be achieved with intensive assistance by educators for students to think, discover new ideas, and find a variety of information required by the students. Other friends who have high creativity must be activated to help students who were less creative, so the learning environment was surrounded by creative people and activities. Educators need to develop an expressive learning atmosphere that respects originality and creativity. The learning process also need to be supported by the atmosphere full of ideas through varied learning activities to provide an opportunity for students to learn and work with people who have high creativity.

According to Fischer et al. [22], learning process in engineering field must begin with self – depiction of the creative process. This means that educators must understand the true meaning of creativity. An understanding of creativity results in more effective learning so that students will have the ability to develop themselves [23]. Therefore, active learning techniques need to be implemented for the students' development through activities that involve and motivate students, make students excited that the students' attention will focus on learning [24]. The generative process on students in an active learning will be formed through the freedom given to students to organize information from textual learning process [17]. Therefore, students can find significance in their learning process and foster a high commitment to achieve learning goals. Learning by doing provides direct experience and knowledge discovery by students. Learning by creating develops students' ability to innovate by seeking and finding new ideas and applying these ideas in the form of products [19]. In addition to reinforcement or mastery of the learning materials, learning by creating is also expected to result in useful products that can be used by others to learn.

4. Conclusions

The implementation of the learning by creating concept in this study was conducted with the following signs: (a) the technology – aided learning is aimed at creating products through the presentation of information in a visual animation format; (b) generative learning process is defined as transforming textual learning into something new; (c) the learning process is carried out by the productive process through creative learning activities to find new ideas and produce products through the provision of adequate time to change the automotive electrical textual material into a visual concept and transform it in the form of animation products; (d) the creation of the animated visual by the fast display of static images in a sequence to create the illusion of motion; and (e) the process of making the animation was conducted with simplified stages of developing ideas, writing scripts , designing, and production.

Products were generally categorized as the quite creative criteria. However, quite a lot of products (44%) did not meet the creative criteria because the products were made of uninteresting and illogical ideas, did not ease the learning process, the message was difficult to understand, there was not element that characterized the motion animation, and careless preparation. The appearance of products was generally included in the quite attractive criteria. Only 12% of the product was categorized as not good because the visualization was not attractive and uninteresting to look at; placement of images, text, and moving elements were imbalanced; and the resulting motion was very rough. Based on the content validity, there was 36% of the products that did not meet the criteria for content validity because the content, animated motion, and the overall contents of visualized electrical system contained some mistakes. Common mistakes made by the students were the lack of the learning material understanding which resulted in the improper concept of product development, and improper visualization. Creative products are the reflection of the creative students who have the creativity in making the products. Creative products have a great visual products appearance and the good content validity as well. Uncreative products have an unattractive appearance and invalid information of the product.

5. References

[1] K. B. A. Jabbar, A. Ong, J. Choy, and L. Lim, 'Effects of experiential-based videos in multi-disciplinary learning', *Australasian Journal of Educational Technology*, vol. 29, no. 4, Art. no. 4, Sep. 2013, doi: 10.14742/ajet.208.

[2] S.-H. Liu, 'Teacher education programs, field-based practicums, and psychological factors of the implementation of technology by pre-service teachers', *Australasian Journal of Educational Technology*, vol. 32, no. 3, Art. no. 3, Jul. 2016, doi: 10.14742/ajet.2139.

[3] A. K. Veira, C. J. Leacock, and S. J. Warrican, 'Learning outside the walls of the classroom: Engaging the digital natives', *Australasian Journal of Educational Technology*, vol. 30, no. 2, Art. no. 2, May 2014, doi: 10.14742/ajet.349.

[4] J. R. Banas and C. S. York, 'Authentic learning exercises as a means to influence preservice teachers' technology integration self-efficacy and intentions to integrate technology', *Australasian Journal of Educational Technology*, vol. 30, no. 6, Art. no. 6, Dec. 2014, doi: 10.14742/ajet.362.

[5] D. F. Yang, J. Catterall, and J. Davis, 'Supporting new students from vocational education and training: Finding a reusable solution to address recurring learning difficulties in e-learning', *Australasian Journal of Educational Technology*, vol. 29, no. 5, Art. no. 5, Nov. 2013, doi: 10.14742/ajet.196.

[6] M. Parkes, C. Reading, and S. Stein, 'The competencies required for effective performance in a university e-learning environment', *Australasian Journal of Educational Technology*, vol. 29, no. 6, Art. no. 6, Dec. 2013, doi: 10.14742/ajet.38.

[7] F. J. S. Grande, H. J. Witt, and J. B. Avalos, 'Evaluation of social competencies in chemical engineering: application and results of the pilot test (academic year 2012-2013)', *JOTSE: Journal of technology and science education*, Jul. 2015, Accessed: Sep. 30, 2021. [Online]. Available: https://raco.cat/index.php/JOTSE/article/view/295441

[8] R. Rabiman, P. Sudira, H. Sofyan, and M. Nurtanto, 'Practical Learning Media in Subject Maintenance of Chassis and Power (MCP) Based Online: Simple Learning Using Videos on YouTube', *International Journal of Interactive Mobile Technologies (iJIM)*, vol. 15, no. 03, Art. no. 03, Feb. 2021, Accessed: Feb. 13, 2021. [Online]. Available: https://www.online-journals.org/index.php/i-jim/article/view/14943
[9] S. Dawkins, M.-E. Ritz, and W. Louden, 'Learning by doing: preservice teachers as reading tutors', *Australian Journal of Teacher Education*, vol. 34, no. 2, Mar. 2009, doi: 10.14221/ajte.2009v34n2.4.
[10] M. O'Brien, 'Fostering a Creativity Mindset for Teaching (and Learning)', *LEARNing Landscapes*, vol. 6, no. 1, Art. no. 1, Jun. 2012, doi: 10.36510/learnland.v6i1.589.
[11] N. Kholifah, H. Sofyan, P. Pardjono, P. Sudira, and M. Nurtanto, 'Explicating the Experience of Beginner Vocational Teachers', *TEM Journal*, vol. 10, no. 2, pp. 719–723, 2021, doi: 10.18421/TEM102-28.
[12] S. Sever, A. Oguz-Unver, and K. Yurumezoglu, 'The effective presentation of inquiry-based classroom experiments using teaching strategies that employ video and demonstration methods', *Australasian Journal of Educational Technology*, vol. 29, no. 3, Art. no. 3, Jul. 2013, doi: 10.14742/ajet.229.
[13] R. H. Kemp, T. Stewart, I. P. W. Fung, and B. Orban, 'Learning by Creating: Letting the Student Do the Work', *Interactive Learning Environments*, vol. 10, no. 2, pp. 121–138, Aug. 2002, doi: 10.1076/ilee.10.2.121.7446.
[14] M. Devlin and J. McKay, 'Teaching students using technology: Facilitating success for students from low socioeconomic status backgrounds in Australian universities', *Australasian Journal of Educational Technology*, vol. 32, no. 1, Art. no. 1, Mar. 2016, doi: 10.14742/ajet.2053.
[15] *Learning with animation: Research implications for design*. New York, NY, US: Cambridge University Press, 2008, pp. ix, 391.
[16] D. Ratnawati, W. Mustafa Kusuma, S. Setuju, M. Nurtanto, and W. Widodo, 'Development of Job Sheet Lathe Machining Practice Based on Animation Video as Interactive Learning Media', *J. Phys.: Conf. Ser.*, vol. 1573, p. 012005, Jul. 2020, doi: 10.1088/1742-6596/1573/1/012005.
[17] A. T. Stull and R. E. Mayer, 'Learning by doing versus learning by viewing: Three experimental comparisons of learner-generated versus author-provided graphic organizers', *Journal of Educational Psychology*, vol. 99, no. 4, pp. 808–820, 2007, doi: 10.1037/0022-0663.99.4.808.
[18] S. R. Daly, E. A. Mosyjowski, and C. M. Seifert, 'Teaching Creativity in Engineering Courses', *Journal of Engineering Education*, vol. 103, no. 3, pp. 417–449, 2014, doi: 10.1002/jee.20048.
[19] D. H. Cropley, J. C. Kaufman, and A. J. Cropley, 'Measuring Creativity for Innovation Management', *Journal of Technology Management & Innovation*, vol. 6, no. 3, Art. no. 3, Aug. 2011, doi: 10.4067/S0718-27242011000300002.
[20] A. Craft, 'The Limits to Creativity in Education: Dilemmas for the Educator', *British Journal of Educational Studies*, vol. 51, no. 2, pp. 113–127, 2003, Accessed: Sep. 30, 2021. [Online]. Available: https://www.jstor.org/stable/3122416
[21] N. Gupta, N. Tejovanth, and P. Murthy, 'Learning by creating: Interactive programming for Indian high schools', in *2012 IEEE International Conference on Technology Enhanced Education (ICTEE)*, Jan. 2012, pp. 1–3. doi: 10.1109/ICTEE.2012.6208643.
[22] S. Fischer, D. Oget, and D. Cavallucci, 'The evaluation of creativity from the perspective of subject matter and training in higher education: Issues, constraints and limitations', *Thinking Skills and Creativity*, vol. 19, pp. 123–135, Mar. 2016, doi: 10.1016/j.tsc.2015.10.002.

[23] B. Lucas, 'A Five-Dimensional Model of Creativity and its Assessment in Schools', *Applied Measurement in Education*, vol. 29, no. 4, pp. 278–290, Oct. 2016, doi: 10.1080/08957347.2016.1209206.

[24] J. Hackathorn, E. D. Solomon, K. L. Blankmeyer, R. E. Tennial, and A. M. Garczynski, 'Learning by Doing: An Empirical Study of Active Teaching Techniques', *Journal of Effective Teaching*, vol. 11, no. 2, pp. 40–54, 2011, Accessed: Sep. 30, 2021. [Online]. Available: https://eric.ed.gov/?id=EJ1092139

Journal of Physics: Conference Series **2111** (2021) 012008 doi:10.1088/1742-6596/2111/1/012008

Development of Hydraulic Cylinder Excavator Learning Media Based on Augmented Reality with Shapr3D

A Setiyawan[1], L C Manggalasari[2], T A Prasetya[3], Towip Towip[4], W Noviansyah[5]

[1]Department of Mechanical Engineering, Universitas Negeri Semarang, Semarang, Indonesia
[2]Department of Technology and Vocational Education, Universitas Negeri Yogyakarta, Indonesia
[3]Department of Mechanical Engineering Education, Universitas Negeri Yogyakarta, Indonesia
[4]Department of Mechanical Engineering Education, Faculty of Teacher Training and Education, Universitas Sebelas Maret Surakarta, Indonesia
[5]Department of Civil Engineering Education, Faculty of Teacher Training and Education, Universitas Sebelas Maret Surakarta, Indones

Email: andrisetiyawan@mail.unnes.ac.id

Abstract. The development process of augmented reality generally requires devices and software with high compatibility so that they are the only obstacles in its development. The purpose of this study was to design and produce an application of learning media on the hydraulic cylinder of excavator based on augmented reality with Shapr3d for prospective students of automotive engineering education and to see the product feasibility level of the test subjects and validators. This learning media development stage consists of five stages: Analysis, Design, Development, Implementation, and Evaluation. For the data method using direct observation and closed questionnaires. The questionnaire of the research results was analysed using Likert scale analysis. The results of the material expert validator with a score of 91.25% are very good category. Expert media design validator 92.5% is included in the very good category with a very proper name. The response to product use from three lecturers of automotive engineering education with a proportion score of 88.34% is in the very good category. The response to the use of the product was from twenty automotive engineering education students with a score proportion of 85.73% in the very good category. This that the application of automatic learning media on excavator hydraulic cylinder based on augmented reality with shapr3d based on Augmented Reality technology that has been developed can be used in elementary schools as a learning media for technical education.

1. Introduction

The government's physical distancing policy to prevent transmission of the covid19 [1]virus caused changes in learning patterns from face-to-face learning to distance education[2]. Along with the rapid development of technology[3], the utilization of technology and its development in education is increasingly advancing and developing, requiring many updates to compensate for these developments[4]. The exponential growth of Information and Communication Technology (ICT) has changed the education systems and contributed to improving the learning process[5][6]. The presence of ICT in the school system will undoubtedly improve the quality and access to better information[7]–[9]. However, scientific advancements that are becoming more advanced in education cannot be used entirely[10]. According to Pribadi, learning media in the form of IT was created to make it easier to communicate facts and awareness to many people[11]. Other teaching media should be used to mitigate problems with this one. Since there are still teaching media such as interactive films, PowerPoint presentations, interactive learning apps, and others that can distribute space-building resources. Augmented Reality is one immersive learning technology that can be used to teach space construction content. Augmented Reality is a device that uses a smartphone to view virtual objects in a real-world setting[12]. Geometry objects are converted into virtual objects using this technology, then projected on a physical object using a smartphone. In hydraulic material in excavator heavy equipment Engineering, Augmented Reality is used to visualize the hydraulic system on the excavator.

In the current era, there is an application in augmented reality modelling that is very easy to do through shapr3d. Shapr3D, launched in March 2016, is a direct modelling tool, which, though developed for professionals, is said to be easy to learn and easy to use. It runs everything through the iPad Pro device, meaning there's no need to push files into the cloud, and it can also be used without an internet connection. Compatible with all major CAD software and ideal for 3D printing jobs, Shapr3D has integrated Parasolid for better geometric modelling and HOOPS Exchange as a leading CAD translation Software Development Kit (SDK)[13]. Shapr3d can model instantly for Augmented Reality.

Developers use this high level of interactivity to create a variety of Augmented Reality-based learning media. Teaching content on standard house examples[14]and human organs are two examples of Augmented Reality-based learning media[15]. Some of the benefits of using Augmented Reality in this broad learning are introducing students to technology in the field of education.

It supports the merdeka curriculum policy that requires the use of IT in learning and supporting government efforts in developing superior human resources for Indonesia's golden generation in 2045, ready to face the industrial age. For all of these benefits, Augmented Reality is an excellent way to educate students about technology at a young age. This is also reinforced by previous studies[16][17], which found that Augmented Reality is suitable for use in school learning because of its high degree of interactivity. Knowing the advantages of using Augmented Reality in the world of education is in line with the purpose of this research, namely the hydraulic cylinder excavator learning media based on augmented reality with shapr3d for prospective students of automotive engineering education. This research was conducted because, generally, Augmented Reality is used to teach historical material, natural sciences, and other subjects. However, there has never been any research related to Augmented Reality as a learning media application, especially for hydraulic excavator material.

2. Methods

This form of research is development research to produce a product. The development procedure consists of five stages: (1) analysis, (2) design, (3) development, (4) implementation/execution, and (5) evaluation/feedback. The development of an adaptive e- earning system is conducted by using the Instructional System Design (ISD) in which typically used for the development of an educational system. Allen (2006) defined that one widely accepted and use disthegeneric "analysis, design, develop, implement and evaluate" model or generally called as an ADDIE model [18][19]. This model is chosen because it has a structured and straightforward development sequence that makes it more accessible. The first step in this research is to analyze needs through observation and interviews; by doing so, more knowledge would be available, allowing the final product to be more efficient and targeted. The design can be defined as the process of implementing various kinds of techniques and principles to define the equipment, process or system in detail[18].

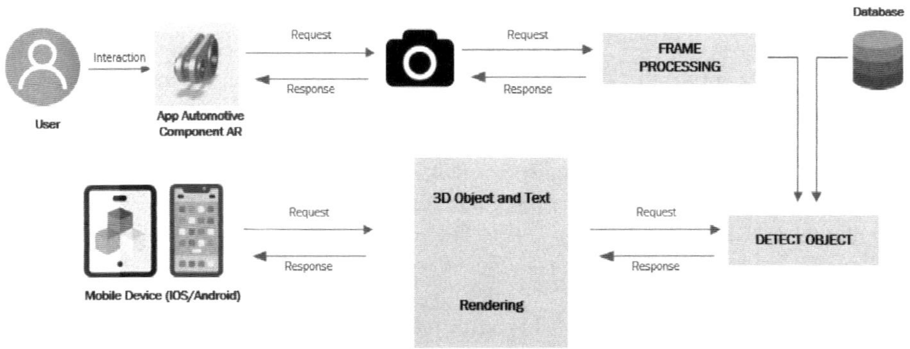

Figure 1. Architecture of AR

The third step is developing an application; at this stage, all previously prepared prototypes are executed until they become a finished product that can be used. This stage includes making the application interface design, drawing for 3d design and extract augmented reality using shapr3d without a marker. After this process is completed, the next step is to implement the application on the student of automotive engineering education.

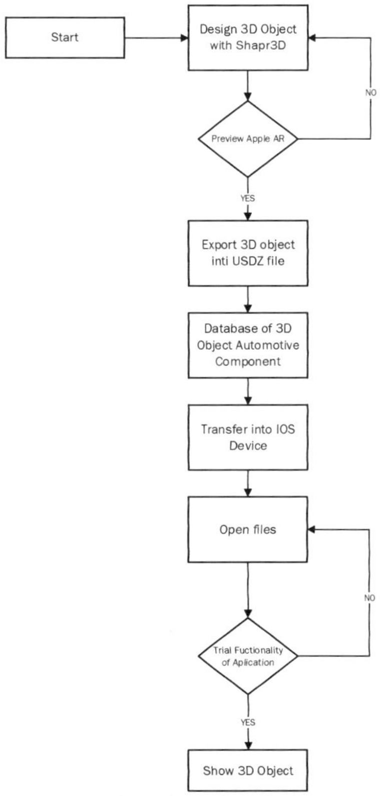

Figure 2. Flowchart development AR

However, before implementation on the subject, the application must be checked on a validator. The goal is to determine how feasible the application is and identify any product flaws to be addressed before it is tested on the subject.

The content expert validator and the validation of learning media design expert are the validators in this report. Although closed questionnaires are used to collect study data from subjects and validators, questionnaires are created using application-based learning media eligibility criteria that adapt previous applicable research, such as the suitability of material content, language, graphic design, colour, ease, and interactivity. The expert judgment test was used to conduct the validity test for this questionnaire. The expert judges would evaluate the questionnaire's suitability as a measuring tool in this research development. The study validator receives the completed questionnaire. In addition, the scores obtained from each validator are evaluated using a Likert scale evaluation analysis to assess the appropriateness of the application submitted.

Table 1. Analysis of Likert Scale Assessment

Score	Information
5	Strongly agree
4	Agree
3	Not agree
2	Disagree
1	Strongly disagree

Data obtained from the validator is analyzed using a Likert scale assessment, and this score will be collected from each of the validators. Furthermore, to be able to make decisions regarding the feasibility of the application using the following decision-making criteria:

Table 2. Indicators of Questionnaire Score Decision

Level of Achievement	Qualification	Information
81-100%	Very Good	Very decent/Valid
61-80%	Well	Eligible/Valid
41-60%	Pretty Good	Inadequate/Less Valid
21-40%	Not Good	Ineligible/invalid
<20%	Very Poor	Very Improper/very invalid

Lecturers and students checked the product at Automotive Engineering education to assess the answer to using the application in learning until the percentage of validator questionnaire scores met the good qualifications. Data processing methods used in the study validator questionnaire were used to calculate scores and make decisions from the instructor and student questionnaires. The evaluation phase is the last step in this research, and it aims to evaluate the products from previous implementations to optimize the research's final product. This stage is completed by reviewing the validator's and test subjects' responses in the questionnaire's comment section. So that it can be determined if the application needs adjustment and can be corrected immediately to increase the quality of the development's performance.

3. Result and Discussion

The research results from this development include product results in the form of an Augmented Reality shooting application. The results of the study also present a description of the validation results in the form of media validation and the results of research that has been carried out. The results of the study also present the results of the pretest and posttest analysis. Needs analysis in this development includes hardware, software and material requirements needed for development. The hardware used is a 7th generation iPad with Apple Pencil.

Furthermore, IOS devices are used for product application implementation. The software used is AutoCAD and Shapr3d. The product design process includes object design and AR design.

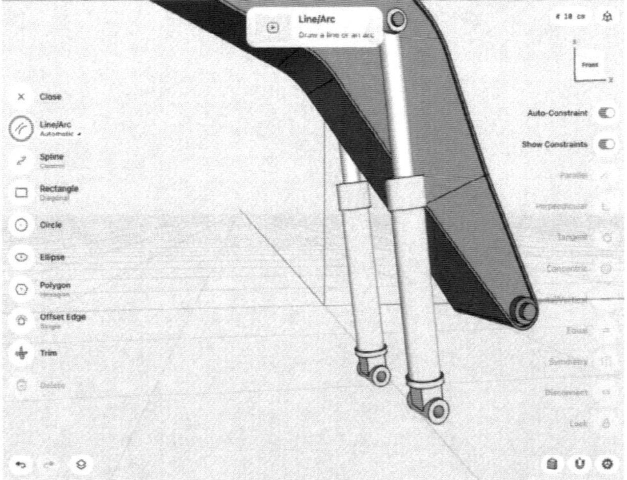

Figure 3. Drwaing 3D Hydraulic Cylinder

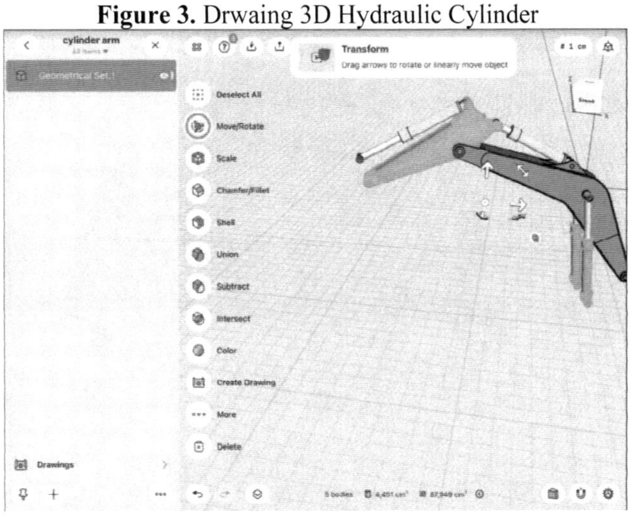

Figure 4. Merging part of Hydraulic Cylinder

Modeling of 3D object to view of all part which merging in 360° and evaluate the part of hydraulic excavator seem like real object in the field. The yellow color shows the part of hydraulic cylinder which easy to indicate by student.

Journal of Physics: Conference Series **2111** (2021) 012008 doi:10.1088/1742-6596/2111/1/012008

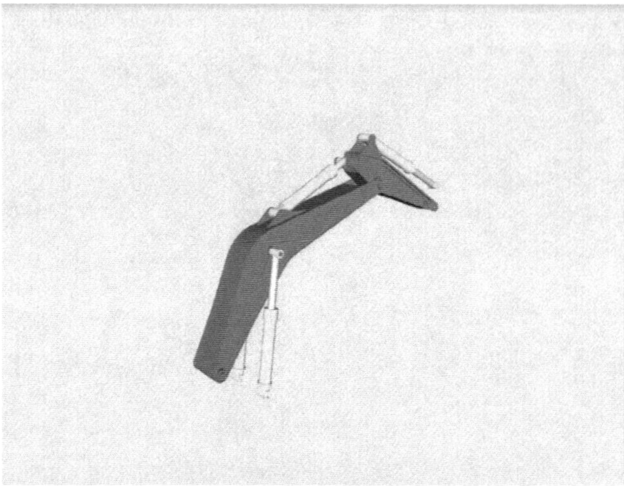

Figure 5. Modeling 3D object

Figure 6. Trial modeling 3D in 360°

Trial of hydraulic excavator AR use Reality composer and IOS device. Camera will detected for the object such as flat object, and if success the model of 3D will appears.

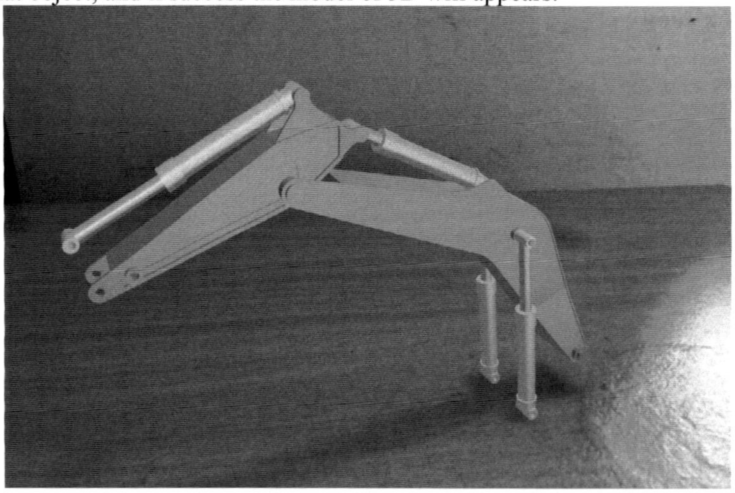

Figure 7. Trial Hydraulic Excavator AR

The following are the results of the validity test of each validator, and this course is a research experiment:

Table 3. Validity test of each validator

Respondent	Percentage	Information
Material Expert Validator	91.25%	Very decent
Media Design Expert Validator	92.50%	Very decent
Lecturer	88.34%	Very decent
Student	85.73%	Very decent

Learning media applications on material based on Augmented Reality technology was developed through five stages, starting from the Analysis, Design, Development, Implementation, and Evaluation stages. The previously designed research design, and product validation by the validator and the test subjects, namely lecturers and students, was carried out at the implementation stage by employing a measuring instrument in the form of a closed questionnaire that had previously been validated by an expert judgment[20]. The certification of the application to the material expert validator involves the material suitability criteria, the design of the material in the application. The application is validated by the content expert using a questionnaire that contains these requirements. The validation results of the application of two material experts were rated on a scale of 5 (excellent) to 4 using a questionnaire (good). So, from the aspect of the material contents included in the criteria very well with a percentage of 91.25% score and is feasible to be applied in higher education as a learning media. The material in this application consists of a hydraulic excavator for the heavy equipment class. The results of the subsequent validation are done by the instructional media design expert and the content expert. Then the confirmation by the media design expert is done using a questionnaire that adapts the feasibility requirements of IT-based media previously done by the media design expert[21]. The appearance of graphic design, colour, interactivity, and ease of operation requirements for media design experts. And, based on the findings of two media design experts' validation, most of the validator scores of the application design aspects ranged between 5 (excellent) and 4 (good).

There is also a "guide" feature that can make it easier for users to operate applications; this is in line with study, which found that ease of operation of applications is critical in developing IT-based media. The findings of the validation conducted on three lecturers in automotive education are presented next. Validation is accomplished by using a questionnaire that adapts the eligibility requirements of IT-based media. The evaluation findings on a scale of 5 (very good) to 4 (poor) based on the results of trials on three teachers who were tested using a questionnaire with elements of the assessment including the contents of the subject, product design, and implementation feature (good). Learning media in the form of this application earned a positive response from elementary school teachers, with a percentage score of 88.34 per cent, which is included in an outstanding certification with very decent knowledge, according to the results of this evaluation.

The final validation result was conducted on students from the automotive engineering faculty of education at Universitas Negeri Semarang to obtain product validation from small groups and obtain feedback on the product's use. This is typically done based on a past investigation by Hidayat [22], which states that item utilization reaction is required when creating an item, particularly for everyday objects that understudies have not attempted. Based on survey information gotten from understudies, the evaluation scores shifted with a extend of scores between 5 (exceptionally great), 4 (excellent), 3 (adequate), and 2 (less). Based on the comes about of this appraisal, a rating score of 85.73% is included in a great capability with exceptionally conventional data. Based on these comes about, the students' reactions to the created application are great, and the application is attainable to be connected in higher instruction as a implies to bolster learning, particularly in automotive engineering.

Referring to the comes about of information investigation gotten in this ponder of validators and test subjects that the application of automotive education media on the hydraulic cylinder of excavator material based on augmented reality with shapr3d innovation is doable to be utilized a learning media with exceptionally great qualifications. The additional significance to past investigation conducted by Hamzah[23] concerning the plan of applications based on Expanded Reality innovation

4. Conclusion

Based on information investigation the comes about of this ponder can be concluded as takes after: (1) The improvement handle incorporates five stages, namely the Analyst (Analysis), Design / Design (Development), Development (Development), Implementation/execution (Implementation), and Evaluation/feedback (Evaluation), this inquire about the show was adjusted from nd past inquire about by who utilized a comparable demonstrate. (2) Augmented Reality application developed is valid/feasible to be used as a media of application of education media on the hydraulic cylinder excavator based on augmented reality with shapr3d innovation with the results of data analysis from the validator and the test subjects as follows: (a) the material expert validator states the application is very feasible with a percentage score of 91.25%, (b) the validator of the media design expert said the application was very feasible with a percentage score of 92.50, (c) three automotive engineering lecturer stated the application was very feasible with a percentage score of 88.34%, (d) fifteen students stated the application was very feasible with the percentage score of 85.73%. Based on the percentage of the study results, the application of automotive education media on the hydraulic cylinder of excavator material based on augmented reality with shapr3d innovation that has been developed can be used in higher education as a learning media. As a result of the findings of this study, it is recommended that lecturers be able to use IT-based media more often to arouse students' interest in learning. Students can also take advantage of today's numerous learning media, such as the application created in this research, which can teach them even outside of the classroom. Finally, researchers and other creators should be able to increase the output of IT-based media that is more eligible both in terms of design and quality of material to increase the diversity of media used in student learning.

5. References

[1] Setiyawan A and Kurniawan A, 2021 The Effect of Pandemic Covid-19 into Internship Activity of Mojokerto Vocational High Schools *JOVES (Journal Vocat. Educ. Stud.* **4**, 1 p. 125–130.

[2] Prasetya T A Harjanto C T and Setiyawan A, 2020 Analysis of student satisfaction of e-learning using the end-user computing satisfaction method during the Covid-19 pandemic *J. Phys. Conf. Ser.* **1700**, 1.

[3] Hariyanto D Mustaqim I and Maruanaya R F, 2021 The Development of Android-based Control System for Reinforcing the Electronic Control Subject *J. Phys. Conf. Ser.* **1737**, 1.

[4] Husaini M, 2014 Pemanfaatan Teknologi Informasi Dalam Bidang Pendidikan (E-Education). *J. Mikrotik* **2**, 1 p. 1–5.

[5] Setiyawan A and Sunyoto, 2020, 50 succes story wirausahawan nasional dan internasional : mata kuliah kewirausahaan : suplemen bahan ajar. .

[6] Setiyawan A Priyanto Prasetya T A and Hastawan A F, 2021 Usability evaluation of assignment and monitoring information learning system of internship students based on SMS Gateway with Raspberry Pi Usability evaluation of assignment and monitoring information learning system of internship students based on SMS G *IOP Conf. Ser. Earth Environ. Sci.*

[7] Setiyawan A Prasetya T A and Hastawan A F, 2021 Usability evaluation of assignment and monitoring information learning system of internship students based on SMS Gateway with Raspberry Pi in *IOP Conference Series: Earth and Environmental Science* **700**, 1 p. 12021.

[8] Setiyawan A, 2020 ASSIGNMENT AND MONITORING INFORMATION SYSTEM OF PRAKERIN STUDENTS BASED ON SMS GATEWAY WITH RASPBERRY PI *VANOS J. Mech. Eng. Educ.* **5**, 1.

[9] Setiyawan A, 2017, Pengembangan Sistem Informasi Penugasan dan Monitoring Siswa

Prakerin Berbasis SMS Gateway dengan Raspberry Pi, Universitas Negeri Yogyakart.
[10] Sawitri, E., Sumiati, M., & Fitriani Y, 2019 Hambatan Dan Tantangan Pembelajaran *Semin. Nas. Pendidik. Progr. Pascasarj. Univ. PGRI Palembang* p. 202–213.
[11] Pribadi B, 2017 *Media & Teknologi dalam Pembelajaran* Jakarta: Kencana.
[12] Azuma R T, 2010 *A Survey of Augmented Reality* Malibu: Hughes Research Laboratories.
[13] Davies S, 2017, Shapr3D integrates Siemens PLM & Tech Soft 3D software to bring 3D modelling capabilities to iPad Pro users, *12 December 2017*. [Online]. Available: https://www.tctmagazine.com/additive-manufacturing-3d-printing-news/shapr3d-siemens-plm-tech-soft-3d-modelling-ipad-pro/.
[14] Pramono A, 2013 Media Pendukung Pembelajaran Rumah Adat Indonesia Menggunakan Augmented Reality **11**, 1 p. 122.
[15] Santoso A, 2013 Rancang Bangun Aplikasi Pembelajaran Organ Tubuh Berbasis Augmented Reality **8**.
[16] Mareta A, 2015, Implementasi Media Ajar Bangun Ruang Berbasis Augmented Reality Pada SMPN 2 Selomerto Kabupaten Wonosobo. Semarang: Jurusan Teknik Elektro, Universitas Negeri Semarang, Universitas Negeri Semarang.
[17] Huda, K., Bagus et al., 2018 Pengembangan Media Pembelajaran Berbasis Android Menggunakan Augmented Reality Pada Materi Bangun Ruang Sisi Datar *J. Pendidik. Mat. dan Sains* **6**, 01 p. 61–69.
[18] Hariyanto D, 2016, An Adaptive E-Learning System based on Student's Learning Styles and Knowledge Level (The Design Step). [Online]. Available: https://www.researchgate.net/publication/323277109_An_Adaptive_E-Learning_System_based_on_Student's_Learning_Styles_and_Knowledge_Level_The_Design_Step.
[19] Allen W C, 2006 Overview and Evolution of the ADDIE Training System *Adv. Dev. Hum. Resour.* **8**, 4 p. 430–441.
[20] Sugiyono, 2017 *Metode Penelitian Pendidikan (Pendekatan Kuantitatif, Kualitatif dan R&D)* Bandung: Alfabeta.
[21] A Fujiastuti, 2018 Pengembangan media pembelajaran retorika berbasis ARCS dengan aplikasi flash *Pros. SAGA* p. 16–24.
[22] I. F. Hidayat Baron, 2015 Implementasi independent t-test pada aplikasi pembelajaran multimedia ragam dan gerak seni tari daerah kalimatan selatan *Kumpul. J. Ilmu Komput. (KLIK)*, **2**, 1 p. 11–21.
[23] Hamzah S, 2019 Pengembangan Media Pembelajaran Perangkat Keras Jaringan Berbasis Augmented Reality Pada Platform Android. *Vokasional Tek. Elektron. dan Inform.* **7**, 3 p. 1–12.

Design and Performance Analysis of 600 W On-Grid Solar Power Plant System based on Maximum Power Point Tracking (MPPT)

Nurhening Yuniarti [1], Alex Sandria Jaya Wardhana [1], Bagas Woro Saputra [1]

[1]Department. of Electrical Engineering, Universitas Negeri Yogyakarta, Indonesia

E-mail: nurhening@uny.ac.id

Abstract. This research aimed to make, test, and analyse the performance of 600 W on-grid solar power plant system based on maximum power point tracking (MPPT). This research used a developmental research method with the ADDIE model that was developed by Robert Maribe Branch. The results obtained data that the average power that was produced and distributed to the PLN grid reached 22.5 W with 95.67% efficiency from the grid-tie inverter usage.

1. Introduction

Electricity is an important element for social, educational, economic and political development of every nation [1]. Electricity need in Indonesia always increases along with the development of technological advances and the increase of public electricity consumption. Based on the data from [2], national electricity production was still dominated by non-renewable energy for 82.64%, consisted of 56.14% coal, 20.2% gas, and 6.3% oil. The usage of renewable energy only contributed for 17.37% from the national electricity production, whereas renewable energy potential is very abundant in Indonesia. Based on [2], the potential of renewable energy reached 442 GW and to date was only implemented for 9.32 GW or less than 3%.

Renewable energy flows are very large in comparison with humankind's use of energy. Therefore, in principle, all our energy needs, both now and into the future, can be met by energy from renewable sources [3]. among all types renewable energy source, solar PV sources can have the biggest impact to electrical power generation in the coming years [4]. Solar energy is one of the renewable energies with high potential of 207.9 GW, although the implementation of the solar power plant in Indonesia was only 0.14 GW or around 0.02%. This is very unfortunate considering the importance of renewable energy usage in industrial revolution 4.0 or Society 5.0 to welcome the Sustainable Development Goals 2030 with the main purpose is the usage of carbon-free energy and environmentally-friendly all over the world. Public's awareness and government's contribution are needed to realize the implementation of renewable energy in Indonesia such as solar power plant rooftop system, off-grid and on-grid, to improve the role of renewable energy in the national electricity mix.

High costs of installation and maintenance as a result of storage units discourage the use of solar Photovoltaic system for power generation. To reduce these costs, Solar PV systems can be installed without storage units alongside conventional power generation systems [5] . Generally, a PV system is categorized into the stand-alone and grid-connected type. In the global market, a large proportion of PV solar power is supplied by the grid-connected type. The conventional current control of grid-connected PV system is to supply the maximum available active power from the PV array to the grid [6].

Based on the introduction, research objectives can be formulated as follows: are to make, test, and analyse the performance of 600 W on-grid solar power plant system based on maximum power point tracking (MPPT).

2. Related Works

2.1. Grid Connected Solar Power Plant System

Solar panels work best in certain weather conditions, but the weather is changing always, so most panels do not operate under ideal conditions [7]. On-grid solar power plan from the PLN network or often called Grid Connected Solar Power Plant System is a power plant system that utilizes solar radiation to produce electricity, where, true to its name, the system is connected to PLN network by optimizing solar energy utilization through the solar module or photovoltaic. This system is also environmentally friendly and free from carbon emission. Grid-connected solar power plant system is also a Green Energy solution for urban communities, both offices and housing, to reduce electricity bill and add values to the owner.

This solar power plant is ruled in [8], on rooftop solar power plant system by consumers of PT Perusahaan Listrik Negara. The On-Grid Solar Power Plant System can be seen in Figure 1.

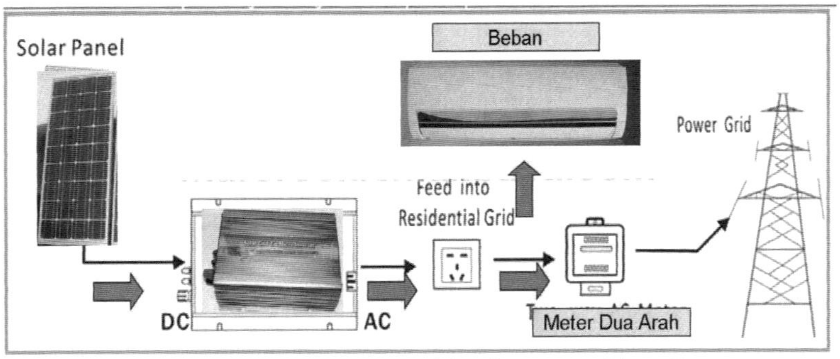

Figure 1 On-Grid Solar Power Plant System

2.2. Grid Tie Inverter

Grid Tie Inverter is a particular type of inverter with different components and specification compared to the standard inverter. A grid tied inverter is a special type of power inverter that converts PV output direct current electricity into alternating current electricity so that one can flow the electricity out into the power grid and actually sell electricity back to the power company [9]. In one-phase system, this inverter has 220-230 VAC output with 50 Hz/60 Hz frequency, pure wave sinus wave, and very identical with PLN network characteristic so it can be transported into the PLN grid. This type of inverter can be used for off-grid and on-grid system and loads with unique characteristic such as AC motor or air conditioning.

Grid-tie inverter is sold in the market with power rating starts from 500 W, 600 W, to up to 1 kW. The specification determines its quality on the grid-tie inverter. The grid-tie inverter quality can be seen from its efficiency, lifetime, and the amount of electricity that can work in high quality

Figure 2 Grid Tie Inverter

2.3. *MPPT (Maximum Power Point Tracking)*

A maximum power point tracking algorithm is absolutely necessary to increase the efficiency of the solar panel as it has been found that only 30-40% of energy incident is converted into electrical energy [10]. Maximum power point tracking or MPPT is a system that is used in the grid tie inverter to improve electricity efficiency produced by the solar cell. Maximum power point tracking works by controlling the current and voltage that is produced by a solar cell so that the produced power is optimized. Several factors that influence the MPPT performance are sensor, an algorithm that is used, and the MPPT speed in creating optimized power.

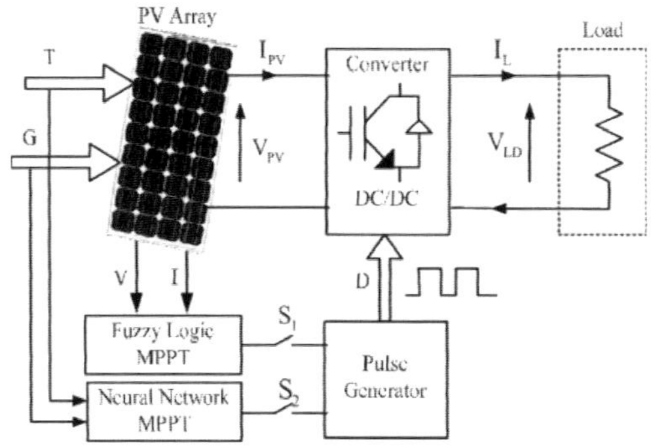

Figure 3 MPPT Schematic Diagram

3. Method

This research used a developmental research method to create design and performance analysis of 600 W on-grid solar power plant system based on maximum power point tracking. The ADDIE approach model that was developed by [11], consists of the below stages:

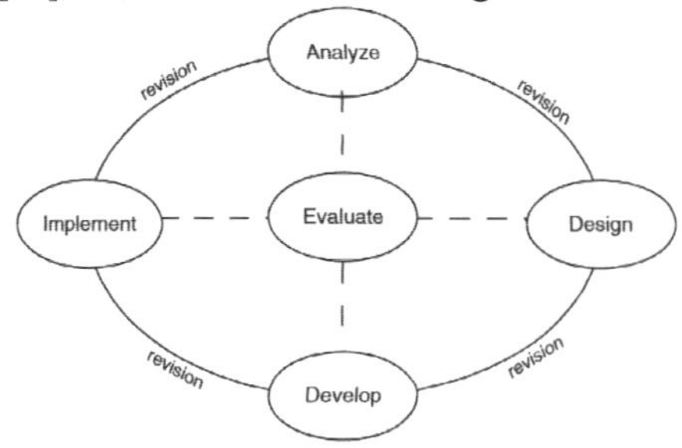

Figure 4. Developmental Stages

Observation and interview related to the performance requirement of on-grid solar power plant system were conducted in the requirement analysis stage. After, the stage was continued with design making of the device and started to compile the required components. The required components among others were 600 W grid-tie inverter, solar panel, digital DC ampere and voltmeter, and digital power meter and other supporters. The design involved box panel design and solar power plant component layout.

After the design stage, the next stage was implementation consisted of tool design and making. All process was done in Mechanical Engineering workshop, Universitas Negeri Yogyakarta.

Figure 5 Device Making Result

4. Findings and Discussions

After the on-grid solar power plant making process was done, the next stage was testing that was performed in the Department of Electrical Engineering Education, Faculty of Engineering, Universitas Negeri Yogyakarta on May 18th, 2020.

Figure 6 Trial Process

Testing was performed without connecting the system to the load because it aimed to analyse the power efficiency that was produced by the system. The testing was done by taking the observation data every 10 minutes from 11:50 to 13:20, considering the optimized solar energy during the hours. Table 1 displays the testing results.

Table 1. 600 W On-Grid System Solar Power Plant Testing Results

Test Result											
	Condition		Photovoltaic		PLN		Grid Tie Inverter Output				
Interval	Luminance (lux)	Temperature (°C)	Voltage (volt)	Current (ampere)	Voltage (volt)	Current (ampere)	Voltage (volt)	Current (ampere)	Power (watt)	Freq (Hz)	PF
11.50	8770	29	43,4	0,56	225	0 (no load)	226,1	0,153	21,794	50	0,63
12.00	7840	29	42,71	0,58	225		228	0,161	23,493	50	0,64
12.10	7350	29	42,49	0,58	225		229	0,153	22,073	50	0,63
12.20	7780	29	42,71	0,58	225		229,1	0,153	22,083	50	0,63
12.30	9860	29	44,48	0,56	225		227,9	0,163	23,403	50	0,63
12.40	7950	29	42,71	0,58	225		228,6	0,153	22,035	50	0,63
12.50	8970	29	43,81	0,56	225		229,1	0,153	22,784	50	0,65
13.00	7860	29	42,71	0,58	225		227,9	0,162	23,259	50	0,63
13.10	7830	29	42,71	0,58	225		228,1	0,153	22,336	50	0,64
13.20	7560	29	42,49	0,59	225		229,2	0,153	22,093	50	0,63

Based on Table 1, all components of the on-grid system solar power plant are well-functioned as proved from the stable electricity and follows the basic theory and manual book from the tool components. The grid-tie inverter connected the solar power plant system that was made with the PLN network. The voltage on average was higher than 2 up to 4 V from the PLN voltage. The readable frequency on the power meter was also similar to PLN frequency of 50 Hz while the average power factor (PF) from the system was 0.63.

The electricity average that was distributed to the PLN reached 22.5 W, if the average sunshine was 8 hours, the system produced up to 180 Wh electricity. From Table 1 above can be analysed the efficiency of the utilized grid-tie inverter using the formulation below:

$$efficiency = \frac{Output\ Power}{Input\ Power} \quad (1)$$

Where the input power was calculated using the formulation below:

$$Input\ Power = V_{sc}\, x\, I_{sc} \quad (2)$$

Table 2. shows the grid-tie inverter efficiency after the analysis

Table 2. Grid Tie Inverter Efficiency Analysis Result

Efficiency calculation		
Input Power (watt)	Output Power (watt)	Efficiency (%)
24,41	22,08	90,46
24,56	23,49	95,67
24,43	22,07	90,35
24,56	22,08	89,92
25,02	23,4	93,54
24,56	22,03	89,73
24,65	22,78	92,45

24,56	23,26	94,71
24,56	22,34	90,95
24,96	22,09	88,51

Based on the above data, the maximum power point tracking system in the grid-tie inverter worked by optimizing the input power in the system to produce the most optimized output power. This statement can be seen from the grid-tie inverter efficiency that reached 95.67% with an average efficiency of 91.63%.

5. Conclusion

The 600 W on-grid solar power plant system was successfully made and tested in the Department of Electrical Engineering Education, Faculty of Engineering, Universitas Negeri Yogyakarta. Based on the testing, the solar power plant system that was made could be connected with PLN network. The average electricity that was distributed to the PLN reached 22.5 W, if the average sunshine was 8 hours, the system could produce the electricity up to 180 Wh. Besides, after the analysis, the grid-tie inverted efficiency reached 95.67% with an average of 91.63% efficiency. For the next research, it can be done by adding the Internet of Things (IoT) for monitoring the power and energy consumption used.

6. References

[1] S. O. Oyedepo *et al.*, "A Study on Energy Demand and Consumption in Covenant University, Ota, Nigeria," *Int. Conf. African Dev. Issues*, pp. 203–211, 2015.
[2] National Energy Council, *Indonesia Energy Outlook (IEO) 2019*. 2019.
[3] P. Holm and J. McIntosh, "Renewable energy - the future for the developing world," *Renew. Energy Focus*, vol. 9, no. 1–2, 2008, doi: 10.1016/S1471-0846(08)70027-1.
[4] M. A. Eltawil and Z. Zhao, "Grid-connected photovoltaic power systems: Technical and potential problems-A review," *Renew. Sustain. Energy Rev.*, vol. 14, no. 1, pp. 112–129, 2010, doi: 10.1016/j.rser.2009.07.015.
[5] A. A. Alabi, A. U. Adoghe, O. G. Ogunleye, and C. O. A. Awosope, "Development and sizing of a grid-connected solar PV power plant for Canaanland community," *Int. J. Appl. Power Eng.*, vol. 8, no. 1, p. 69, 2019, doi: 10.11591/ijape.v8.i1.pp69-77.
[6] T. Vu, "A Study of Grid-Connected PV System with Power Control Structure," no. December, pp. 3–5, 2018.
[7] A. M. Ali, S. S. Mustafa, and A. H. Mutlag, "Design optimization of solar power system with respect to temperature and sun tracking," *Al-Sadiq Int. Conf. Multidiscip. IT Commun. Tech. Sci. Appl. AIC-MITCSA 2016*, no. March, pp. 317–321, 2016, doi: 10.1109/AIC-MITCSA.2016.7759956.
[8] Kementerian Energi dan Sumber Daya Mineral (KESDM), "Permen ESDM Nomor 13 Tahun 2019 Perubahan Nomor 49 Tahun 2018," 2018.
[9] M. Nahid Hossain, T. K. Routh, A. H. Bin Yousuf, M. M. Asasduzzaman, M. Iqbal Hossain, and U. Husnaeen, "Design and development of a grid tied solar inverter," *2012 Int. Conf. Informatics, Electron. Vision, ICIEV 2012*, no. May 2019, pp. 1054–1058, 2012, doi: 10.1109/ICIEV.2012.6317344.
[10] A. B. Jain and C. Sharma, "Maximum Power Point Tracking Techniques : A Review," no. May, 2014.
[11] Robert Maribe Branch (2009). Instructional Design: The ADDIE Approach. USA Springer Science Business Media.

Development Mobile Application Android Based for Sensor and Transducer

Y I Hatmojo[1] and S M Azis[2]

[1,2] Electrical Engineering Education Study Program, Faculty of Engineering, Universitas Negeri Yogyakarta, Indonesia

Email: yuwono_indro76@uny.ac.id

Abstract. The aims of this research are: (1) producing a mobile applications android based for sensor and transducer field of study on vocational education; (2) know the quality of the product mobile applications android based for sensor and transducer field study on vocational education. The method used is the Software Development Life Cycle with the Waterfall model. The stages in Waterfall model are: (1) requirements definition; (2) system and software design; (3) implementation and unit testing; (4) integration and system testing; (5) operation and maintenance. The black box test results show the performance of Sensor application software can all function properly. While the results of the compatibility test show the application can run on various android systems (at least Android 4.2).

1. Introduction

Education is the most important foundation in supporting the progress of a country. Education is an activity to distribute knowledge, values and skills through an educational institution [1]. Education will be more focused if an institution has a focus on certain fields that are applied to students, one of which is vocational education. Vocational education has a main focus on equipping students with certain skills and abilities to be ready to enter the world of work [2]. In Indonesia, Vocational High Schools are vocational education at the secondary level that must produce graduates who are in accordance with skill in the industry needed [3]. Broadly speaking, the purpose of Vocational High School is to improve the skills of students in accordance with the field of expertise following the development of science, technology and art, as well as preparing a professional attitude at work. Steps to achieve these goals require appropriate and innovative learning methods and learning media.

According to [4] media is anything that is useful for delivering messages or information in the process of learning activities. Media has the meaning as an intermediary for a message from the source to the recipient of the message. According to [5] without media, the communication process will not run and learning will not be able to run optimally. Referring to the explanations of some experts above, learning media are all forms or things that function as knowledge intermediaries between educators and students so that learning objectives can be achieved optimally. According to [6] teachers and students have challenges and opportunities in the world of education if they can use media and technology efficiently and appropriately. The technology that has great opportunities in the world of education is the smartphone.

Learning media that using smartphones is known as mobile learning. According to [7], mobile learning can be run on various types of devices, such as smartphones, laptops, tablet PCs, media players, and game consoles. [8] 'mobile learning' allows students to learn learning materials anytime and anywhere. Students are allowed to access and manage learning materials without time restrictions. Mobile learning can overcome the limitations of learning media because it can visualize components and tools that are difficult to get or expensive. According to [9] mobile learning has three main functions in learning, as an addition, a complement, and or a substitute. Mobile learning in addition, where students are free to choose and use it, students who use these devices will have more knowledge than those who do not use them. Function as a complement, it's means that the material in mobile learning is structured to strengthen the material in conventional learning. Mobile learning as a substitute, aims to enable students to access learning materials according to their time and daily activities. Mobile learning along with the development of the world of education, has begun to develop both in terms of learning media and learning systems. According to [10] in a publication on Statista.com, smartphone users in Indonesia in 2020 reached 196.7 million users. From the exposure to the data, it can be used as an opportunity factor for the use of smartphones in learning activities. However, there are still many schools especially Vocational High School that have not utilized smartphones as an optimal learning medium. Based on observations during distance learning in the Sensor and Transducer course at SMK SMTI Yogyakarta, the researchers found that the learning process was only through WhatsApp groups and the teacher only delivered material in the form of power point files. The limited learning media and supporting modules in the learning process resulted in a lack of interest in learning and student learning activity. This results in less effective learning so that the absorption of knowledge from each student is not optimal.

Therefore, researchers will develop an Android-based mobile application for the field of Sensors and Transducers in vocational education. Researchers in this development focus on basic competencies about proximity sensors, photoelectric sensors, and liquid level sensors that are adapted from the subjects of the Sensors and Transducers. The purpose of this study was to determine the design of an Android-based mobile application for the field of sensor and transducer studies in vocational education. Researchers will develop learning media using research and development methods with the Waterfall model.

2. Research Method

Researchers developed this application software using the 'Waterfall Model' or Software Life Cycle [11]. Figure 1 is an instructional waterfall models, in this study the focus is on making and functional testing of Android-based software applications.

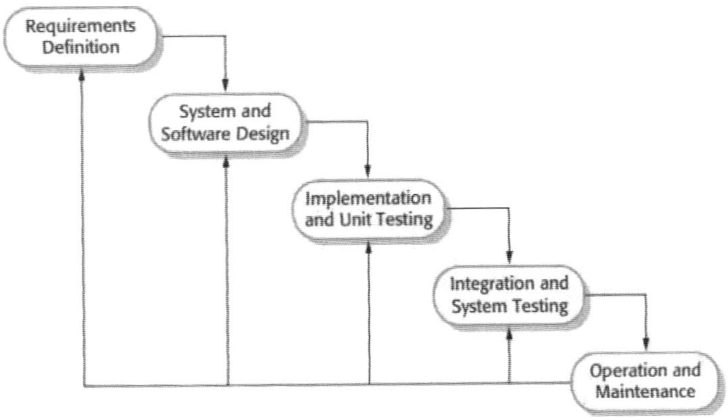

Figure 1. Waterfall Model

The following are the steps of development software application using waterfall model.

2.1. Requirements Definition

This step aims to find out the problems, objectives, and application needs through observations in the subjects of sensors and transducers in the Mechatronic Engineering department, SMK SMTI

Yogyakarta. Interviews were conducted with subject teachers and students who taking sensor and transducer subjects. According to subject teachers, Sensor and Transducer learning media are still limited and have not been able to increase student activity in learning, so that the absorption of knowledge from each student is not optimal. Situation analysis, analyzing the state of learning activities and students during learning. Media analysis is seen from the learning media used by teachers in teaching. While the analysis of supporting data is based on an analysis of learning tools such as syllabus, job sheets, and supporting modules. After that, the results obtained are defined and use as general application specifications.

2.2. System and Software Design
This stage is an android-based sensor and transducer application software planning. The steps taken are designing the application layout and information architecture. The process of making the layout design is using vector-based graphic design software, namely CorelDraw X7. All media elements in the application are made by the author himself. Application design is made by considering the needs and results of the Requirement Definition stage.

2.3. Implementation and Unit Testing
At this stage the results of software requirements and design will be realized in the form of a program into a mobile application, by processing the elements of the media used. The media elements in this application are text, images, videos, animations, and simulations. In this android-based sensor and transducer application software, there are Simulation media elements that are rarely found in other applications. The simulation aims for users to interact directly with the media, so that students interest and activeness in learning will increase. The process of managing media elements using Corel Draw X7 and Adobe Animate software with Android SDK version 32.0.

2.4. Integration and System Testing
At this stage, the completed program will be integrated and tested. Testing is divided into two stages, Black box test and Compatibility Test. The black box test in making an application is carried out to determine the performance or functionality in terms of some aspects. Black box testing in this development is reviewed from the interface aspect and function suitability [12]. While the Compatibility test according to ISO 25010 [13] rules, this test aims to determine the performance of the system if it is run from one device with different Android versions and screens.

2.5. Operation and Maintenance
At this stage, the Android-based sensor and transducer application software is directly operated and maintained. This stage makes it possible to make corrections for errors in each of the previous stages. This stage is usually the longest life cycle phase of an application software.

3. Result and Discussion

3.1 Stage of Requirements Definition
The results obtained at this stage through observations and interviews about the conditions and needs of users, are:

(1) The limitations of learning media and supporting modules used by teachers in the subject of Sensors and Transducers. There is no learning media that is used specifically for Sensor and Transducer subjects.
(2) In the process of Distance Learning (PJJ), the delivery of material by the teacher is still monotonous and less interesting. So that the interest and activeness of students in learning is still low, this causes the absorption of knowledge from each student to be not optimal.
(3) All students already have smartphones with most of them using the Android operating system, but teachers only use smartphones to discuss through WhatsApp Groups and sometimes do Video Conferences.

3.2 Stage of System and Software Design

The results at this stage are in the form of designing the layout and information architecture of an Android-based mobile application for the sensor and transducer field. The layout design describes the placement and content of each page. Making this layout design using CorelDraw X7 software, the layout design is shown in Figure 2. Every page and menu in the application software needs a layout design, so that application content can be neat and consistent.

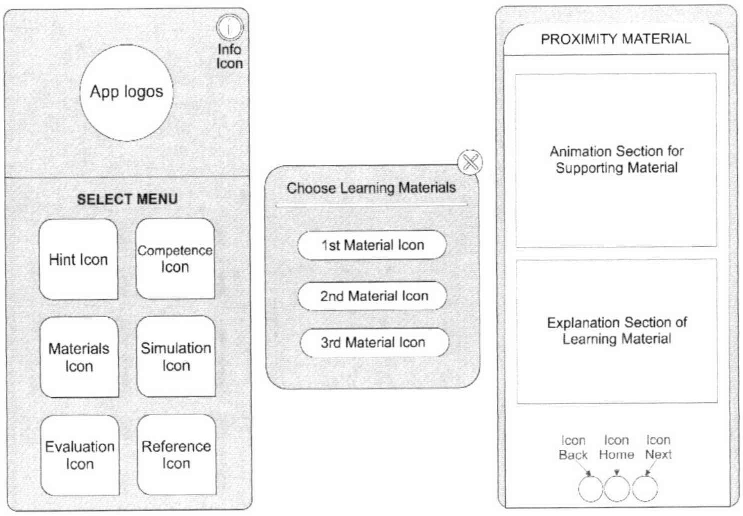

Figure 2. Layout Design

Information architecture (IA) shows an outline of information in a mobile application, starting from the main menu, sub-menu, sub sub-menu, etc. The main purpose of IA was to make it simple and structured to navigate. The information pattern is in the form of a hierarchy, as shown in Figure 3. The top order of the hierarchy is the main menu in the application, there are 7 main menus in the mobile application for this sensor and transducer field. The next hierarchy is the content and features contained in each menu.

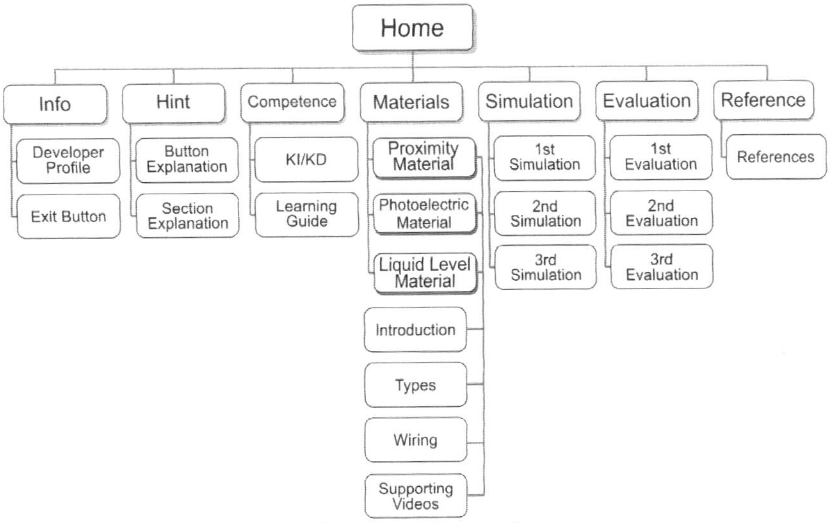

Figure 3. Information Architecture

3.3 Stage of Implementation and Unit Testing

Media elements in this mobile application are text, images, videos, animations, and simulations. The initial stage of creation is to create an interface that refers to layout planning and create drawings as needed. Making interfaces and image components using vector-based Corel Draw X7 software, as shown in Figure 4. The designs that have been created are saved in .png format so that the image has a transparent background and is easy to combine with other media elements.

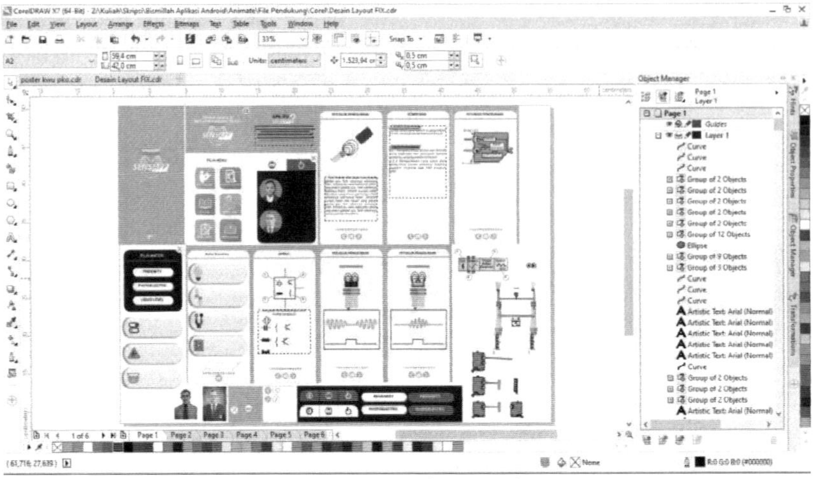

Figure 4. Interface Design

After designing the interface components and images, then processing various media elements according to the design stage. In this stage used Adobe Animate CC 2017 software to program and process all media elements, as shown in Figure 5. The programming language used in the Adobe Animate CC 2017 software is Action Script 3.0 which is a development of the previous version of Adobe Flash software. The file is exported using Android SDK version 32 as a builder program, so that it can be used on Android.

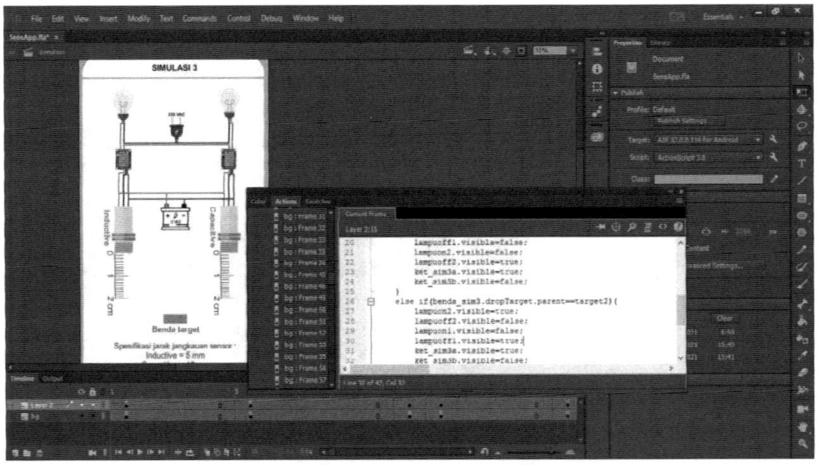

Figure 5. Processing Media Elements

The developed mobile application consists of 7 main menus, namely: info menu, hint menu, competency menu, material menu, simulation menu, evaluation menu, and reference menu. The menus are displayed on the main page that can be selected by the user, the display of the mobile application is shown in Figure 6. The info menu *("info")* contains the developer bio and a button to exit the application. The hint menu *("petunjuk")* explains the function of the buttons and the content of each part of the application. The competency menu *("kompetensi")* explains the learning objectives and competencies in the application. The material menu *("materi")* page contains several media elements, namely text, images, animation, sound and video. The simulation menu *("simulasi")* is used to support the material, in this menu the user can interact directly with the media by selecting and sliding objects to be simulated. The evaluation menu *("evaluasi")* contains multiple choice questions as a test of user understanding of the material in the mobile application. The reference menu *("referensi")* contains a bibliography of teaching materials. This learning media has a final file form in *.apk format that can be run on smartphones with the Android operating system.

Figure 6. Interface Mobile Application

3.4 Stage of Integration and System Testing

3.4.1 Black Box Testing

Black box testing in making application software is carried out to determine the performance of the Android-based Sensor learning media in terms of some aspects. Black box testing in this development is viewed from the aspect of the interface and the suitability of the function according to Pressman. Table 1 is the result of the test data obtained. The data obtained are in the form of two nominals between 0 or 1, the number 0 represents "Not Working" and the number 1 means "Working". The meaning of "Not Working" is that the test step fails or does not show results, and "Working" means that the test results match the statement under test.

Table 1. Black Box Test Data

No.	Statement	Description	
		Working	Not Working
Interface Aspects			
1.	The Sensor application software logo animation is displayed when the application is run	1	
2.	The main page of the application displays seven menus	1	
3.	The material page displays a choice of 4 sub menus on each material topic	1	
4.	Videos are presented in the supporting video sub menu in the material menu	1	

No.	Statement	Description	
		Working	Not Working
5.	Questions and answer choices are presented for each evaluation option	1	
6.	The score results on the questions are displayed at the end of the evaluation session	1	
Aspects of Functional Suitability			
7.	A pop up in the form of developer profile info is displayed if the info icon is pressed	1	
8.	The help page is displayed if the hint icon is pressed	1	
9.	The competency page is displayed if the competency icon is pressed	1	
10.	The material selection pop up is displayed if the material icon is pressed	1	
11.	The simulation page is displayed if the simulation icon is pressed	1	
12.	The evaluation page is displayed if the evaluation icon is pressed	1	
13.	The reference page is displayed if the reference icon is pressed	1	
14.	Goes to the main page when the home button is pressed	1	
15.	Returns to the previous page if the previous button is pressed	1	
16.	Returns to the next page if the next button is pressed	1	
17.	Animation or video will run if the play button is pressed	1	
18.	Animation or video will stop if stop button is pressed	1	

The statement on the interface aspect shows that the appearance of each menu presented is in accordance with the statement, and the function suitability aspect is used to test the buttons and commands in the application. Table 1 shows the recapitulation of the scores obtained from the black box test, the interface aspect gets an average score of 1, while the function suitability aspect gets an average score of 1. Based on the data obtained, it can be concluded that the performance of the Android-based Sensor application software can all function properly.

3.4.2 Compatibility Test

Researchers conducted compatibility testing by installing, running, and installing application software on various versions of the Android operating system and screen sizes. The selection of smartphone devices used in the test is based on different android versions and screen sizes.

Table 2. Compatibility Test Data

No	Device	Screen Resolution	Android Version	Successful	Failed
1	Asus Zenfone C	480 x 854	4.4.2	✓	
2	Xiaomi 5A	1080 x 1920	7.1.2	✓	
3	Asus Max M1	720 x 1280	8	✓	
4	Asus Max Pro M1	1080 x 2160	9	✓	
5	Realme 5	1080 x 2340	10	✓	

Based on the data obtained as shown in Table 2, it can be concluded that the compatibility of the Android-based Sensor application software can all function properly on five different devices and with different screen resolutions. This conclusion means that the Android-based Sensor application software can be used for various versions of Android with a minimum specification of Android version 4 which has a screen resolution of 480 x 854.

3.5 Stage of Operation and Maintenance

Maintenance includes repairing and monitoring the application if errors are found during the development stage. In the implementation and unit testing stages, the researcher conducted several experiments, especially when testing the Simulation menu. The program code on the Simulation menu must be adjusted to the animation that will be displayed, then the application software needs to be tested via debug software or directly installed on a smartphone.

4. Conclusion

Research on the development of android-based mobile applications for sensor and transducer field studies in vocational education has the conclusion that the design of sensor learning media uses the Waterfall development model, the stages in this development starting from the definition of requirements, system and software design, implementation and unit testing, integration and system testing, operation and maintenance. The final result is a file with *.apk format that can be run on smartphones with the Android operating system. This Android-based Sensor mobile application consists of 7 main menus, namely the info menu, hint menu, competency menu, material menu, simulation menu, evaluation menu, and reference menu.

References

[1] D. dkk Siswoyo, *Educational Sciences,* First. Yogyakarta: UNY Press, 2013.
[2] H. Sofyan, *Vocational Technology Education*. Yogyakarta: UNY Press, 2018.
[3] Regulation of the Minister of Education and Culture, "Regulation of the Minister of Education and Culture Number 34 of 2018 concerning National Standards for Vocational High School Education/Vocational Madrasah Aliyah." 2018.
[4] A. Arsyad, *Learning Media*, 1st ed. Depok: PT RajaGrafindo Persada, 2019.
[5] Daryanto, *Learning Media has a very important role in achieving learning objectives*. Yogyakarta: Gava Media, 2016.
[6] H. Praherdhiono, "Technology in Learning Media," 2017. https://teknologipendidikan.org/teknologi-pada-media-pembelajaran-2/ (accessed Mar. 12, 2021).
[7] I. Aripin, "Mobile Learning Concepts and Applications in Biology Learning," *J. Bio Educ.*, vol. 3 Nomor 1, 2018.
[8] M. Ally, *Mobile Learning Transforming the Delivery of Education and Training*. Athabasca University: AU Press, 2009.
[9] H. Kurniawan, "Mobile Learning Media Using Android (Case Study: Information Systems Department IIB Darmajaya)," *Explor. – J. Sist. Inf. dan Telemat.*, vol. Volume 8, p. 48, 2017.
[10] H. Nurhayati, "Number of smartphone users in Indonesia from 2015 to 2025," *statista.com*, 2021. https://www.statista.com/statistics/266729/smartphone-users-in-indonesia/ (accessed Mar. 19, 2021).
[11] I. Sommerville, *Software Engineering: Ninth Edition*. USA: Pearson Education, 2011.
[12] R. S. Pressman, *Software engineering: A practitioner's approach*. New York: McGraw-Hill, 2010.
[13] ISO/IEC, "ISO 25010," 2011. https://www.iso.org/obp/ui/#iso:std:iso-iec:25010:ed-1:v1:en:en (accessed Jun. 14, 2021).

Development and Effectiveness of Drone as a Learning Media in Islamic Boarding School

I W Djatmiko[1], S Yatmono[2], and A C Nugraha[3]

[1, 2, 3]Department of Electrical Engineering Education, Faculty of Engineering Universitas Negeri Yogyakarta, Indonesia

Email: [1]istanto_wj@uny.ac.id

Abstract. Nowadays, drone technology is not only be used in commercial businesses but can also be utilized in a teaching-learning process, including in Islamic boarding school. This article will be discussed the development, performance, feasibility, and effectiveness of drone technology to be utilized as a learning media at *Pesantren Hidayatullah Yogyakarta* that is one of Islamic boarding school. The developed drone was the multicopter with a multi-rotor type using four rotors. This study was conducted with the development research approach developed by Alessi and Trollip and the descriptive research approach. This drone was applied in teaching-learning process at *Pesantren Hidayatullah* Yogyakarta. Data were obtained through drone's testing and observation, interviews, questionnaires for the drone's affectivity in teaching-learning process. The results of this study can be known that the drone performance is categorized as "Very Good" in terms of functionality and usability aspects. The drone feasibility in the term of material quality and its usefulness is included in the "Very Feasible" category. The drone as learning media has had significantly impact when it's applied in learning process in extracurricular activities.

1. Introduction

Today's technological and scientific innovations continue to grow along with the flow of globalization to produce supporting devices. All of these technologies are highly dependent on electronic devices that require electricity to operate them. One of the rapidly developing technologies today is intelligent robot technology [1]. Intelligent robot technology has various types such as the types of land robots, water robots, and air robots. Drone or Unmanned Aerial Vehicle (UAV) is one type of intelligent robot that can be operated manually or automatically which is well known in a wider community. The development of robot technology has had many impacts and benefits in life nowadays, especially in the industrial and business sectors. The industrial and business sectors have taken advantage of the development of robotic technology to support the expansion of their activities [2] because production costs can be reduced so that product prices become cheaper and speed up a production process more efficiently [3]. In addition to the industrial and business sectors, the development of robotic technology also has an impact and benefit on an education sector.

The education sector can take advantage of this robot technology as a learning media or a learning aid to convey material in teaching and learning process activities. Learning media in the form of robots can be used to redesign a classroom atmosphere to be able to produce high-level thinking skills for students [4]. In addition, this learning media is also useful for equating students' perceptions and strengthening a material to be delivered by the teacher [5]. Besides being applied to academic learning activities, learning media in the form of robots can also be applied to non-academic learning activities, such as extracurricular activities. The application of robots in extracurricular learning activities can foster cognitive skills and teamwork so that students can express ideas that are valuable for their lives in the future [6]. Therefore, this robot technology, such as a drone, can be designed to be used as a learning media in schools.

Pesantren or Islamic Boarding School *Hidayatullah* Yogyakarta is one of an educational institution that organizes religious and general science-based learning activities that provide at the levels of *Madrasah Ibtidaiyah* level, which is equivalent to an elementary school, *Madrasah Tsanawiyah* which is equivalent to a junior high school, and *Madrasah Aliyah* which is equivalent to a Senior High School. Students or usually be called as *"santri"* are also provided with life skills learning activities that aim to train students' skills and expertise. Drone technology is one of the latest technologies that need to be introduced to them, especially its contribution as a learning media in extracurricular activities at *Pesantren Hidayatullah* Yogyakarta.

1.1. Drone or Unmanned Aerial Vehicle (UAV)

Drone or UAV is an aircraft system without a crew or pilot inside it [7]. Drones are usually controlled by an operator remotely using a remote control, although now various types of drones can be operated autonomously based on coordinates points or programming codes. Currently, drones have been widely applied in various fields because drones are safer when operated and their costs are relatively cheaper [8]. Nowadays, drones are operated in real-time monitoring [9]. Real-time monitoring can be done using drones because most drones are currently equipped with a high-resolution camera so that real-time monitoring activities can be carried out properly. In addition, a drone has quite complex components, such as a brushless motor, propeller, battery, etc. [15]. Components of a drone can be seen in Figure 1.

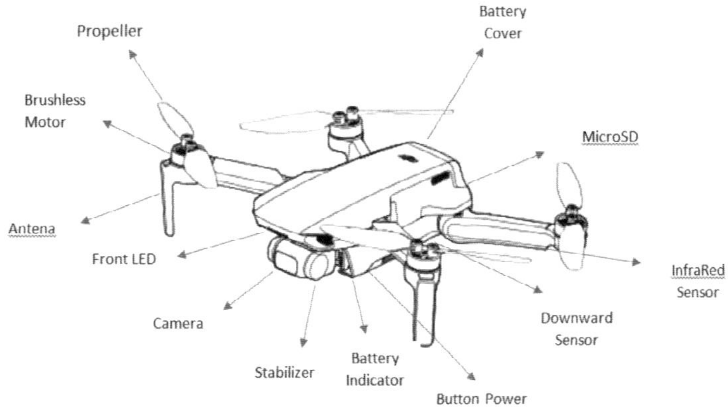

Figure 1. The Drone's Components

All of those drone components are very easy to buy or obtain in today's market with various types and advantages. Therefore, in this study, using a drone on the market was chosen, it's and then modified to suit the study objectives. The drone was selected taking some consideration such as the price and technical capabilities of the drone that adapted to the study objectives. In terms of price, the drone should affordable prices, while in terms of technical capabilities, the drones must have an ability to record images and videos in good quality, because this drone was used for environmental monitoring and also as a learning media at *Pesantren Hidayatullah*. Referring to Stefano's study, drones with high recording resolution were the right choice to be used as a mapping or monitoring tool [13]. In addition, the technical capabilities of the drone should also be considered, such as the ease of control so that it can be operated by students easily.

These drone modifications were performed by adding a camera and a transmitter and receiver video. These additional devices were chosen so that the interface development can be carried out easily for this drone. The drone's interface development was carried out to suit the needs of this study's objectives. After the drone's interface development was completed, the next activities determined the coordinates of several paths that will be passed by the drone. Examples of the coordinate points that have been taken at *Pesantren Hidayatullah* can be seen in Table 3. After all drone development was completed, the activity continues with the development of drone-oriented learning modules that had been developed to facilitate the delivery of material regarding drone operations.

1.2. Drone as a Learning Media

Learning media is needed by teachers as a tool in a learning process either in a classroom or in online learning. Learning media is a component of learning resources or physical devices that contain instructional materials for students that can stimulate students to learn [12]. Learning media currently has a function as a tool utilized by teachers to deliver material in their learning process [16] so that learning media has an important role in determining the effectiveness and efficiency of achieving learning objectives [17]. Therefore, it is not easy to choose suitable learning media to be applied in a learning process.

Some aspects need to be considered in choosing instructional media, such as suitability for learning objectives, capabilities and characteristics of the media, duration of media procurement, and the price of media itself [18]. Another aspect that needs to be considered in choosing learning media is the novelty of media technology. Learning media with the latest technology is proven to have a positive effect on students in terms of improving performance and their level of interactivity among students [19]. Therefore, learning media with the latest technology began to be implemented in a learning process as a learning media. Drones are one of the learning media that utilize modern technology.

Nowadays, robots are starting to be used in learning as media, including flying robots or drones, which are applied in mapping or monitoring lessons. Drones have many advantages, especially in terms of time flexibility, wider coverage area, and detailed work results [20]. In addition, drones can also be performed for programming learning media as an automatic delivery tool based on predetermined coordinate points [21].

Drones are also currently being developed as learning media by considering students' characteristics. Based on the study [22] students will not only be able to fly drones, but also students must be able to develop understanding deeply of drones and ethics in flying these drones. Furthermore, students not only learn skills but also they can create interdisciplinary knowledge, problem-solving abilities, and hands-on abilities by utilizing drones in the learning process.

1.3. Drone DJI FLY as a Learning Media

DJI FLY is an interface application with a simple and very intuitive design created by dji.com. This application is compatible with various types of drones such as Mavic Mini, Mavic Air 2, DJI Mini 2, DJI FPV, DJI Air 2S, and DJI Mini SE. This application can work on iOS and Android platforms with application version V 1.4.8 [23]. This application is also applied to fly and control drones, for example, the DJI Mini 2 drone. DJI FLY can order the drone to take pictures or videos with some techniques embedded in the system. The system is named the active track feature. This feature allows the drone to take video with some automatic movement. In addition, the DJI FLY can also control a drone's flight manually with pilot control. However, these video recordings with manual techniques may have lower quality than those with automatic video recordings. Therefore, the operation of the DJI FLY requires the ability of a pilot who is able to fly a drone so that good quality recordings are obtained as operated with automatic features. This condition is necessary for learner pilots, including *"santri"* at *Pesanten Hidayatullah.*

The drone learning module had been developed on how to manually fly a drone for "*santri*" and operate it. This drone learning module was developed based on the DJY FLY application which has been modified according to the characteristics of students. The drone learning module contains four materials, include basic drone's theory, drone's construction, drone's operation, and drone's maintenance. In addition, this study also developed two research instruments to determine the performance and feasibility of drones as learning media. The grids of these two instruments can be seen in Table 2. Furthermore, to know the effectiveness of drones in learning, two drone comprehension instruments were developed that applied before and after learning or training activities.

Table 1. The Performance and Feasibility Grids of Drone as a Learning Media

Performance of Drone's Media		Feasibility of Drone's Material	
Aspects	Reviewed Studies	Aspects	Reviewed Studies
Functionality	Navigation accuracy	Material Quality	Material suitability
	Component's functionality		Material equipment
Usefulness	Media benefits		Material collapse
	Media effectiveness		Suitability with student characteristics
		Usefulness	Make material delivery easier
			Support learning process

2. Research Method

This study was carried out using the development approach developed by Alessi and Trollip [10] to produce drone as learning media and test its effectiveness. This study was conducted in three stages, including the planning, design and development stages as shown in Figure 1.

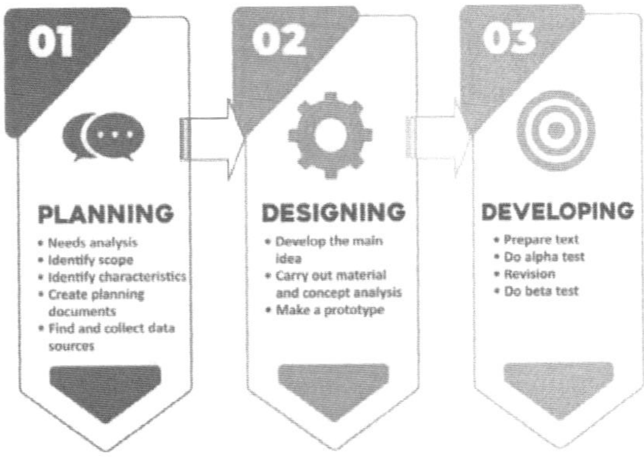

Figure 2. The Drone's Studies Stages

As explained in [10], the steps of each stage can be described as follows: in the planning stage, researchers plan the need for drones that will be used as learning media by considering several aspects, such as determining the scope of drones as media, identifying students' characteristics, determining obstacles, determining development costs, making planning documents, making learning module, determining, and selecting materials, collecting data, and submitting user approvals. Designing stage was related to several activities such as collecting materials, designing drone performance tests, and drone feasibility tests in terms of functionality and the meaning of drones as learning media, its effectiveness. These steps were taken in this stage include developing ideas, compiling an analysis of drone test concepts, compiling a description of drone learning programs, preparing drone prototypes, creating flowcharts and storyboards, and submitting approval to users. The developing stage was assembling a drone that has been designed into a learning media product so that it was ready to be tested. This stage is carried out with several steps including preparing a drone prototype, determining coordinates, preparing learning modules, conducting training to students, conducting alpha tests, revising products, conducting beta tests, revising final products.

The results of designing drones as learning media are obtained through performance and feasibility tests carried out by experts who have some experience in the field of drone technology. Furthermore, the effectiveness test of drones as a learning medium was carried out through training for 30 students at the Madrasah Aliyah level at *Pesantren Hidayatullah*. The effectiveness test was conducted to determine the level of students' understanding before participating in the training (comprehension 1) and after completing the training (comprehension 2). The used instrument to determine the effectiveness of drones can be known through the coefficient results of Pearson's product-moment validity test of 0.482 to 0.885, while the coefficient of reliability test results with Cronbach's alpha obtained is 0.946. Data were analyzed using a descriptive approach which was presented through photo illustrations and graphics.

3. Research Results and Discussion

3.1. Development of Drone as Learning Media

As explained above, the DJI Mini 2 drone was chosen as a learning media in this study. The DJI Mini 2 drone was modified by adding a camera and a video transmitter and receiver to obtain monitoring recordings related to these study objectives. The changes to the specifications of the DJI Mini 2 Drone after modifications are made as shown in Table 2.

Table 2. Specifications of Modified Drone DJI Mini 2

No.	Specifications	Remark
1.	Weight	<249 g
2.	Dimension	245mm x 289mm x 56mm
3.	Sensor System	Downward system
4.	Distance range	4 km
5.	Maximum flight time	31 minutes
6.	Camera	12 MP
7.	Video Resolution	4.0K, 2.7K
8.	Battery Capacity	5200mAh

Furthermore, the design of the Graphical User Interface or interface had done so that it can display objects that can convey information and represent the results for users. This interface is a system of interactive visual components for computer software. The design of this drone application interface is carried out to make it easier for students to know and control drones. Designing the drone application interface was applied by making storyboards to facilitate the further development process. The storyboard design of the drone interface can be seen in Figure 3.

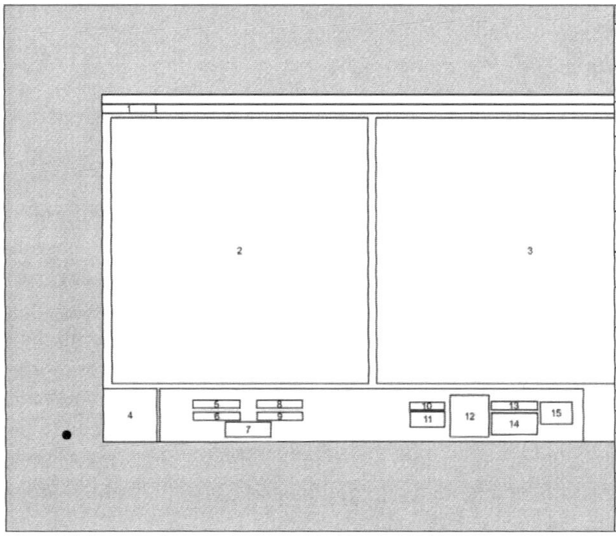

No	Descriptions	Remark
1	Resolution	Recording video resolution setting
2	Video Display	Displaying video recording in real time
3	Coordinate Display	Displaying drone coordinates in real time
4	University Logo	UNY's logo display
5	Longitude	Displaying longitude coordinates
6	Latitude	Displaying latitude coordinates
7	Port Connection	Port Connectivity
8	Port Options	List of available port options
9	Baud rate Options	List of available baud rate options
10	Data Check	
11	Delete	
12		
13	Camera Options	List of available camera options
14	Camera Connection	Camera connectivity
15	Save	Recording data storage

Figure 3. The Storyboard of Interface Display

After the storyboard was completed, the development of the drone application interface was carried out related to the storyboard design that had been made previously. Visualization of the drone application interface is shown in Figure 4.

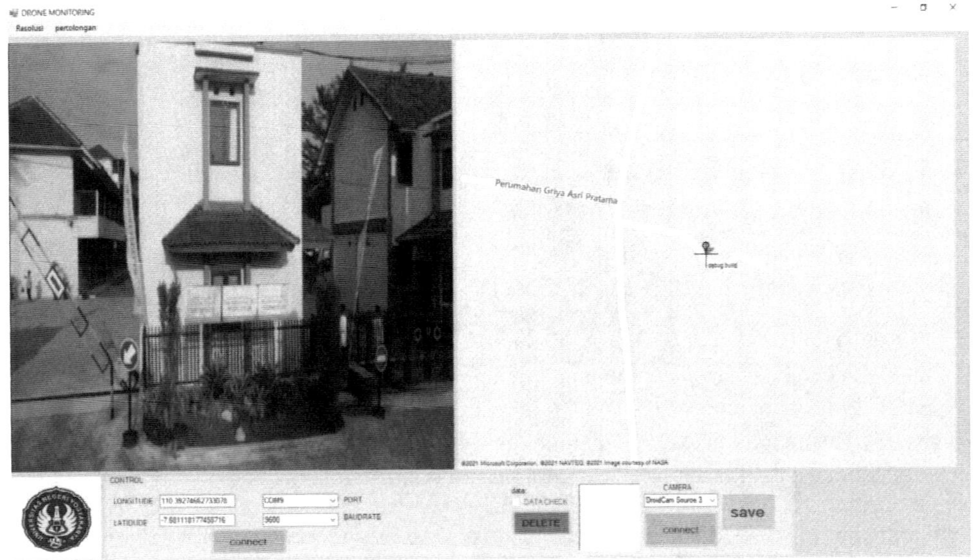

Figure 4. The Display of the Drone Application Interface at *Pesantren Hidayatullah*

Establishing coordinate points was the drone utilization as a learning media for monitoring in *Pesantren Hidayatullah* environment. Coordinates setting is performed to determine routes or paths that will be passed by the drone. The results of defining the coordinate points at *Pesantren Hidayatullah* can be presented in Table 3.

Table 3. Drone's Waypoint

Waypoint	Names of Coordinate Point	Latitude	Longitude
H	Home	-7.68086000530349	110.392019748688
1	South-left point	-7.68150626606614	110.39218939840794
2	South-right point	-7.681825575005529	110.39334811270237
3	Lower-east point	-7.681319863962482	110.39340376853943
4	Center-east point	-7.680899877461258	110.39342824369669
5	Upper-east point	-7.680740388807714	110.39310302585363
6	North-right point	-7.680390842632617	110.39322573691607
7	North-center point	-7.680345654132719	110.39271980524063
8	North-left point	-7.680219392122235	110.39239089936018

Based on these coordinate points are used to plan the drone flight path around *Pesantren Hidayatullah* complex as shown in Figure 5.

Figure 5. Drone Flight Path around *Pesantren Hidayatullah* Complex

Finally, the final stage of developing drones as learning media is the development of drone learning module that are applied in extracurricular learning in *Pesantren Hidayatullah*. The results of designing drones as learning media are carried out through five main activities. Firstly, the activity of carrying out a needs analysis was carried out by reviewing several studies related to the application of drones in learning. Secondly, the activity identifies the scope by looking at what materials can be developed and collaborated with the application of drones as learning media. Third, activities to identify the characteristics of *santri* and extracurricular activities at *Pesantren Hidayatullah* can collaborate in applying a drone as learning media. Fourthly, making instruments for performance testing, feasibility tests, and learning media in the form of drones and training modules. Finally, the activity of testing the effectiveness of drones as a learning media through learning or training activities.

Based on the first, second, and third activity stages above, the development of a drone learning module had been compiled which contains five chapters covering the basic theory of the drone, drone components construction, drone operating, and drone maintenance. This drone learning module was delivered in theoretical and practical learning with a learning duration of 8 hours. In addition, the development of instruments on understanding this drone was also elaborated to determine the students' comprehension level toward drone learning before and after participating in this learning process. These drone comprehension instruments contain 20 questions respectively related to the material contained in the drone learning module. Furthermore, the instrument for drone media performance, the instrument for the feasibility of drone materials, and the effectiveness of a drone as learning media are also prepared with the completed grids of each of these instruments described in the further discussion section.

3.2. Performance of Drone as Learning Media

The media performance test was carried out to determine the performance quality of the modified drone. The drone media performance was tested by two experts who have expertise in this field. The tested aspects in this media performance test consist of two aspects which include the functionality aspect and the usability aspect. The functionality aspect consists of two reviews covering navigation accuracy and component functionality, while the usability aspect consists of two studies covering media meaning and media effectiveness as described in Table 1.

The results of the drone media performance test can be seen that the quality of this media is in the "Very Good" category (94.16%) that is relevant to the reference [11]. It means that the performance of this drone media can be utilized as a learning media. In addition, the result of this study is also in line with the study result referring to [15] that the result of the media performance test score of at least 82.10% can be stated to be included in the very feasible category.

3.3. Feasibility of Drone as Learning Media

The feasibility test of the material is carried out by knowing the quality level of the drone material and the learning module separately. The feasibility of this drone material had been tested by two experts who have a comprehensive mastery of drone material. The aspects tested in the feasibility test of this material consist of two aspects which include material quality and material usefulness aspects as described in Table 1. The material quality aspect contains four studies which include material suitability, material equipment, material collapse, and suitability with student characteristics aspects. The usability aspect contains two studies which include making material delivery easier and supporting the learning process.

The results of the drone material feasibility test can be seen that the quality of this drone material is in the "Very Feasible" category (90.27%) that is relevant to the reference [11]. It means that the feasibility of this drone material can be applied to a learning process activity. This study result was reinforced by study results that refer to [14] which stated the result of the material feasibility test score of at least 82.50% can be stated to be included in the very feasible category.

3.4. Effectiveness of Drone as Learning Media

The beta test is carried out by measuring the effectiveness of a drone when it's applied in the learning process which was carried out through the Wilcoxon Signed Ranks Test before participating in the training and after completing the drone's training. The results of the beta test were shown that a drone as a learning media has a significant level of effectiveness (significant value of 0.000). This means that a drone as a learning media can increase a students' level of understanding in a learning process. This level of effectiveness was seen through the achievement of students' understanding of four main materials during drone training, including basic theory of the drone, drone components constructions, drone operating, and drone maintenance. Overall, the level of understanding of students before training (comprehension 1) and after training (comprehension 2) of the drone as a learning media is shown in Figure 6.

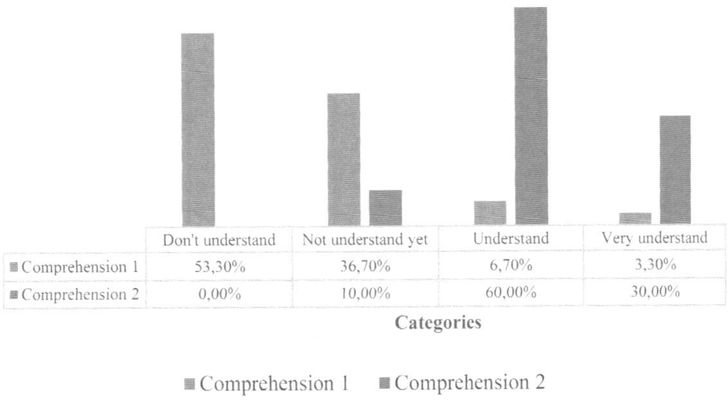

Figure 6. The Overall Comprehension of Drone's Knowledge and Skills

The descriptions of the increment in the level of understanding achievement on the topics of the basic theory of drones, the drone components constructions, the drone operating, and the drone maintenance were reviewed in the dominant category when before and after completion of drone training. Each of those topics can explain that the student's comprehension of the basic theory of drones in the "very understand" category is increasing from 16.70% to 93.30%. The students' comprehension of the drone components construction in the "very understand" category is raised from 6.70% became 50.00%. The students' comprehension of the drone operating in the "very understand" category is a rise from 3.30% became 33.3%. The students' comprehension of the drone maintenance in the "understand" category is ascended from 6.70% became 33.3%.The results of this study are in line with the results of research [22] which states that the use of media in learning has a positive effect on independent learning and more innovative thinking to students and is also relevant according to references [23] that stated utilizing learning media will enhance a learning process which in the end can improve quality and achieved learning outcomes.

4. Conclusion

The results of this study can be seen that drones can be developed and applied as a learning media for *"santri"* in Madrasah Aliyah level at *Pesantren Hidayatullah.* The results of the alpha test on drone media showed that the performance level of the drone was included in the "very feasible" category (94.16%) and also included in the "very good" category (90.27%) when viewed toward the media feasibility. Furthermore, the results of the beta test can determine the effectiveness of drones as a learning media that can significantly improve the quality of drone learning at *Pesantren Hidayatullah.* The results of this study can be concluded that drones as learning media can be applied in the learning process.

5. References

[1] Darwis R, dkk. 2019. *Perancangan Robot PemadamApi Hexapod.*JurnalAL-Azhar Indonesia Seri Sains dan Teknologi, Vol. 5, No. 1, p 1-4.
[2] Susan A, &Novianti W. 2019. *Benefits of Technology for Business.* INCITEST 2019, Vol. 662, p 032036, doi: 10.1088/1757-899X/662/3/032036.
[3] Wilburn K. M, & Wilburn H. R. 2018. *The Impact of Technology on Business and Society.* Global Journal of Business Research, Vol. 12, No. 1, p 23-39.
[4] Hopson M. H, dkk. 2002. *Using a Technology-Enriched Environment to Improve Higher-Order Thinking Skills.* Journal of Research on Technology in Education, Vol. 34, No. 2, p 109-119.
[5] Puspitarini Y. D., & Hanif M. 2019. *Using Learning Media to Increase Learning Motivation in Elementary School.*Anatolin Journal of Education, Vol. 4, No. 2, p 53-60.
[6] Alimisis, D. 2013. *Educational Robotics: Open Question and New Challenges.* Theme in Science and Technology Education, Vol. 6, No. 1, p 63-71.
[7] Saputra D.R.H. &Pramujati B. 2013. *RancangBangun Prototype Unmanned Aerial Vehicle (UAV) denganTiga Rotor.*Jurnal Teknik Pomits, Vol. 2 No. 1, p 47-52.
[8] Barreto J, dkk. 2021. *Drone-Monitoring: Improving the Detectability of Threatened Marine Megafauna.* MDPI, Vol. 5, No. 14, p 1-14.
[9] Witayangkurn A. 2011. *Real-Time Monitoring System Using Unmanned Aerial Vehicle Integrated with Sensor Observation Service.* International Archives of the Photogrammetry, Remote Sensing and Spatial Information Sciences, DOI: 10.5194/isprsarchives-XXXVIII-1-C22-107-2011.
[10] Alessi & Trollip. 2001. *Multimedia for learning: Methods and Development.* Massachusetts: A Pearson Education.
[11] Arikunto S. & Jabar C. S. A. 2018. Evaluasi Program Pendidikan: PedomanTeoritisPraktis Bagi Mahasiswa dan Praktisi Pendidikan. Jakarta. SinarGrafikan Offset.
[12] Arsyad, Azhar. 2016. *Media Pembelajaran.* Jakarta: Raja GrafindoPersada.
[13] Stefano Andrew. 2020. *Pemanfaatan Drone dalamPemetaan Kontur Tanah.*Buletin LOUPE, Vol. 16, No. 02, p 32-41.
[14] Sukir, dkk. 2019. *Kelayakan E-Modul Berbasis Android untukMendukungPembelajaranAplikasi PLC sebagaiPengendaliMesinPengisi dan Penutup Botol Otomatis di Smk.*JurnalEdukasiElektro, Vol. 3, No. 2, p 88-98.
[15] Muliady&Subagya E. J. 2019. *SistemPemetaan Udara MenggunakanPesawat Fixed Wing.* TESLA, Vol. 21, No. 1, p 26-35.
[16] Puspitarini Y. D., dkk. 2019. *Using Learning Media to Increase Learning Motivation in Elementary School.* Anatolian Journal of Education, Vol. 4, No. 2, p 53-60.

[17] Miftah M. 2013. *Fungsi dan Peran Media Pembelajaransebagai Upaya PeningkatanKemampuanBelajarSiswa.*Jurnal KWANGSAN, Vol. 1, No. 2, p 95-105.
[18] Chotib S. H. 2018. *Prinsip Dasar PertimbanganPemilihan MediaPembelajaran.*Jurnal PGMI, Vol 1, No. 2, p 109-115.
[19] Sakat A. A. 2012. *Educational Technology Media Method in Teaching and Learning Progress.* American Journal of Applied Sciences, Vol. 9, No. 6, p 874-878.
[20] Utomo B. 2017. *Drone untukPercepatanPemetaanBidang Tanah.* MKG, Vol. 18, No. 2, p 146-155.
[21] Hakim M. L., dkk. 2020. *The Effects of Vertical and Horizontal Distance on The Performance of QR Code Detection System.* ICE-ELINVO 2020, Vol. 1737, p 012029.
[22] Chen C, dkk. 2018. *Exploring the Learning Effectiveness of "The STEAM Education of Flying and Assembly of Drone".* 2018 Seventh International Conference of Educational Innovation through Technology (EITT). DOI 10.1109/EITT.2018.00021
[23] Sudjana, N. 2013. Dasar-dasar Proses BelajarMengajar (Cetakan ke-13). Bandung: SinarBaruAlgesindo.

Effect of Mesh Topologies to Predict Cooling Effect on Evaporative Cooling Duct

A Jamaldi[1], Sarjito[2], A S Nurrohkayati[3], N T Atmoko[4]

[1,4]Department of Mechanical Engineering, Sekolah Tinggi Teknologi Warga Surakarta, Indonesia.
[2]Department of Mechanical Engineering, Universitas Muhammadiyah Surakarta, Indonesia.
[3]Department of Mechanical Engineering, Universitas Muhammadiyah Kalimantan Timur, Indonesia.

Email: agus.jamaldi@sttw.ac.id

Abstract. This paper examines the effect of different mesh types on a numerical study of evaporative cooling in the chimney. This research is a follow-up study from previous research. The test specimen used is an evaporative chimney design with the addition of a nozzle arrangement in it. The main focus of this research is the study of mesh refinement, namely by applying structured mesh during the simulation process. Three types of mesh with different levels of fineness were used for the specimens. They are coarse (mesh A), medium (mesh B), and fine (mesh C). In addition to differences in mesh, research was also carried out with variations in the level of relative humidity (RH). The RH levels used are 5, 10, and 15%. Two main parameters of evaporative cooling performance are airflow distribution and temperature drop in the chimney. Method for measuring the distribution of airflow and temperature drop in the chimney by making five planes with different heights. The results showed that the simulation with mesh B produced a good agreement data with previous studies than mesh A and C. The RH level that generated the most optimal cooling is found at 5% RH.

Keywords: chimney, evaporative cooling, mesh, numerical study, temperature.

1. Introduction

The comfort level of a building is influenced by several aspects. Among those aspects, one of them is thermal comfort related to air circulation inside it. Nowadays, there are many devices employed to generate good air circulation and environmentally friendly such as windows installation as natural ventilation in the building [1] [2] [3] [4], heating, ventilation and air conditioning (HVAC) [5], and evaporative cooling tower which used to catch airflow around the building [6]. A multi-stage down-draft evaporative cool tower (DECT) is an air conditioning mechanism in buildings with catch the wind flows that pass through the building [7] [8] [9]. DECT is very environmentally friendly and does not require electricity to use [8]. Another technology developed by the researchers is an energy-efficient and pollution-free cooling system, called indirect evaporative cooling (IEC) [10], wind tower [11], and passive cooling system [12].

To create a more optimal cooling effect at the evaporative cooling tower modified by adding other components such as water spray by the nozzle [13]. This water spray gives a better cooling effect. The nozzle addition as a component of the evaporative cooling has been carried out by many previous researchers both experimentally and computationally [14] [15] [16]. Georges and Buchlin [17] carried experiments using a single nozzle on evaporative cooling in the chimney. A simulation study was conducted by Gant [18] by using a single nozzle in the water misting process. The study focused to observed the thermal energy and momentum transfer that occurs in the surrounding air. The results of Gant's research were identical to those of Georges and Buchlin. Experiments study also done by Pearlmutter about evaporative cooling with the addition of a water spraying system in the chimney [19] [20]. The results show that the most optimal cooling occurred when the water spray has a little intensity with smooth misting.

The cooling effect is also influenced by the number, type, and nozzles arrangement [21] [22] [23] [24]. Tambur and Guetta [25] conducted research using two types of nozzles, namely BETE PJ32 and TF6. The orifice diameter of each nozzle has been modified by the total modified of it is 16. The results obtained that the PJ32 nozzle produces a better spray than TF6 at the same surface area per unit volume. Sarjito [26] researched the number of nozzles and the arrangement of the nozzle positions for the most optimal cooling effect in the evaporative chimney. The simulation research carried by Sarjito et al. [27] focused on the differences of number and nozzles arrangement to the cooling effect in the evaporative chimney. The results showed that the nozzle configuration with 8 nozzles at the top of the chimney and 3 nozzles with a height of 3.5 m from the bottom of the chimney produced the optimal temperature reduction and mass flow rate. The humidity level also influences of evaporation cooling effect [23].

Referred to research conducted by Sarjito et al. [27], in this study, a refinement mesh study was carried out by applying a structured mesh to evaluate the cooling effect in the evaporative chimney. Two main parameters of evaporative cooling performance are airflow distribution and temperature drop in the chimney. The results of this study are then compared with the founding data carried by Sarjito et al. [27].

2. Methods

The research was conducted in two steps. The first step is to make the geometry of the evaporative chimney, according to Sarjito's research [27]. The next step is the meshing process of the specimens using the Gambit meshing tools. Then the simulation process is carried out by adopting the structured mesh that has been made before.

2.1. Making the Geometry

This process is carried out to ensure that the test specimens conform to the geometry used in Sarjito's research. The specimens used in this study are tubular, as shown in Figure 1. Chimney diameter (D) = 3 m, height (Lo) = 4 m, nozzle diameter (do) = 0.00625 m, nozzle length (lo) = 0.05 m, the injection centre is located at 0.01 m from the tip of the nozzle with the nozzle spray angle (θ) is 30 degrees.

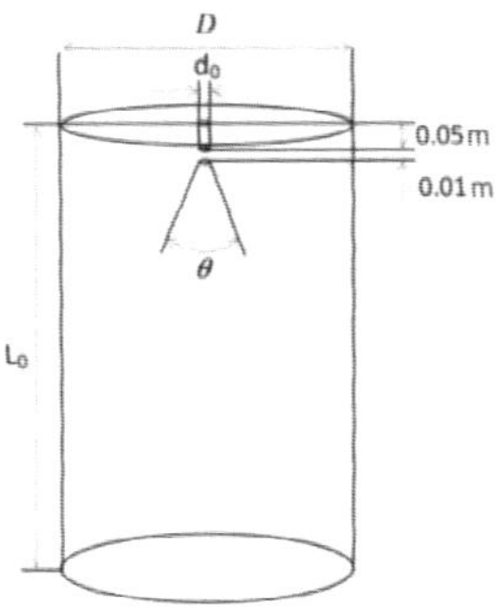

Figure 1. The dimensions of evaporative chimney. [27]

Journal of Physics: Conference Series **2111** (2021) 012012 doi:10.1088/1742-6596/2111/1/012012

2.2. Meshing Process

The structured mesh is created using the Gambit meshing tool. Three types of mesh i.e coarse mesh (mesh A), medium (mesh B), and fine (mesh C), were used in this study. The differences between each mesh are shown in Figure 2. The characteristics of each mesh model are shown in Table 1. The use of structured mesh is the main focus of this research. The use of structured mesh provides the same size for each element so that the number of nodes and cells is less than the unstructured mesh. Structured and unstructured meshes have their advantages and disadvantages. The superiority of structured mesh is that we can determine the size of each element so that each cell tends to have the same size. While the limitation of structured mesh is that the meshing process takes a long time compared to unstructured mesh. Hereafter the superiority of unstructured mesh is that the meshing process is so simple. This is because the geometry of the element is automatically determined by the system. While the limitation is that the meshing results have different element sizes and produce many cells when used in complex shapes.

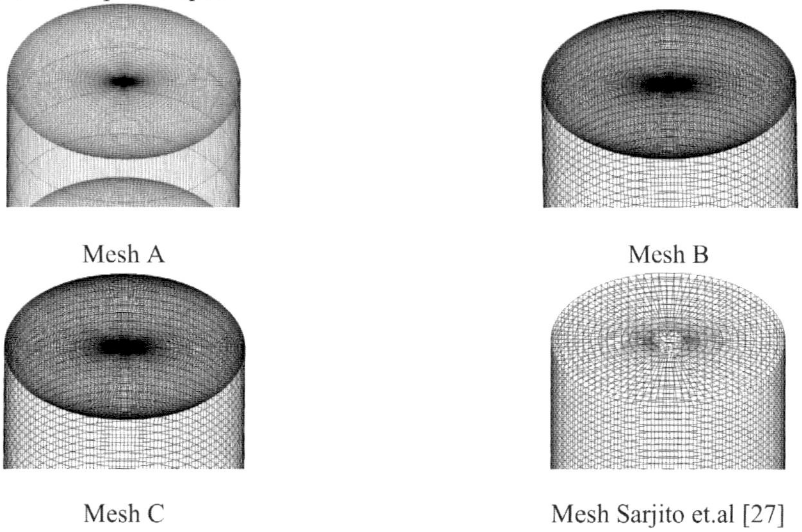

Figure 2. Meshing.

Table 1. Mesh characteristic

No	Type mesh	Nodes	Elements
1	Mesh A	63005	50000
2	Mesh B	104091	100000
3	Mesh C	327626	312500
4	Sarjito et.al	348068	2016345

2.3. Domain and Boundary Conditions

The domain and boundary conditions used in this study refer to research conducted by Sarjito et al. [27]. The top of the chimney is "opening" with the intake air temperature of 30°C, on the chimney cover is "wall-free slip adiabatic", and at the bottom of the chimney is "outlet" with a pressure of 0 Pa. The exiting water velocity from the nozzle is 21.57 m/s with a temperature of 10°C. In this simulation k-epsilon, the turbulent model was applied, which represent the average turbulent flow condition the domain and boundary conditions of the test specimen shown in Figure 3.

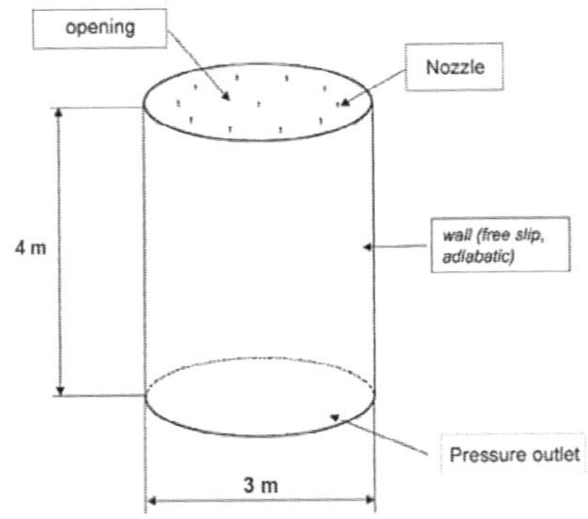

Figure 3. Domain and boundary conditions. [27]

2.4. Nozzle Configuration
The nozzle configuration used is 8 nozzles at the top of the chimney and 3 nozzles with a height of 3.5 m from the bottom of the chimney. The nozzle configuration is shown in Figure 4.

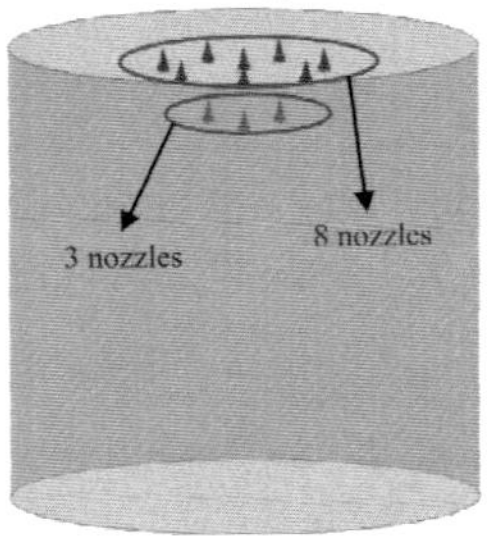

Figure 4. Nozzles configuration.

2.5. Measurement Position
A total of five planes with different heights were used as measurement positions to analyze the airflow distribution and temperature drop in the evaporative chimney. The five measurement positions are shown in Figure 5.

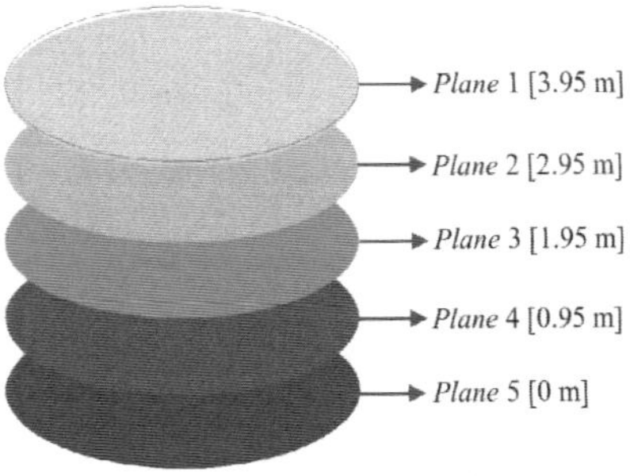

Figure 5. Plane position.

3. Results and Discussion

3.1. Mesh Validation
The mesh validation is shown in Figure 6. The simulation results used in mesh B produce consistent data with the results conducted by Sarjito et al. [27]. It is shown by the identical trend graph, while the

simulation results are under prediction from previous studies. The discrepancy between the prior research was 0.093%.

The simulation results using mesh A and C do not match with the reference data by the previous researcher. In measurement positions, 1 and 2 occurred "under-prediction". At measurement positions 4 and 5 occurred "over-prediction" only at measurement position 3 has the same value as the reference data. The difference in simulation results using mesh A and C against the previous studies was 0.094% and 0.083%, respectively. Based on the results, it shows that the difference in level mesh smoothness affected the simulation results. Furthermore, mesh B was used to analyze the cooling performance in the evaporative chimney.

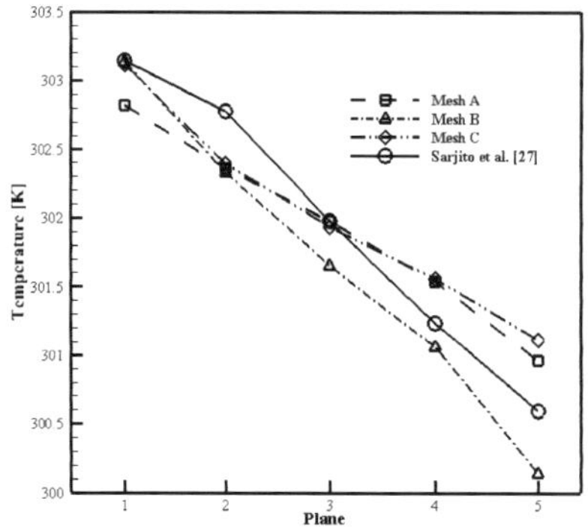

Figure 6. Temperature distribution.

3.2. *Airflow Velocity Distribution in the Chimney*

Figure 7 shows the velocity distribution of the airflow in the chimney at each measurement position. Mesh B gave consistent data with reference data conducted by Sarjito et al. [27]. Similar trends occurred at positions 1 to 4 only at position 5 occur striking diversity. The discrepancy with prior research is 20.55%.

The simulation results with mesh A and C produce a similar trend. The results obtained an "under-prediction" to the reference data. The discrepancy to the reference data is 24.27% (mesh A) and 28.96% (mesh C).

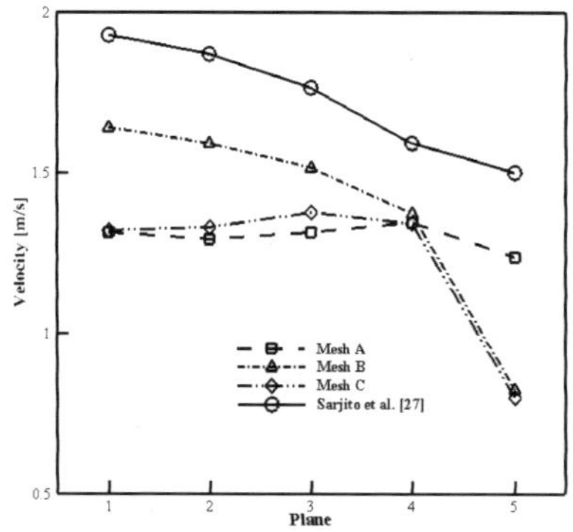

Figure 7. Velocity distribution.

3.3. *The Effect of RH level on Evaporative Cooling Performance*

The difference in RH level affect the evaporative cooling performance. In this study, the RH levels used 5, 10 and 15%. RH can be calculated using the following equation.

$$RH = \frac{P(H_2O)}{P'(H_2O)} \times 100\% \quad (1)$$

Table 2 shows the data from the calculation of RH levels with variations in temperature differences.

Table 2. RH level calculation.

Temperatur (°C)	RH 5%	RH 10%	RH 15%
0	0.0002	0.0004	0.0006
5	0.0003	0.0005	0.0008
10	0.0004	0.0008	0.0011
15	0.0005	0.0010	0.0016
20	0.0007	0.0014	0.0022
25	0.0010	0.0019	0.0029
30	0.0013	0.0026	0.0039

Figure 8 shows the effect of RH differences on the temperature distribution inside the evaporative chimney. Based on the results, the differences in RH levels affect the temperature distribution. RH level of 5% with the highest temperature of 303.14 K (29.99°C) and the lowest temperature of 300.49 K (27.34°C), there was a decrease in temperature in the lowest plane of 2.65°C. RH level of 10% with the highest temperature of 303.14 K (29.99°C) and the lowest temperature of 300.59 K (27.44°C), there was a decrease in temperature in the lowest plane of 2.55°C. RH level of 15% with the highest temperature of 303.14 K (29.99°C) and the lowest temperature of 300.69 K (27.54°C), there was a decrease in temperature in the lowest plane of 2.45°C. Based on this analysis, it can be seen that the cooling effect decreases with increasing RH. This is because the higher RH indicates the water vapour content in the air is high. Hence the impact of water spray from the nozzle is less influential because there is a natural temperature drop of air in the chimney.

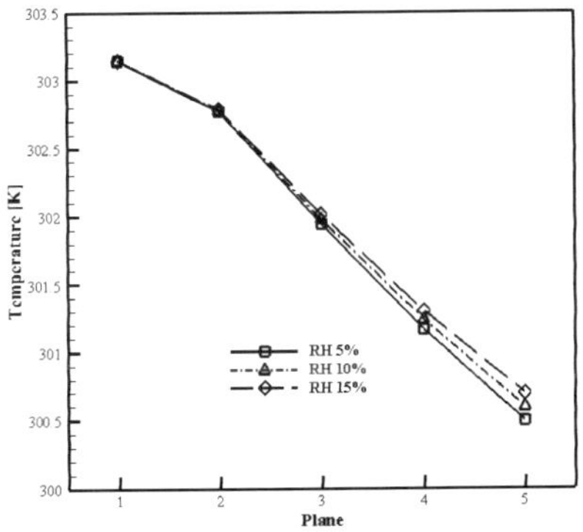

Figure 8. Effect of RH

4. Conclusion

The mesh refinement study on cooling simulations in the evaporative chimney has been completed. The temperature drop and airflow cooling distribution in the evaporative chimney are the focus of this research. The three types of mesh produced different simulation results. Based on the validation results, mesh B provides consistent data with the results of previous studies, both in temperature drop and airflow distribution in the chimney. Of all the mesh types used, it can be concluded that mesh B produces good agreement results to the previous data compared to mesh A and C. The study of mesh refinement on evaporative cooling in the chimney gave consistent results with the data of previous researchers. In the future, this research can be developed by varying the position and direction of the nozzle spray, i.e. horizontal, diagonal or setting a certain angle.

References

[1] Calautit J K, Aquino A I, Shahzad S, Nasir D, and Hughes B 2017 *Energy Procedia* **105** 2865-70
[2] Koranteng C, Essel C, and Nkrumah J 2015 *Int. Adv. Res. J. Sci. Eng. Technol.* **2** 10
[3] Aldawoud A 2017 *Adv. Build. ENERGY Res.* **11** 1 67–86
[4] Amos-Abanyie S, Akuffo F O, and Kutin-Sanwu V 2013 *Sci. World J.* **2013**
[5] Hughes B R, Calautit J K, and Ghani S A 2012 *Appl. Energy* **92** 606–27
[6] Badran Ali A 2003 *Energy Build.* **35** 1031–35
[7] Erell E, Pearlmutter D, and Etzion Y 2007 *Sol. Energy* **82** 420–29
[8] Vakiloroaya V, Samali B, Fakhar A, and Pishghadam K 2014 *Energy Convers. Manag.* **77**, 738–54
[9] Calautit J, O'Connor D, Shahzad S, Calautit K, and Hughes B 2019 *Energy Procedia* **158** 3125–30
[10] Liu Y, Yang X, Li J, and Zhao X 2018 *Energy*
[11] Bahadori M N 1985 *Sol. Energy* **15** 119–29
[12] Givoni B 2011 *Sol. Energy* **85** 1692–26
[13] Bowman N T, Eppel H, Lomas K J, Robinson D, and Cook M J 2000 *Indoor Built Environ.* **9** 284–90
[14] Baptista da Silva A C S, Bellini da Cunha Neto J A, and Lamberts R 2005 *Build. Simul.*
[15] Farnham C, Nakao M, Nishioka M, Nabeshima M, and Mizuno T 2011 *Procedia Environ. Sci.* **4** 228–38
[16] Ketan N A, Aya H, and Jun T 2019 A simplified numerical model for evaporative cooling by water spray over roof surfaces *Appl. Therm. Eng.* j.applthermaleng.2019.114514
[17] St-Georges M and Buchlin J M 1994 *Int J Multiph. Flow* **20** 6 979–92
[18] Gant S E 2006 *Health & Safety Laboratory*, Harpur Hill, Buxton Derbyshire.
[19] Pearlmutter D, Etzion Y, and Erell E 2006 *The 23rd Conference on Passive and Low Energy Architecture,* Geneva, Switzerland
[20] Pearlmutter D, Erell E, and Etzion Y 2008 *Sol. Energy* **82** 430–40
[21] Sun T, Huang X, Chen Y, and Zhang H 2020 *Energy Build.* **214**
[22] Delele M A, Schenk A, Tijskens E, Ramon H, Nicolaï B M, and Verboven P 2009 *J. Food Eng.* **91** 228–39
[23] Taha Al-Zubaydi A Y and Hong G 2019 *Appl. Therm. Eng.* **151** 283–93
[24] Stefano De Antonellis, Joppolo C M, Liberati P, Milani S, and Molinaroli L 2016 *Energy Build.*
[25] Tambur Y and Guetta S, "Optimizing the Design and Operation of the Sprayers in the Tower," Faculty of Aerospace Engineering, Technion - Israel Institute of Technology.
[26] Sarjito 2013 *Appl. Mech. Mater.* **315** 835–40
[27] Sarjito, Siswanto W A, Jamaldi A, and Yao Y 2018 *Int. J. Mech. Prod. Eng. Res. Dev. IJMPERD* **8** 6 81–90

Journal of Physics: Conference Series **2111** (2021) 012013 doi:10.1088/1742-6596/2111/1/012013

Fuzzy Time Series for Forecasting Railway Passengers in Indonesia

Muhammad Fatih Rizqon[1] and Handaru Jati[2]

[1,2] Department of Electronics and Informatics Engineering Education, Postgraduate Program, Yogyakarta State University, Indonesia

Email: [1]fatihrizqon@student.uny.ac.id

Abstract. Some fuzzy time series models have their own advantages and disadvantages. In addition, these models sometimes are complex and claimed to have better forecasting result than each other. The suitable model for forecasting depends on a wide variety of considerations. The models proposed by Chen (1996) applied simplified arithmetic operations and claimed more efficiency than before. The model proposed by Chen was introduced in 1996 and still exists in several previous studies. This research aims to forecast the number of railway passengers in Indonesia using the fuzzy time series. In addition, this research also evaluates the forecasting results based on mean absolute error (MAE) and mean absolute percentage error (MAPE). The results showed the forecasting results in this research has accuracy for 86.6%.

1. Introduction

Public transportation is one of the favors to support mobility across cities, provinces, and islands in Indonesia. The railway is one of the most preferred public transportation in Indonesia which is currently managed and operated by the Indonesian Railway Company (PT. Kereta Api Indonesia). According to the data in [1] shows the railway passengers in Indonesia has increased every year from 2006 to 2019. The highest passengers occurred in July 2019 which has reached 39035 passengers. The increase of passengers is correlated with the quality of services provided by the Company. Based on a survey in 2018, Indonesian Railway Company gained the highest Customer Satisfaction Index (CSI) score of 4.08 followed by aircraft with a score of 4.05, shuttle by 3.54, and bus with CSI score of 3.52 [2]. Indonesian Railway Company offers various class services such as sleeper class, priority class, executive class, business class, and economy class to support passenger's mobility.

However, in April 2020 the railway passengers have significantly decreased to 5898 due to COVID-19 Pandemic. The number of passengers in May 2020 decreased by 5484 and became the lowest in the last decade. Therefore, in terms of business it is important to analyze the movement of historical data to determine what to do in the future. In this case, it is possible to determine the departure schedule by considering the supply and demand to operate Railway at a certain time by using forecasting method. There are various forecasting methods that can be used such as CNN Backpropagation, Regression, Moving Average, Fuzzy Time Series, etc.

This research aims to forecast the dataset of railway passengers by using fuzzy time series proposed by Chen (1996) [3]. Fuzzy time series is one of the forecasting methods which firstly introduced by Song & Chissom in 1993 [4]. There is various development of fuzzy time series models that are used to forecast a set of historical data. However, some of these models have their own advantages and disadvantages. In addition, these models sometimes are complex and claimed to have better forecasting results than each other. The suitable models in forecasting depends on a wide variety of considerations [5]. Chen found in the proposed model by Song & Chissom used complicated procedures [3]. Therefore, Chen applied simplified arithmetic operations and claimed more efficiency than before. The model proposed by Chen was introduced in 1996 with many developments afterward. This model still exists in several previous studies in [6]–[8]. Based on the comparison forecasting model results in [6] showed that the model proposed by Chen has better results than the model proposed in [9]. In addition, this research also evaluates the forecasting results based on the mean absolute error (MAE) and mean absolute percentage error (MAPE) which commonly used to evaluate forecasting results based on related studies [10], [11].

2. Literature Review

2.1. Fuzzy Time Series

Song & Chissom was introduced fuzzy time series by using the concept of fuzzy sets in [12] as the basis for calculations where the value of historical data are represented by fuzzy sets [4]. Fuzzy sets theory proposed by [12] has good achievement both in theory and practice that allow to make decisions, plans, smart systems, artificial intelligence, etc. Song & Chissom explained their basic forecasting procedure are used the historical data patterns and form it to the next data (forecasting value). Fuzzy time series has 4 characteristics: (1) fuzzy time series is a dynamic process; (2) fuzzy sets as the object; (3) the universe of discourse (*U*) for the fuzzy sets are subsets of *R*; and (4) fuzzy time series doesn't require conventional time series method [4]. Let the universe of discourse (*U*), $U = \{u_1, u_2, ..., u_n,\}$, then a fuzzy set *A* of *U* are determined as follow:

$$A = f_A\,(u_1)/u_1 + f_A\,(u_2)/u_2 + \ldots + f_A\,(u_n)/u_n \qquad (1)$$

where f_A is the membership function of *A*, then $f_A : U \to [0,1]$. Membership function is a curve that defines the mapping of input/output points into their membership between 0 and 1. Song & Chissom applied the model of $A_i = A_{i-1}{}^{\circ} R$ to forecast the enrollment of students in University of Alabama, where A_{i-1} is the enrollment of the year $i-1$ terms of a fuzzy set, and the systematic procedures proposed by Song & Chissom show as follows:

Step 1: Defining the *U* (universe of discourse) for the fuzzy sets are subsets of *R*
Step 2: Collecting the historical linguistic values
Step 3: Defining the fuzzy sets (fuzzification)
Step 4: Set up the fuzzy logic relationship (FLR)
Step 5: Summarizing the FLR
Step 6: Applying the value of the data to the model and determine the output for the forecasted value
Step 7: Defuzzification

Chen found in [4] required a large number of calculations to obtain the fuzzy relationship. Chen proposed a new model by using simpler arithmetic operations rather than the complicated max-min composition operations presented in [4]. There are two definitions of fuzzy time series that proposed in [3]. The first definition, let $Y(t)$ (t = …, 0, 1, 2, …) is a subset of *R* (real numbers) be the universe of discourse (*U*) on fuzzy sets $f_i(t)$ $(i = 1, 2, \ldots)$. The input sometimes referred from the universe of discourse (*U*). Then $F(t)$ is called a fuzzy time series on $Y(t)$ $(t = \ldots, 0, 1, 2, \ldots)$. $F(t)$ represented as linguistic variable, and $(i = 1, 2, \ldots)$ as possible linguistic value of $F(t)$, where $f_i(t)$ $(i = 1, 2, \ldots)$ regarded as fuzzy sets. The values of $F(t)$ sometimes can be different at certain times due to the fact that the universe of discourse can be different at certain times. When $F(t)$ only caused by $F(t-1)$ then this relationship represented as $F(t-1) \to F(t)$, where $F(t-1)$ can be viewed as past time. The second definition, let $F(t)$ is a fuzzy time series and if $F(t-1)$ and $F(t)$ has unlimited elements, then $F(t)$ is called a time-variant fuzzy time series. Otherwise, if $F(t-1)$ and $F(t)$ has limited elements it called a time-invariant fuzzy time series.

2.2. Measuring the Forecasting Results

2.2.1. Mean Absolute Error (MAE). MAE is one of the common methods to evaluate model performance because it's summing the total of absolute error and dividing it by the total of records (n) [13]. MAE would be good if the result is closer to 0 and it depends on the size of data. MAE can be obtained by using the following equation:

$$MAE = \frac{\sum_{t=1}^{n}|A_t - F_t|}{n} \quad (2)$$

2.2.2. Mean Absolute Percentage Error (MAPE). MAPE is one of options for measuring the error of forecasting results. MAPE is easy to understand because it represents the error in percentage [14]. Lewis's (1982) interpreted MAPE if the value under 10% regarded as "highly accurate"; 11% - 20% regarded as "a good forecast"; 21% - 50% is "acceptable"; and >51% is an "inaccurate forecast" [15]. MAPE can be obtained through this equation:

$$MAPE = \frac{100}{n}\sum_{t=1}^{n}\left|\frac{A_t - F_t}{A_t}\right| \quad (3)$$

3. Methods

This research applied the proposed models in [3] to avoid the max-min composition operations, because it would be complicated when the fuzzy relation R is very big as used in [4].

3.1. Determining the Universe of Discourse

The operations started by determining the maximum of data as D_{max} and the minimum of the data as D_{min}. The universe of discourse (U) can be obtained by using the following equation:

$$U = [D_{min} - D_1 \,;\, D_{max} + D_2] \quad (4)$$

where the D_1 and D_2 are independent positive numbers. Furthermore, the operation of the forecasting process is as follows:

3.2. Forming Intervals

The forming of intervals is to divide the set of the U that has been obtained previously into several intervals with the same distance. In determining the intervals of U, the number and length of intervals should be determined first by using the following equation:

$$Intervals = 1 + 3.3 * log10(n) \quad (5)$$

where n is the amount of the dataset. The length of intervals can be obtained by using this equation:

$$Length\ of\ Intervals = \frac{D_{max} - D_{min}}{Intervals} \quad (6)$$

3.3. Fuzzification and Forming Fuzzy Logic Relationship (FLR)

Fuzzification is a process to convert the input from *crisp* to *fuzzy value* (linguistic variable) which presented as fuzzy sets with a respective membership function. The fuzzification is carried out based on the determined intervals, from the initial data then grouped according to the number of formed intervals. FLR indicates the association between the fuzzy time series and the composing operator. FLR is a fuzzy logic that has a relationship with a series of data members that have been assigned from the historical data to the forecasted data.

3.4. *Fuzzy Logic Relationship Group (FLRG)*

FLRG is formed by dividing all the derived FLRs into several groups based on the current states of the enrollments of FLR.

3.5. *Defuzzification and Forecasting*

The Defuzzification process is used to determine the forecasting value based on the results of FLRG by using 3 rules as follows:

Rules 1: If there is only one logical relation of fuzzy in a series of fuzzy logic relationship, for example $A_1 \rightarrow A_j$*, then the forecasting value defined as* $F(t) = A_j$.

Rules 2: If there is more than one relationship in a series of fuzzy logic relationship, for example $A_i \rightarrow A_{j1}, A_{j2}, \ldots A_{jn}$*, then the forecasting value should be the average of the midpoint from* $A_{j1}, A_{j2}, \ldots A_{jn}$ or defined as $F(t) = \overline{A_{j1}, A_{j2}, \ldots A_{jn}}$.

Rules 3: If there is no relation on A_j *there is nothing like* ($A_j \rightarrow \#$)*, then the forecasted value defined as* $F(t) = A_j$.

4. Results

4.1. *Determining the Universe of Discourse*

The training data in this research are monthly train passenger from January 2006 to July 2021 that available in [1]. The minimum and maximum data (5484 and 39035) are used to determine the *U* by using equation (4). The values of $D_1 = 4$, and $D_2 = 5$ are obtained arbitrarily to get the appropriate results of the D_{min} and D_{max}. Therefore, the universe of discourse in this research defined as $U = [5480 ; 39040]$.

Table 1. Railway Passengers in Indonesia January 2006 – July 2021

	January	February	March	April	May	June	July	August	September	October	November	December
2006	11828	11931	13314	12909	13575	13203	14433	13255	13436	14290	13631	13614
2007	13960	10969	13409	14415	15232	15104	16454	15419	15033	15866	14391	15084
2008	15027	14378	16071	15711	16363	17010	17887	17108	15879	17337	15973	15332
2009	14494	13869	17132	16775	17824	18143	18385	17527	17281	17281	16778	17581
2010	17424	15207	16992	16832	16988	17259	17680	16477	17301	16908	16469	17733
2011	16891	14890	16978	16441	17522	17265	18132	14846	16921	16461	16179	16811
2012	16283	15490	17090	16746	17771	18062	18309	17056	16368	17127	15773	16104
2013	14900	14594	15826	16000	16113	17301	20245	19423	19738	20534	19919	21417
2014	21092	19998	22836	21908	22988	23840	22500	23199	23593	24923	24356	26275
2015	24676	22790	27267	26565	27910	27562	27612	27796	27549	28718	27669	29831
2016	28358	26510	28617	28435	30703	29159	28831	29588	29516	30263	29690	32150
2017	30949	27342	32170	31502	33745	30723	34310	33791	32498	35070	34361	36807
2018	34717	31278	35875	35754	35482	33030	36800	35190	34504	36236	35298	37965
2019	35122	31899	35751	35809	35102	35090	39035	35189	35221	36448	35877	37463
2020	34130	32283	23425	5898	5484	9290	12238	12774	11429	11937	13722	13515
2021	11901	11479	14284	14864	14857	14556	5747	-	-	-	-	-

4.2. *Forming the Interval*

Forming the interval data started by determining the number of intervals by using equation (5) and the length of interval by using equation (6). The determined number of intervals are 8 with the length of the interval is 4195. The result of the formed interval are shown in Table 2.

Table 2. Formed Interval Data

Interval	Bottom	Median	Top
A0	5484	7582	9679
A1	9679	11777	13874
A2	13874	15972	18069
A3	18069	20167	22264
A4	22264	24362	26459
A5	26459	28557	30654
A6	30654	32752	34849
A7	34849	36947	39044

4.3. *Fuzzification and Forming the Fuzzy Logic Relationship (FLR)*

The fuzzification process was carried out referred on the intervals that have been determined from the initial data, then grouped according to Table 2. The formed Fuzzification and their relationships are shown in Table 3.

Table 3. Fuzzification Results

No.	Data	Fuzzification	FLR
1	11828	A1	-
2	11931	A1	A1 → A1
3	13314	A1	A1 → A1
4	12909	A1	A1 → A1
5	13575	A1	A1 → A1
6	13203	A1	A1 → A1
...	...	...	...
180	13515	A1	A1 → A1
181	11901	A1	A1 → A1
182	11479	A1	A1 → A1
183	14284	A2	A1 → A2
184	14864	A2	A2 → A2
185	14857	A2	A2 → A2
186	14556	A2	A2 → A2
187	5747	A0	A2 → A0

Table 3 shows the first data is 11828 which is fuzzified as A1 and the last data is 5747 which is fuzzified as A0 which is in accordance with the formed intervals in **Table 2** . The FLR are formed based on previous fuzzification, for example in the first data fuzzified as A1, and then for the second data was also fuzzified as A1. Therefore, the first formed FLR is A1 → A1 as well as next.

4.4. *Fuzzy Logic Relationship Group (FLRG)*

Table 2 contains 8 intervals from A0 to A7 with the distance of interval by 4195. In addition, Table 3 presents the fuzzification results and established relationship between each data. All established relationship then grouped as in Table 4.

Table 4. The Fuzzy Logic Relationship Group (FLRG)

Group	Fuzzification	FLRG
1	A0 →	A0, A1
2	A1 →	A1, A2
3	A2 →	A0, A1, A2, A3
4	A3 →	A2, A3, A4
5	A4 →	A0, A3, A4, A5
6	A5 →	A5, A6
7	A6 →	A4, A5, A6, A7
8	A7 →	A6, A7

4.5. *Defuzzification and Forecasting*

The next step is the defuzzification and determine the forecasting value based on the rules that presented in section 3.5. The defuzzification and forecasting results are shown in Table 5. The forecasting value are obtained from the average of median (midpoint) from the grouped FLR. For example, the A0 → A0, A1 where the midpoint of A0 is 7582, and the midpoint of A1 is 11777 (See Table 2). Therefore, the forecasting results obtained from the group of A0 is 9679.

Table 5. Defuzzification and Grouping Forecasted Value to the FLRG

Group	Fuzzification	FLRG	Median	Forecasting Value
1	A0 →	A0, A1	7582, 11777	9679
2	A1 →	A1, A2	11777, 15972	13874
3	A2 →	A0, A1, A2, A3	7582, 11777, 15972, 20167	13874
4	A3 →	A2, A3, A4	15972, 20167, 24362	20167
5	A4 →	A0, A3, A4, A5	7582, 20167, 24362, 28577	20167
6	A5 →	A5, A6	28577, 32752	30654
7	A6 →	A4, A5, A6, A7	24362, 28577, 32752, 36947	30654
8	A7 →	A6, A7	32752, 36947	34849

4.6. Forecasting Results

In summary, the forecasting results collectively are shown in Table 6.

Table 6. Forecasting Results

No.	Date	Data	Fuzzification	FLRG	Forecasting Results
1	January 2006	11828	A1	A1 → A1, A2	-
2	February 2006	11931	A1	A1 → A1, A2	13874
3	March 2006	13314	A1	A1 → A1, A2	13874
4	April 2006	12909	A1	A1 → A1, A2	13874
5	May 2006	13575	A1	A1 → A1, A2	13874
6	June 2006	13203	A1	A1 → A1, A2	13874
...	...	...	...	...	...
180	December 2020	13515	A1	A1 → A1, A2	13874
181	January 2020	11901	A1	A1 → A1, A2	13874
182	February 2020	11479	A1	A1 → A1, A2	13874
183	March 2021	14284	A2	A2 → A0, A1, A2, A3	13874
184	April 2021	14864	A2	A2 → A0, A1, A2, A3	13874
185	May 2021	14857	A2	A2 → A0, A1, A2, A3	13874
186	June 2021	14556	A2	A2 → A0, A1, A2, A3	13874
187	July 2021	5747	A0	A0 → A0, A1	13874
188*	August 2021	-	-	-	9679

**Forecasted result*

While this research was conducted, the available data in Table 1 started from January 2006 to July 2021. Based on the calculations, the railway passengers in August of 2021 approximately will be around 9679 passengers because of the previous relationship was A0 which means A0 → A0, A1 has forecasting value for 9679. Overall forecasting results are presented in Figure 1. The solid line represents the actual data, while the dotted line represents the forecasting results.

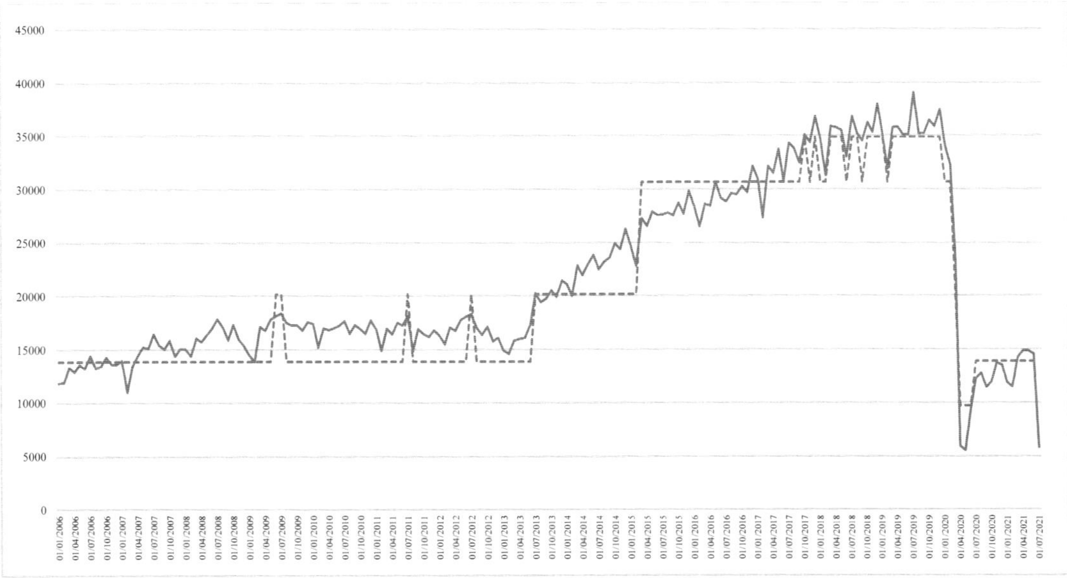

Figure 1. Overall Forecasting Results

4.7. Forecasting Evaluation

Forecasting evaluation is an important stage to find the performance of forecasting method. It would be 'good' if the forecasting results are similar or close with the actual data, and/or has low of errors value. The MAE and MAPE were applied to evaluate the forecasting results based on the error value obtained in this research.

Table 7. Forecasting Evaluation

No.	Date	Data	Forecasting Results	Absolute Error	Absolute Percentage Error
1	January 2006	11828	-	-	-
2	February 2006	11931	13874	1943	0.163
3	March 2006	13314	13874	560	0.042
4	April 2006	12909	13874	965	0.075
5	May 2006	13575	13874	299	0.022
6	June 2006	13203	13874	671	0.051
…	…	…	…	…	…
180	December 2020	13515	13874	152	0.011
181	January 2020	11901	13874	359	0.027
182	February 2020	11479	13874	1973	0.166
183	March 2021	14284	13874	2395	0.209
184	April 2021	14864	13874	410	0.029
185	May 2021	14857	13874	990	0.067
186	June 2021	14556	13874	983	0.066
187	July 2021	5747	13874	682	0.047
		Total		**433957**	**25**

Table 7 shows that the total absolute error in this research is 433957, and the total absolute percentage error is 25. Therefore, the MAE and MAPE obtained in this research are 2333 and 0.134 (13.4%) respectively.

5. Conclusion
The results showed that the forecasting model in this research has mean absolute error (MAE) for 2333 and mean absolute percentage error (MAPE) for 13.4% which means has a good forecast based on Lewis's (1982) interpretation. Therefore, the overall forecasting accuracy is 86.6%. In addition, based on forecasting calculation the railway passengers in August of 2021 will be around 9679 passengers.

6. References
[1] Badan Pusat Statistik, "Badan Pusat Statistik," *Badan Pusat Statistik*, 2021. https://www.bps.go.id/indicator/17/72/1/jumlah-penumpang-kereta-api.html (accessed Jul. 20, 2021).
[2] PT Kereta Api Indonesia, "Report Survey Kepuasan Pelanggan KAI 2018," *PPID PT KAI*, 2019. https://ppid.kai.id/layanan-berkala-view?_tokenizer=Njc= (accessed Jul. 01, 2021).
[3] S.-M. Chen, "Forecasting enrollments based on fuzzy time series Shyi-Ming," *Fuzzy Sets Syst.*, vol. 81, pp. 312–319, 1996, doi: 10.1007/11925903_25.
[4] Q. Song and B. S. Chissom, "Fuzzy time series and its models," *Fuzzy Sets Syst.*, vol. 54, no. 3, pp. 269–277, Mar. 1993, doi: 10.1016/0165-0114(93)90372-O.
[5] C. Chatfield, "What is the 'best' method of forecasting?," *J. Appl. Stat.*, vol. 15, no. 1, pp. 19–38, 1988, doi: 10.1080/02664768800000003.
[6] Arnita, N. Afnisah, and F. Marpaung, "A Comparison of the Fuzzy Time Series Methods of Chen, Cheng and Markov Chain in Predicting Rainfall in Medan," *J. Phys. Conf. Ser.*, vol. 1462, no. 1, 2020, doi: 10.1088/1742-6596/1462/1/012044.
[7] R. W. Ningrum, B. Surarso, Farikhin, and Y. M. Safarudin, "Forecasting Jakarta composite index (IHSG) based on chen fuzzy time series and firefly clustering algorithm," *J. Phys. Conf. Ser.*, vol. 983, no. 1, 2018, doi: 10.1088/1742-6596/983/1/012055.
[8] Zaenurrohman, S. Hariyanto, and T. Udjiani, "Fuzzy time series Markov Chain and Fuzzy time series Chen & Hsu for forecasting," *J. Phys. Conf. Ser.*, vol. 1943, no. 1, 2021, doi: 10.1088/1742-6596/1943/1/012128.
[9] C. H. Cheng, T. L. Chen, H. J. Teoh, and C. H. Chiang, "Fuzzy time-series based on adaptive expectation model for TAIEX forecasting," *Expert Syst. Appl.*, vol. 34, no. 2, pp. 1126–1132, 2008, doi: 10.1016/j.eswa.2006.12.021.
[10] U. Khair, H. Fahmi, S. Al Hakim, and R. Rahim, "Forecasting Error Calculation with Mean Absolute Deviation and Mean Absolute Percentage Error," *J. Phys. Conf. Ser.*, vol. 930, no. 1, 2017, doi: 10.1088/1742-6596/930/1/012002.
[11] W. Wang and Y. Lu, "Analysis of the Mean Absolute Error (MAE) and the Root Mean Square Error (RMSE) in Assessing Rounding Model," *IOP Conf. Ser. Mater. Sci. Eng.*, vol. 324, no. 1, 2018, doi: 10.1088/1757-899X/324/1/012049.
[12] L. A. Zadeh, "Outline of a New Approach to the Analysis of Complex Systems and Decision Processes," *IEEE Trans. Syst. Man. Cybern.*, vol. SMC-3, no. 1, pp. 28–44, 1973, doi: 10.1109/TSMC.1973.5408575.
[13] C. J. Willmott and K. Matsuura, "Advantages of the mean absolute error (MAE) over the root mean square error (RMSE) in assessing average model performance," *Clim. Res.*, vol. 30, no. 1, pp. 79–82, 2005, doi: 10.3354/cr030079.
[14] P. M. Swamidass, Ed., "MAPE (mean absolute percentage error)MEAN ABSOLUTE PERCENTAGE ERROR (MAPE)," in *Encyclopedia of Production and Manufacturing Management*, Boston, MA: Springer US, 2000, p. 462.
[15] C. D. Lewis, *Industrial and business forecasting methods: A practical guide to exponential smoothing and curve fitting*. Butterworth-Heinemann, 1982.

Internet of Things-based Hydroponic: Literature Review

Alan Hadinata[1], Mashoedah[2]

[1,2]Graduate Program of Electronics and Informatics Engineering Education, State University of Yogyakarta

E-mail: alanhadinata.2020@student.uny.ac.id[1]

Abstract. Hydroponics is a method of growing crops without using soil, with the benefits of controlling the environment and nutrients, conserving water, and reducing labor. The applied technology is used to improve results that have consistency. This research was conducted with the aim of knowing the IoT devices and platforms used in the development of IoT-based hydroponics. Data obtained from the website garuda.ristekbrin.go.id and www.sciencedirect.com from 2016 to 2021. This study is a literature review using the PRISMA method. This method is used for literature review using a systematic and structured basic framework. The results of this study indicate that the variables measured are Temperature, Potential Hydrogen (pH), Total Dissolve Solid, Water Temperature, Humidity, UV, Carbon Dioxide, Soil Moisture and Electrical Conductivity. The device used is the ESP8266, Arduino, and Raspberry PI with the MySQL, Thingspeak, Firebase, Domoticz, and Wyliodrin IoT Platforms.

1. Introduction

Farming is a human habit that has existed since time of immemorial. This farming activity has also changed along with the development of technology. Changes in farming activities due to technological developments exist in various aspects. Starting from the purpose of the farming activity itself, changes in plant types to changes in planting methods. One of them is the method of farming using hydroponics.

Hydroponics is a method of growing crops without using soil as a growing medium, but using nutritious water or materials containing nutrients [1]. Hydroponics itself comes from the Greek "Hydroponos" where hydro means water and ponos means power [2]. With the hydroponic planting method, the use of environmental control, nutrition, and water conservation can be directly carried out and periodically observed by farmers.

This hydroponic method can be done vertically in a narrow space such as a yard or terrace so it does not require a large space. The requirements for hydroponic growing media are: (1) water absorbant, (2) air circulation, and (3) cheap [2].

The advantages of cultivating plants using this hydroponic method include: (1) producing products that are superior in quality and quantity to conventional cultivation, (2) lower plant pests and diseases, and (3) easily control the nutrients provided [3].

Although there are many advantages and conveniences created by the hydroponic method, in the conditions of the covid-19 pandemic, there are still many farmers that use conventional methods of farming. This is because the hydroponic method is considered the same as the conventional farming method which must be controlled directly and manually. Therefore, we need a technology-based hydroponic system that can advance agriculture by increasing quantity and optimizing yields and having better quality and making it easier for farmers to control and care for their plants.

Based on developing technology, all aspects of life adapt by applying the latest technology that can provide convenience, including agriculture. Internet of things technology can be applied to this hydroponic system to provide convenience to farmers in carrying out their duties in the midst of the COVID-19 pandemic. The conveniences obtained from the application of the IoT in hydroponic farming activities include monitoring and controlling activities as well as plant care that can be done anywhere and anytime.

2. Methods

This research is a systematic review research using the Preferred Reporting Items for Systematic Reviews and Meta-analyses (PRISMA) method. PRISMA is a literature review method using a systematic and structured basic framework. The procedure of this research method consists of several stages, namely: 1) Preparing the background and research objectives, 2) Formulating research questions, 3) Searching for literature, 4) Selecting literature according to the criteria, 5) Practical screen, 6) Data extraction, and 7) Data synthesis [4].

2.1. Object of the Research

In this study, the object are research journals on IoT-based hydroponic methods based on the following reasons:

1. There are several additions to the IoT system in the hydroponic farming method that can assist farmers in monitoring and remotely controlling the plants being cultivated.
2. An IoT-based hydroponic system that is implemented using a specific microcontroller.
3. IoT-based hydroponics has a variety of protocols that can be customized as desired.

2.2. Method

2.2.1. Research question. Based on the background and object of the research that has been submitted, here is the questions of the research

RQ1. What variables are used in IoT-based Hydroponics research?
RQ2. What systems or devices are used in IoT-based hydroponics?
RQ3. what are the disadvantages of a platform implemented on an IoT-based hydroponic method?

2.2.2. Search Process. The process of searching for journal articles using the website garuda.ristekbrin.go.id and www.sciencedirect.com. the data search process in this case uses keywords related to hydroponics and the internet of things.

2.2.3. Inclusion and Exclusion Criteria. After carrying out the journal search step, the next procedure is to make a selection to meet the inclusion and exclusion criteria. This stage aims to determine appropriate data used in research based on the following criteria:

1. Data with a time span of 2016-2021.
2. The data comes from the website garuda.ristekbrin.go.id and www.sciencedirect.com.
3. Data related to hydroponics and IoT.

2.2.4. Quality Assessment. Data that has gone through the inclusion and exclusion criteria selection process will then be evaluated according to the quality assessment criteria as follows:

QA1: Does the journal list the Variables used for measurement?
QA2: Does the journal list the tools used to develop the system?
QA3: Does the journal list the Platform used for the IoT system?

2.2.5. Data Collection. Journal data that obtained from the garuda.ristekbrin.go.id and www.sciencedirect.com website and has met the criteria and quality assessment, will be extracted to answer research questions referring to RQ1, RQ2, and RQ3.

2.2.6. Data Analysis. Based on the results obtained at the data extraction stage, the next procedure is to perform data analysis which includes an explanation of the following points:

1. Variables measured in the development of an IoT-based Hydroponic system.
2. Devices used in the development of IoT-based Hydroponic systems.
3. The IoT platform used in the development of an IoT-based Hydroponic system.

3. Result and Discussion

3.1. Search and Selection Process

Based on the search process using the keywords hydroponics and internet of things on the garuda.ristekbrin.go.id website page, 16 (sixteen) and only 7 (seven) journals met the following inclusion and research induction criteria.

Table 1. Data of *Internet of Things-bases Hydroponic* Journal Research Result on the website garuda.ristekbrin.go.id that Match the Criteria

No.	Title	Author	Journal Name	Year
1	Advanced Encryption Standard (Aes) 128 Bit Untuk Keamanan Data Internet Of Things (Iot) Tanaman Hidroponik	Roiya Ravida, and Heru Agus Santoso	Jurnal Resti (Rekayasa Sistem Dan Teknologi Informasi)	2020
2	Optimalisasi Pertumbuhan Pada Sayuran Hidroponik Nutrient Film Technique Dengan Metode Fuzzy Logic Berbasis Internet Of Things	Akhmad Wahyu Dani	Jurnal Teknologi Elektro	2020
3	Rancang Bangun Aplikasi Mobile Berbasis Internet Of Things Untuk Pemantauan Nutrisi Tanaman Selada Hidroponik	Muhammad Rizki Juanda	Jurnal Komputer, Informasi Teknologi, Dan Elektro	2020
4	Penggunaan Internet Of Things (Iot) Untuk Pemantauan Dan Pengendalian Sistem Hidroponik	Yodi Setiawan, Harlianto Tanudjaja, and Sandra Octaviani	Tesla: Jurnal Teknik Elektro	2018
5	Pengembangan Greenhouse Fodder Jagung Hidroponik Menggunakan Kontrol Logika Fuzzy (Klf) Sebagai Pengendali Suhu Dan Kelembaban Terintegrasi Dengan Sistem Internet Of Things	Mohammad Mufti Fajar, Muhammad Aziz Muslim, and Purwanto	Jurnal Mahasiswa Teub	2019
6	Sistem Nutrisi Tanaman Hidroponik Berbasis Internet Of Things Menggunakan Nodemcu Esp8266	Andi Heryanto, Jian Budiarto, and Sirojul Hadi	Jurnal Bumigora Information Technology (Bite)	2020
7	Perancangan Sistem Monitoring Hidroponik Nutrient Film Technique (Nft) Berbasis Internet Of Things (Iot) Menggunakan Web Server Thingspeak	Primadoly Nababan, Trias Andromeda, and Yosua Alvin Adi Soetrisno	Transient: Jurnal Ilmiah Teknik Elektro	2020

From the website www.sciencedirect.com obtained 99 (ninety nine) titles of scientific writing using the keyword hydroponic internet of things. Then from the 99 (ninety nine) titles, 50 (fifty) titles are obtained which are research articles. After the review, 20 (twenty) journals of research articles were found that were accessible and the journals met the inclusion criteria and research induction as follows.

Table 2. Data of *Internet of Things-bases Hydroponic* Journal Research Result on the website www.sciencedirect.com that Match the Criteria

No.	Title	Author	Journal Name	Year
1	Hydroponic and Aquaponic Farming: Comparative Study Based on Internet of things IoT technologies	Ibtissame Ezzahoui, Rachid Ait Abdelouahidb , Khaoula Taji , and Abdelaziz Marzak	Procedia Computer Science	2021
2	Monitoring rocket (Eruca sativa) growth parameters using the Internet of Things under supplemental LEDs lighting	Jolan Jamal, Sadoon Azizi, Alireza Abdollahpouri, Nasser Ghaderi, Behrooz	Sensing and Bio-Sensing Research	2021

		Sarabi, Arturo Silva-Ordaz, and Victor M. Castano-Meneses.		
3	Effect of different root lengths for retaining freshness of hydroponic lettuce	Rui Suo, Wenqi Wang, Yidong Ma, Longsheng Fu, and Yongjie Cui	Journal of Agriculture and Food Research	2021
4	The Internet of Things at the service of tomorrow's agriculture	Younes Abbassi And Habib Benlahmer.	Procedia Computer Science	2021
5	Physical and mechanical properties of hydroponic lettuce for automatic harvesting	Wenqi Wang, Yidong Ma, Longsheng Fu, Yongjie Cui, and Yaqoob Majeed	Information Processing in Agriculture	2020
6	Fully Automated Hydroponic System for Indoor Plant Growth	Vaibhav Palande, Adam Zaheer, and Kiran George	Procedia Computer Science	2018
7	Architecture design of monitoring and controlling of IoT-based aquaponics system powered by solar energy.	Taji Khaoula, Rachida Ait Abdelouahid, Ibtissame Ezzahoui, and Abdelaziz Marzak	Procedia Computer Science	2021
8	Investigation on environment monitoring system for a combination of hydroponics and aquaculture in greenhouse	Shanhong Zhang, Yu Guo, Shuai Li, Zhixin Ke, Huajian Zhao, Jinqi Yang, Yang Wang, Daoliang Li, Liang Wang , Wenhua Yang, and Zhili Zhang.	Information Processing in Agriculture	2021
9	CARMEN: A framework for the verification and diagnosis of the specification of security requirements in cyber-physical systems	Ángel Jesús Varela-Vaca, David G. Rosado, Luis E. Sánchez, María Teresa Gómez-López, Rafael M. Gasca, and Eduardo Fernández-Medina	Computers in Industry	2020
10	Monitoring system of environmental variables for a strawberry crop using IoT tools	Amaya Diaz Juan Carlos, Luzmila Rojas Estrada, Cardenas-Ruiz, Cesar Augusto, Ariza-Colpas Paola Patricia, Piñeres-Melo Marlon Alberto, Ramayo González Ramón Enrique, Morales-Ortega, Roberto César, Ovallos- Gazabon David Alfredo, and Collazos-Morales Carlos Andrésa	Procedia Computer Science	2020
11	A Design of IoT-based Monitoring System for Intelligence Indoor Micro-Climate Horticulture Farming in Indonesia	Emil Robert Kaburuana, Riyanto Jayadi, and Harisno	Procedia Computer Science	2019
12	A 3.5-tier container-based edge computing architecture	Ching-Han Chen, and Chao-Tsu Liu	Computers and Electrical Engineering	2021
13	Enhancing aquaponics management with IoT-based Predictive Analytics for efficient information utilization	Divas Karimanzira, and Thomas Rauschenbach	Information Processing in Agriculture	2019
14	How Green Can the Lettuce Be? A Case Study on Greening Initiatives in Supply Chains and Sustainable ERP Systems	Slaman, M & Haddara, M.	Procedia Computer Science	2019
15	Urban surface uses for climate resilient and sustainable cities: A catalogue of solutions	Silvia Croce, and Daniele Vettorato.	Sustainable Cities and Society	2021
16	The Digitisation of Agriculture: a Survey of Research Activities on Smart Farming	Manlio Bacco, Paolo Barsocchi, Erina Ferro, Alberto Gotta, and Massimiliano Ruggeri	Array	2019
17	Proposed Collective ICOM-based post-scarcity/post-capital networked communities	S. Mason Dambrot	Procedia Computer Science	2021
	Food, meaning-making and ontological uncertainty: Exploring 'urban foraging' and productive landscapes in London	Marcus Nyman	Geoforum	2019
18	'Aha' moments in the water-energy-food nexus: A new morphological scenario method to accelerate sustainable transformation	Claire Hoolohan, Carly McLachlan, and Alice Larkin	Technological Forecasting & Social Change	2019
19	Adaptation and development pathways for different types of farmers	L.C. Stringer, E.D.G. Fraser, D. Harris, C. Lyon, L. Pereira, C.F.M. Ward, and E. Simelton	Environmental Science and Policy	2020
20	Enhanced approaches to automated monitoring environmental quality in non-isolated thermodynamic system	Ondrej Kainz, František Jakab, Miroslav Michalko, Miroslav Hudák, and Rastislav Petija	IFAC-PapersOnLine	2019

3.2. Quality Assesment Result

The journals that have been obtained were selected according to the quality assessment criteria referring to QA1, QA2, and QA3, so that the data in Table 3 and 4. is obtained below.

Table 3. Data of the *Internet of Things-based Hydroponic* Journal quality assessment from website garuda.ristekbrin.go.id.

Journal Number	QA1	QA2	QA3	Result
1	√	√	X	X
2	√	√	√	√
3	√	X	√	X
4	√	√	√	√
5	X	X	X	X
6	√	√	√	√
7	X	X	X	X

Based on the data table 3 from the quality assessment results, it is known that there are 3 journals that meet the quality assessment criteria, namely journals number 2, 4, and 6. Meanwhile, two journals did not meet the quality and there were two journals that could not be accessed completely so that it was not possible to know their eligibility to be used in this study.

Table 4. Data of the *Internet of Things-based Hydroponic* Journal quality assessment from website www.sciencedirect.com

Journal Number	QA1	QA2	QA3	Result
1	X	X	X	X
2	√	√	√	√
3	X	X	X	X
4	X	X	X	X
5	X	X	X	X
6	√	√	√	√
7	√	√	√	√
8	√	X	√	X
9	X	X	√	X
10	√	√	√	√
11	√	√	√	√
12	X	√	√	X
13	√	X	√	X
14	X	X	X	X
15	X	X	X	X
16	X	X	X	X
17	X	X	X	X
18	X	X	X	X
19	X	X	X	X
20	√	√	√	√

Meanwhile, based on table 4 of the quality assessment results, it is known that there are 6 journals that meet the quality assessment criteria, namely journals number 2, 6, 7, 10, 11, and 20. Meanwhile, 14 journals do not meet the quality to be used in this study.

3.3. Data Analysis

After conducting a quality assessment, 9 journals that met the quality were then extracted to answer research questions referring to RQ1, RQ2, and RQ3.

3.4. Measured Variables
The results of the extraction from the 9 journals containing the variable points measured in the use of the IoT in the hydroponic planting method can be presented in Table 5.

Table 5. Data of the Variable Points Measured in the Journal of *Internet of Things-based* Hydroponic.

No.	Variable	Total
1.	Temperature	8
2.	Potential Hydrogen	7
3.	Total Dissolve Solid	2
4.	Water Temperature	2
5.	Humidity	5
6.	UV	3
7.	Carbon Dioxide	3
8.	Soil Moisture	3
9.	Electrical Conductivity	4

Based on the data in Table 5, there are 8 journals that analyze the temperature variable, 7 journals that analyze the potential Hydrogen (pH) value variable, 2 journals that analyze the total dissolve solid variable, 2 journals that analyze the water temperature variable, 5 journals that analyze the humidity variable, 3 journals that analyzed the UV variable, 3 journals that analyzed the carbon dioxide variable, 3 journals that analyzed the soil moisture variable, and 4 journals that analyzed the electrical conductivity variable.

3.5. Device Used
In the journal analysis process that refers to the research question RQ2, it was found that there are three devices used, namely ESP8266, Arduino, and Raspberry PI.

3.5.1. ESP8266. ESP 8266 is an additional module for microcontroller devices such as Arduino that can be connected to wifi. ESP8266 has three wifi modes including station, access point, and Both [5]. Besides being able to connect to wifi, this device also has a USB to Serial communication chip so that programming data can use a micro usb data cable.

3.5.2. Arduino. Arduino is an open source electronic kit that has the main component, namely the AVR type microcontroller chip. Using Arduino, users can easily create programs that can be erased and reprogrammed. Users can create a system using this Arduino that can use sensors and actuators as interactions with the environment [7].

3.5.3. Raspberry PI. Raspberry PI is a single board computer that is equipped with functions like a computer in general. This Raspberry PI uses a system on an ARM chip mounted on a card-sized pcb board. The operating systems that can be installed on this device are RaspbianOS and Arch Linux [8].

3.6. IoT platform used
The data obtained through the IoT device is then forwarded through the Wifi module to an IoT platform. An IoT platform is a system consisting of hardware and software used to manage IoT devices. An IoT platform is defined as middleware and infrastructure that allows users to interact with smart objects [9]. IoT platforms are useful for storing, displaying and analyzing data originating from IoT devices.

Based on the extraction of selected journals, each journal in his research uses three different IoT platforms. The different platforms used in each journal consider the advantages and disadvantages of each existing platform according to needs.

3.6.1. MySQL. MySQL is software that provides a SQL database management system or DBMS (database management system) and can be accessed by multi-users [10]. The advantages of the MySQL platform include being free, good database capabilities, using SQL language, being able to be accessed simultaneously, being easily accessible via the internet and guaranteed security factors. There is a drawback of this platform, namely the absence of transaction processing facilities [11].

3.6.2. Thingspeak. Thingspeak is an open source platform and provides IoT applications and application programming interfaces (APIs) as data storage taken from devices using hyper text transfer protocol (HTTP) over the internet network [5]. The advantages of this platform are that it is free, easy to create data visualizations using line charts, can be connected to hardware such as Arduino microcontrollers and raspberry pi, and is relatively easy for new users. The drawback is that there are several features that require users to upgrade their accounts to get data limits for API channels and create custom charts [12].

3.6.3. Firebase. Google has a cloud and backend service called. This platform provides Real Time Database or NoSQL. The storage method uses the JSON format. Firebase can store and synchronize data using a database to each connected client/device. The updated data will be stored in the cloud and share the latest data directly to the client [13]. This platform has several advantages, namely firebase cloud messaging which can send messages and notifications for android, IOS and the web, firebase analytics can provide information about application usage and user engagement, and realtime database provides direct database services and APIs to synchronize application data throughout client and stored on the firebase cloud. The drawbacks are that if a database is updated by a user with large enough data it can affect its performance, the JSON storage format so that data migration cannot be done easily, and there are performance problems from the server provided by firebase [14].

3.6.4. Domoticz. Domoticz is an open source automation system. This platform allows inspection, monitoring, and control of third devices and can be easily accessed using a browser on the user's desktop or mobile device. This platform system uses C++ machine language. The domoticz display uses a web browser to enter the assigned ip address and add a port at the end, usually 8080. The panels on the web display are tailored to the device [15].

3.6.5. Wyliodrin. Wyliodrin is an IoT platform that supports the connection of multiple third-party systems. The platform can be used easily for data transmission and is very practical to implement on devices that are used to view variables. Wyliodrin is a programming and device control service accessible from a browser. Users can write and run programs and save them wherever the platform is [16].

4. Conclusion

The results of the analysis and discussion show that almost all research publications related to hydrophonics analyzed measure the same variables, namely Temperature and Hydrogen potential (pH), and use the module device, namely ESP 8266, Arduino, and Raspberry PI. While the IoT platforms implemented are Mysql, Thingspeak, Firebase, Domoticz, and Wyliodrin adjusted with the advantages and disadvantages of each platform.

References

[1] Izzuddin A 2016 Wirausaha Santri Berbasis Budidaya Tanaman Hidroponik *DIMAS* **16** 351–66
[2] Singgih M, Prabawati K and Abdulloh D 2019 Bercocok Tanam Mudah dengan Sistem Hidroponik NFT *Januari* **03** 21–4
[3] Sumartono G H and Sumarni E 2013 Pengaruh Suhu Media Tanam Terhadap Pertumbuhan Vegetatif Kentang Hidroponik di Dataran Medium Tropika Basah *Agronomika* **13** 1–9
[4] Moher D, Liberati A, Tetzlaff J, Altman D G, Altman D, Antes G, Atkins D, Barbour V, Barrowman N, Berlin J A, Clark J, Clarke M, Cook D, D'Amico R, Deeks J J, Devereaux P

Journal of Physics: Conference Series **2111** (2021) 012014 doi:10.1088/1742-6596/2111/1/012014

J, Dickersin K, Egger M, Ernst E, Gøtzsche P C, Grimshaw J, Guyatt G, Higgins J, Ioannidis J P A, Kleijnen J, Lang T, Magrini N, McNamee D, Moja L, Mulrow C, Napoli M, Oxman A, Pham B, Rennie D, Sampson M, Schulz K F, Shekelle P G, Tovey D and Tugwell P 2009 Preferred reporting items for systematic reviews and meta-analyses: The PRISMA statement *PLoS Medicine* **6** 1–6

[5] Setiawan Y, Tanudjaja H and Octaviani S 2018 Penggunaan Internet of Things (IoT) untuk Pemantauan dan Pengendalian Sistem Hidroponik *TESLA* **20** 196–207

[6] Mesquita J, Guimarães D, Pereira C, Santos F, Almeida L and Almeida L 2018 Assessing the ESP8266 WiFi module for the Internet of Things *CISTER-TR-181106* pp 784–91

[7] Louis L 2016 Working Principle of Arduino and Using it as a Tool for Study and Research *International Journal of Control, Automation, Communication and Systems* **1** 21–9

[8] Sachdeva P and Katchii S 2014 *A Review Paper on Raspberry Pi* vol 3818

[9] Vogel B, Dong Y, Emruli B, Davidsson P and Spalazzese R 2020 What is an open IoT platform? Insights from a systematic mapping study *Future Internet* **12** 1–19

[10] Dhika H, Isnain N and Tofan M 2019 Manajemen Villa Menggunakan Java NetBeans dan MySQL *Jurnal IKRA-ITH Informatika* **3** 104–10

[11] Supriyanti W 2013 Rancang Bangun Aplikasi Sistem Pendukung Keputusan Penerima Beasiswa dengan Metode SAW *Citec Journal* **1** 67–75

[12] Kusuma H A, Anjasmara R, Suhendra T, Yunianto H and Nugraha S 2020 An IoT Based Coastal Weather and Air Quality Monitoring Using GSM Technology *Journal of Physics: Conference Series* vol 1501 (IOP Publishing)

[13] Dani A W 2021 Optimalisasi Pertumbuhan Pada Sayuran Hidroponik Nutrient Film Technique Dengan Metode Fuzzy Logic Berbasis Internet of Things *Jurnal Teknologi Elektro* **11** 1–10

[14] Gupta S and Kapoor B 2016 Firebase In APP Development *International Research Journal of Engineering and Technology (IRJET)* **3** 180–1

[15] Kolanur C B, Banakar R M and Rajneesh G 2021 Design of IoT based Platform Development for Smart Home Appliances Control *Journal of Physics: Conference Series* vol 1969 (IOP Publishing Ltd)

[16] Juan Carlos A D, Estrada L R, Cesar Augusto C R, Paola Patricia A C, Marlon Alberto P M, Ramón Enrique R G, Roberto César M O, David Alfredo O G and Carlos Andrés C M 2020 Monitoring system of environmental variables for a strawberry crop using IoT tools *Procedia Computer Science* vol 170 (Elsevier B.V.) pp 1083–9

Stress Distribution Analysis of the Rectangular and Star-Blade for Plastic Crusher Machine Using Finite Element Analysis

Nalendro Mataram[1], Sigiet Haryo Pranoto[1,*], Rizqi Ilmal Yaqin[2], Anis Siti Nurrohkayati[1], Noer Aden Bahry[1], Nuruddin Wahyu Eko Saputro[1], Sabaruddin Syach[1]
[1]Mechanical Engineering Department, Faculty of Science and Technology, Universitas Muhammadiyah Kalimantan Timur, Indonesia.
[2]Ship Machinery, Maritime and Fisheries Polytechnic of Dumai.

Email: shp904@umkt.ac.id

Abstract. Recycling plastic waste can help reduce plastic waste, which is the world's largest problem. The first solution is a plastic crusher machine, which converts plastic packaging into pellets, especially for plastics made from polyester, which is often used in bottle packaging. An integral part of the crushing machine is the crushing blade, which determines the design and analysis of the blade. The finite element analysis method is widely used in engineering analysis. Its results can provide useful information for the analysis of manufacturing processes. A rectangular blade and a star-shaped blade are the two types to consider. The analytical results obtained on the blade of the crusher machine in the rectangular shape that was given loads by 200 N, 400 N, and 600 N are 1.285 x 10^3 N/m^2, 2.570 x 10^3 N/m^2, and 3.855 x 10^3 N/m^2, respectively. We obtained the displacements are 1.080 x 10^{-6} mm, 2.160 x 10^{-6} mm, and 3.241 x 10^{-6} mm, respectively. The maximum von mises stress result on consecutive star-shaped blades are 8.890 x 10^2 N/m^2, 1.778 x 10^3 N/m^2, and 2.667 x 10^3 N/m^2. The displacement obtained are 1.211 x 10^{-6} mm, 2.422 x 10^{-6} mm, and 3.633 x 10^{-6} mm. The results of this analysis indicate that for the star blade, the shape is better than for the rectangular blade at the same time, and for the star blade, the stress is smaller than for the rectangular blade. Based on this simulation, the safety factor is 15, which means that it is more than 1, which means that it is safe.

Keywords: Displacement, Von misses stress, Blade shape, Finite element analysis, the Safety factor

1. Introduction

Indonesia has high biodiversity and coral reefs. Unfortunately, Indonesia is estimated to be the second-largest contributor to ocean plastic pollution after China [1]. PET (Polyethylene terephthalate) is a plastic material used for making bottles, this material is widely accepted because it has advantages such as being disposable, cheap, and lightweight. Plastic crusher machine serves to destroy or change plastic bottles into smaller pieces that can be recycled into new products [2]. The blade is one of the most important components in a plastic crusher machine, the suitability of the design in its various variations greatly affects the durability or the results of the crusher machine [3]. Finite element analysis (FEA) is one of the techniques of the Finite Element Method, this technique can analyze a design, with a CAE program such as Abaqus. This technique has advantages such as reducing the number of design prototypes produced [4]. This method can also see the results of von mises stress, deformation, and fluid flow to heat transfer in the design, to optimize the design [5]. The renewal and purpose of this research are to determine and compare which two blade shapes have good resistance by looking for the von misses stress and displacement of the two shapes.

2. Method

The research aims are to compare the two blades, rectangular and star-shaped. The modeling of the two blades design was carried out using CAD (Computer-Aided Design). Rectangular blade and star blade design modeling is attached in Figure 1. CAD method using Autodesk Inventor

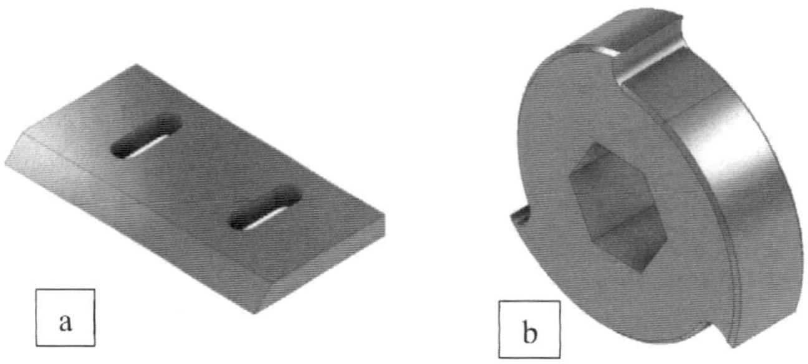

Figure 1. (a) *Rectangular blade* design, (b) *Star blade* design make by Autodesk Inventor

2.1 Finite Element Analysis

The way FEA works is by dividing objects that have been designed into finite small parts or commonly called meshing. These small parts will be analyzed and the results will be combined again to get the overall results of the objects that have been designed. FEA can be used to find von mises stress, displacement, and even heat transfer [6]. In this research, FEA will be used to find von mises stress and displacement, here also using tetrahedral type meshing [7]. The meshing application is attached in Figures 2. FEA method using Abaqus program and Autodesk Inventor

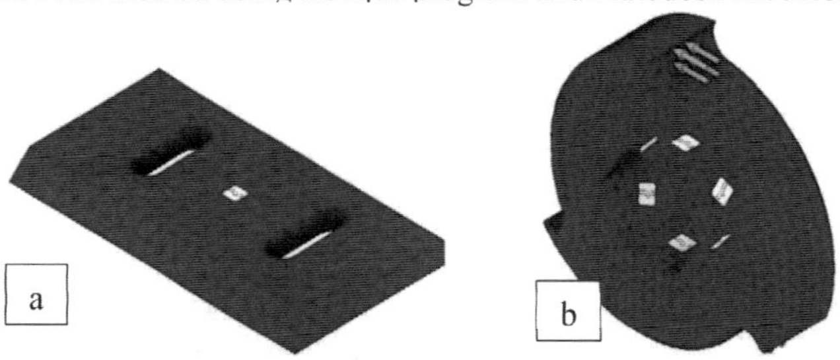

Figure 2. (a) Meshing in rectangular blade, (b) Meshing in Star Blade make by Autodesk Inventor

The meshing obtained from the CAE program has 1336 nodes on the rectangular blade, while the star blade has 3226 nodes. The elements contained in the rectangular blade are 639, while for the star blade there are 1868 elements.

2.2 Von Mises Stress and Displacement

Von mises stress calculation on the elements of Finite Element Analysis given by

$$\begin{Bmatrix}\sigma_{xx}\\ \sigma_{yy}\\ \sigma_{zz}\end{Bmatrix} = \frac{E}{(1+v)(1-2v)}\begin{Bmatrix}(1-v)\varepsilon_x + v\varepsilon_y + v\varepsilon_z\\ v\varepsilon_x + (1-v)\varepsilon_y + z\varepsilon_z\\ v\varepsilon_x + v\varepsilon_y + (1+v)\varepsilon_z\end{Bmatrix} \quad (1)$$

Where E is the young modulus for each material used in the blade design, and ε is the strain within each design [8]. The strain itself is found from displacement

$$\varepsilon = \frac{\Delta l}{l_0} = \frac{l-l_0}{l_0} \quad (2)$$

Where Δl is the difference between the total length and the initial length of the geometry of each design [9].

2.3 Safety factor

Safety factor It is a factor of the ability of an object when it receives a load that has been determined from the outside, this load can be compressive or tensile [10]. The safe value for the safety factor itself is 1, if the safety factor value obtained is less than one, then the design is considered unsafe [11].

$$Safety\ factor = \frac{\sigma_{Maximum}}{\sigma_{Allowable}} \quad (3)$$

Where $\sigma_{Maximum}$ is the maximum stress that a design or object can fracture, while $\sigma_{Allowable}$ is the stress that is allowed in a material [12]. Each material has a different allowable stress [13].

3. Result and Discussion

Following are the results of the von mises stress and displacement in both blade shapes. The analysis was carried out using the CAE program, called Abaqus. Abaqus can be used for complex analytical modeling, analysis, monitoring and visualizing simulation.

3.1 Von mises stress

The process is analyzed with the load magnitude, the load magnitude is 200 N, 400 N, and 600 N. Given different magnitudes to know the stress distribution of the blades in different load magnitudes.

3.1.1 Rectangle Blade Shape

The following is an analysis of the von mises stress on a rectangular blade, given a load of 200N, 400N, and 600N. The analysis results are attached in Figure 3, Figure 4, and Figure 5

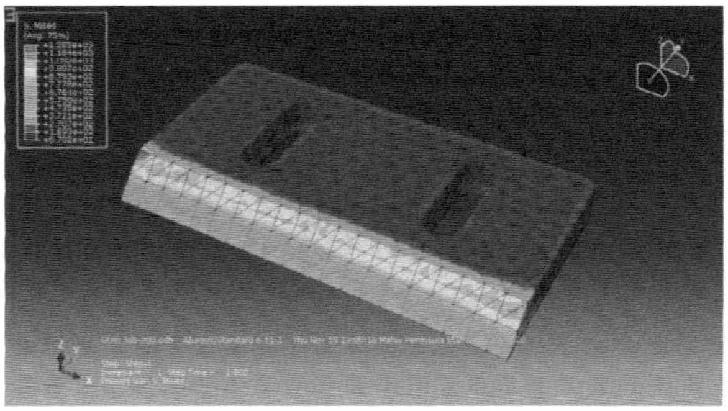

Figure 3. Result of rectangle blade analysis at 200 N load

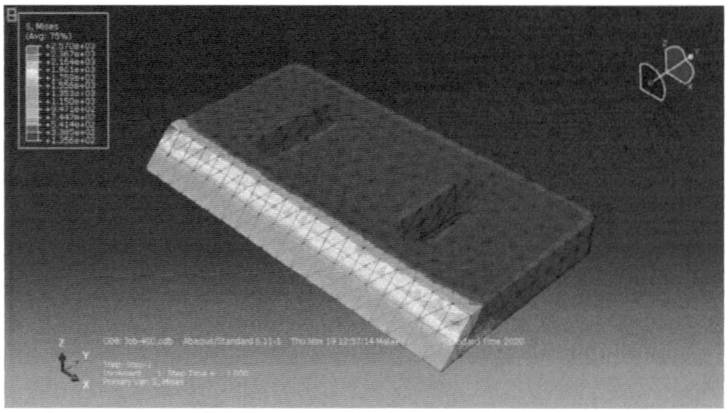

Figure 4. Result of rectangle blade analysis at 400 N load

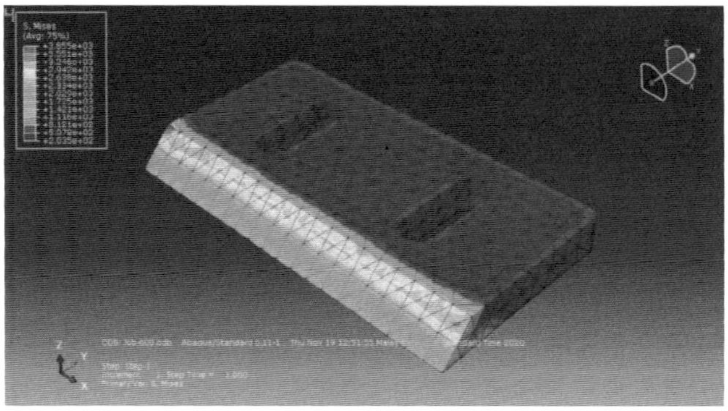

Figure 5. Result of rectangle blade analysis at 600 N load

The maximum result that occurs in a rectangular blade at 200 N load is $1.285\ x\ 10^3$ N/m^2 and the average von mises stress is $6.764\ x\ 10^2$ N/m^2. The maximum result at a load of 400 N is $2.570\ x\ 10^3$ N/m^2 and the average stress is $1.353\ x\ 10^3$ N/m^2. The maximum result at a 600 N load is $3.855\ x\ 10^3$ N/m^2 and the average stress is $2.029\ x\ 10^3$ N/m^2.

3.1.2 Star Blade Shape

The following is an analysis of the von mises stress that occurs in a star blade, given a load of 200N, 400N, and 600N. The analysis results are attached in Figure 6, Figure 7, and Figure 8.

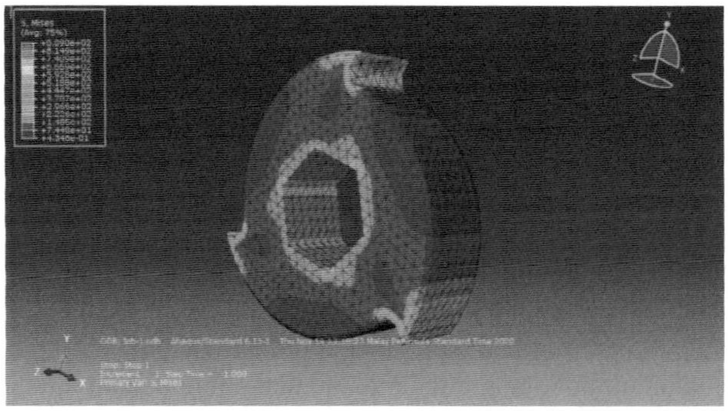

Figure 6. Results of star blade analysis at 200 N load

Journal of Physics: Conference Series **2111** (2021) 012015 doi:10.1088/1742-6596/2111/1/012015

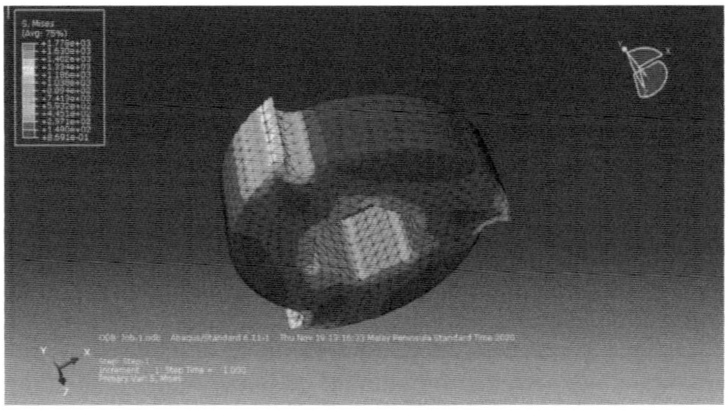

Figure 7. The results of star blade analysis at 400 N load

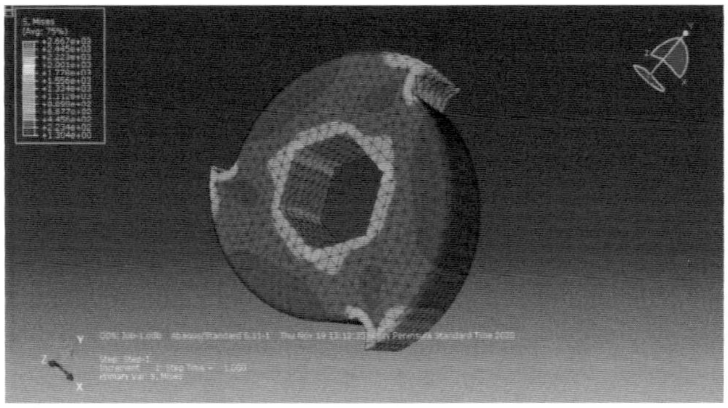

Figure 8. The results of star blade analysis at 600 N load

The maximum result that occurs in a rectangular blade at 200 N load is $8.890 \, x \, 10^2$ N/m^2 and the average von mises stress is $4.447 \, x \, 10^2$ N/m^2. The maximum result at a load of 400 N is $1.778 \, x \, 10^3$ N/m^2 and the average stress is $8.894 \, x \, 10^2$ N/m^2. The maximum result at a 600 N load is $2.667 \, x \, 10^3$ N/m^2 and the average stress is $1.334 \, x \, 10^3$ N/m^2.

3.2 Displacement

The process is analyzed with a load magnitude, as the load to be pressed by the blade, the magnitude is 200 N, 400 N, and 600 N. Given different magnitudes to know the distribution of deformation that occurs the star blade and rectangle blade under different load magnitudes.

3.2.1 Rectangle Blade Shape

The following is an analysis of the displacement that occurs in a rectangular blade, given loads of 200 N, 400 N, and 600 N. The analysis results are attached in Figure 9, Figure 10, and Figure 11.

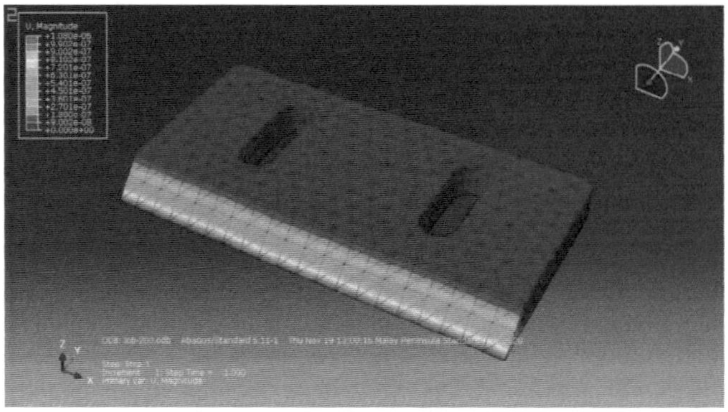

Figure 9. Result of rectangle blade analysis at 200 N load

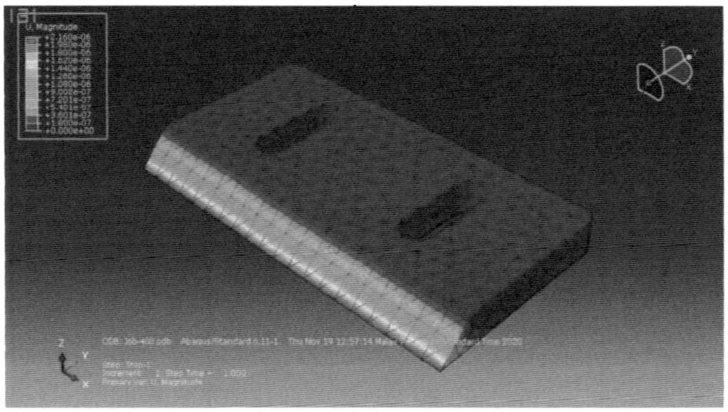

Figure 10. Result of rectangle blade analysis at 400 N load

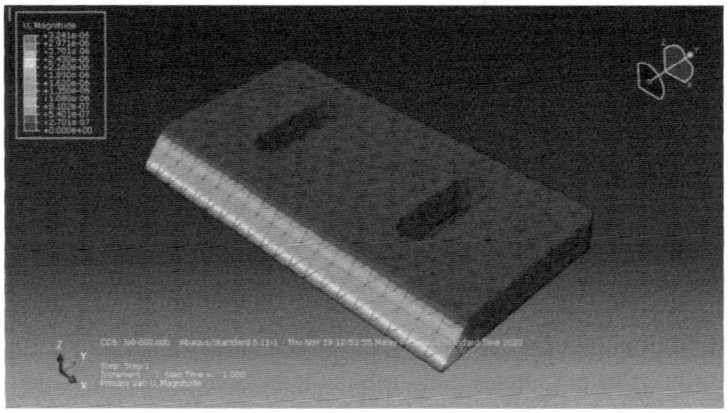

Figure 11. Result of rectangle blade analysis at 600 N load

The maximum result that occurs in a rectangular blade at 200 N load is 1.080 x 10^{-6} mm and the average displacement is 5.401 x 10^{-7} mm. The maximum result at a load of 400 N is 2.160 x 10^{-6} mm and the average is 1.080 x 10^{-6} mm. The maximum result at a 600 N load is 3.241 x 10^{-6} mm and the average is 1.620 x 10^{-6} mm.

3.2.2 Star Blade Shape

The following is an analysis of the displacement that occurs in a rectangular blade, given a load of 200N, 400N, and 600N. The analysis results are attached in Figure 12, Figure 13, and Figure 14.

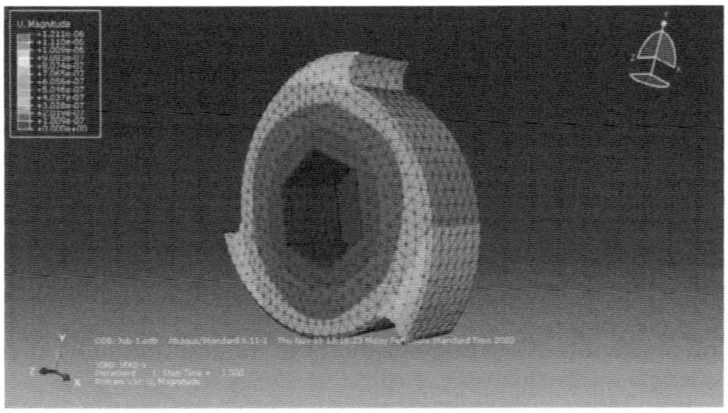

Figure 12. The results of star blade analysis at 200 N load

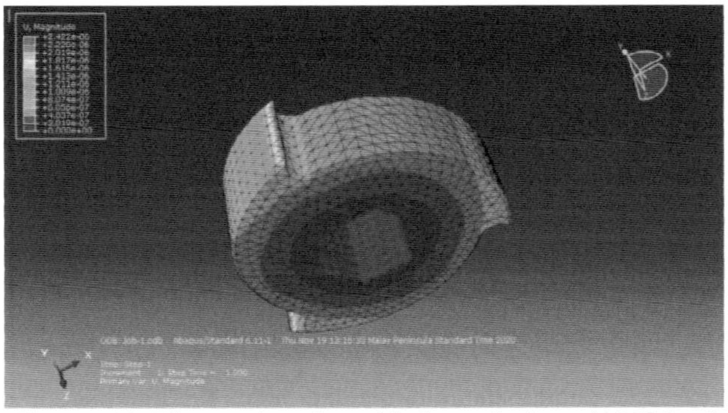

Figure 13. The results of star blade analysis at 400 N load

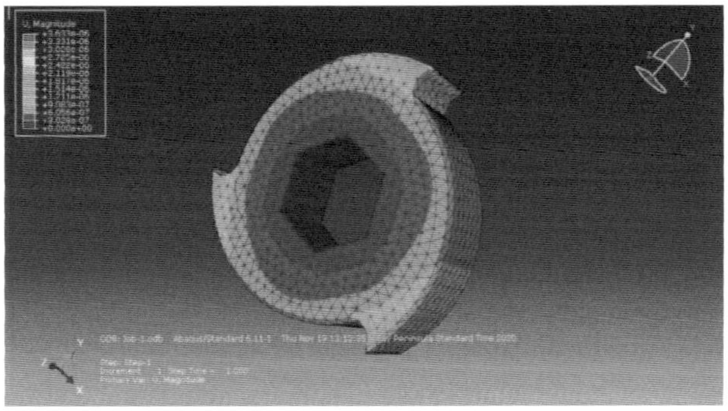

Figure 14. The results of star blade analysis at 600 N load

The maximum result that occurs in Star Blade at 200 N load is 1.211 x 10^{-6} mm and the average displacement is 6.050 x 10^{-7} mm. The maximum result at a load of 400 N is 2.422 x 10^{-6} mm and the average displacement is 1.211 x 10^{-6} mm. The maximum result at a 600 N load is 3.633 x 10^{-6} mm and the average is 1.817 x 10^{-6} mm.

3.3 Blade Shape Comparison

After analyzing using the CAE program, the comparison of the two-blade shapes will be compared between the maximum value and the average value that occurs in each design. The following is the attachment in table 1,2,3 and 4.

Table 1. Comparison of Average Von Mises Stress

Load	**Rectangular Blade**	**Star Blade**
200N	$6.764 x 10^2$ N/m^2	$4.447 x 10^2$ N/m^2
400N	$1.353 x 10^3$ N/m^2	$8.894 x 10^2$ N/m^2
600N	$2.029 x 10^3$ N/m^2	$1.334 x 10^3$ N/m^2

Table 2. Comparison of Maximum Von Mises Stress

Load	**Rectangular Blade**	**Star Blade**
200N	$1.285 x 10^3$ N/m^2	$8.890 x 10^2$ N/m^2
400N	$2.570 x 10^3$ N/m^2	$1.778 x 10^3$ N/m^2
600N	$3.855 x 10^3$ N/m^2	$2.667 x 10^3$ N/m^2

Table 3. Comparison of Average Displacement

Load	**Rectangular Blade**	**Star Blade**
200N	$5.401 x 10^{-7}$ mm	$6.050 x 10^{-7}$ mm
400N	$1.080 x 10^{-6}$ mm	$1.211 x 10^{-6}$ mm
600N	$1.620 x 10^{-6}$ mm.	$1.817 x 10^{-6}$ mm

Table 4. Comparison of Maximum Displacement

Load	**Rectangular Blade**	**Star Blade**
200N	$1.080 x 10^{-6}$ mm	$1.211 x 10^{-6}$ mm
400N	$2.160 x 10^{-6}$ mm	$2.422 x 10^{-6}$ mm
600N	$3.241 x 10^{-6}$ mm	$3.633 x 10^{-6}$ mm

3.4 Safety Factor

This research uses a compressive load with a maximum value of 1000N. The CAD program reads both blade shapes having a safety factor value of 15, attached to Figure 15 and Figure 16

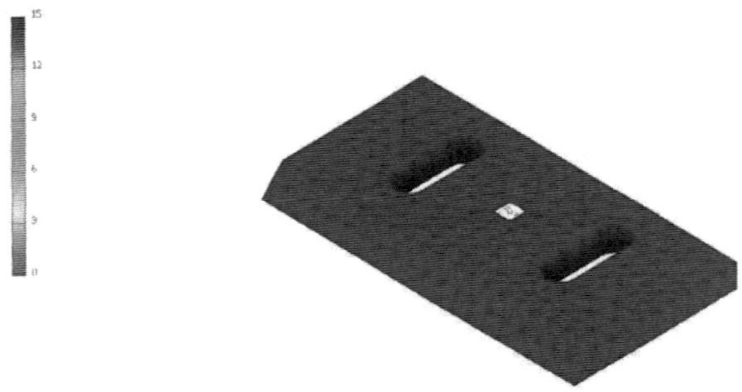

Figure 15. *Safety Factor* Rectangular Blade shape

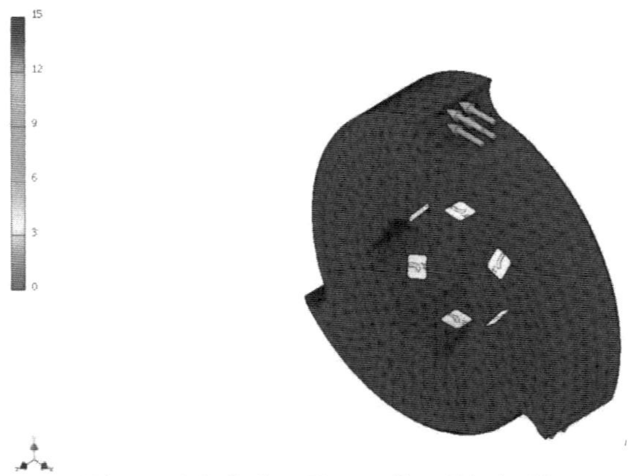

Figure 16. *Safety Factor* Star-Blade Shape

4. Conclusion

The results of the analysis conducted using CAD and CAE programs found that the blade had increased von mises stress and displacement on a load 200N, 400N, and 600N. Rectangle blade has a spread of von mises star higher than the blade, with the highest value is 3.855 x 10^3 N/m^2 while the highest value that occurs in the star blade is 2.667 x 10^3 N/m^2. Areas of high concentration (red color) are classified as few in both blade designs. The star blade has a higher displacement value than the rectangle blade, the maximum value is 3.663 x 10^{-6} mm, and the maximum value of the rectangle blade of 3.241 x 10^{-6} mm. The difference in the displacement of the two blades is small so it can be concluded that the two blades have almost the same displacement value. The displacement that occurs in both blades is close to 0 so it can be ignored, and there is no significant deformation. The design of these two blades has a relatively safe factor, with a 1000N load, a safety factor of 15 is found, this value is more than 1, which means that the blade design is safe for crushing plastic.

References

[1] A. A. Phelan, H. Ross, N. A. Setianto, K. Fielding, and L. Pradipta 2020 *Ocean plastic crisis Mental models of plastic pollution from remote Indonesian coastal communities*, ed S B Borrelle (Canada : PLOS ONE) p 3

[2] C. P. Yepes, P. M. R. Angel, and P. M. G. José *2018 Analysis by Means of the Finite Element Method of the Blades of a PET Shredder Machine with Variation of Material and Geometry* vol 11 (Colombia: Contemporary Engineering Sciences) p 4116

[3] I. Aniekan, and O. Ikechukwu 2017 *Design of Used PET Bottles Crushing Machine for Small Scale Industrial Applications* vol 3 (Nigeria: INTERNATIONAL JOURNAL of ENGINEERING TECHNOLOGIES) pp 157-158

[4] O. Adeyefa, O. Oluwole 2011 *Finite Element Analysis of Von-Mises Stress Distribution in a Spherical Shell of Liquified Natural Gas (LNG) Pressure Vessels* vol 3 (Nigeria: Scientific Research Engineering) pp. 1012-1017

[5] S. N. Ali, and A. Hardianto 2019 *Finite Element Simulation of Acceleration Pedal On XX Vehicle* vol 4 (Indonesia: Journal of Mechanical Engineering and Mechatronics) pp. 45-50

[6] N. Dantulwar, R. Maske, and J. Patel 2017 *Finite Element Analysis of Ball Valve Assembly for Earthquakes* vol 5 (India: International Conference on Ideas, Impact and Innovation in Mechanical Engineering) pp.1460-1463

[7] M. V. Sereshk and M. Salimi 2011 *Comparison of finite element method based on nodal displacement and absolute nodal coordinate formulation (ANCF) in thin shell analysis* (Isfahan: INTERNATIONAL JOURNAL FOR NUMERICAL METHODS IN BIOMEDICAL ENGINEERING) pp.1185-1190

[8] M. Yeh. and C. Wang 2017 *Stress analysis of composite wind turbine blade by finite element method* (Hsinchu: 5th Asia Conference on Mechanical and Materials Engineering) pp. 1-4
[9] X. Shen 2016 *The Research of Crusher Blade Carrier Shaft Based on Balancing Test and Dynamic Analysis*, Applied Mechanics, Mechatronics and Intelligent Systems pp. 79-83
[10] M. Szulc and I. Malujda 2015 *Method of determination of safety factor on example of selected structure* (Poland: The 20th International Conference Machine Modeling and Simulation, Elsevier) pp. 50-51
[11] J. C. Musto 2015 *The safety factor: case studies in engineering judgment* (Milwaukee: Mechanical Engineering Department, Sage Journal) p. 289
[12] S. Qiang and M. Liu 2014 *A New Safety Factor Prediction Model for Mass Concrete Surface Cracking in Early Age* (London: Mathematical Problems in Engineering Volume 2014, Hindawi Publishing Corporation) p. 2
[13] R. D. Beldar and S. Komble 2018 *Mechanical Design of Shell and Tube Type Heat Exchanger as per ASME Section VIII Div.1 and TEMA Codes for Two Tubes* vol 8 (Indonesia: International Journal of Engineering and Technical Research (IJETR), Neliti) p. 2
[14] S. H. Pranoto and M. Mahardika 2018 *Design and Finite Element Analysis of Micro Punch CNC Machine Modeling for Medical Devices* vol 1941 (Semarang: AIP Conference Proceedings) p. 4

Journal of Physics: Conference Series **2111** (2021) 012016 doi:10.1088/1742-6596/2111/1/012016

Oxygen Availability Control and Monitoring System in Hospitals using IoT

Nikko Aji Bayu Nugraha[1], Mashoedah[2]

[1,2]Post Graduate Program of Electronic & Informatics Engineering Education, Faculty of Engineering, Universitas Negeri Yogyakarta, Indonesia

E-mail: [1]*nikkoaji.2020@student.uny.ac.id*; [2]*mashoedah@uny.ac.id*

Abstract. Medical Gas is an important component in the treatment of patients with COVID-19 disease. Medical gas is used to help COVID-19 patients to reduce the effects of respiratory disorders by providing oxygen ventilators to patients. With the surge in typical COVID-19 sufferers as of July 2021, the need for Oxygen in hospitals is getting higher. This is when the control and monitoring of medical gases in hospitals are late because the integrated system is very dangerous. Therefore, a system that can be used to control and monitor medical gases in hospitals that are integrated and automated and can be monitored by the Government and Medical Gas Producers. This is useful for anticipating the lack of availability of Medical Gas in hospitals. In this study, the system used IoT systems as a base in delivery and control. This system uses Arduino as a minimum system that reads the press sensor on the Medical Gas tube and regulates the valve. The data obtained is then sent to the Local Server to be processed and delivered to the Hospital Officer. Local Servers also send the data to cloud servers to be monitored by the government funds of several medical gas producers. This design can help in the process of controlling and monitoring Medical Gases in hospitals in hopes of minimizing the risk of delays in supplying medical gases to hospitals.

1. Introduction

Lately in Indonesia and all countries are fighting the Covid-19 Pandemic. This resulted in the hospital being filled with patients with the disease. Patients suffering from this disease mostly experience respiratory problems. To reduce the effects of respiratory disorders, the medical treatment carried out is to provide oxygen ventilators to patients. It is also increasing the need for Medical Gas for oxygen ventilators for patients with CoVid-19. It is as revealed by Hermawan [1] that Covid-19 patients in Java and Bali soared, thus requiring oxygen of 2,620 tons per day. This surge in patients resulted in insufficient oxygen supply in the hospital. This is as revealed by Novelino [2] that the oxygen supply for the treatment of Covid-19 patients in D. I. Yogyakarta is dwindling. Even Dr. Sardjito General Hospital clarified that in the duration of 1 day only on Sunday, July 4, 2021, 63 patients at this Hospital died after the hospital lacked oxygen supply [3].

From these problems, control and monitoring are needed to find out the supply of Oxygen in the Hospital in Real-Time and can be accessed by all parties who are concerned in the provision of Oxygen Medical Gas. This is also in line with the Ministry of Health's program that the government continues to strive for and monitor the availability of Oxygen, following reports from each hospital related to Oxygen stock [4]. With the development of IoT technology at this time, the problem should be helped by this technology.

2. Research Methods

In this study, the research method used is the study of literature. Literature is a method of collecting data from libraries, reading, and recording, and managing research materials [5]. In addition, literary studies are a mandatory activity when conducting research to develop theoretical aspects and aspects of practical benefits. Literature studies are conducted by researchers as a basis in the research process. The rationale is used to find and create the basis of theory, frame of mind, and the initial hypothesis in research. This is expected so that the researcher can group, allocate organize, and use the variety of libraries in their fields. This circuit study is also used by researchers to have a broader and deeper understanding of the problems to be studied. Researchers did a literature study at the time after they determined the research topic and established the formulation of the problem until before they plunged into the field to collect the necessary data [6].

3. Research Results

In several journals analyzed, the development of control systems is used to detect medical gas pressure in hospitals. As in the research conducted by Nur Hudha Wijaya [7] about the design of a tool used for monitoring the pressure of medical gas installations with digital systems. This tool is used to perform medical gas monitoring that can show low and high-pressure indicator information displayed on LCD screens equipped with a buzzer. The study succeeded in developing an automation system that can warn officers to know the level of pressure of medical gases that are being installed in the medical gas system at the hospital. But in the study conducted by Nur Hudha Wijaya [7] is only a monitoring system with buzzer signs for alarms centrally in the hospital. It still requires action from the officer to manually.

In addition to the research mentioned above, the study conducted by Atika Hendryani [8] also made the Medical Gas automation system at the Hospital in the form of a manifold system equipped with pressure monitoring for the automatic exchange of Oxygen gas cylinders using pressure sensors and microprocessors. The research results in an automation system for the automatic exchange of active gas cylinders on *manifolds*. Unfortunately, this system only discusses how a working manifold valve automatically moves from the Primary tube to the Secondary tube. With this research, it can be redeveloped with a touch of IoT technology for control and monitoring processes. The following is a block diagram design of the system developed by Atika Hendryani.

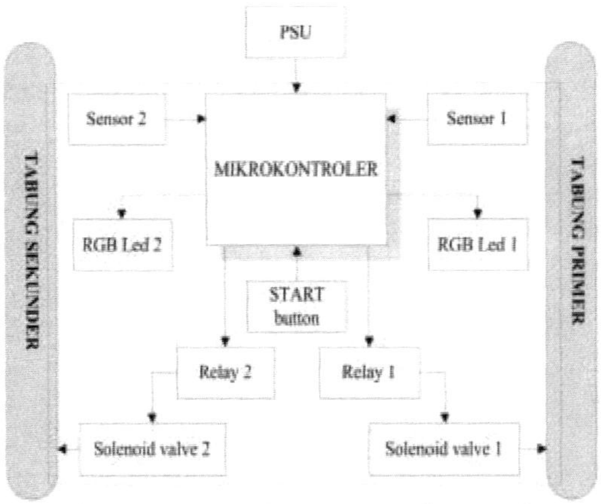

Figure 1. Block Diagram Automatic system exchange of gas cylinders [8]

The use of IoT in Medical Gas systems in Hospitals has been alluded to by Miasih [9] in his Research entitled*"Development of Oxygen Volume Monitoring as a Basis for Determining Tariffs with IoT-Based Real-Time".* In his research, IoT was used to set regulators and tariffs. In this system, take data from the gas flow meter sensor to be processed with a microcontroller to produce a tariff value that is then displayed on the LCD and also WEB The advantages of this system can facilitate the calculation of rates and monitoring of Medical Gas in hospitals in a real case.

The use of IoT in Medical Gas monitoring systems in Hospitals was also implemented by Basuki Rahmat [10] in his research entitled*"Telemonitoring Medical*Gases". In the study, IoT was used to monitor the availability of medical gases and inform officers using mobile phone communication networks. From the study, the use of a phone when used to send virgins *in real-time* will cost a lot of money. Therefor in this research, the network system is used for the process of sending data using the internet network.

Lately, the concept of IoT has been widely used in the environment around us. However, IoT has been used on many devices to support daily life, the Internet of Things does not yet have a standard definition [11]. Therefore, based on some of the studies mentioned above, this study will develop the design of a centralized medical gas system control and monitoring system that can be monitored by various parties concerned with the IoT base. This system is expected to reduce the risk of delays in providing medical gas reserves. This is because the design of the development of this system aims so that every hospital, government, and medical gas producer can monitor the needs of medical gases at the hospital.

4. Discussion

4.1. Solenoid Valve Control

Based on research conducted by Atika Hendryani [8], the design developed in this study is a valve system that works automatically to move to a Secondary tube when the Primary tube is at a predetermined minimum pressure. The following are the valves and sensors in the Hospital Medical Gas installation system.

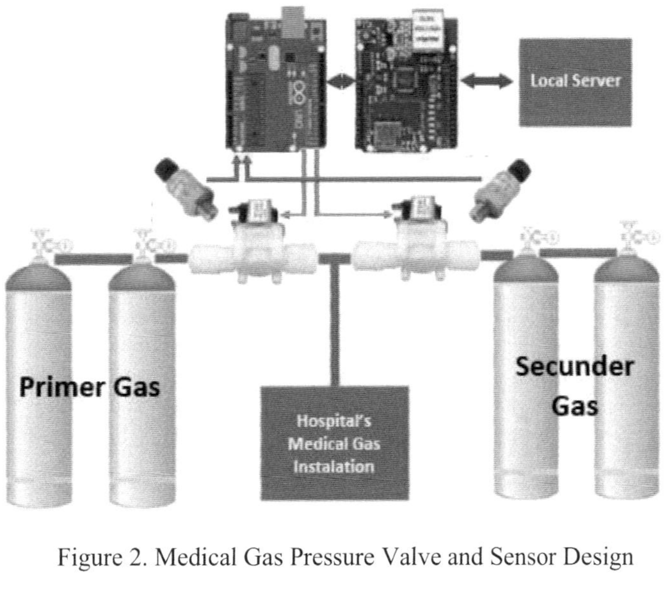

Figure 2. Medical Gas Pressure Valve and Sensor Design

In the design, Medical Gas is divided into two, namely Primary Gas used to supply Medical Gas at Hospital Installations, while secondary gases are used as backup and turned into Primary Gas when primary gas was previously at minimum pressure.

The Medical Gas controller system uses an Arduino microprocessor to process the pressure sensor data and the position of the opened valve. This Arduino in addition to processing data from the sensor is also tasked with sending the data to the server with *an Ethernet Shield device*. The data sent to the server is in the form of large primary and secondary gas puzzles and the position of the valve opened.

4.2. Server

The server on this system has 2 parts, namely the local server and the cloud server. The Local Server is used by the Hospital to monitor the condition of the Medical Gas in the Hospital. This local server is tasked with processing data sent from Arduino which is then displayed on the computer for monitoring hospital officers. The data displayed from the computer is in the form of large pressures on primary and secondary gas cylinders and can provide warnings when there is an empty gas cylinder. In addition to monitoring, the Local Server is also used by the Officer to set the valve to work automatically or by the Officer's command.

Unlike the local server, cloud servers are useful for intermediaries in monitoring the condition of medical gases in hospitals. This server is used jointly by outside parties of the Hospital concerned in this case such as the Government and medical gas producers. This aims to monitor the condition and supply of Medical Gas at the Hospital so that there are no delays in *the re-stock of* Medical Gases such as the case that occurred at Dr. Sardjito Hospital [3]. The following is the design of the server system on the Hospital Medical Gas Control and Monitoring System.

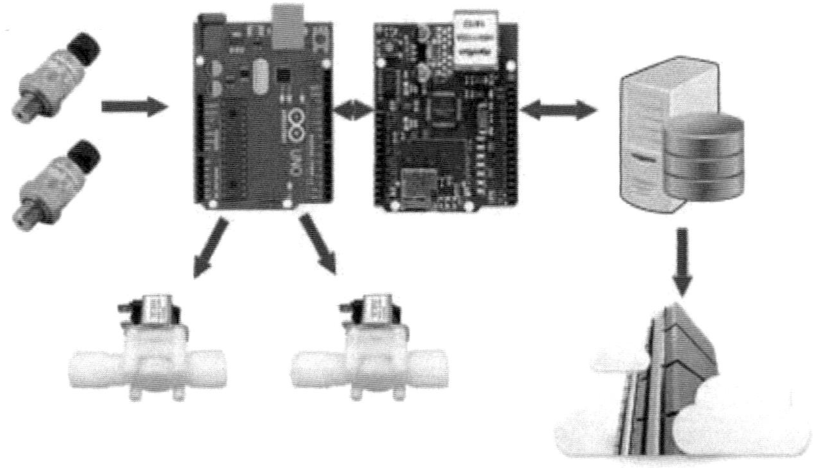

Figure 3. Server System Design

4.3. Monitoring System

Monitoring on this system is divided into 2 parts, namely the first part and the second part. In the first part, monitoring is carried out in the form of pressure monitoring on the primary tube and secondary tube and can control the valve module fully in this system. Monitoring in the first part is carried out by hospital officials.

The second part is different from the first part. In the second part, this is only used to monitor the availability of Medical Gas in the Hospital. This monitoring is carried out by parties concerned in this case the Government and Medical Gas Producers. The following is the design of the monitoring system for both parts.

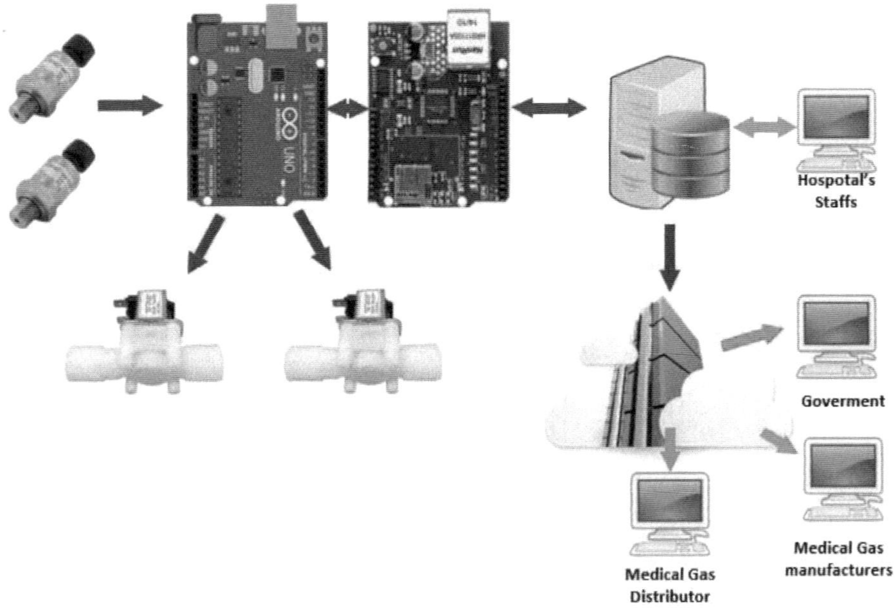

Figure 4. Monitoring System Design

5. Conclusion

The conclusion of this study is based on several journals that have reviewed the design of Oxygen Availability Control and Monitoring Systems in Hospitals that can use IoT systems. This design can help in the process of controlling and monitoring Medical Gases in hospitals in hopes of minimizing the risk of delays in supplying medical gases to hospitals.

This study is only limited to making the design of the Oxygen Availability Control and Monitoring System in the Hospital. Therefore, it is necessary to review and further research so that this system can be realized. More research is needed to identify the minor needs needed to create a well-integrated system.

References

[1] H. Handaka, 09 July 2021. [Online]. Available: https://www.tribunnews.com/nasional/2021/07/10/pasien-covid-19-melonjak-jawa-bali-butuh-oksigen-2620-ton-per-hari-produksi-cuma-1400-ton. [Accessed 25 Agust 2021].

[2] A. Novelino, 23 June 2021. [Online]. Available: https://www.cnnindonesia.com/nasional/20210623145229-20-658349/stok-oksigen-di-yogyakarta-mulai-langka. [Diakses 25 Agust 2021].

[3] A. B. Tamtomo, 5 July 2021. [Online]. Available: https://www.kompas.com/tren/read/2021/07/05/173000165/-klarifikasi-kehabisan-oksigen-63-pasien-rsup-dr-sardjito-meninggal-dalam?page=all. [Accessed 25 Agust 2021].

[4] G. LOTULUNG, 04 July 2021. [Online]. Available: https://nasional.kompas.com/read/2021/07/04/15125391/kementerian-kesehatan-terus-upayakan-dan-monitor-ketersediaan-oksigen?page=all. [Accessed 25 Agust 2021].

[5] M. Zed, Literature Research Methods, Jakarta: Yayasan Obor Indonesia, 2008.

[6] H. Darmadi, Educational Research Methods, Bandung: Alfabeta, 2011.

[7] B. U. I. K. Nur Hudha Wijaya, "Medical Gas Pressure Monitoring at Hospital Medical Gas Installations," *Medika Teknika,* 2019.

[8] V. N. N. D. Atika Hendryani, ". Design of Manifold with Pressure Controller for Automatic Exchange of Oxygen Gas Cylinders in Hospital," *TEKNIK,* 2021.

[9] B. G. I. A. K. Miasih, "Development of Oxygen Volume Monitoring as a Basis for Determining Tariffs with Real Time Equipped with IOT-Based Safety Regulators (Oxygen Volume as Tariff Determinants)," *Doctoral dissertation Poltekkes Kemenkes Surabaya,* 2020.

[10] B. Rahmat, "TELEMONITORING GAS MEDIS," *Prosiding Seminar Nasional Widya Husada 1,* 2019.

[11] A. Wijaya dan M. Rivai, "IoT-Based Irrigation System Monitoring and Control Using Banana PI," *Journal Teknik ITS,* 2018.

Information System for Disaster Mitigation Using Google Data Traffic

Handaru Jati[1], Muhammad Izzuddin Mahali[1], Satriyo Agung Dewanto[1]

[1,2,3]Electronics and Informatics Engineering of Education Department, Universitas Negeri Yogyakarta, Indonesia

E-mail: handaru@uny.ac.id, izzudin@uny.ac.id, satriyoad@uny.ac.id

Abstract. The Head of the Data Information Center and Public Relations of the National Disaster Management Agency predicts the potential for forest fires in 2018 will increase. The areas that are burned are border areas, both provincial/district/city borders because border areas are poorly supervised. This study aims to (1) Develop an Information System that can be used for Disaster Mitigation Using Google Data Traffic (2) Ensure that the developed Information System is suitable for use for Disaster Mitigation Using Google Data Traffic to obtain accurate traffic data. Google Data Traffic is a giant database used by the Google company to provide real-time traffic information through the Google Maps application. Google Maps uses multiple iterations of satellite cameras and uses Google Earth to map traffic including congestion information and the shortest possible path alternatives. The development method chosen is the Rational Unified Process (RUP). This research uses several methods in data collection including observation, interviews, and questionnaires. Interviews were conducted with prospective users of the application, namely firefighter drivers and people in disaster-prone areas. Questionnaires are used to provide a set of questions to respondents. In the determination test, the recall and precision values are calculated from the data generated by the developed system. While the function test uses descriptive data analysis techniques with the percentage feasibility formula. This research produces an application that can assist firefighters in carrying out their daily duties to be more effective and efficient and the test results on this application show that it is declared feasible in terms of efficiency, performance and compatibility.

1. Introduction

The Head of the Information Data Center and Public Relations of the National Disaster Management Agency predicts the potential for forest fires in 2018 will increase. The areas that are burned are border areas, both provincial/district/city borders because border areas are poorly supervised. One of the primary causes of deforestation in Indonesia is fire. [1]. Forest and peat fires occur almost every year in Indonesia, with some fires causing significant damage. For example, the 1997/1998 fire incident, which is regarded as Indonesia's worst forest fire, destroyed more than 11 million ha of Indonesian forest. [2].

The number of hotspots causing forest and land fires in several parts of Indonesia has increased. According to a report by the National Disaster Management Agency (BNPB), based on satellite monitoring, the increasing number of hotspots only occurred within a week. Initially on July 27, there were only 173 points, and then it jumped to 239 points until Sunday, July 30, 2017. Six Sumatera and Kalimantan provinces (South Sumatera, Jambi, Riau, West Kalimantan, Central Kalimantan, and South Kalimantan) experienced extremely poor air quality as a result of the haze, with PM10 concentrations exceeding 500 mg/m3 recorded at several measuring stations. [3]

Land and forest fires have had a wide-ranging impact, affecting not only transportation and health, but also wildlife habitat. Several Indonesian endemic, endangered species, such as the orangutan and the Sumatran tiger, are forest dependent; thus, forest degradation due to fires directly threatens their existence and indirectly contributes to their population decline. [4]. On the other hand, Information Systems in today's modern era have a very important role in presenting information and become the backbone of almost all companies and organizations. Information systems are interconnected components that work together to collect, process, store, and disseminate information in an organization to support decision making, coordination, control, analysis, and visualization. [5]. For many researchers and engineers, the use of sensor networks to monitor natural hazards such as wildfires has become a special research topic. [6]. Of course, along with the advancement of internet technology, the information systems used today are computer-based. In addition, smartphone technology is also growing rapidly. So that people can access information systems not only through computers, but also through smartphones whose marketing is currently mushrooming in all circles of society. The advantages of smartphones compared to ordinary cellphones are that they have an operating system, web and internet applications, touch screens, and message processing features. Meanwhile, in Indonesia, the statistics of information and communication technology (ICT) adoption are more than promising. Internet penetration increased significantly over the last three years, from 15% in 2014 to 51% in 2017, making Indonesia the country with the highest growth in Internet users, more than five times the global average between 2016 and 2017.[7]

The role of Android phones is certainly very important in today's daily life, both at home and when we work outside the home. The scope of location when using an Android phone is also wide, it can be at home, office or outside the office. Various kinds of features are offered in android mobile devices. Applications that use the internet as a navigation base are also widely developed by service providers with the aim of making it easier for every driver to find their way with the help of digital maps. Various kinds of features are also embedded in internet-based navigation applications, for example features to determine the level of congestion on a road. This feature can be used by all drivers to avoid getting stuck in traffic jams.

Meanwhile, to obtain accurate traffic data, the research team intends to use Google Data Traffic. Google Data Traffic is a giant database used by the Google company to provide Realtime traffic information through the Google Maps application. Google Maps uses multiple iterations of satellite cameras and uses Google Earth to map traffic including congestion information and the shortest possible path alternatives. The traffic flow we see on Google Maps is a very accurate real-time display of the number of Android phones currently available. All traffic data including roads, buildings, restaurants, motorcyclists and so on will be stored in Google Traffic Data.

IoT is defined by the International Telecommunication Union as a global infrastructure for the information society enabling advanced services by interconnecting things based on existing and evolving interoperable information and communication technologies [8]. There is a lot of interest in developing disaster management systems that can save lives and property while reducing the need for costly economic investments. In order to build an effective monitoring infrastructure, information must be gathered from a variety of sources. IoT technology has seen significant success in this context. [9]. In recent years, innovative real-time monitoring and disaster warning systems have emerged, based on the emerging IoT paradigm, in which things (sensors) are globally interconnected. WSNs have been widely used as part of the Internet of Things to monitor natural disasters in remote and inaccessible areas. [10].

Departing from the various backgrounds above, the research team felt the need to build an information system that can be used for disaster mitigation, especially fire disasters. This information system can be accessed through computers, laptops, and of course Android-based smartphones. Fire victims can immediately send requests for assistance through this system, so that their position can be immediately tracked by the system which will then be followed by the process of picking up and evacuating firefighters. This information system works by applying IoT technology. It is hoped that this system can help people who need emergency assistance and of course firefighters can also be more alert in picking up fire victims.

2. Discussion

2.1. Inception Stage

At this stage, observation and interview techniques were used. Researchers conducted interviews with several residents affected by the fire disaster in the Daerah Istimewa Yogyakarta and surrounding areas. The results of interviews with the following components:

1. Early report to related unit / fire service
   Reports from related parties in this case are fire service units, there are many cases of fires that occur but lack of fast information, so that fires cannot be handled properly. Meanwhile, this is only limited to using cable telephones that are used to related parties.
2. The length of time for firefighters to arrive at the location of the fire disaster
   The duration of the fire department is constrained because of the large number of motorized vehicles on the road, thus preventing fire engines from reaching the fire destination.
3. Stuck in a traffic light
   Due to the lack of coordination between related parties and the police, traffic light jams often occur which causes fire trucks to not arrive at the location of the fire on time.
4. Map of fire hazard
   The map will be there automatically showing where the fire occurred as well as showing the closest route to the fire scene so as to make time shorter and fires can be extinguished quickly. The results of observations and interviews are needs analysis in the form of specifications needed in the development of Fire Disaster Mitigation Applications.

2.2. Elaboration Stage

Figure 1 depicts how Google Traffic data was developed in detail. The Disaster Mitigation Information System Development System makes use of Backend Service Firebase services such as Realtime Database, Authentication, REST Support, Data Analysis, and SaaS Services [11]. The use of Firebase allows communication between the User and the fire department. The user interacts with the Disaster Mitigation Application system using an application on a smart phone. This application has several main features including (1) SignUp/Login (2) User verification (3) Fire disaster report (4) Upload Photos (5) Upload Location using GPS. Users can use this application together. The application will send updated vehicle location data to Firebase using an internet connection and a server that accommodates calculations in the cloud data by using the GPS found on the Smartphone [12].

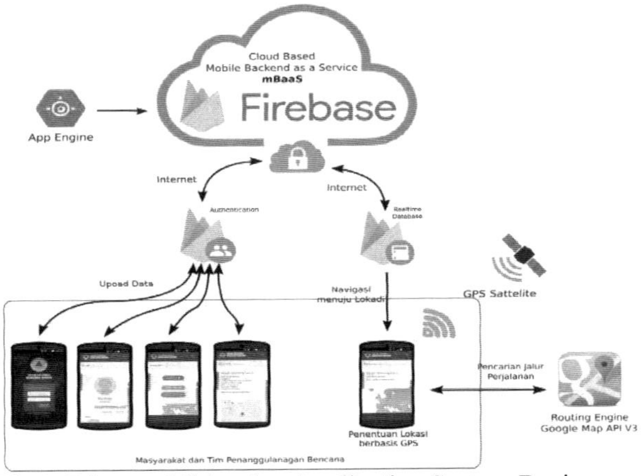

Fig 1. Disaster Mitigation Application System Design

2.3. Construction

2.3.1. Firebase

The use of Firebase as a Mobile Backend as a Service is a step that helps researchers. Firebase as a service that provides various services in one place. There are 2 firebase services used in this research, namely Realtime Database and Authentication. The Firebase Realtime Database which is used as a storage place for all data will be accessed by the user through the Fire Disaster Mitigation application and will be accessed by all application users together as shown in figure 2.

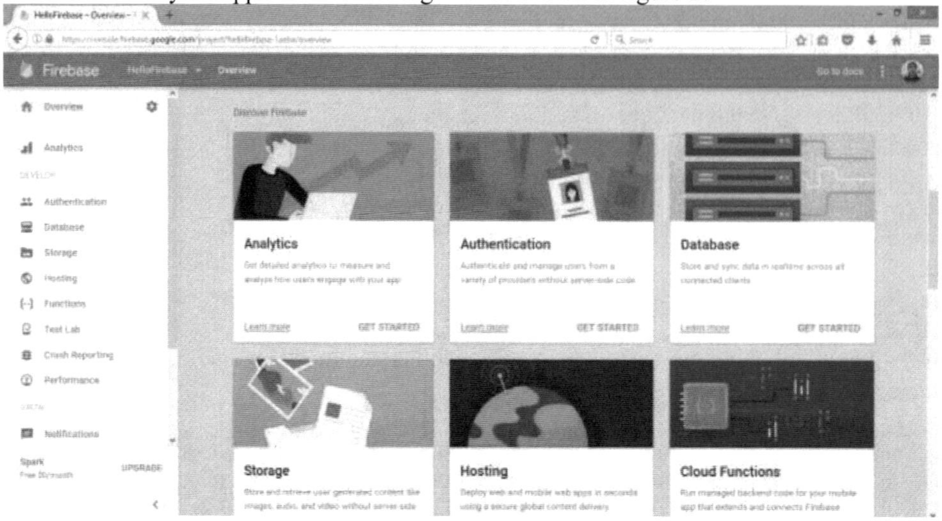

Figure 2. Dashboard Firebase

Firebase communication using android utilizes authentication and realtime database. In the development of Android-based Fire Disaster Mitigation, the firebase configuration is centered on one configuration file in the form of a google-services.json configuration file which must be downloaded on the firebase server and included in the android project. By using this configuration file, Smart Traffic Apps applications can interact with Firebase Services both Authentication and Realtime Database.

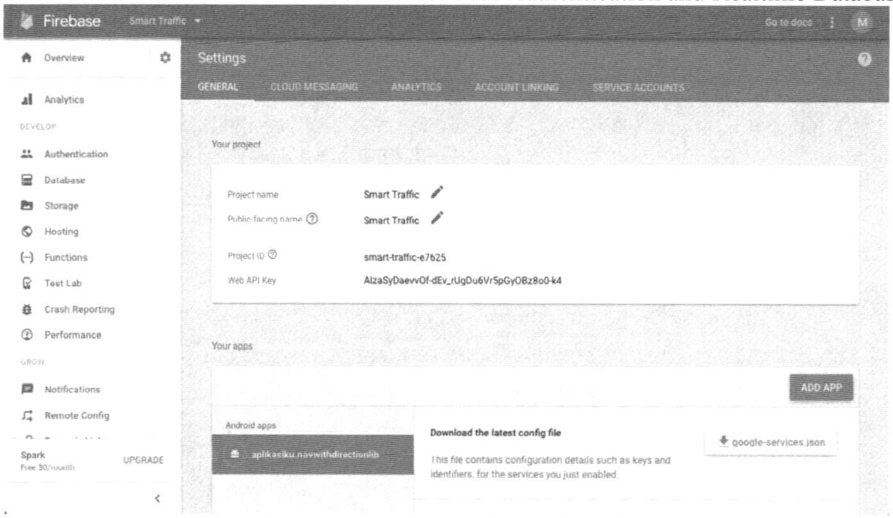

Figure 3. Firebase Configuration File

2.3.2. Android Application

Android-based Disaster Mitigation application was developed by using Android Studio software. In the development of Disaster Mitigation application that are connected to Firebase authentication and realtime databases function as Mobile Backend as a Service. The authentication service on firebase is used for account registration and user management systems. This service will handle user registration data and is used for system login. The smart traffic light application features an account registration and login system.

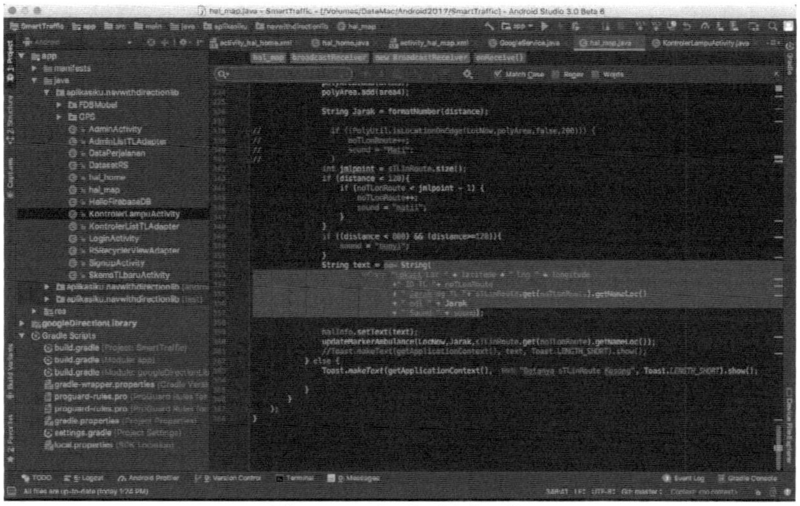

Figure . Android Studio

Realtime Database service as data storage base for communication bridge with smart sensor. The realtime database on firebase has a data structure in JSON (Java Script Object Notation) format as shown in Figure .

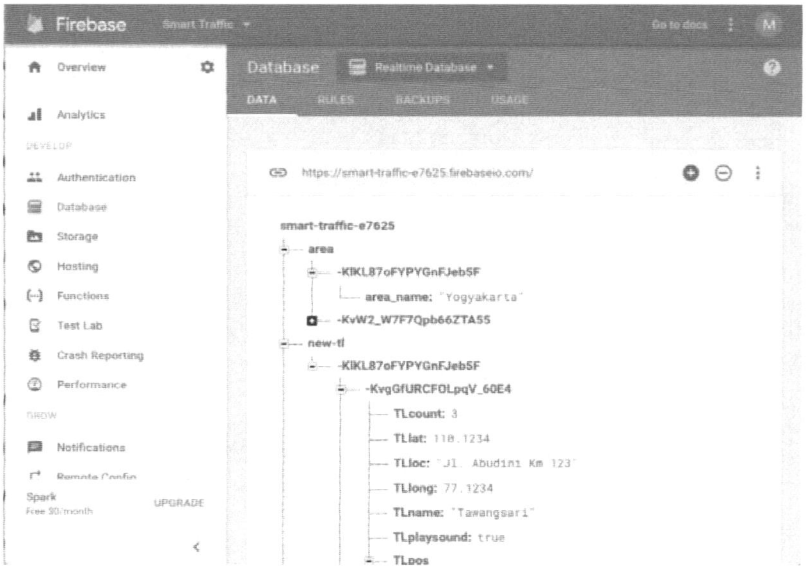

Figure . Firebase Database

The Disaster Mitigation application uses the digital map feature to search for travel routes. The digital map used is the Google Maps API V2 service which already supports programming in Android Studio. Android Studio uses the library 'com.google.maps.android:android-maps-utils:0.4+' to access services in Google Map API V2. The features used in this library include:

1. Travel Route Search
2. Determination of the ordinance of the travel route
3. Calculations rarely use point coordinates
4. Making a Marker Map
5. Displaying Google Map in Application

The Google Map API V2 is used by the Disaster Mitigation application to access Google digital maps via KEY. The key will be obtained after we register through a Google developer account and access the Google Developer Console page as shown in Figure .

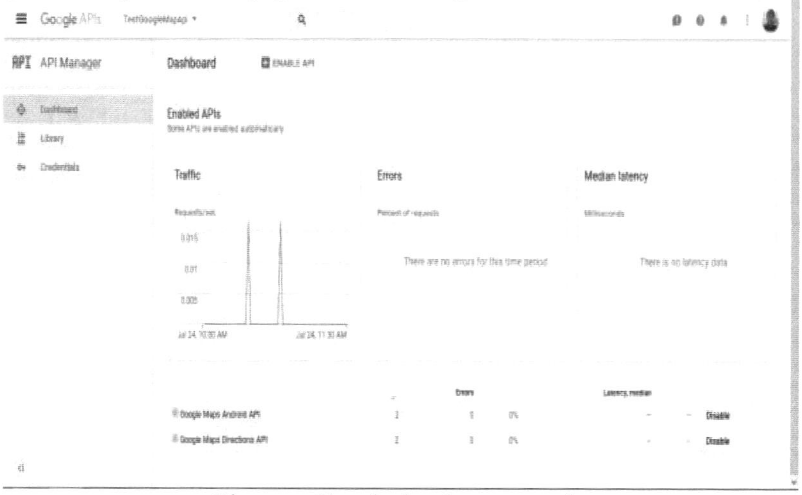

Figure . Google developer console

Communication between The Disaster Mitigation application and Google Map API V2 uses a retrofit library. This library is used to read JSON data and parse information from JSON data. Firebase's Realtime database also uses the JSON format for its data storage.

2.3.3. *Transition*

The development and testing stages are included in the transition stage. The Disaster Mitigation android application, also known as the Disaster Mitigation application, was created with the help of the Android Studio software and the Java programming language. The Disaster Mitigation application has the following features:

1. SignUp, Login and Verify
2. Filling in and Details of User Profile info
3. Latest disaster info
4. Disaster Info Details
5. Disaster data input form
6. Photo upload form
7. Location search form using GPS

2.3.4. *Application*

Disaster Mitigation has a service for registering user accounts, this aims to make it easier to add application users. User account data and stored in MbaaS Authentication Firebase The user must always be connected to the internet when using the Fire Disaster Mitigation application. When the Disaster Mitigation application is launched for the first time, the user is prompted to enter his or her login username and password, as shown in Figure . If the user does not have an account, the user can register through the account registration shown in the image below.

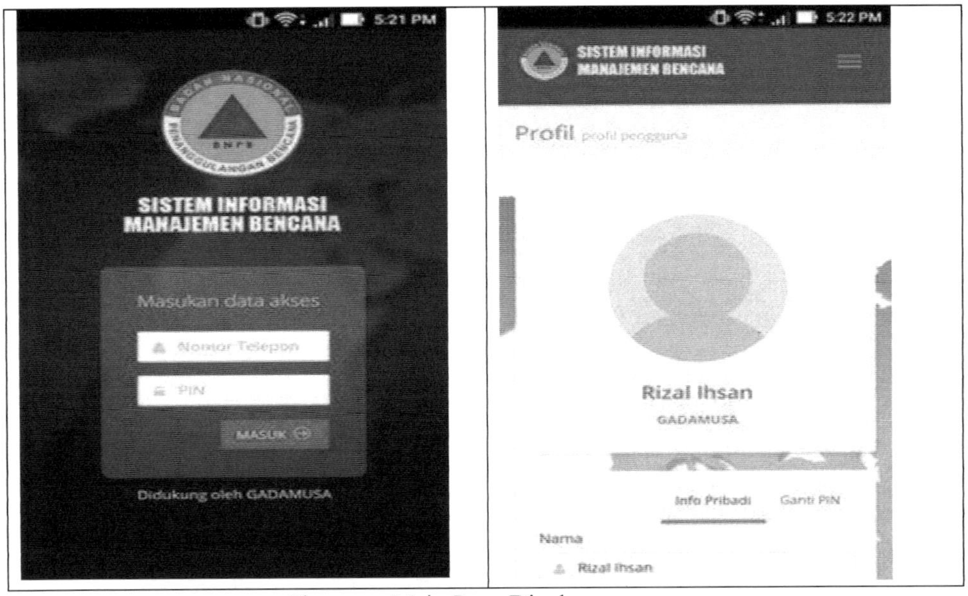

Figure . Main Page Display

The main page of the Disaster Management Information System contains information on all user data required by the operator, such as name, address, email, mobile number, and the current position of the user. The user's location will always be monitored by the operator through the facilities provided by GPS. In this study, the researcher took a sample of user data from various location points to determine how fast the information was from the user to the operator.

The next page contains information about disasters that will occur in the near future, as shown in Figure 9, there was a flash flood in the city of Garut. The news is linked by software that is connected by the BMKG (Meteorology and Geophysics Agency) from all over Indonesia, the data is displayed on the page randomly.

The user profile on the next page will be displayed automatically if the user has been previously verified by the relevant operator. Disaster Management Information System is a system that combines Android Application with Hardware Embedded System. The hardware embedded system referred to in this study is a heat sensor. This application has one feature to manually configure into sensor hardware tools. This facility is needed to check system performance so that everything runs smoothly so that the time needed for firefighters to reach the scene can be faster and more precise. The system also provides a menu to select the location of the fire sensor in the database list. Settings regarding configuration can also be done by selecting manual or automatic configuration. In the "disaster details" menu, the application provides details related to the disaster such as the type of disaster, the location of the disaster, and even the level of the disaster. The data displayed is very useful for users to estimate or just knowledge if at any time a similar disaster occurs that can be experienced by users anywhere and anytime.

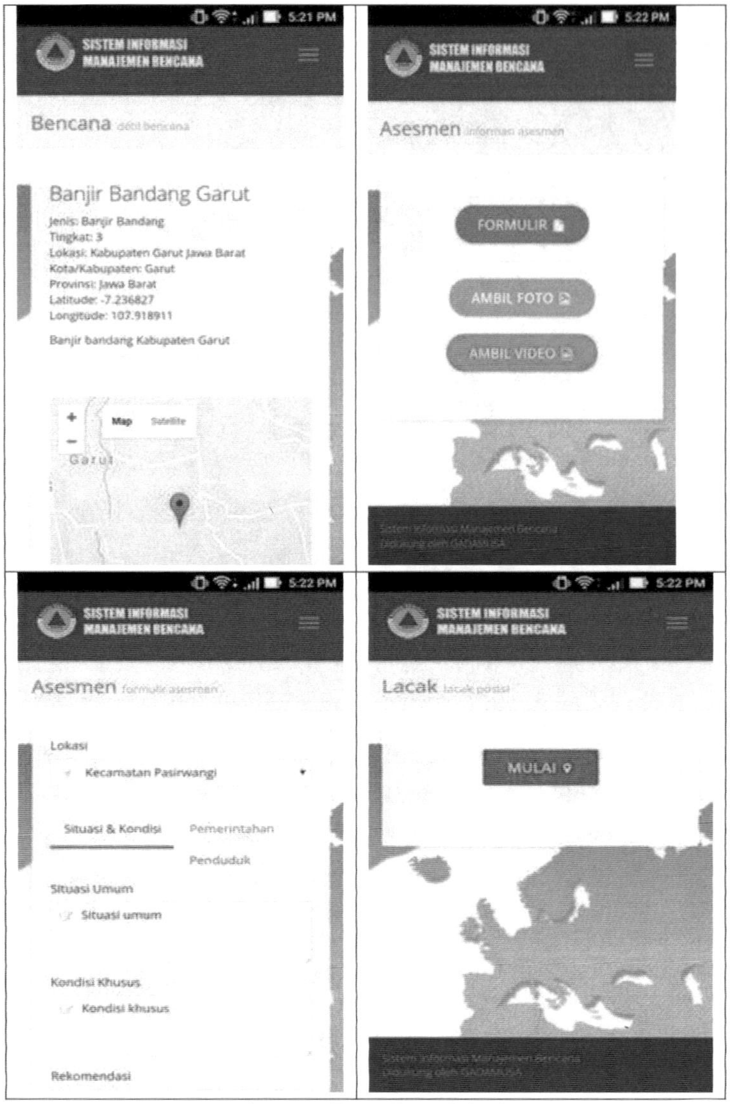

Figure 8. Form Page Display

On the "form" page, there are several fields that must be filled in, namely location, current situation and conditions, general situation, special conditions and recommendations. These data are used by operators to make reports that occur to superiors which will then be used to determine what actions will be taken by the disaster management agency on the basis of reports that occur.

On the "Track" page, the operator will find the position of the reporter using the GPS device that has been installed in the application. This page is very important to determine the coordinates of the complainant as well as to find the closest way to a fire or other disaster.

This disaster mitigation application provides a facility to see the travel route that will be passed by firefighters in the closest route between the fire department to the fire scene. The algorithm in this disaster mitigation system makes it possible to send data to the nearest traffic light so that the green light will adjust to be passed by the fire department. This system, in addition to speeding up the fire extinguisher to the scene of a fire, is also designed to report incidents that are catastrophic in nature. The journey of a fire engine has different characteristics and levels of density. The characteristic in question is when the path in the direction of the fire engine gets a turn of the red traffic light, can the fire engine exit the queue using the contra flow path or the fire engine cannot take the lane because there is a roadblock at that location. Based on testing the accuracy and speed of data transmission, it is found that the threshold distance of the map display in this disaster mitigation application will always update the location of different fire engines and are made to use the icon. The distance between the fire engine and the nearest traffic light will be shown on the icon when the application is used in an emergency as shown in Figure 9.

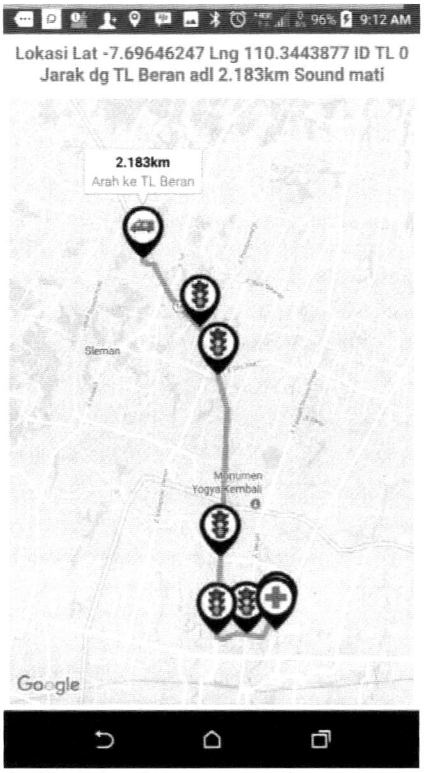

Figure 9. Firefighter's journey to the scene of the fire

3. Testing

Testing consists of 2 kinds, namely:

3.1. Performance efficiency test

This test uses an android application in the form of test android. This application will display the memory usage that will be used by the device. If a memory leak occurs, the application will display information related to what is needed.

3.2. Compatibility

This test is carried out by direct observation at the stage of installing or installing and in running the application on an Android screen on a different screen size.

The results of these two tests are presented in table 1 below

Table 1. Performance of Application

No	CPU	Memory	Information	Compatibility
1	Samsung J7 Pro	110	Usage does not cause memory leak	Success
2	LG Nexus 6	153	Usage does not cause memory leak	Success

3	Sony Experia Z2	131	Usage does not cause memory leak	Success
4	Nokia 5	145	Usage does not cause memory leak	Success

From the two tests, it can be concluded that the test can be concluded that it is feasible to be used as a disaster application to monitor and even cope based on data related to disaster management related to the accuracy and speed of data communication.

Based on the results of research and discussion, this research can be concluded as follows: This research produces an application that can assist firefighters in carrying out their daily duties to be more effective and efficient by using the latest technological developments and also to reduce the number of victims who fall due to fire disasters. The application is built using the Rational Unified Process (RUP) development method, where the risks and errors found will be corrected in several iterations so as to produce a good architecture and high-quality applications. RUP consists of several stages, namely Inception, Elaboration, Construction, and Transition. At each stage in the RUP, iterations of the business modeling, requirements, analysis & design, implementation, test, deployment, configuration & change management, project management, and environment processes are carried out.

The test results on this application show that it is declared feasible in terms of efficiency, performance and compatibility. All memory data shows above the normal threshold of 100.

4. References

[1] D. Tsujino, R., Yumoto, T., Kitamura, S., Djamaluddin, I., & Darnaedi, "History of forest loss and degradation in Indonesia," *Land use policy*, vol. 57, pp. 335–347, 2016.

[2] A. P. Tacconi, L., & Vayda, "Slash and burn and fires in Indonesia: A comment," *A comment. Ecol. Econ.*, vol. 56, no. 1, pp. 1–4, 2006.

[3] D. V. Crippa, P., Castruccio, S., Archer-Nicholls, S., Lebron, G. B., Kuwata, M., Thota, A., & Spracklen, "Population exposure to hazardous air quality due to the 2015 fires in Equatorial Asia," *Sci. Rep.*, vol. 6, no. 1, pp. 1–9, 2016.

[4] A. T. Linkie, M., Martyr, D. J., Holden, J., Yanuar, A., Hartana and N. Sugardjito, J., & Leader-Williams, "Habitat destruction and poaching threaten the Sumatran tiger in Kerinci Seblat National Park, Sumatra.," *Oryx*, vol. 37, no. 1, pp. 41–48, 2003.

[5] P. Kenneth, C., LAUDON, L., & JANE, *Management Information Systems: Managing the Digital Firm*. PEARSON, 2019.

[6] C. T. Kung, H. Y., Hua, J. S., & Chen, "Drought forecast model and framework using wireless sensor networks," *J. Inf. Sci. Eng.*, vol. 22, no. 4, pp. 751–769, 2006.

[7] W. A. Social, "Digital in 2017 global overview. January," 2017. .

[8] L. Rose, K., Eldridge, S., & Chapin, "The internet of things: An overview," *internet Soc.*, vol. 80, pp. 1–50, 2015.

[9] G. Atzori, L., Iera, A., & Morabito, "The internet of things: A survey," *Comput. networks*, vol. 54(15), no. 15, pp. 2787–2805, 2010

[10] H. Chen, D., Liu, Z., Wang, L., Dou, M., Chen, J., & Li, "Natural disaster monitoring with wireless sensor networks: A case study of data-intensive applications upon low-cost scalable systems," *Mob. Networks Appl.*, vol. 18, no. 5, pp. 651–663, 2013.

[11] A. Daramas, S. Pattarakitsophon, K. Eiumtrakul, T. Tantidham, and N. Tamkittikhun, "HIVE: Home Automation System for Intrusion Detection," in *Proceedings of the 2016 5th ICT International Student Project Conference, ICT-ISPC 2016*, 2016, pp. 101–104, doi: 10.1109/ICT-ISPC.2016.7519246.

[12] B. B. P. Rao, P. Saluia, N. Sharma, a Mittal, and S. V Sharma, "Cloud computing for Internet of

Things & sensing based applications," *Sens. Technol. (ICST), 2012 Sixth Int. Conf.*, pp. 374–380, 2012, doi: 10.1109/ICSensT.2012.6461705.

Journal of Physics: Conference Series **2111** (2021) 012018 doi:10.1088/1742-6596/2111/1/012018

Experimental Analysis of Heat Transfer on the Tubular Tube Heater with a Variable of Twisted Tape Inserts and Wire Coil to Prevent the Icing Contaminated Tailplane Stall

G Marausna[1], F S Reza[1], and F Jayadi[1]
[1]Sekolah Tinggi Teknologi Kedirgantaraan, DI. Yogyakarta, Indonesia

Email: gaguk.marausna@sttkd.ac.id

Abstract. An anti-icing system has the purpose of protecting the leading edge of the tailplane from contamination during aircraft flight. An anti-icing system in a turboprop aircraft employs heaters that use electrical energy as their power source or heat generated by the bleed air. Protecting the tailplane from contamination is preventing the aircraft stall from occurring that triggers dangerous flight conditions. The aluminum prototype tailplane is assembled with a variable of twisted tape insert and wire coil. The twisted tape insert comes in three different geometries with twist ratio T3 = 9.3; T4 = 7; and T5 = 5.6, as well as a wire coil with fixed geometry. This study shows the best heat transfer rate occurs in T3 with a value of 33.90 W. The consequence of this condition is an decrease in pressure drop that occurs. Twisted 3 has the greatest pressure drop when compared to other geometry, with an average value of 4.72 Pa.

1. Introduction

An aircraft is a vehicle in the air that requires stability during flight. The heater on the tailplane is important for this purpose because it prevents icing formation caused by low temperature exposure. Because of the icing effect of changing the shape of the leading edge shape, an icing-forming or contaminated tailplane is dangerous. Airflow around the tailplane becomes turbulent and triggers flow separation to bring the aircraft into a stall condition. Aircraft at supercooled condition up to -40 °C induced liquid water droplet and formation of ice on leading-edge of the wing that can prevent by the electro-thermal anti-icing[1]. Another technic to prevent the formation of ice is the anti-icing system is the piccolo tube anti-icing is commonly used in commercial aircraft [2]. By adding a heater on the leading edge of the tailplane, icing formation can be prevented. Airflow around the tailplane becomes smooth and prevents separation at a low angle of attack.

The heater is operated by transferring high energy from a source to the air on the tailplane surface, especially on the leading edge of the tailplane. The heater can employ sources from bleed air, exhaust gas, or electrical device to produce heat sources. The tubular tube has a simple geometry that accommodates the purpose of a pipeline that contains fluid at a higher temperature than the surrounding environment for heaters that use bleed air. Due to easy manufacturing, the tubular tube is the best choice as a heater. The problem came from heat transfer tubular tubes have a tendency to low heat transfer. Many methods can be applied to enhance heat transfer, such as adding a flow disturber. The flow was disrupted when any shape such as a twisted insert or wire coil was added to the tubular tube.

Maradiya reported many heat transfer enhancement techniques were developed such as twisted tape insert, helical insert, winglet twisted tape, wire coil insert, perforation, fin, rib and grove, roughness, duct shape, and nanofluid [3]. This method results in material and cost reduction using in device operational by increasing heat transfer surface or modify flow into the turbulent region. The twisted insert can be applied with one or more as reported in He's investigation [4]. From the thermal-fluid dynamics viewpoint, more than one twisted insert applied is not favorable due to one twist has a performance efficiency coefficient larger than two twisted inserts. Patil reported that an experimental study of laminar flow inside a square duct resulted in a low heat transfer enhancement due to a change in twist ratio [5].

2. Experimental setup

A twisted plate with dimensions of 420 mm length, 15 mm width, and 0.75 mm thickness. Twist inserted into a wire coil with a diameter 2 mm, pitch 10 mm, and tubular tube with a diameter of 19 mm and 435 mm length as shown in figure 1. In previous research, Patil conducted a small twist ratio of 2.66, 3.55, and 4.01 were connected in series. The larger twist ratio with uniform ratio needed to be evaluated in this research to conduct the heat transfer characteristic at a larger twist ratio in transitional flow. The ratio between length and width (twist ratio) for each geometric is T3 = 9.3; T4 = 7; and T5 = 5.6.

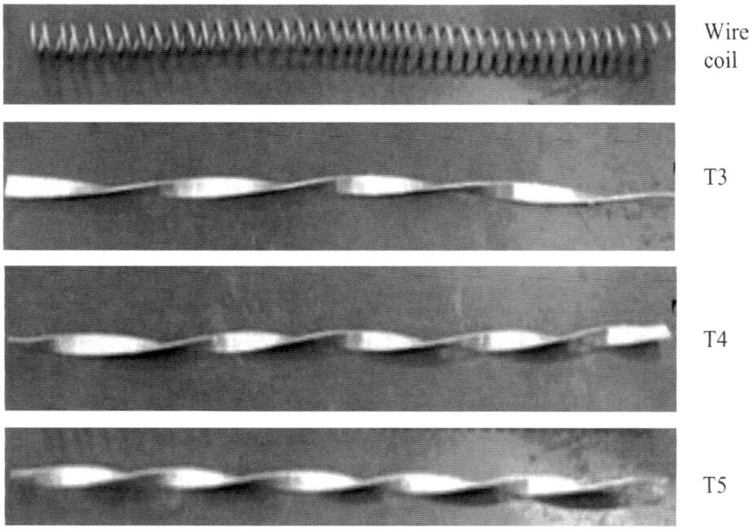

Figure 1. Three twisted tape geometries with a wire coil inserted on the outside.

Figure 2 shows a schematic diagram of an apparatus with a tubular tube inserted into a prototype wing made of a 0.75 mm aluminum plate. The wing has a chord of 580 mm, 140 mm thickness, 400 mm width. To prevent leading-edge temperature drop due to cooling by water spray from the nozzle at low temperature, hot air from the heat gun is transferred to a tubular tube as a heater. Water is sprayed on the leading edge to simulate a drop in temperature caused by the wing being exposed to air during flight. Experimental results were obtained by setting the heat gun temperature within the range of 50 °C, 60 °C, 70 °C, 80 °C, 90 °C, 100 °C as inlet temperature for each twisted geometric tested.

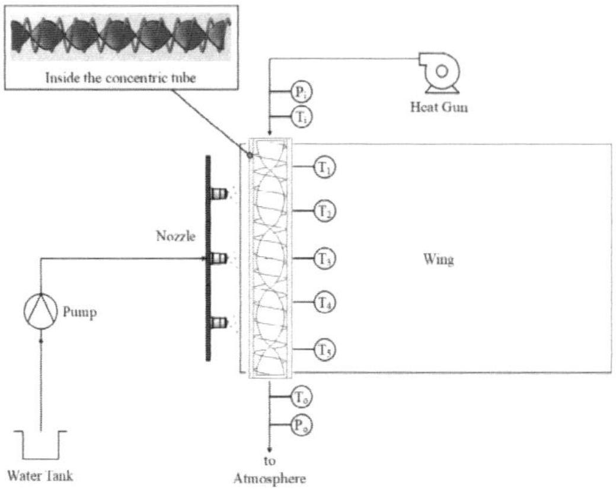

Figure 2. Schematic diagram of test apparatus of heat transfer enhancement by using twisted tape insert and wire coil.

3. Experimental calculations

In this study, twisted tape insert and wire coil were investigated using three different twist ratios to evaluated thermal characteristic tubular heater in a turbulent flow region. Thermal characteristics of a heater are determined by Reynold number, Nusselt number, convective heat transfer, heat transfer rate, friction factor, and pressure drop using equation (1-6) below [6]. A dimensionless quality is called Reynolds number to determine the flow regime of the fluid:

$$Re = \frac{\rho V D}{\mu} \tag{1}$$

where ρ is the density, V is the velocity, D is the diameter, and μ is dynamic viscosity. Dimensionless convection heat transfer coefficient is called Nusselt number for turbulent flow in tube from Dittus-Boelter equation with approach turbulent flow assumption:

$$Nu = 0.23\, Re^{0.8}\, Pr^{0.4} \tag{2}$$

where Pr is Prandtl number. The convective heat transfer coefficient calculated from Nusselt number:

$$h = \frac{Nu\, k}{D} \tag{3}$$

where k is thermal conductivity of working fluid. Heat transfer rate must be supplied by flowing fluid:

$$\dot{Q} = \dot{m}\, Cp\, \Delta T \tag{4}$$

where $\dot{m}$ is mass flow rate of working fluid, Cp is specific heat of working fluid, and ΔT is temperatur between diffferent inlet and outlet. Pressure drop is calculated from friction factor below which is valid for $2300 < Re < 4500$:

$$f = 3.03 \text{ x } 10^{-12}\, Re^3 - 3.67 \text{ x } 10^{-8}\, Re^2 + 1.46 \text{ x } 10^{-4}\, Re - 0.151 \tag{5}$$

$$\Delta P = f \frac{L}{D} \rho \frac{V^2}{2} \tag{6}$$

where f is friction factor, ΔP is pressure drop, and L is length of tube.

4. Results and discussion

The tubular tube is a simple geometry that can be used for the heater to prevent leading edge icing. Flow inside the tubular tube has the best heat transfer performance while in a turbulent region. To accomplish this, a twisted insert and wire recoil are installed in a tubular tube and tested at various temperatures to determine the Reynolds number. The Reynolds number is an important parameter to determine flow characteristics in heat transfer. Figure 3 shows T3 has the largest Reynolds number compared to T4 and T5. It is beneficial to enhance flow to the turbulent regime by reducing the number of twists from 5 to 3, causing inertia force more dominance than viscous force. Reducing Reynold number due to adding twist number followed by reducing Nusselt number as shown in figure 4. A Nusselt number greater than one indicated that convection rather than conduction dominated flow inside the tubular tube.

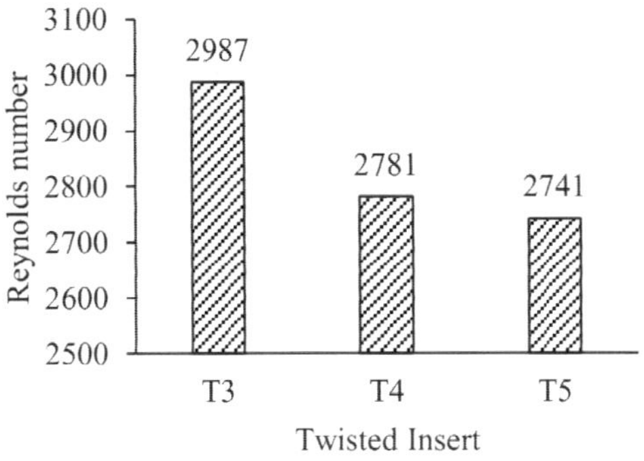

Figure 3. Effect twisted insert geometry to Reynolds number.

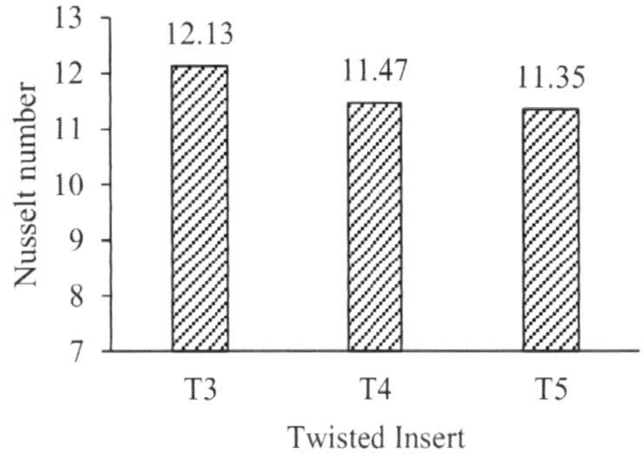

Figure 4. Effect twisted insert geometry to Nusselt number.

To increase heating on the leading-edge using T3 is a better choice than the other two due to the fact T3 has large heat transfer as shown in figure 5 and follows with convective heat transfer coefficient as shown in figure 6. It is shown that the Reynolds number affects heat transfer rate. Increasing Reynolds number will increase the heat transfer rate from the heater to the leading-edge during use to prevent icing forming.

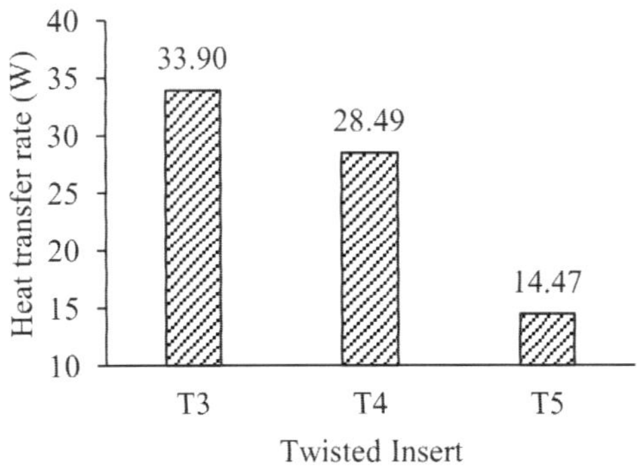

Figure 5. Effect twisted insert geometry to heat transfer rate.

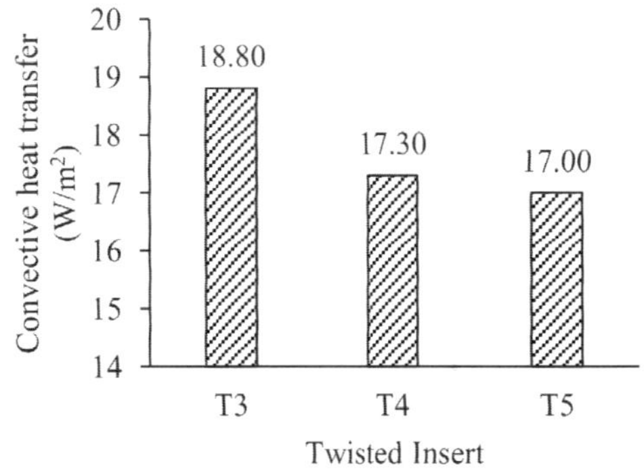

Figure 6. Effect twisted insert geometry to convective heat transfer coefficient.

The largest friction factor is about 7.1% has a significant effect on fluid flow followed by pressure drop as shown in figure 7-8. It came from a heater that was inserted into the 400 mm wide wing prototype in this study. Hot air flows inside the heater for a short period due to short piping for the heater is used. This study compared a heater with and without a wire coil related to Reynolds number and convective heat transfer. Figure 9-10 shown the wire coil affects reducing Reynold number and convective heat transfer coefficient. Reynolds number and convective heat transfer coefficient for the heater with and without a wire coil to Reynolds number have different tendencies. For the heater with a wire coil, increasing twist ratio results in increasing Reynolds number and convective heat transfer coefficient. For the heater without a wire coil, increasing twist ratio results in decreasing Reynolds number and convective heat transfer coefficient. For that reason, removing a wire coil is the best choice for increasing heat-transfer characteristics in this study.

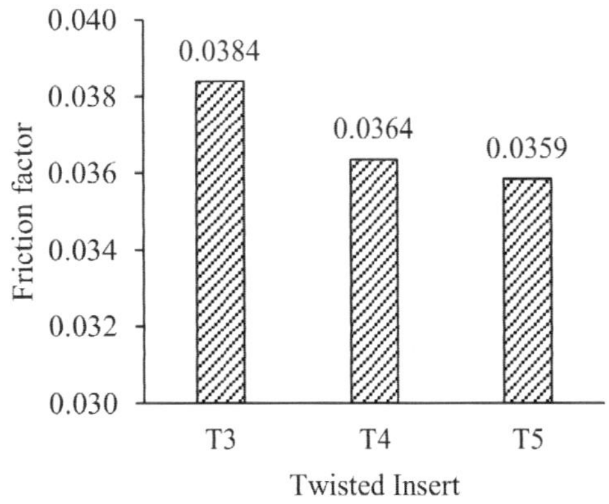

Figure 7. Effect twisted insert geometry to friction factor.

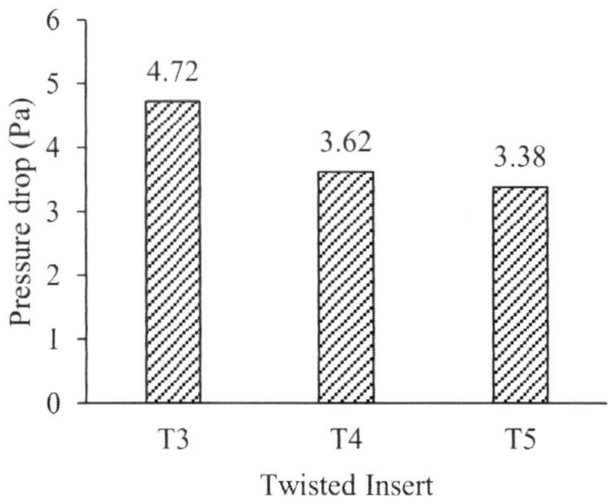

Figure 8. Effect twisted insert geometry to pressure drop.

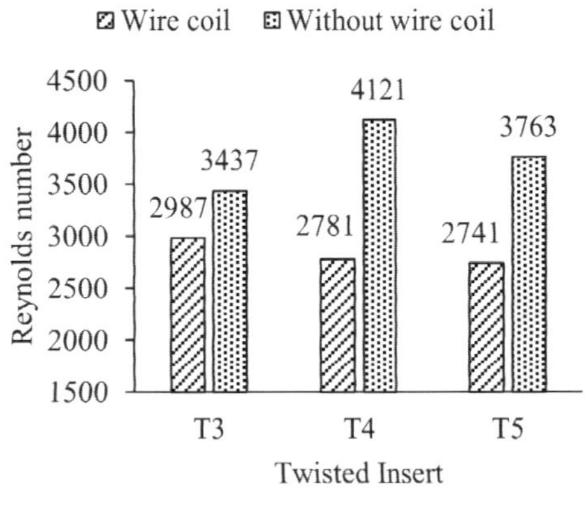

Figure 9. Comparison of a heater with and without a wire coil to Reynolds number.

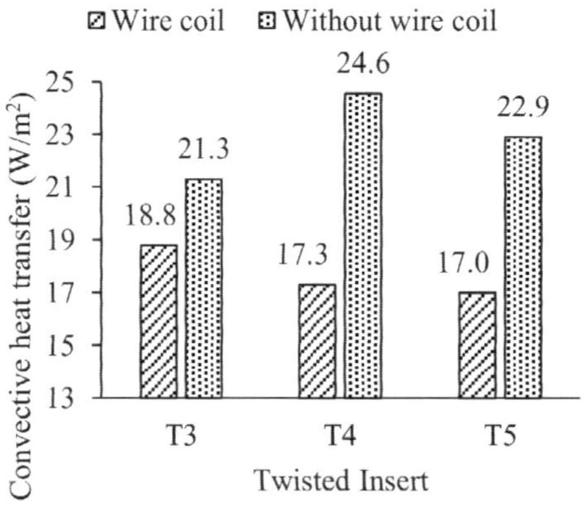

Figure 10. Comparison of a heater with and without a wire coil to convective heat transfer coefficient.

5. Conclusions

The purpose of this study to analyze perform of wire coil as flow disturbed combine twisted insert. Applied wire coil into twisted insert leads to better performance at high twist ratio where T3 has largest Reynolds number and convective heat transfer. T3 has the greatest pressure drop when compared to other geometry, with an average value of 4.72 Pa where its related to Reynolds number of the combination wire coil and twisted insert. Removing wire coil is more favorable due to increasing twist ratio where T4 has the largest Reynolds number and convective heat transfer in turbulent flow. In transitional flow, increasing twist ratio has pros to heat transfer characteristics, more heat transferred to leading-edge conducted to prevent the formation of ice. Future work to evaluate laminar and fully turbulent flow is needed to study the effect of the combination wire coil and twisted insert.

Nomenclature

Cp	specific heat (J/kg·K)
D	diameter (m)
f	friction factor
h	convective heat transfer coefficient (W/m^2·K)
L	length (m)
k	thermal conductivity (W/m·K)
Nu	Nusselt number
$\dot{m}$	mass flow rate (kg/s)
Pr	Prandtl number
$\dot{Q}$	heat transfer rate (W)
Re	Reynolds number
V	velocity (m/s)
ΔP	pressure drop (Pa)
ΔT	temperature different (°C)
μ	dynamic viscosity (kg/m·s)
ρ	density (kg/m^3)

References

[1] A. Carozza, F. Petrosino, and G. Mingione, "Numerical procedure for electro-thermal anti-icing system simulation coupling internal thermal analysis and external multi-physic code," IOP Conf. Ser. Mater. Sci. Eng., vol. 1024, p. 012010, Jan. 2021, doi: 10.1088/1757-899X/1024/1/012010.

[2] M. H. Y. Syed, M. A. Ismail, Q. Azam, P. Rajendran, and N. M. Mazlan, "Simulation study of the effect of anti-icing on the nacelle lip-skin material," IOP Conf. Ser. Mater. Sci. Eng., vol. 370, p. 012011, May 2018, doi: 10.1088/1757-899X/370/1/012011.

[3] C. Maradiya, J. Vadher, and R. Agarwal, "The heat transfer enhancement techniques and their thermal performance factor," *Beni-Suef Univ. J. Basic Appl. Sci.*, vol. 7, no. 1, pp. 1–21, 2018, doi: 10.1016/j.bjbas.2017.10.001.

[4] W. He, D. Toghraie, A. Lotfipour, F. Pourfattah, A. Karimipour, and M. Afrand, "Effect of twisted-tape inserts and nanofluid on flow field and heat transfer characteristics in a tube," *Int. Commun. Heat Mass Transf.*, vol. 110, p. 104440, Jan. 2020, doi: 10.1016/j.icheatmasstransfer.2019.104440.

[5] S. V. Patil and P. V. V. Babu, "Heat transfer and pressure drop studies through a square duct fitted with increasing and decreasing order of twisted tape," *Heat Transf. Eng.*, vol. 35, no. 14–15, pp. 1380–1387, Oct. 2014, doi: 10.1080/01457632.2013.877318.

[6] Y. A. Cengel and A. J. Ghajar, *Heat and mass transfer: Fundamentals & applications*, 5th ed. New York: McGraw-Hill Education, 2015.

Journal of Physics: Conference Series **2111** (2021) 012019 doi:10.1088/1742-6596/2111/1/012019

The Design of Mini CNC Assembly Simulator Based on Virtual Reality Toolkit (VRTK)

C A J Malik[1] and S Hadi[2]

[1,2] Master Education Program of Electrical Engineering, Faculty of Engineering, Universitas Negeri Yogyakarta , Indonesia

Email: choirulabdul.2020@student.uny.ac.id[1]

Abstract. Information and communication technology development today has provided the latest innovation in didactic world. One of such innovation is the usage of virtual reality (VR) as learning media. The application of virtual reality in learning environment started to be adopted because VR is considered more capable to explain learning material when compared with other learning material. Moreover, learning with VR has the capability for student to learn independently anytime and anywhere. One of VR utilization in learning environment is to use it in a Mini CNC machine assembly simulator. The learning media is meant to be used to improve student skill in assembling a Mini CNC machine independently. This research uses Research and Development method with *waterfall* model which has several stages which are communication, planning, modelling, construction, and deployment. However, this research only aims up to construction stage, because this research focuses on media functionality test. The media black box research has shown that the media is capable to function as expected.

1. Introduction

Education in higher education currently experiencing a big challenge. This is because technology advancement in higher education cause education in higher education must be capable to prepares student in accordance to recent development. Technology advancement today has reached the 4th era which known as 4.0 industrial era. 4.0 Industry era had been introduced for the first time in Germany during Hannover Fair in 2011 [1]. This Industry era aims to increase competitiveness between countries which utilized digital technology [2].

One of the uses of digital technology in didactic world is the use as learning media. The availability of learning media today is still limited in and tends to lack in development, especially for schools or higher education which situated in remote areas [3]. This limit of media usage causes the learning process to be less than optimal. This situation is getting worse with COVID-19 pandemic that makes learning must be done online and independently by student.

Due to that, learning media development must be done so that learning process can continue. One of the learning media which could accommodate that is learning media development with virtual technology like virtual reality which based on Virtual Reality Toolkit (VRTK). Learning media based on virtual technology has several benefits during online learning process nor independently. This is due to virtual technology is capable to introduce practical knowledge towards student without the need to be present in classroom [4] or in other words, media can be used anytime and anywhere by its user or student [11]. Moreover, virtual learning media can reduce the risk of work accident in practical learning and of course it also reduces expense [5], so that it is very appropriate to be used as student's independent learning media, especially for CNC Maintenance and Repairs Practice course. Virtual learning media application like simulator in afore mentioned course, could give more experience towards student on maintenance practice up to build a CNC, so that the purpose of leaning competence can be done even without the needs to be present in a classroom.

1.1. Virtual Reality Toolkit (VRTK)

Virtual Reality or abbreviated as VR is a term to describe virtual environment which can be manipulated by its user in real time, this virtual environment is the results of computer processing which projected in a device strapped in one head, a computer monitors or projection screen [6]. Whereas Virtual Reality Toolkit known as VRTK is a library that contains method and component that could work with Virtual Technology. VRTK library consists of numbers of solution or order, like movement in VR, interaction with object like touching and catching, and 2D and 3D control like button, lever, and another object [7]. VRTK can ease users in creating a game or simulator application because of built-in program in one library. One of program or feature in VRTK library is Snap Zone. Snap Zone has the principle to integrate an object with another object according to initial object placement setting. Because of that, VRTK is very compatible to be used to create an application or game with the type of simulation. Because of that, it can be concluded that virtual reality is a virtual environment produced by computer image which can be done by manipulating said image in accordance to user need. While VRTK itself is a library which contain methods and components to create virtual reality.

1.2. Mini CNC

CNC machine or the abbreviation of Computer Numerical Controlled is a machine that controlled by computer with numeric language in accordance to ISO standard [8], while a Mini CNC is a CNC machine that smaller in size. Today CNC machine has become manufacture tool which often used in industry. This is because CNC machine has higher accuracy in comparison with conventional machine in general [9]. Because of that, several schools and higher education today opened expertise program that has relation with CNC machine operation. However, as time goes by, not only CNC machine operating skill, but the skill to make and repair CNC machine is also opened in several schools or higher education even though it is only with a Mini CNC machine. Because of that, manufacture and maintenance simulation learning media of CNC machine must be developed in the future.

2. Research Method

This research uses the method of Research and Development with waterfall model developed by Pressman [10] which has several stages which are communication, planning, modelling, construction, and deployment. Waterfall model is started with communication stage which consist of observation and interview. Communication stage is useful to dig information that supports later development process. The next stage is planning done by planning time and flow of to be made learning media development process, so that media development can be controlled and does not need a long time to be done. Modeling stage is done by designing product in accordance to data which has been collected through the stages before, so that product that will be made will correspond with research needs. Construction stage is done by writing programming code, so that learning media can function in accordance to program code that has been made. Deployment stage is to find out the feedback of user toward learning media that has been made, so that the media is suitable with its user characteristic. However, deployment stage is not carried out in this research, because this research is only focuses on media functionality test only.

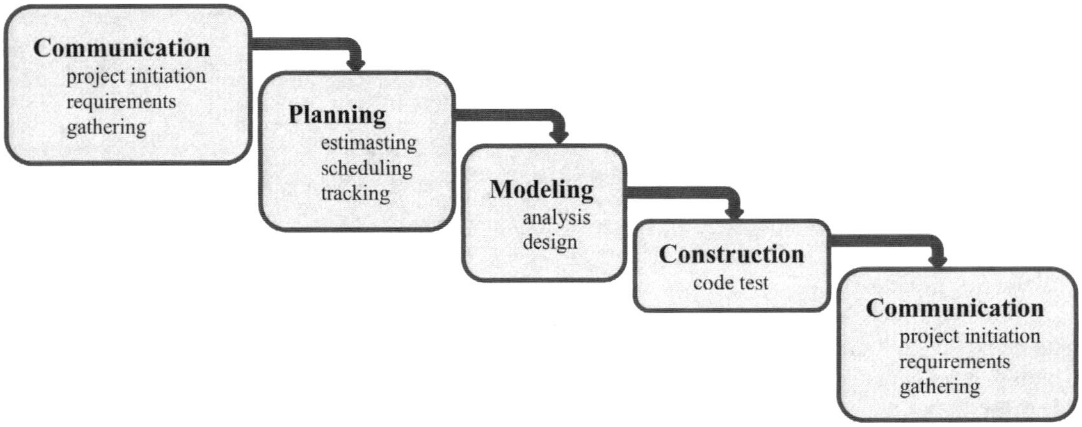

Figure 1. *Waterfall Scheme* [10]

3. Research Results and Discussion

This research is started with communication stage which consists of observation and interview. Observation is done by doing needed competency identification related to Mini CNC machine assembly. Competency Identification is done by studying existing document like lesson plan and other supporting document, while interview is done by asking several questions towards lecturer that teach CNC Maintenance and Reparation course in Electrical Engineering Department to get data which will be used in media development process.

Planning stage is done by determining which application will be used in developing media, like 3D design application for virtual reality development. 3D design application which will be used is Unity 3D. Furthermore, in planning stage preparation time of manufacture is also done so the media development does not waste a lot of time.

Modeling stage is started by designing program flow diagram to ease development process (Figure 2), then it is proceeded with designing the content of media by using 3D design application up to designing the interface of learning media in game engine application. Moreover, instrument design of black box test is also carried out in this stage. Designing stage of the media is done by creating 3D design of Mini CNC machine by using Blender application then it is converted into .FBX or .OBJ file type so that it can be processed in game engine application (Figure 3). After designing the content of media is done, the next activity is the modelling stage which design user interface of the media in accordance to existing 3D design (Figure 4).

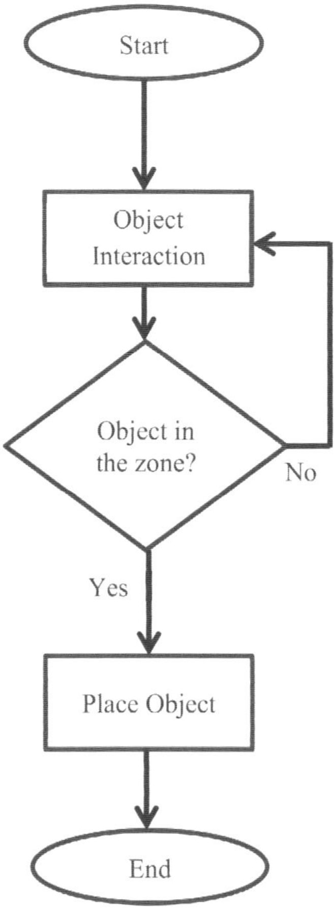

Figure 2. Programming Flowchart

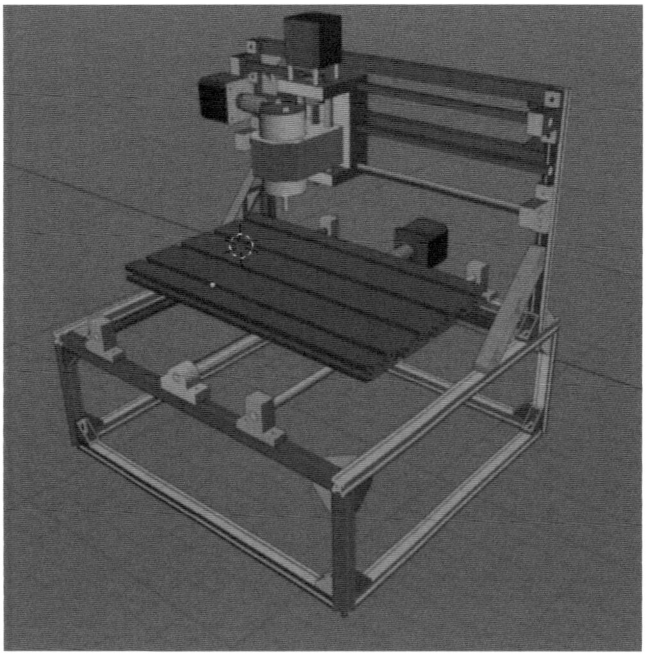

Figure 3. Mini CNC 3D Design

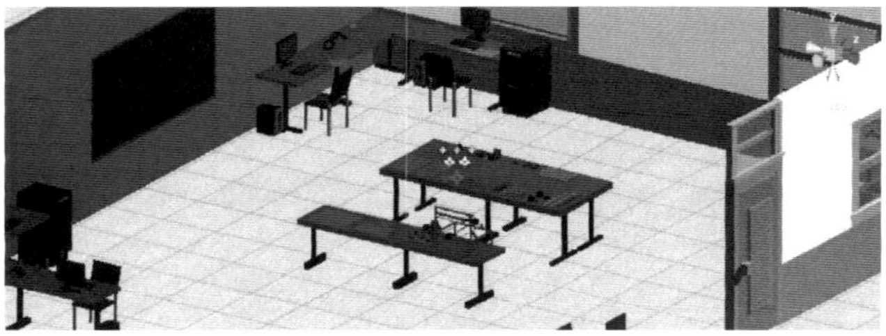

Figure 4. Media User Interface

Construction stage is done by programming component content of the media which has been made in modelling stage to be integrated with each other until an artificial environment (virtual reality) is created which will be used during Mini CNC machine assembly practice. Media programming is done by using Unity 3D application by utilizing Virtual Reality Toolkit (VRTK) feature. Programming is started by inputting "InteractableSnapZone" towards work shape (Figure 5). InteracableSnapZone function as laying place of object which will be integrated with another object. After the process of inputting InteractableSnapZone succeed, the next step is to change the shape of InteractableSnapZone to becomes certain object so that the laying area is easier to be identified (Figure 6), after that apply the same towards every object which will be integrated (Figure 7).

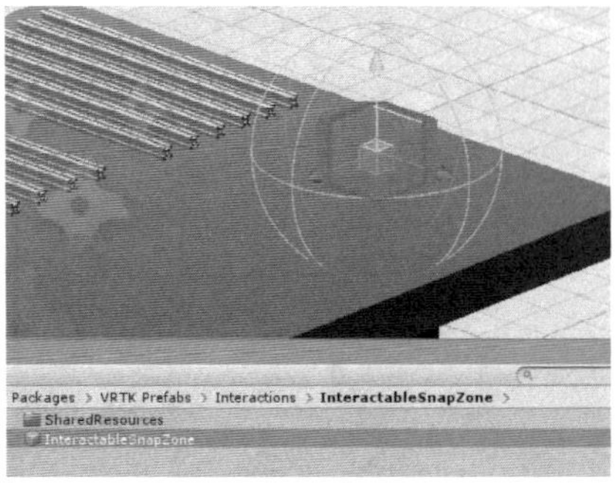

Figure 5. Insert InteractableSnapZone

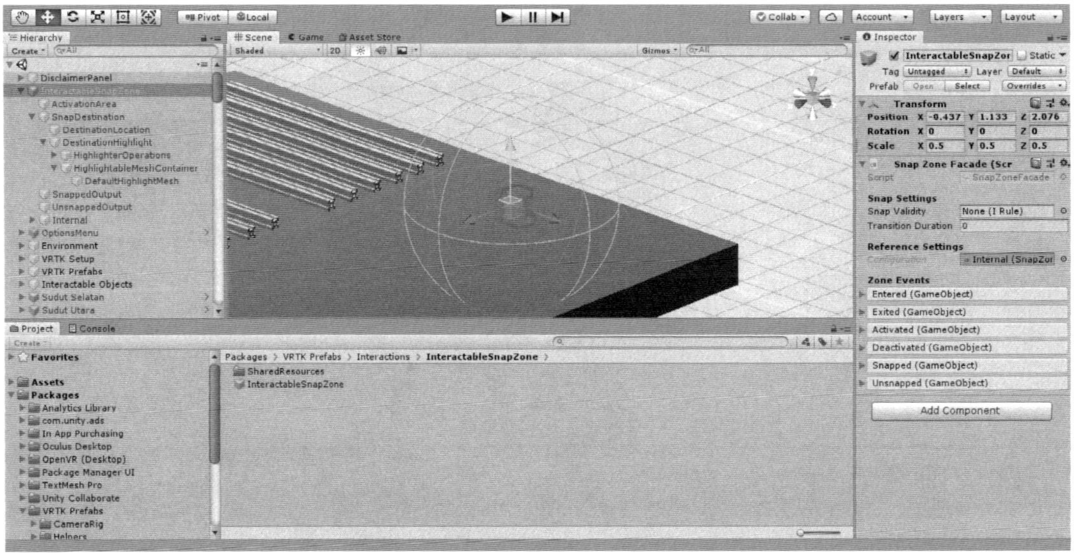

Figure 6. DefaultHighlightMesh Setting

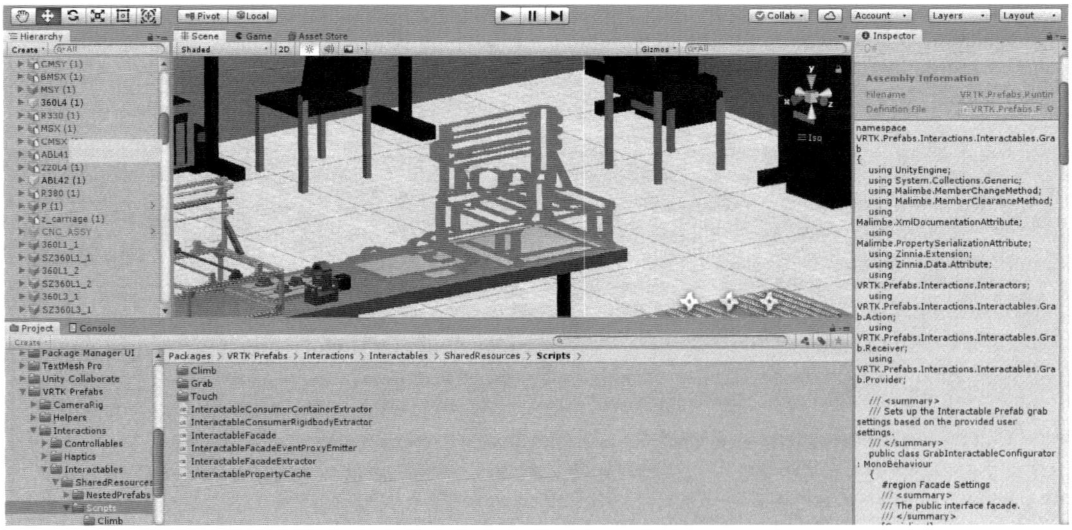

Figure 7. Virtual Reality Toolkit Final Setting

Besides that, in modelling stage a black box test is also done to understand the functionality of media that has been made. Black box functionality test has shown that the system is capable to be working as expected as shown on Table 1. According to Table 1. it can be seen that the media main screen view is working properly. Moreover, user's movement command and connectivity between media components can work properly.

Table 1. Black Box Test Result

Num-ber	Scenario	Expected Result	Result	
			Appro-priate	Not Appropriate
1.	Clicking RIG camera appearance	Media main screen is shown virtually	√	-
2.	Pressing W button on Keyboard	VR user will move forward	√	-
3.	Pressing S button on Keyboard	VR user will move backward	√	-
4.	Pressing D button on Keyboard	VR user will move to the right	√	-
5.	Pressing A button on Keyboard	VR user will move to the left	√	-
6.	Dragging the mouse to the right	VR user will turn right	√	-
7.	Dragging the mouse to the left	VR user will turn left	√	
8.	Dragging the mouse to the top	VR user will look up	√	-
9.	Dragging the mouse to the bottom	VR user will look down	√	-
10.	Clicking scroll on mouse	Activates teleport command	√	-
11.	Clicking Right button on mouse	Activates picking object command	√	-
12.	Putting object according to its place	Object can interact with each other	√	-

4. Conclusion
This research resulted in a learning media in the form of Mini CNC Assembly simulator. This media development is done by using waterfall model with four steps which are communication, planning, modelling, and construction. Media is developed by using Virtual Reality Toolkit or VRTK. VRTK is used to develop learning media, especially for movement command or connectivity between objects. Mini CNC Simulator learning media can be used in online learning or independent learning by student. This is proved by black box test towards media which has shown that every system that has been built could function appropriately. However, to find out feasibility and effectivity level of media, further research is needed to be done.

5. References
[1] Sawitri D. (2019). *Revolusi Industri 4.0. Big Data Menjawab Tantangan Revolusi Industri 4.0.* Jurnal Ilmiah Maksitek, Vol. 4, No. 3, p 1-9

[2] Prasetyo H & Sutopo W. (2018). *Industri 4.0: Telaah Klasifikasi Aspek dan Arah Perkembangan Riset.* Jurnal Teknik Industri, Vol. 13, No. 1, p 17-26

[3] E Prianto, dkk. (2019). *Mini CNC Design to Increase Students' Programming and Control CNC*

Competencies. Journal of Physics: Conference Series, Vol. 1413, No. 1, p 012006. doi:10.1088/1742-6596/1413/1/012006

[4] Sholihin M, Sari R.C., Yuniarti N, Ilyana S. (2020). *A New Way of Teaching Business Ethics: The Evaluation of Virtual Reality-Based Learning Media.* The International Journal of Management Education, Vol. 18, No. 3, p 100428

[5] Coles T.R., dkk. (2011). *The Role of Haptics in Medical Training Simulators: A Survey of the State of The Art.* IEEE Transactions On Haptics, Vol. 4, No. 1, p 51-66

[6] Mandal, S. (2013). *Brief Introduction of Virtual Reality & its Challenges*. International Journal of Scientific & Engineering Research, Vol. 4, No. 4, p 304–309.

[7] Saravas V, dkk. (2020). *The Development Of Educational Application With Virtual Reality Placing Objects System Using Snap Zone Technology.* The 11th IEEE International Conference on Dependable System, Servieces and Technologies, DOI: 10.1109/DESSERT50317.2020.9125023

[8] Amala M & Widyanto S.A. (2014). *Pengembangan Perangkat Lunak Sistem Operasi Mesin Milling CNC Trainer.* Jurnal Teknik Mesin S-1, Vol. 2, No. 3, p 204-210

[9] Pramana F.Y. & Rameli M. (2012). *Pengatura Kecepatan Spindle pada Retrofit Mesin Bubut CNC Menggunakan Kontroler PI Gain Scheduling.* Jurnal Teknik POMITS, Vol. 1, No. 1, p 1-5

[10] Pressman. 2010. *Software Engineering : a Practitioner's Approach.* New York. McGraw-Hill

[11] Triatmaja A. K., dkk. (2021). *Virtual Laboratorium Teknik Digital berbasis Mobile Virtual Reality.* Jurnal Edukasi Elektro, Vol. 05, No. 1, p 1-13.

Innovation Media Learning: Online Project-Based Learning (O-PBL) on Drawing Competence in Automotive Engineering Using Video on YouTube

Haryanto[1], Wahyu Mustafa Kusuma[2], Farid Mutohhari[3], Muhammad Nurtanto[4], Suyitno Suyitno[5]

[1]Department of Electrical Engineering Education, Yogyakarta State University, Yogyakarta, Indonesia
[2,3]Postgraduate Program of Technology and Vocational Education, Yogyakarta State University, Yogyakarta, Indonesia
[4]Department of Mechanical Engineering Education, Universitas Sultan Ageng Tirtayasa, Indonesia
[5]Department of Automotive Engineering Education, Universitas Muhammadiyah Purworejo, Indonesia

E-mail: haryanto@uny.ac.id

Abstract. This study aims to develop and analysis the feasibility of a video-assisted project-based learning model on YouTube in automotive engineering drawing subjects. Research development using ADDIE design which is simplified into 4 steps. The developed model was validated by material experts, media and learning practitioners as well as user perception tests. Data were collected through a questionnaire with a product feasibility questionnaire and a perception questionnaire using 5 Likert scales. The results of developing a project-based learning model assisted by YouTube videos contain planning procedures, syntax implementation, and assessment mechanisms that are packaged in a learning model guide book. The results of the validation test obtained a very high level of feasibility, both from material experts, media experts and learning practitioners. Likewise, the student response test showed a very good response to the developed model. The video-assisted project-based learning model on YouTube in automotive engineering drawing subjects is very feasible to be applied in learning.

1. Introduction

The essence of vocational education is to provide work competencies to students through their learning [1], [2]. Entering the 21st century, vocational learning must be transformed to suit the needs of competence in today's world of work [3], [4]. Improving creativity skills, critical thinking and problem solving are some of the focuses of current vocational learning. Various efforts have been made to improve these skills, such as the development of models, media and learning resources that are continuously being carried out. Project-based learning is one of the appropriate learning models to be used in current vocational learning. Project-based learning models have a positive influence in shaping and improving students' 21st century skills [5]–[7]. Through results-oriented learning products will foster critical thinking skills, creative and problem solving. However, in its implementation, subject competencies need to be adjusted to the concept of the model.

Automotive engineering drawing is one of the vocational education subjects that can apply a project-based learning model. The output of student learning outcomes in the form of image products is the basic reason for implementing this learning [8]. Nevertheless, the complexity of learning automotive engineering drawings is a serious obstacle in achieving learning competencies. The procedure for drawing 2D and 3D objects with various projection models is a competency that must be mastered by students [9], [10]. In addition, the cleanliness of the image, the level of precision, the clarity of various symbols and drawing lines are important aspects in the assessment. The complexity of automotive engineering drawing learning competencies becomes a separate difficulty for students in learning it. Various efforts made to overcome these problems have not been optimal. In addition, the low learning motivation of students also affects students' learning difficulties, so that the quality of the results of student drawing projects is still low [11], [12].

Students' learning difficulties in learning automotive engineering drawings are increasing due to the COVID-19 pandemic conditions [13]. Face-to-face learning is transformed into online learning using the internet [14]. Researchers conducted observations and interviews with teachers and students in 6 vocational high schools. Productive teachers expressed difficulties in teaching and facilitating students during learning, especially when practicing drawing [15]. Teachers find it difficult to determine media and learning resources that can support project-based learning easily and practically [16]. Meanwhile, students complained about the lack of knowledge and literacy to carry out effective online learning to achieve learning competencies. Thus, effective learning media innovations are needed that can overcome project-based learning problems during the covid19 pandemic.

Project-based learning in automotive engineering drawing subjects can be done online using media assistance in the form of video tutorials on YouTube. YouTube is one of the social media platforms that provides video resource services, including practical learning videos [17]. The ease and practicality of access are the advantages of the YouTube platform as a media to support project-based learning [18]. Videos on YouTube can be accessed anywhere and anytime by students. The need for smartphone ownership is the basic reason for making it easier for students to learn effectively and efficiently [19]. The teacher will act as a facilitator and direct learning resources on YouTube that are relevant and in the context of learning automotive engineering drawings. Meanwhile, students through YouTube will learn independently to make automotive engineering drawing projects according to the instructions and achievement of learning competencies .

Based on the description above, this research will focus on developing a project-based learning model in automotive body learning which is carried out online with the help of video media on YouTube. The research will be conducted in 6 vocational high schools in Sleman-Indonesia. The research questions are: (1) How is the process of developing project-based learning models in automotive body learning assisted by video media on YouTube? and (2) How is the feasibility of a project-based learning model in automotive body learning assisted by video media on YouTube?

2. Methodology

This research is a research and development that adopts the ADDIE model developed byWalter Dick [20]. This development model consists of five simplified stages without the fifth stage, namely evaluation. In the first stage (analyzing), the researcher analyzes the development needs from the perspective of students and teachers. Then the second stage (design), the design of a project-based learning model assisted by video learning media on YouTube is carried out. The video media used include video tutorials and animated videos related to technical drawing procedure tutorials and image construction animations from various types of views. Simulation of animated videos and tutorial videos using the YouTube platform, then modified according to learning needs. At this stage the expert validates the content and media to determine the feasibility before being tested on users. After that, in the third step (development), the product starts to be developed to completion. After the product is ready, then proceed with the implementation of project-based learning assisted by video media on YouTube to get student feedback about the product being developed.

The trial design consisted of a product feasibility test involving material and media experts, and a perception test and students. The trial respondents involved 8 university lecturers, 120 students of class X and 16 teachers who were divided into 4 vocational schools in Sleman regency. The distribution of test subjects is shown in table 1 below.

Table 1. Distribution of test Respondents

No	Characteristics	Sum (n)	Percentage (%)
1	*Expert Validation Respondents University lecturer*		
	1.1. Expert of material	4	2,94
	1.2. Expert of media	4	2,94
2	*Expert of practitioner Public school teachers and Private school teachers*		
	2.1. SMK N 1 Ngaglik	4	2,21
	2.2. SMK N 1 Seyegan	4	2,21
	2.3. SMK Maarif 1 Sleman	4	2,21
	2.4. SMK Muhammadiyah Prambanan	4	2,21
3	*User Public school students and Private school students*		
	3.1. SMK N 1 Ngaglik	30	22,06
	3.2. SMKN 2 Depok	38	27,94
	3.3. SMK Maarif 1 Sleman	25	18,38
	3.4. SMK Muhammadiyah Prambanan	27	19,85
	Total	**144**	**100**

Data collection was carried out using a questionnaire technique to determine development needs, and validation techniques to determine feasibility based on expert and user assessments. The data collection instrument used a 5 Likert scale questionnaire and an expert validation sheet consisting of a material expert validation sheet, media expert, learning practitioner expert and user. The grid of validation instruments for material experts, media experts and users is shown in table 2 below.

Table 2. Instrument validation based on characteristics and aspects

Characteristics	Aspects	Item (n)	Distribution
Need assessment	Use of learning media	4	1 – 4
	Use of learning model	4	5 – 8
	Student learning outcomes	4	9 – 12
	Development support	4	13 – 16
Expert of Material	Suitability of the material	9	1 – 9
	Syntax suitability	5	10 – 16
	Language and format	5	17 – 21
Expert of Media and Practitioner	Video output	4	1 – 4
	Consistency	3	5 – 7
	Context suitability	8	8 – 15
	Ease of use	3	16 – 18
User/Student	Video output	4	1 – 4
	Consistency	3	5 – 7
	Ease of use	3	8 – 10
	Language suitability	2	11 – 12

Data analysis techniques are carried out in determining categories at the stage of needs analysis, and validation of experts and users. The categories in the needs analysis include five categories, namely VN "very need"; N "needs"; FN "fairly need; NN "not need"; and VNN "very not need". Meanwhile, the categories at the validation stage are divided into five categories, namely VF "very feasible; F "feasible"; FF "fairly feasible"; NF "not feasible", and VNF "very not feasible". The number of respondents determines decision making, the rating scale chosen, and the question items for each fixed variable. The data were analyzed descriptively using a quantitative approach based on the formula shown in table 3 below.

Table 3. Interval Score of Feasibility

Interval Score	**Category**	
	Need Assessment	**Validation/Respon**
$M > Mi + 1{,}5\ Sdi$	Very Need	Very Feasible/ Very Good
$Mi + 0{,}5\ SDi < M \leq Mi + \ 1{,}5\ SDi$	Need	Feasible/Good
$Mi - 0{,}5\ SDi < M \leq Mi + \ 0\ SDi$	Fairly Need	Fairly Feasible/Fairly Good
$Mi - 1{,}5\ SDi < M \leq Mi + (- \ 0{,}5\ SDi$	Not Need	Not Feasible/ Not Good
$M \leq Mi - 1{,}5Sdi$	Very Not Need	Very Not Feasible/Very Not Good

Source: Mardapi (2012) [21]

3. Result and Discussion

3.1. Analysis

The first stage, the researcher conducted a development needs analysis through a questionnaire containing 16 statements with 5 answer options, namely strongly agree, agree, disagree, disagree, and strongly disagree. The contents of the questionnaire consist of model problems and those that have been applied, learning outcomes problems, and support for the development of project-based learning models assisted by YouTube video media. The results of the needs analysis, obtained data from 20 respondents which is shown in table 4 below.

Table 4. Results of the needs analysis

Respondent	Mean	Percentage	Median	Mode	Std. Dev	Min	Max
Teachers	64,92	82,66%	64.50	64	4,581	61	76
Students	64,50	81,15%	64,00	64	4,252	60	76

The average value of 64.50 for students is in the "very need" category, so it can be concluded that students need and support the development of project-based learning models assisted by video media on YouTube. Meanwhile, the average score of 64.92 for teachers is in the "very need" category, so it can be concluded that teachers need and support the development of project-based learning models assisted by video media on YouTube.

3.2. Design

The focus on design is to design by integrating videos on YouTube into project-based learning syntax. Researchers explored video sources which included animated videos and automotive engineering drawing tutorial videos. The video that will be used for student learning is adjusted to the material and competencies to be achieved. Video is also adjusted to the quality with minimum criteria of 1280x720 resolution (720p) and 25 fps with a max file of 255MB. After obtaining a draft list of videos that are eligible to be integrated in project-based learning on automotive engineering drawing subjects,

After the draft list of videos is collected, then the preparation of learning guidebooks, learning tools and worksheets is carried out. The manual contains the implementation of the developed model. The integration of the videos in the book in the form of links and images (clickbait) that will be displayed in the learning syntax. The manual is also equipped with procedures and assessment instruments to measure the achievement of learning competencies.

3.3. Development

This development stage is carried out to produce a needs-based learning model to support online-based-learning projects. The development process is carried out by focusing on learning syntax that is integrated with video tutorials on YouTube. The syntax in learning automotive engineering drawings developed is adapted to the context. The learning syntax is packaged into a manual to make it easier to implement. The following display cover and table of contents of the developed model manual is shown in Figures 1-4 below.

Figure 1. The cover of Guidebook

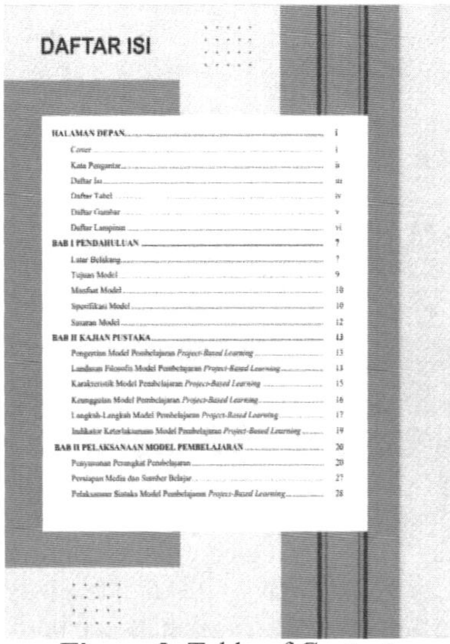

DAFTAR ISI

HALAMAN DEPAN ... i
Cover ... i
Kata Pengantar ... ii
Daftar Isi ... iii
Daftar Tabel ... iv
Daftar Gambar ... v
Daftar Lampiran ... vi
BAB I PENDAHULUAN ... 7
Latar Belakang ... 7
Tujuan Model ... 9
Manfaat Model ... 10
Spesifikasi Model ... 10
Sasaran Model ... 12
BAB II KAJIAN PUSTAKA ... 13
Pengertian Model Pembelajaran *Project-Based Learning* ... 13
Landasan Filosofis Model Pembelajaran *Project-Based Learning* ... 13
Karakteristik Model Pembelajaran *Project-Based Learning* ... 15
Keunggulan Model Pembelajaran *Project-Based Learning* ... 16
Langkah-Langkah Model Pembelajaran *Project-Based Learning* ... 17
Indikator Keterlaksanaan Model Pembelajaran *Project-Based Learning* ... 19
BAB II PELAKSANAAN MODEL PEMBELAJARAN ... 20
Penyusunan Perangkat Pembelajaran ... 20
Persiapan Media dan Sumber Belajar ... 27
Pelaksanaan Sintaks Model Pembelajaran *Project-Based Learning* ... 28

Figure 2. Table of Contents

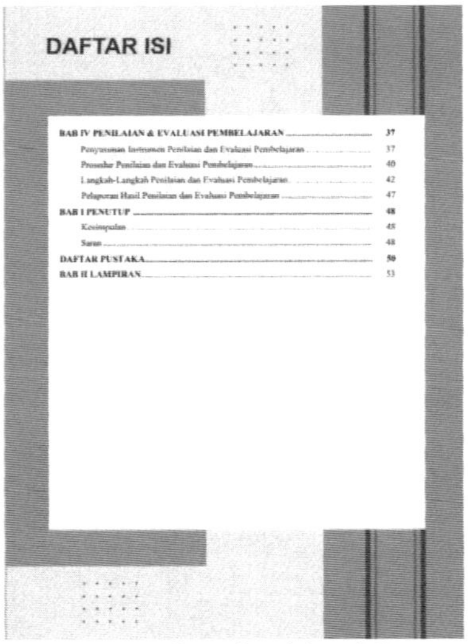

DAFTAR ISI

BAB IV PENILAIAN & EVALUASI PEMBELAJARAN ... 37
Penyusunan Instrumen Penilaian dan Evaluasi Pembelajaran ... 37
Prosedur Penilaian dan Evaluasi Pembelajaran ... 40
Langkah-Langkah Penilaian dan Evaluasi Pembelajaran ... 42
Pelaporan Hasil Penilaian dan Evaluasi Pembelajaran ... 47
BAB I PENUTUP ... 48
Kesimpulan ... 48
Saran ... 48
DAFTAR PUSTAKA ... 50
BAB II LAMPIRAN ... 53

Figure 3. Continued Table of Contents

Figure 4. Author Information

After the product has been developed, the next step is to conduct a feasibility validation test by asking for an expert assessment. Experts involved include material experts, media experts, and learning practitioners. Material experts and media experts appointed to provide product feasibility assessments are Yogyakarta State University lecturers with academic positions of lector and head lector. Meanwhile, learning practitioners who conduct product feasibility assessments are productive SMK teachers who have been certified. The following table 5 is a summary of the results of the validation carried out by material experts, media, and learning practitioners.

Table 5. Result of Product Feasibility

Expert	Aspect	Score	Category
Material	Suitability of the material	4,22	Feasible
	Syntax suitability	4,40	Very Feasible
	Language and format	4,40	Very Feasible
Media	Video output	4,25	Very Feasible
	Consistency	4,00	Feasible
	Context suitability	4,00	Feasible
	Ease of use	4,67	Very Feasible
Practitioner	Video output	4,67	Very Feasible
	Consistency	4,33	Very Feasible
	Context suitability	4,33	Very Feasible
	Ease of use	4,67	Very Feasible

Based on the results of material validation, it can be explained that the value of the suitability of the material to the learning context is in the "appropriate" category. The value of the suitability of the learning steps, language and format is in the "very feasible" category. Then the results of media validation, obtained the feasibility value for video output and ease of use into the "very feasible" category. The feasibility value on the consistency and suitability of the context is included in the "adequate" category. Meanwhile, the assessment carried out by learning practitioners, the results obtained on video output, consistency, context suitability and ease of use are in the very feasible category. Thus, it can be concluded that the feasibility of the developed product is in the "very feasible" category to be used and implemented.

3.4. Implementation

After being declared feasible to be used and implemented, the next step is to implement the product developed in learning. Implementation is done to find out the responses or responses from students related to the product being developed. Implementation is carried out in 2 stages, namely the first stage is carried out on a limited sample and the second stage is carried out on a wider sample. The implementation of the first phase involved 52 students from 2 SMKs. The results of the responses from students in the first stage are shown in table 6 below.

Table 6. Limited Scale Assessment Results

No	Aspect	Score	Category
1	Video output	4,00	Good
2	Consistency	3,67	Good
3	Ease of use	4,67	Very Good
4	Language suitability	4,00	Good

The results of the user response analysis in the first stage of implementing the developed product were obtained in 4 aspects. Video output, consistency, and language suitability fall into the "good" category. While the aspect of ease of use received an "excellent" rating. Thus, it can be concluded that in the implementation of the first stage, a good response was obtained for the product being developed. After the first stage is completed, then the second stage is implemented with a wider sample. The implementation of the second phase involved 120 students in 4 SMKs. The results of the response test in the implementation of the second stage are presented in table 7 below.

Table 7. Large Scale Assessment Results

No	Aspect	Score	Category
1	Video output	4,50	Very Good
2	Consistency	4,33	Very Good
3	Ease of use	4,67	Very Good
4	Language suitability	4,50	Very Good

The results of the user response analysis in the second stage of the implementation of the developed product were obtained in 4 aspects. All aspects including video output, consistency, ease of use and language suitability fall into the "very good" category. Thus it can be concluded that in the first stage of implementation the response was very good. The results of the comparison of user response tests in the first and second stages are shown in the following chart.

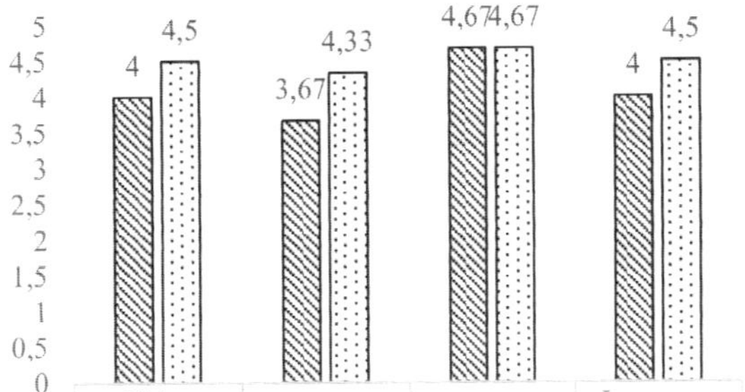

	Video Output	Consistency	Ease of Use	Language Suitability
Limited Scale	4	3,67	4,67	4
Large Scale	4,5	4,33	4,67	4,5

Figure 5. User Response Test Chart

The video-assisted project-based learning model on YouTube is feasible to use after getting validation from experts. In addition, the results of the development are also received and get a good response from users. The feasibility of the model to be applied in automotive engineering drawing subjects in SMK is influenced by several factors. Increased motivation and interest in learning as a result of using a learning model that integrates video-based learning media on YouTube is an important factor [22], [23]. In addition, through the integration of videos from YouTube, students can choose their own learning resources according to their respective interests [24].

In addition, the problems faced during online learning during the current COVID-19 pandemic provide a positive value for the development carried out. This learning model can be done online, so it can provide the flexibility and flexibility of independent learning anywhere and anytime [25]. The teacher acts as a learning facilitator to provide instruction and literacy on how to implement online project-based learning. Meanwhile, students are given the freedom to work on projects with the help of video tutorials and animated videos on YouTube [26].

The easy accessibility of technology such as smartphones and the YouTube platform is a distinct advantage of developing this model. Students have the opportunity to develop their abilities independently and explore resources according to their interests [27]. Thus, through independent learning using digital sources, especially YouTube, it can form learning creativity and students' critical thinking power [28]. In addition, through learning to make projects, self-efficacy can be formed by itself [29].

4. Conclusion

The development of video-assisted project-based learning models from Youtube can be applied to automotive engineering drawing subjects. Model development is carried out by positioning YouTube as a source of student learning in working on automotive engineering drawing projects. Various characteristics of animated videos and video tutorials are adapted to the context and achievement of automotive engineering drawing learning. The entire procedure for implementing the learning syntax is packaged in a manual. The developed model received high feasibility assessment results from material experts, media and learning practitioners. In addition, a good positive response to the developed model was obtained from users. Thus, the project-based learning model assisted by YouTube videos is feasible to be applied in vocational high schools, especially in automotive engineering drawing subjects.

5. References

[1] S. Billett, *Vocational Education: Purposes, Traditions and Prospects*. Dordrecht: Springer, 2011. doi: 10.1007/978-94-007-1954-5.

[2] L. Clark and C. Winch, *Vocational Education : International Approaches, Developments and Systems*. New York: Routledge, 2007.

[3] P. Sudira, *Paradigma Baru Pembelajaran Vokasional Era Revolusi Industri 4.0*. Yogyakarta: UNY PRESS, 2020.

[4] M. Nurtanto, H. Sofyan, P. Pardjono, and S. Suyitno, 'Development model for competency improvement and national vocational qualification support frames in Automotive technology', *International Journal of Evaluation and Research in Education (IJERE)*, vol. 9, no. 1, Art. no. 1, Mar. 2020, doi: 10.11591/ijere.v9i1.20447.

[5] D. Probowati, R. S. Iswari, and S. Sukaesih, 'The Influence of Project Based Creative Problem Solving Toward Creative Thinking Ability on Circulation System', *Journal of Biology Education*, vol. 9, no. 2, pp. 167–177, 2020, doi: 10.15294/jbe.v9i2.22430.

[6] I. W. Santyasa, N. K. Rapi, and I. W. W. Sara, 'Project based learning and academic procrastination of students in learning physics', *International Journal of Instruction*, vol. 13, no. 1, pp. 489–508, 2020, doi: 10.29333/iji.2020.13132a.

[7] O. Kizkapan and O. Bektas, 'The effect of project based learning on seventh grade students' academic achievement', *International Journal of Instruction*, vol. 10, no. 1, pp. 37–54, 2017, doi: 10.12973/iji.2017.1013a.

[8] Z. Furqon and J. Pramono, *Gambar Teknik Otomotif*, II. Yogyakarta: Penerbit Andi, 2017.

[9] F. E. Glesecke, S. Lockhart, M. Goodman, and C. M. Johnson, *Technical Drawing with Engineering Graphics*, 15th ed. New York: Macmillan Publishing Co, Inc, 2016.

[10] M. Nurtanto, H. Sofyan, and P. Pardjono, 'E-Learning Based AutoCAD 3D Interactive Multimedia on Vocational Education (VE) Learning', *Journal of Engineering Education Transformations*, vol. 34, no. 1, Art. no. 1, Jul. 2020, doi: 10.16920/jeet/2020/v34i1/147793.

[11] A. El-Adl and H. Alkharusi, 'Relationships between self-regulated learning strategies, learning motivation and mathematics achievement', *Cypriot Journal of Educational Sciences*, vol. 15, no. 1, pp. 104–111, 2020, doi: 10.18844/cjes.v15i1.4461.

[12] J. C. Tu and K. H. Chu, 'Analyzing the relevance of peer relationship, learning motivation, and learning effectiveness-design students as an example', *Sustainability (Switzerland)*, vol. 12, no. 10, pp. 1–26, 2020, doi: 10.3390/SU12104061.

[13] E. M. Onyema *et al.*, 'Impact of Coronavirus Pandemic on Education', *Journal of Education and Practice*, vol. 11, no. 13, pp. 108–121, 2020, doi: 10.7176/jep/11-13-12.

[14] E. Liguori and C. Winkler, 'From Offline to Online: Challenges and Opportunities for Entrepreneurship Education Following the COVID-19 Pandemic', *Entrepreneurship Education and Pedagogy*, vol. 3, no. 4, pp. 346–351, 2020, doi: 10.1177/2515127420916738.

[15] F. Mutohhari, P. Sudira, and M. Nurtanto, 'Automotive Engineering Drawing Learning: Effective Online Learning Using Autocad Application', *Journal of Education Technology*, vol. 5, no. 2, pp. 1–11, 2021, doi: 10.23887/jet.v5i2.33197.
[16] M. Nurtanto, N. Kholifah, A. Masek, P. Sudira, and A. Samsudin, 'Crucial problems in arranged the lesson plan of vocational teacher', *International Journal of Evaluation and Research in Education (IJERE)*, vol. 10, no. 1, Art. no. 1, Mar. 2021, doi: 10.11591/ijere.v10i1.20604.
[17] R. Rabiman, P. Sudira, H. Sofyan, and M. Nurtanto, 'Practical Learning Media in Subject Maintenance of Chassis and Power (MCP) Based Online: Simple Learning Using Videos on YouTube', *International Journal of Interactive Mobile Technologies*, vol. 15, pp. 130–145, 2021, doi: https://doi.org/10.3991/ijim.v15i03.14943.
[18] S. Moghavvemi, A. Sulaiman, N. I. Jaafar, and N. Kasem, 'Social media as a complementary learning tool for teaching and learning: The case of youtube', *International Journal of Management Education*, vol. 16, no. 1, pp. 37–42, 2018, doi: 10.1016/j.ijme.2017.12.001.
[19] E. T. Maziriri, P. Gapa, and T. Chuchu, 'Student perceptions towards the use of youtube as an educational tool for learning and tutorials', *International Journal of Instruction*, vol. 13, no. 2, pp. 119–138, 2020, doi: 10.29333/iji.2020.1329a.
[20] W. Dick, 'The dick and carey model: Will it survive the decade?', *ETR&D*, vol. 44, no. 3, pp. 55–63, Sep. 1996, doi: 10.1007/BF02300425.
[21] D. Mardapi, *Pengukuran, Penilaian, dan Evaluasi Pendidikan.* Yogyakarta: Mitra Cendekia Press, 2012.
[22] H. Mujianto, 'Pemanfaatan Youtube Sebagai Media Ajar Dalam Meningkatkan Minat Dan Motivasi Belajar', *Jurnal Komunikasi Hasil Pemikiran dan Penelitian*, vol. 18, no. 4, pp. 355–361, 2019, doi: 10.1177/1069072710374496.
[23] P. Namubiru Ssentamu, D. Ng'ambi, E. Bagarukayo, R. Baguma, H. Mutambo Nabushawo, and C. Nalubowa, 'Enhancing Student Interactions in Online Learning: A Case of Using YouTube in a Distance Learning Module in a Higher Education Institution in Uganda', *Higher Education Research*, vol. 5, no. 4, pp. 103–117, 2020, doi: 10.11648/j.her.20200504.11.
[24] J. E. Klobas, T. J. McGill, S. Moghavvemi, and T. Paramanathan, 'Compulsive YouTube usage: A comparison of use motivation and personality effects', *Computers in Human Behavior*, vol. 87, no. 12, pp. 129–139, 2018, doi: 10.1016/j.chb.2018.05.038.
[25] Z. Yaacob and N. H. M. Saad, 'Acceptance of YouTube as a Learning Platform during the Covid-19 Pandemic: The Moderating Effect of Subscription Status', *TEM Journal*, vol. 9, no. 4, pp. 1732–1739, 2020, doi: 10.18421/TEM94-54.
[26] A. A. Jaffar, 'YouTube: An emerging tool in anatomy education', *Anatomical Sciences Education*, vol. 5, no. 3, pp. 158–164, 2012, doi: 10.1002/ase.1268.
[27] J. Balakrishnan and M. D. Griffiths, 'Social media addiction: What is the role of content in YouTube?', *Journal of Behavioral Addictions*, vol. 6, no. 3, pp. 364–377, 2017, doi: 10.1556/2006.6.2017.058.
[28] D. R. Rizaldi, E. Nurhayati, and Z. Fatimah, 'The Correlation of Digital Literation and STEM Integration to Improve Indonesian Students' Skills in 21st Century', *International Journal of Asian Education*, vol. 1, no. 2, pp. 73–80, 2020, doi: 10.46966/ijae.v1i2.36.
[29] S. T. Steinemann *et al.*, 'Potentials and pitfalls of increasing prosocial behavior and self-efficacy over time using an online personalized platform', *PLoS ONE*, vol. 15, no. 6, pp. 1–12, 2020, doi: 10.1371/journal.pone.0234422.

Utilization of Electronic Waste for Energy-saving Lamp Circuits

Zamtinah[1], I Mustaqim[2], HS Pramono[3]

[1,2,3]Department of Electrical Engineering Education, Yogyakarta State University, Yogyakarta, Indonesia

Email: zamtinah@uny.ac.id

Abstract. Indonesia is one of the largest electronic consumer countries in the world. The negative impact unless the increase in the use of electrical energy, there is also an increase in electronic waste or e-waste. If electronic waste (e-waste) is not managed properly, it will cause its own problems, especially environmental health problems. One alternative to overcome these problems is the use of e-waste as a raw material for a series of Waste-Based Energy Saving Lamps, abbreviated as SLHE BBL. This study aims to develop SLHE BBL prototypes base on electric and electronics waste, and test the performance of SLHE BBL prototypes.The SLHE BBL method developed by research and development adapted from Borg & Gall with the following stages: First, needs analysis and product planning; Second: Initial product development, validation and initial product revision. Third: Product development, testing, product revision. Fourth: Finalization dan Dissemination. The research instrument consisted of an observation sheet, luxmeter, voltmeter, ammeter, multimeter, watt meter. The collected data was then analyzed quantitatively and descriptively.The results of the research are as follows: 1) waste that can be recycled into raw materials for SLHE BBL include: resistors, capacitors, LED lights on computer mice or electronic component indicators, used goods casings, adapters, diodes, cables, TV's control remote; 2) Electrical and mechanical design of SLHE BBL can be assembled into prototype of SLHE BBL; 3) The performance of the SLHE BBL prototype is that the lamp can be adjusted in colour and light intensity; the measured electrical parameters are within the range in accordance with the provisions of electrical standards such as Perasyaratan Umum Instalasi Lstrik (PUIL), International Electrical Commisionin (IEC), or National Electrical Code (NEC)

1. Introduction

The volume of electronic waste generated worldwide in 2019 was roughly 54 million metric tons. Several factors such as increased spending power and the availability of electronics have fuelled e-waste generation in recent decades, making it the fastest growing waste stream worldwide. This trend is expected to continue, with projections showing that by 2030, annual e-waste generation worldwide will have increased by approximately 30% as shown in Figure 1 [1]. According to [2], On March 6, 2021, an article from Suara.com stated that it was estimated that Citation is not good

Indonesia would produce 2 million tons of e-waste in 2021, and this is the most in Southeast Asia. Annual e-waste production in Indonesia is projected to increase to 3.2 million tons in 20 years. Based on the Material Flow Analysis (MFA) method [3], the accumulation of electronic waste generated from households in Indonesia in 2025 can reach 622,000 tons. Based on the MFA model, the total e-waste generated (2015-2025) is estimated at 3.75–4.98 million tons. The negative impact unless the increase in the use of electrical energy, there is also an increase in electronic waste or e-waste. If e-waste is not managed properly, it will cause its own problems, especially environmental health problems. One alternative to overcome these problems is the use of e-waste as a raw material for a series of Waste-Based Energy Saving Lamps, abbreviated as SLHE BBL.

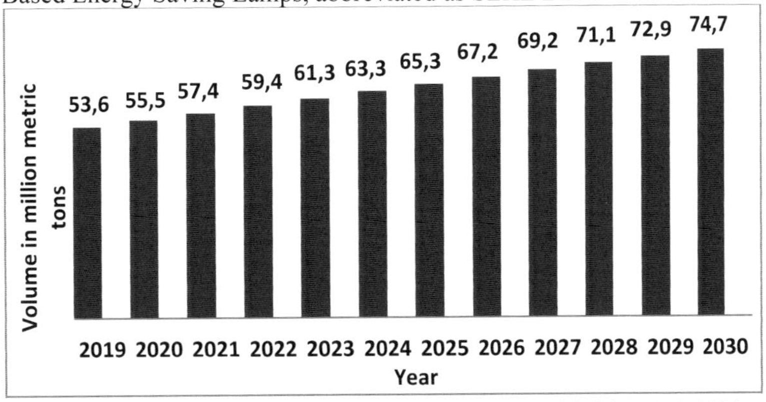

Figure 1 Projected electronic waste generation worldwide 2019-2030
(Source: https://www.statista.com/statistics/517449/large-household-appliances-ewaste-uk)

The waste used in this research is electrical and electronic waste. Electrical and electronics waste, for example, used tubes lamp, other electrical components that are not used, electronic waste, which components can still be utilized. The Energy Saving Lamp Circuit Made from Waste with remote control (here in after abbreviated as SLHE BBL) developed in this study was inspired by the working principle of the remote control on a TV set. The analogy is that if the remote control is used on a TV set to adjust the sound volume, in this study it is used to adjust the intensity or dimness of the light. Furthermore, the function to change the TV channel, in this study is used to change the colour of the lamp.

Energy saving which is meant in this study is saving in electricity consumption, meaning that compared to other lamps with the same electrical power, the intensity of light produced by SLHE BBL is greater. Another savings found in the SLHE BBL is the use of a remote control as a regulator of the intensity and colour of the lamp that can be adjusted according to needs. In other words, the SLHE BBL lamp can be dimmed or brightened, and the colour of the light can be changed as needed, for example, for studying, the bright white colour is selected and the electrical power is set to 10 watts, for sleep, blue or green is selected and the power is reduced to 5 watts.

The SLHE BBL prototype is different from similar lamp products that have ever existed [4]. For example, PT Philip Indonesia produces a lamp called "Tornado Easy Scene" which is equipped with a remote control, but the remote control on the lamp only functions as a substitute for a switch and light intensity regulator, not yet equipped with a lamp colour setting and only one colour is available (usually milky white colour).

SLHE BBL is also different from similar studies that have been carried out by previous authors, for example the study conducted by [5], [6], [7], and [8]. The study entitled "Design of a light intensity regulator with remote control for incandescent lamps" [5], has limitations on the type of lamp that is controlled not the type of energy-saving lamp, namely incandescent lamps that can cause global warming. Another weakness is that the remote control device is only able to replace the function of the switch and light intensity regulator, the control sensor used (infrared) has a limited range. The weakness of Zulfahmianuddin's study is also found in the study of [6] entitled "Home lighting control system with Remote Control based on the ATMEGA8535 Microcontroller utilizing radio waves", and [7] entitled "SiMBeR as a load regulator for household electrical appliances automatically based on

ATMEGA16". The advantage of the study compared to other studies is the use of radio waves whose control is able to penetrate walls, so that lights can be controlled from outside the building (walls) [8].

The AS unit of luminous intensity is the candle-power, abbreviated cp (cd), and normally represented by the letter "I". An ordinary wax candle has a luminous intensity horizontally of approximately one candlepower. The candela and candlepower have the same magnitude [9]. This is consistent with Mullin & Smith (2002) which state that the luminous intensity of a source, when expressed in candelas, is the candlepower (cp) rating of the source.

According to [10], lumen (lm) is the amount of light received in a unit of time on a unit area at a unit distance from point source of one candela. If the units are in SI, then the unit area is one square meter, and the unit distance is one meter, thus a one-candela source produce one lumen on each square meter (one lux) for a total of 12.57lumen. Referring to [6], one lumen of luminous flux, produce an illumination of one foot candle (fc). Illumination is normally represented by the letter "E". Illumination is the density of luminous power, expressed in terms of lumen per unit area, stated mathematically:

$$E = \frac{\phi}{F} \quad \text{(1)}$$

Where:
E = illumination (lux)
Φ = quantity of light (lumen)
F = square meters of area (m^2)

The unit of electric power is watt (W). In AC circuit impedance is compressed of resistance and reactance (AC resistance of inductance and capacitance) and causes a phase difference between voltage and current. This phase difference is represented by and angle, the cosine of which is called cos θ or power factor, abbreviated "pf". The AC power equation is:

$$W = V \,.\, I \,.\, \cos\theta \quad \text{(2)}$$

Where: W = electric power (watt)
V = voltage (volt)
I = current (ampere)
Cos θ = power factor

2. Method

The method used is R & D developed by [11] with various modifications as shown in Figure2.

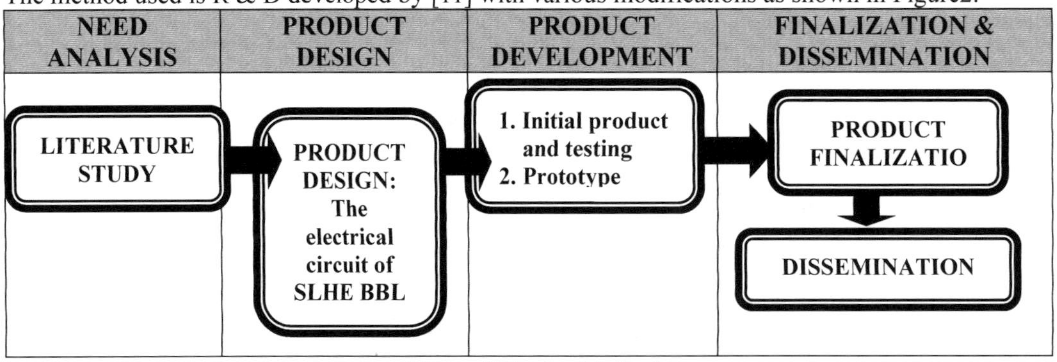

Figure 2 Design Method

The instruments used to measure the electrical parameters of the BBL SLHE are: a multi meter to measure current, voltage, and e-waste connections; wattmeter to measure lamp power; lux meter to measure lamp illumination as shown in Table 1.

Table 1: Measurements instrument

Instruments	To measure:
Multi meter	Electric current
	Voltage
	E-waste connection
Watt meter	Lamp power
Lux meter	lamp illumination

3. Results and Discussion

Based on laboratory tests, observations and testing of e-waste components, it can be identified several components that can still be recycled into basic materials or components of SLHE BBL, as follows: resistor, capacitor, LED lamps, for example the one on a computer mouse, casing of used goods, such as syrup bottles, packaged drinks, perfumes, broken incandescent lamps, used adapters for example from radio, TV and other equipment, diode, cable, remote control casing, e.g. TV remote control.

Furthermore, based on the identification of e-waste, the required circuit can be determined, namely: 1) a power supply circuit that requires components of transformers, diodes, resistors, and capacitors; 2) The system circuit, which consists of resistors, connectors, and connectors; 3) Rectifier, which requires a diode, power supply, resistor, capacitor, and cable; 4) LED circuit.

Figure 3 show the remote control circuit uses an infrared receiver with a working frequency of 38 kHz. The shape of the pulse issued by the sensor is the opposite of the pulse received from the remote (inverting pulse). The sensor output pulses are directly forwarded to the microcontroller.

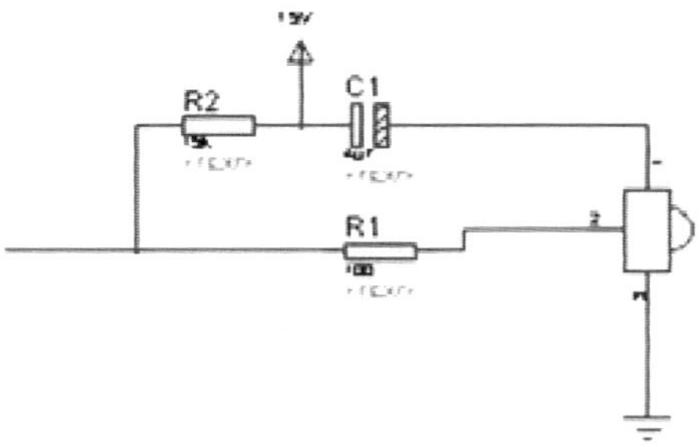

Figure 3 Remote Control circuits

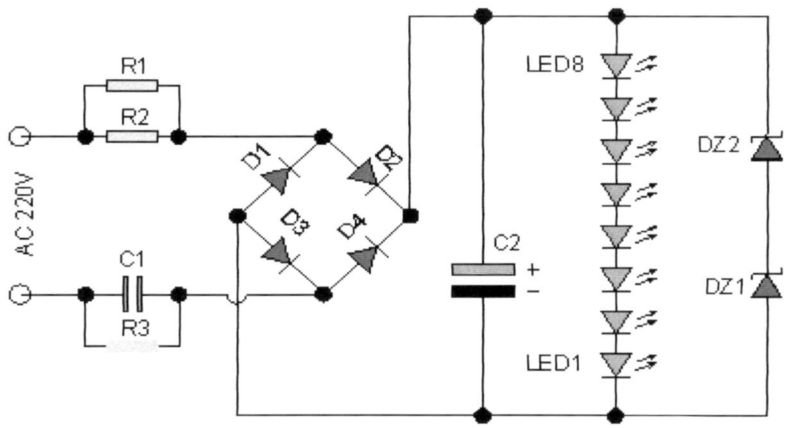

Figure 4 LED Lamp Circuit Schematic

Figure 4 adopting Hartono [12], this circuit consists of several sets of colours, namely white, red, green and blue. Performance that is measured from the LED circuit is the illumination, power, voltage, and current. The microcontroller in this study serves to translate the pulses sent via the remote into numerical data by reading the length of each wave. The value of the translated number is then used to change the PWM value of the microcontroller output. Changes in the PWM value result in changes in the intensity of the LED lights as shown in Figure 5.

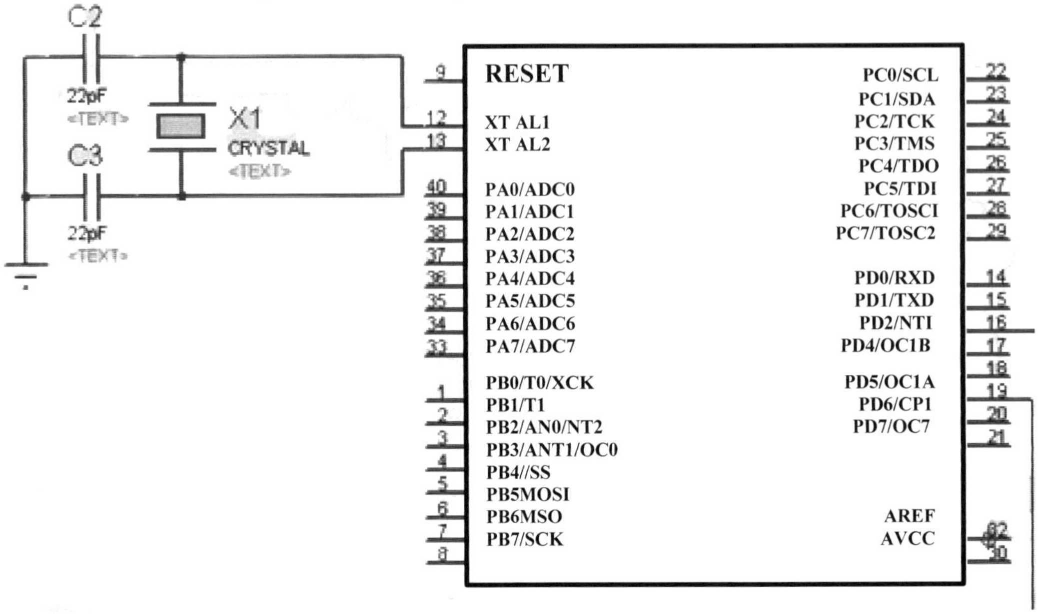

Figure 5 Microcontroller circuits

The output of the microcontroller is limited to a voltage of 5 volts and a maximum current of 40mA, while the LED used requires a voltage of 32 volts. Therefore we need a device that can control a voltage of 32 volts with a microcontroller output voltage of only 5 volts. The control device used is a transistor. The PWM connection of the microcontroller output is connected to the base of the transistor, so that it can adjust the on-off transistor. The on-off transistor will produce an average voltage that varies according to the ratio of the on and off time as shown in Figure 6.

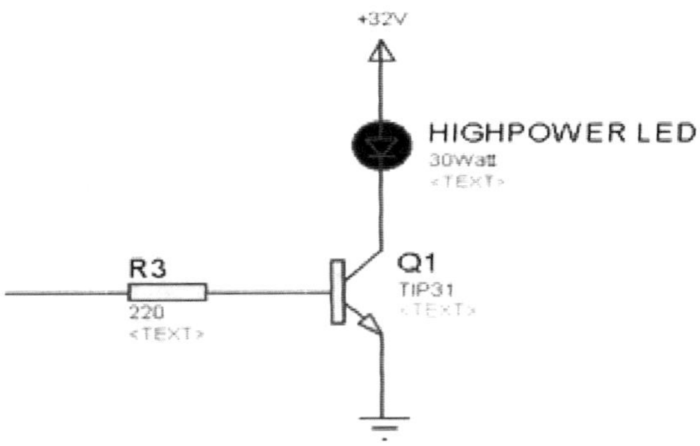

Figure 6 LED control circuits

Table 2 shows the reference parameters based on catalog and analysis for Performance Testing of SLHE BBL.

Table 2. Reference parameters

Parameters	Standard size	Parameters	Standard size
Illumination (E)	80 lux	Voltage (V)	220 volt
Lamp power (W)	40 watt	Cos θ	0.56
Current (I)	323 Ma	Temperature	10° – 40° C

The performance of this study was tested through measurements of illumination, electric power, current, voltage, wiring system testing, and mechanical testing. Figure 7, 8, 9 and 10 show the results of testing the performance of SLHE BBL.

The lighting system on the LED is quite good because it is gradual and does not get hot easily. At a voltage of 220V, the light intensity on the LED does not decrease. It is still consistent, for white lights, the higher the voltage, the intensity light is higher. Likewise, the higher the current and the higher the electric power, the stronger the lighting of a lamp. As for coloured lights (red, green, and blue), the higher the power, current, and voltage, the stronger the lighting will be. This is according to the character of the colour, the redder, the greener, and the bluer, the darker or stronger the light is, the smaller it is.

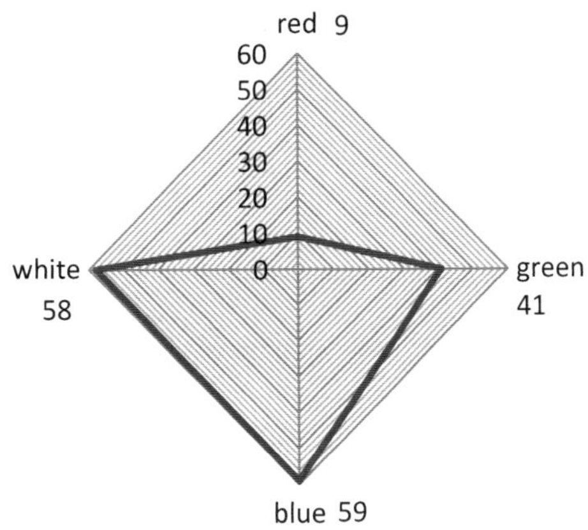

Figure 7 Illumination of lamp color in lux

Figure 7 show that the blue light produces the largest illumination of 59 lux, while the red light has the lowest illumination. Referring to Table 1, the ideal lamp illumination is 80 lux. Based on the observation data, the blue light illumination is lower than the ideal illumination. This is due to the depreciation factor. This is in accordance with formula 1.

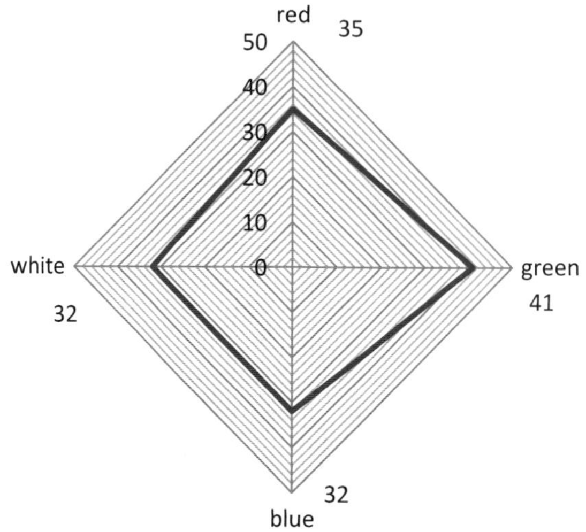

Figure 8 Electric power of lamp color in watt

Based on Figure 8, the largest lamp power is the green lamp and the smallest is the blue and white lamp. This is identical to the illumination produced by the two lamps. If compared with Table 1, the lamp power from the observation is lower than the lamp power from the analysis. This is in accordance with formula 2.

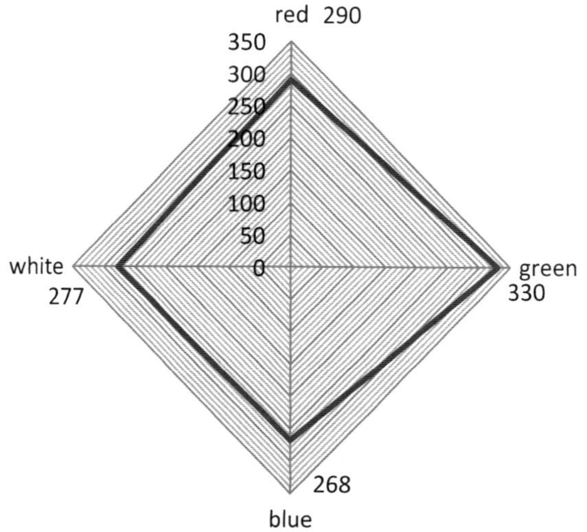

Figure 9 Electric current of lamp color in mA

Based on Figure 9, the largest electric current is the green lamp and the smallest is the blue and white lamp. This is identical to the illumination produced by the two lamps. If compared with Table 1, the lamp power from the observation is lower than the lamp power from the analysis. This is in accordance with formula 2.

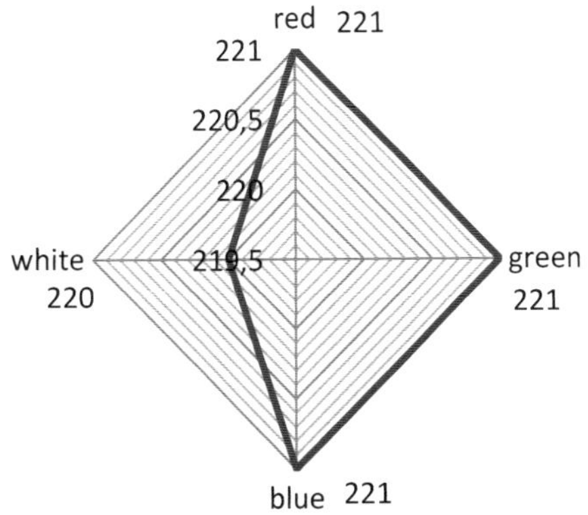

Figure 10 Electric voltage of lamp color in volt

Based on Figure 10, the largest voltage is the red, green, and blue lamp and the smallest is the white lamp. If compared with Table 1, the lamp voltage from the observation is higher than the lamp power from the analysis. This is in accordance with formula 2.

Observational data shows that all test units have provided indicators that they can work according to design. Several points tested for the electrical system were at the input of the power supply, at the input of the microcontroller, at the output of the microcontroller, at the input LED, and at the output LED. The results of this test also give the result that electrically, all the wiring is good.

Furthermore, the mechanical test showed that all test points had met the requirements, except for the finger test and the isolation test. In the finger test, the SLHE BBL circuit does not meet the requirements because the circuit is still open, has not been packaged or included in the casing, so there is still the possibility that the electric current circuit is still open or can endanger safety because if it is touched the human body can electrocute.

4. Conclusion

The design and prototype of the SLHE BBL has been electrically and mechanically tested and the result is that the lamp circuit can be completed as planned.

It is necessary to make improvements to the design of the circuit, assembly or manufacturing circuit, as well as to continue with testing the temperature or heat generated by the SLHE BBL, whether it is still within reasonable limits or still causing excessive heat; The lamp casing is adjusted to the size of the LED lamp to withstand the heat or temperature caused by the high power LED lamp.

References

[1] https://www.statista.com/statistics/1067081/generation-electronic-waste-globally-forecast/

[2] https://www.suara.com/tekno/2021/03/06/190500/peluang-bisnis-baru-limbah-elektronik-indonesia-bernilai-rp-200-triliun?page=all

[3] Rochman, F. F., Ashton, W. S., & Mochamad. (2016). E-waste, money and power: Mapping electronic waste flows in Yogyakarta. *Environmental Development*.

[4] ----------, (2011) *Lampu Philip Tornado Easy Scene.* www.Philip.co.id.

[5] Zulfahmianuddin (2007) *Perancangan Alat Pengatur Intensitas Cahaya dengan Kendali Jaraj Jauh untuk Lampu Pijar.* Laporan Skripsi Jurusan Teknik Elektro Universitas Tanjungpura Pontianak

[6] Ifit Khusnul Arif (2007) *Sistem Penerangan Rumah dengan Remote Control Berbasis Mikrokontroler ATMEGA8535 Memanfaatkan Gelombang Radio.* Laporan Proyek Akhir Program Studi Teknik Elektro Fakultas Teknik UNY

[7] Agusalim (2010). *Perancangan Lampu Belajar LED Hemat Energi Berbasis ATTiny2313.*

Laporan Proyek Akhir Jurusan Pendidikan Teknik Elektro Fakultas Teknik UNY

[8] Bambang Purwanto (2011). SiMBeR Sebagai Alat Pengatur Beban Peralatan Listrik Rumah Tangga Secara Otomatis Berbasis ATMEGA16. Laporan Proyek Akhir Program Studi Teknik Elektro Fakultas Teknik FT UNY

[9] Stein, B., Reynolds, J.S., McGuinnes, W.J (1986). *Mechanical and electrical equipment for buildings, seventh edition vol 2.* New York: John & Wiley, Inc.

[10] Mullin, R.C. & Smith, R.L. (2002). *Electrical wiring commercial eleventh edition.* New York: Delmar Publishers, Inc.

[11] Borg,W.R & Gall,M.D (2003) Educational Research an Introduction Seventh Edition. New York: Longman, Inc.

[12] Hartono (2009). *"lampu tidur LED".* available from http://www.gemar-elektronika.co.cc/merancang-lampu-led.html. 21 Desember 2009. Accessed April 20, 2013

Analysis of the Impact of Distance and Temperature in Determining Transfer Power Using Laser Applications

I A Rahardjo[1], S Muslim[1] , F Wadjdi[1], M Rifan[1], A Sunawar[1], M Subekti[1], Sakirah[2]

[1]Electrical Engineering Education Program, Universitas Negeri Jakarta, Rawamangun, Indonesia
[2]Electrical Engineering, Taruna Bangsa of Vocational High School, Bekasi, Indonesia

Email: [1]imam_ar@unj.ac.id

Abstract: This article aimed to analyze the impact of distance and temperature in determining transfer power using laser applications. This study used instruments consisting of a power meter, photovoltaic, laser, and DC power supply. This research was using experimental method. The impact of distance and temperature in determining transfer power using laser applications was concluded based on the results of calculation analysis, measurements of photovoltaic output power, distance, temperature, and laser input power. The impact of distance and temperature in determining transfer power using laser applications showed that the probability value or p-value was lower than the error/alpha level (a) (2.207E-18 <0.05). Likewise, the calculated F value is greater than the significance F (15290048>4.68386E-27). With the applicable provisions in testing the hypothesis, that the conclusion in this article rejected the null hypothesis (H0 : The distance and temperature have not an impact in determining transfer power using laser applications), and accepted the alternative hypothesis (H1:The distance and temperature have an impact in determining transfer power using laser applications).

Keywords: *Impact, Distance, Temperature, Power, Laser Applications*

1. Introduction

The concept of wireless power transmission using the application of lasers with photovoltaic arrays is known as laser power beaming [1-4]. Laser power beaming can be an alternative for charging electric vehicles because it is able to transfer power to the receiver wirelessly, making it more convenient and safer [5], space applications [6]. The photovoltaic module circuit through the photovoltaic effect converts sunlight into electricity. Sunlight with a wavelength that carries photon energy excites electrons to move from the negative semiconductor to the holes in the positive semiconductor. In the photovoltaic effect that uses sunlight, not all solar wavelengths are absorbed by the photovoltaic module. This is because sunlight consists of various wavelengths or polychromatic. Sunlight with these various wavelengths or polychromatic is different from lasers that have one wavelength or monochromatic. So that the wavelength of the laser light on the laser power beaming will adjust to the wavelength of the p-n junction arrangement of photovoltaic cell materials or vice versa. [7].

The monochromatic laser beam passes through the atmosphere to the surface of the photovoltaic module in the concept of wireless power transmission, then the efficiency or output power of the photovoltaic module illuminated by the laser will be influenced by reducing factors the weakening of the light irradiation by way of the atmosphere. [8-11]. Imam Arif Rahardjo et al have conducted research showing that the temperature affects the irradiation of laser light through the atmosphere to the photovoltaic module [12]. Because distance impacts the temperature transfer of a substance. So, in this article will explain the analysis of the impact of distance and temperature in determining transfer power using laser applications.

2. Method

This study used instruments consisting of a power meter, photovoltaic, laser, and DC power supply. This research was using experimental method. The electric power generated by the power supply is used to activate the beam in laser applications. The light from the laser is captured by photovoltaics to be converted into electricity. The influence of the laser beam received by the photovoltaic from the laser in the form of changes in distance and changes in temperature is observed, measured using an energy meter, and analyzed for changes in the electrical power produced by photovoltaics.

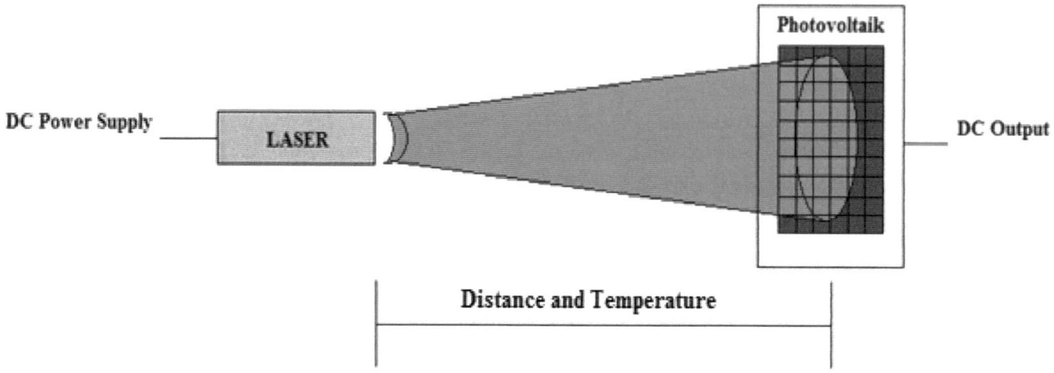

Figure 1. Research Method

The research methods was shown in Figure 1. The impact of distance and temperature in determining transfer power using laser applications was concluded based on the results of calculation analysis, measurements of photovoltaic output power, distance, temperature, and laser input power.

To analyze the the impact of distance and temperature in determining transfer power using laser applications using the applicable provisions in testing the hypothesis. Where the null hypothesis (H0) will be rejected and the alternative hypothesis (H1) will be accepted, when the probability value (p-value) is lower than the error/alpha level (a). On the other hand, the null hypothesis (H0) will be accepted and the alternative hypothesis (H1) will be rejected , when the probability value (p-value) is greater than the error/alpha level (a). Likewise, the null hypothesis (H0) will be rejected and the alternative hypothesis (H1) will be accepted , when the calculated F value is greater than the significance F. On the other hand, the null hypothesis (H0) will be accepted and the alternative hypothesis (H1) will be rejected , when the calculated F value is lower than the significance F.

Null hypothesis (H0) : "The distance and temperature have not an impact in determining transfer power using laser applications"

Alternative hypothesis (H1) : "The distance and temperature have an impact in determining transfer power using laser applications"

3. Results

Table 1 ANOVA shows the results of the impact of distance and temperature in determining transfer power using laser applications.

Table 1. ANOVA of the Impact of Distance and Temperature in Determining Transfer Power using Laser Applications

	df	*SS*	*MS*	*F*	*Significance F*
Regression	2	1,1E-10	5,5E-11	15290048	4,68386E-27
Residual	8	2,87769E-17	3,597E-18		
Total	10	1,1E-10			

	Coefficients	*Standard Error*	*t Stat*	*P-value*
Intercept	0,005216047	1,34641E-05	387,40397	2,207E-18
Distance	-0,000102308	2,6713E-07	-382,99027	2,419E-18
Temperature	8,6313E-10	9,19447E-10	0,9387489	0,3753332

Based on Table 1 ANOVA of the impact of distance and temperature in determining transfer power using laser applications, it shows that the probability value (p-value) is lower than the error/alpha level (a) (2.207E-18<0.05). Based on rules in hypothesis testing as explained in the research method, it means the null hypothesis (H0: "The distance and temperature have not an impact in determining transfer power using laser applications") will be rejected. And the alternative hypothesis (H1: "The distance and temperature have an impact in determining transfer power using laser applications") will be accepted.

Likewise, the calculated F value is greater than the significance F (15290048 > 4.68386E-27). Based on rules in hypothesis testing as explained in the research method, it means the null hypothesis (H0: "The distance and temperature have not an impact in determining transfer power using laser applications") will be rejected. And the alternative hypothesis (H1: "The distance and temperature have an impact in determining transfer power using laser applications") will be accepted.

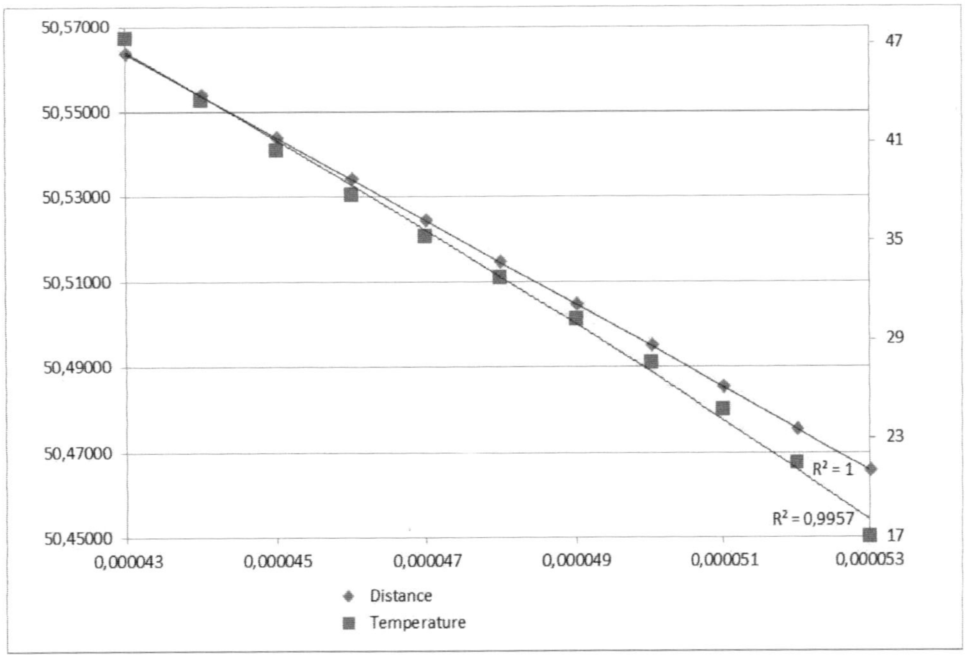

Figure 2. Impact of Distance and Temperature in Determining Transfer Power

Based on Figure 2 above, the lower of distance and temperature are, the higher energy that occurs is.

4. Conclusion

The results of the analysis shows that the probability value (p-value) is lower than the error/alpha level (a) (2.207E-18<0.05). Likewise, the calculated F value is greater than the significance F (15290048 > 4.68386E-27). Based on rules in hypothesis testing as explained in the research method, it means the null hypothesis (H0: "The distance and temperature have not an impact in determining transfer power using laser applications") will be rejected. And the alternative hypothesis (H1: "The distance and temperature have an impact in determining transfer power using laser applications") will be accepted.

Acknowledgements

The authors would like to thank International Conference Committe of ice-elinvo 2021 which has accepted this article to be published.

References

[1] Pollnau, M., & Eichhorn, M. (2020). *Spectral coherence, Part I: Passive-resonator linewidth, fundamental laser linewidth, and Schawlow-Townes approximation. Progress in Quantum Electronics, 100255.* doi:10.1016/j.pquantelec.2020.100255.

[2] G. A. Landis, "Photovoltaic receivers for laser beamed power in space," *J. Propuls. Power*, vol. 9, no. 1, pp. 105–112, 1993, doi: 10.2514/3.11491.

[3] J. H. LEE and E. J. CONWAY, "Power Laser Beaming and Applications in Space," *Le J. Phys. IV*, vol. 01, no. C7, pp. C7-715-C7-719, 1991, doi: 10.1051/jp4:19917194.

[4] S. Chhawchharia, S. K. Sahoo, M. Balamurugan, S. Sukchai, and F. Yanine, "Investigation of wireless power transfer applications with a focus on renewable energy," *Renew. Sustain. Energy Rev.*, vol. 91, no. September 2017, pp. 888–902, 2018, doi: 10.1016/j.rser.2018.04.101.

[5] Y. Rathod, L. Hughes, Y. Rathod, and L. Hughes, "ScienceDirect ScienceDirect Simulating the charging of electric vehicles by laser Simulating the charging of electric vehicles by laser," *Procedia Comput. Sci.*, vol. 155, no. 2018, pp. 527–534, 2019, doi: 10.1016/j.procs.2019.08.073.

[6] Zhong, M.-L., Li, Y.-Z., Mao, Y.-F., Liang, Y.-H., & Liu, J. (2017). *Coupled optic-thermodynamic analysis of a novel wireless power transfer system using concentrated sunlight for space applications. Applied Thermal Engineering, 115, 1079–1088.* doi:10.1016/j.applthermaleng.2017.01.052

[7] Y. Yuan and C. Wu, "Thermal analysis of fi lm photovoltaic cell subjected to dual laser beam irradiation," *Appl. Therm. Eng.*, pp. 1–8, 2015, doi: 10.1016/j.applthermaleng.2015.01.054.

[8] H. Baig *et al.*, "Conceptual design and performance evaluation of a hybrid concentrating photovoltaic system in preparation for energy," *Energy*, vol. 147, pp. 547–560, 2018, doi: 10.1016/j.energy.2017.12.127.

[9] J. Cubas, S. Pindado, and M. Victoria, "On the analytical approach for modeling photovoltaic systems behavior," *J. Power Sources*, vol. 247, pp. 467–474, 2014, doi: 10.1016/j.jpowsour.2013.09.008.

[10] J. Mukherjee, S. Jarvis, M. Perren, and S. J. Sweeney, "Efficiency limits of laser power converters for optical power transfer applications," vol. 264006, 2013, doi: 10.1088/0022-3727/46/26/264006.

[11] N. J. Butt, R. A. Roberts, and S. S. Patnaik, "Laser diode optical output dependence on junction distance for high- power laser systems," *Opt. Laser Technol.*, vol. 125, no. November 2019, p. 106019, 2020, doi: 10.1016/j.optlastec.2019.106019.

[12] I. A. Rahardjo and F. Wadjdi, "The effect of temperature on energy transfer capacity by laser," *J. Phys. Conf. Ser.*, vol. 1402, no. 4, 2019, doi: 10.1088/1742-6596/1402/4/044013.

Economic Analysis Development and Provider Elective Decision Model on Auto-ID Technology Investment

Anis Siti Nurrohkayati[1], Iwan vanany[2]

[1]Mechanical Engineering, Universitas Muhammadiyah Kalimantan Timur, Indonesia
[2]Industrial Engineering, Institut Teknologi Sepulun Nopember, Surabaya, Indonesia

Email: asn826@umkt.ac.id

Abstract. In the last few years, the warehousing system use Auto Identification technology to manage the warehouse operation. Auto-ID is often used to obtain the information about amount, location, content, and condition of part or product. However, some industries still use manual warehouse management. Manual warehouse management systems usually cause some problem like inventory shrinkage and stock out. The investment and application of Auto-ID technology on the warehouse system can reduce the effect of the manual system. The use of Auto-ID technology in warehousing systems has a significant effect on the reduction of human error, the reduction of loss of goods, the accuracy of inventory stock, and can handle the product in large quantities, the speed of acceptance, and improve the delivery process. In this research, the cost and benefit analysis was carried out to assess the Auto-ID technology investments. The cost of inventory shrinkage reduction, labor and stock out, and the increase of productivity and maintenance on the spare parts warehouse management system used to analyse the cost benefit. To determine the expected NPV value in Auto-ID technology investments used Monte Carlo analysis. Furthermore, a company's evaluation for selecting an Auto ID provider is a significant aspect. The evaluation of decision making for selecting Auto ID vendors is based on three criteria. This is referred to as many criteria decision making. Vendor performance, service after installation, and system attributes are the criteria employed in the decision-making model. Evaluation of Auto-ID vendor selection is done by using AHP. Based on the results of AHP analysis for the selection of Auto-ID suppliers, it was found that supplier PT C was the main priority (39.3%) for best suppliers, and the last priority was PT B (13.8%).

Keywords: Cost-benefit analysis, Provider Decision Making, RFID, Barcode

1. Introduction

In recent years, Auto-ID (automatic identification) technology is a popular technology in manufacturing or industrial companies, especially in warehousing, purchasing, or goods delivery systems. Auto-ID technology provides information about human, animal, and product data. There are several types of Auto-ID technology. Barcode and RFID (Radio frequency identification) technology are two types of Auto-ID that are often used in the manufacturing or service industries. Barcode technology is a cheaper technology, but barcodes can only be used in simple problems and cannot be reprogrammed [1]. RFID is often referred to as a smart card. RFID technology systems provide information via radio waves. At this time, many manufacturing companies have implemented RFID technology. RFID technology is widely used in the mobile phone and cordless telephone industrial sectors [1]. In addition, although RFID technology is a new technology, many companies realize the potential of RFID to deal with complexity globally and to be able to balance cost and performance in the supply chain [2].

RFID technology has been identified as one of the top 10 technologies in the last 21 years [3]. In 1999 Wal-Mart implemented RFID to replace the existing barcode system [4]. Wal-Mart uses RFID to track inventory. In addition, RFID is also used to obtain information about the location, content, location, and condition of goods or products. Several community service centers such as hospitals have also implemented Auto-ID technology. With the aim of reducing costs and improving patient and service safety, several health care centers are implementing RFID systems to track patient data, medications, mattresses, blood, and medical history [5]. RFID technology is also widely applied to manufacturing companies, especially in the supply chain. The implementation of RFID is growing very quickly since several companies have succeeded in implementing RFID in logistics activities.

Some studies only focus on the general description of RFID or the application of RFID in the fashion, service, retail, manufacturing, healthcare, and automotive industries [2]. This is because RFID technology can minimize inventory errors or inventory inaccuracy in retail companies [6]. Constraints such as inaccuracies in warehousing operating systems that often exist can be minimized by implementing the right information technology. Although the cost of implementing or investing in information technology is quite expensive, with this information technology, companies can reduce losses and increase profits or company performance. Companies can invest in Auto-ID (RFID and/barcode) information technology in the company's operating system. Investment in Auto-ID technology is expected to reduce management operating costs and also make a good contribution to business processes [7]. However, some studies also say that RFID technology does not provide many benefits even with good planning.

Ustundag [5] conducted an economic analysis to increase the profits from the use of Auto-ID in companies. The increase in profit is the sum of reductions in labor costs, inventory costs, shrinkage costs, and revenue increases. However, in the research conducted by Ustundag (2010), calculations have not been carried out to reduce stock-out costs. The cost of stock out is one of the costs of the warehousing system that can affect the company's production process. The costs included in the cost of stock out are backorder costs, lost sales opportunity costs, and lost profit costs. The result of stock out is lost sales, and customer dissatisfaction occurs because customers cannot get the goods they need. With the help of RFID technology, companies can always monitor the amount of stock available in the warehouse. RFID technology can provide an alert when the amount of stock available in the warehouse is at a minimum. Srivastava says that reducing stock-outs can increase company profits and customer satisfaction [8]. Therefore, in this study, a model was developed to get an increase in company profits by taking into account the reduction of stock-out costs.

Based on research conducted by [5] it recommends making decisions to evaluate RFID investments. The research conducted by Ustundag has not provided a solution for selecting the best RFID supplier according to the company's conditions. The selection of suppliers or suppliers is one of the success factors for manufacturing companies. According to Ghoddsypur [9] supplier selection is an important decision-making problem for smooth business or industrial processes and company competitiveness. Based on Cebi strategy and operational factors such as tangible and intangible factors must be considered in the analysis of supplier selection decisions. Meanwhile, Amid in [9] said that the selection of suppliers must include quality and quantity factors.

The complexity of the decision problem is not only caused by imperfections or information uncertainty but the many factors that influence the choices, with various selection criteria in conflicting situations is another factor that can cause the complexity of making a decision. With so many alternatives in conflicting situations, the multi-criteria decision-making method can be used to get the optimal solution [10]. The analytical hierarchy process (AHP) method is the method most often used to solve multi-criteria problems. The hierarchical approach was developed based on the theory of preference measurement by conducting pairwise comparisons for all quantitative and qualitative decision criteria [10]. The purpose of using AHP is to determine the weight of the relative importance of each criterion; then, this criterion is used as a basis for reference for evaluating the relative determination of decision alternatives.

This study aims to conduct an economic analysis of Auto-ID investment in the ROL (re-order level) spare part warehouse system and the decision to select the best RFID supplier according to the company's conditions. Auto-ID technology investment in the spare part warehouse system is expected to help warehouse staff in the warehouse management process. So far, the warehouse management system still uses a manual system. Inventory shrinkage and spare part shortages are caused by the manual system. Spare parts that are in the warehouse beyond the time limit of use are categorized as inventory shrinkage. The amount of inventory shrinkage and stock out resulted in equipment breakdown, and the maintenance process did not run smoothly. Thus, the production level cannot reach the targeted production quantum, and the company does not get the maximum profit.

2. Method

In this research, cost and benefit analysis are used for the economic analysis of Auto-ID investment, and supplier selection is used for the selection of Auto-ID solution providers. The economic analysis carried out is a cost-benefit analysis of the Auto-ID investment in the ROL spare part warehouse system. The total benefit is the sum of the reduction of stock-out costs, inventory shrinkage costs, labor costs in the ROL spare part management system. In addition, it is hoped that reducing the number of ROL spare part stock-outs and the length of maintenance time can increase the amount of production, so as to increase the total profit company too. By investing in Auto-ID, it is hoped that the company will be able to reduce costs in the warehousing system and also increase the amount of production, which will affect the increase in total profit or company profits.

Misinformation on the number of spare parts available is often one of the causes of end-users being unable to carry out the maintenance process on time. Data on the number of spare parts at the end-user and the actual number in the warehouse are often not the same. When the spare part data at the end user is available, when a request is made for the spare part that is needed, it is not available in the warehouse. This is because demand forecasting is uncertain. The sudden demand in large quantities causes the warehouse to have a shortage of spare part inventory, so when the required spare parts are not available, the end-user must wait to be able to carry out the repair or maintenance process. Therefore, in this study, the fuzzy rules-based system method was used to calculate the percentage increase in orders for spare parts and productivity increases for the total profit from the integration of Auto-ID technology. The NPV and IRR values are used to assess investments for the next 5 years.

Furthermore, after an economic analysis of the Auto-ID investment has been carried out, the Auto-ID supplier is selected. Auto-ID supplier selection decisions aim to assist company managers in choosing the best Auto-ID suppliers according to company conditions. The AHP method is used to make decisions on the best Auto-ID supplier to choose.

2.1 Auto-ID Investment Analysis

Cost and benefit analysis (cost and benefit analysis) is used to perform economic analysis in this research. Cost and benefit analysis was also carried out in research conducted by Ustundag [5]. The benefits of investing in Auto-ID technology used in this study are based on previous literature reviews. According to Ustundag [5], RFID implementation can reduce inventory costs, shrinkage, labor, and lost sales. However, Srivastava [8] also stated that the implementation of RFID could affect the minimization of stock out, inventory shrinkage and also affect correct pricing. Figure 4.1 shows the benefit structure of implementing Auto-ID technology. When the number of stock items in the warehouse is small, the possibility of stock-outs will be even greater. Likewise, with lost production and inventory shrinkage. The greater the possibility of stock-outs lost production and inventory shrinkage, the greater the costs that must be incurred.

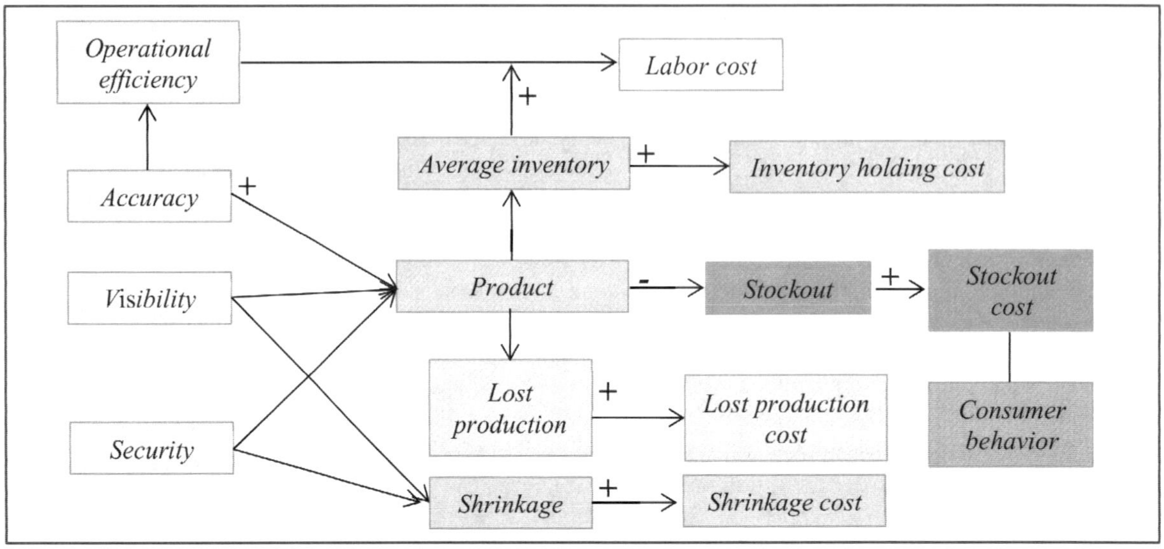

Figure 1. The benefit of Auto-ID Implementation

2.2 *Auto-ID Investment Model Analysis*

In this study, the economic analysis model of Auto-ID technology investment used refers to the mathematical model used by Ustundag [5]. Ustundag conducted a cost-benefit analysis which was also carried out in this study. Ustundag used the FRBS method and Monte Carlo simulation to calculate NPV. In this study, in addition to calculating the reduction of labor costs and inventory shrinkage, it will also calculate the reduction of stock-out costs.

Inventory shrinkage is an item that has been in the warehouse for a long time and has not been used (dead inventory). With the inventory shrinkage, the company must incur storage costs and also the cost of loss because the item is considered unused or lost in the ROL spare part warehouse management system, which is now still manual. Jobs such as calculating the number of ROL spare parts, searching for, and routinely checking the number of ROL spare parts are still done manually by employees. This causes high labor costs because workers need a long time to do these jobs. By investing in Auto-ID, it is expected to reduce labor costs because it can minimize worker time in doing ROL spare part warehouse management jobs.

Stock-out costs are costs caused by the unavailability of goods when needed. The consequences of stockouts can cause the production process to stop. This is because when the factory requires ROL spare parts during maintenance, the required ROL spare parts are not available. In this study, the occurrence of stock out of ROL spare parts will result in large losses to the company's system, one of which can cause a decrease in productivity levels. By reducing the cost of stock out can increase the company's profit. This is because, with the availability of the number of ROL spare parts and the fulfillment of end-user demands, the maintenance process can run smoothly so that productivity increases. Furthermore, the basic model created is used to analyze the benefits obtained by the company by investing in Auto-ID technology in the warehouse system.

2.3 *Proposed Model Analysis*

In this study, the costs of Auto-ID are divided into three areas such as hardware cost (C_H), middleware cost (C_M), software cost (C_S), and service cost (C_Z). The Auto-ID benefits are broken down into two parts: cost reduction and increase productivity and maintenance. The total value of the benefit is calculated by using Eq. (1), which is developing a model of Ustundag [11]. The cost reduction is computed considering the increased spare part order (Q'), cost/unit (c), and cost reduction rate (r). Cost reductions are calculated in Eqs. (2) – (4).

$$T_B = \Delta C_{IS} + \Delta C_{lab} + \Delta C_{sto} + \Delta P - C_{tag.n} \quad (1)$$

$$\Delta C_{IS} = Q' x \, c_{IS} \, x \, r_{IS} \quad (2)$$

$$\Delta C_{lab} = Q' x \, c_{lab} \, x \, r_{lab} \quad (3)$$

$$\Delta C_{sto} = Q' x \, c_{sto} \, x \, r_{sto} \quad (4)$$

Where ΔC_{IS} is a reduction of inventory shrinkage cost, ΔC_{lab} is a reduction of labor cost, ΔC_{sto} is a reduction of stock-out cost, ΔP is production and maintenance increase, and $C_{tag.n}$ is tag cost in year-n.

The increase spare part order (Q') and productivity and maintenance increase (ΔP) are calculated by:

$$Q' = Q(\mu, \sigma) x \, (1 + d) \quad (5)$$

$$\Delta P = p \, x \, Cprod \quad (6)$$

Where Q is the actual demand for spare parts, d is the increase order of spare part ROL, p is increased productivity (%), and C_{ord} is production cost per year.

The NPV and IRR [12] value of the total Auto-ID investment is calculated using Eq.7 and Eq.8, respectively. The NPV and IRR value is used to estimate the likely financial outcome of the investment. The NPV Monte Carlo simulation is utilized in this study to reduce the random variable's error probability.

$$\mathrm{NPV} = -(C_H + C_M + C_S + C_Z) + \sum_{n=1}^{t} \frac{T_B}{(1+i)^n} \quad (7)$$

$$\mathrm{IRR} = -I_0 + \sum_{t=n}^{n=1} \frac{CF_t}{(1+IRR)^t} \quad (8)$$

Where I used the term "discount rate" as an index.

The rules established for the increase in spare part ROL orders are shown in Table 1.

Table 1. The Rules of Increase Orders Spare Part ROL

Rule		Delivery Time		Type Accuracy		Amount Accuracy		Increase Order
1	If	Short	AND	High	AND	High	THEN	High
2	If	Short	AND	High	AND	Medium	THEN	Medium
3	If	Medium	AND	Medium	AND	High	THEN	Medium
4	If	Medium	AND	High	AND	High	THEN	High
5	If	Long	AND	Medium	AND	Low	THEN	Low
6	If	Low	AND	Low	AND	Low	THEN	Low

2.4 Selection of Auto-ID solution supplier

In this study, vendor selection decisions the Auto-ID refers to the research that has been conducted by Sari (2013) [13]. Figure 1 shows the hierarchical structure of the decision model. The model aims to help managers of a company for deciding the best provider of Auto-ID.

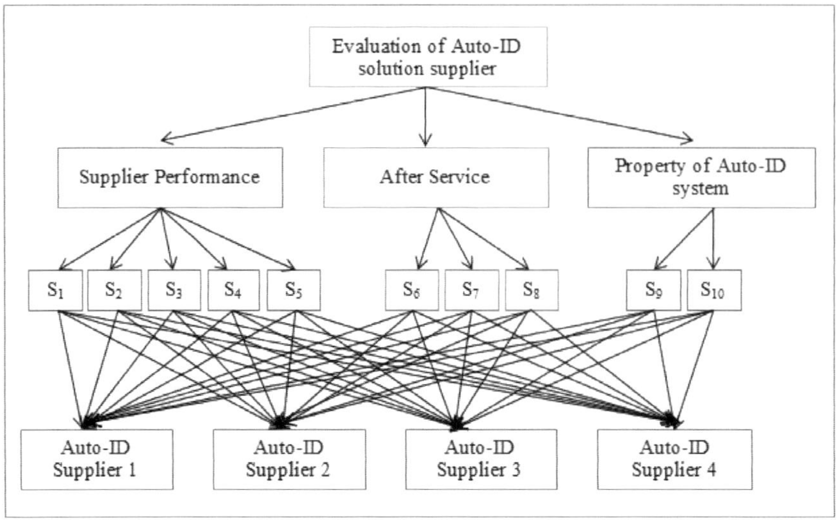

Figure 2. Hierarchy structure of provider elective decision

3. Result

3.1 The benefit of Auto-ID Investment

By using a 10% discount rate and an average ROL demand for critical spare parts of 66,654 units with a standard deviation of 80776%, the total benefit value is Rp 1,462,764,326.60 for RFID investment and Rp 269,305,292.29 for barcode investment. With the same discount rate (10%), NPV RFID (3,659,844,069.75) > NPV barcode (913,723,450.85, but the percentage value of IRR RFID (29%) < IRR barcode (55%). with RFID investments with a faster payback time compared to RFID investments.

Table 2. Auto-ID Cost Benefit Investment

	RFID	*Barcode*
Total benefit (Rp)	1,462,642,632.83	269,614,104.90
NPV (Rp)	3,659,237,181.61	915,270,294.29
IRR (%)	29.00%	55.49%

With the same expenditure for inventory shrinkage, labor, and stock-out costs for both types of technology, but the amount of value reduced after investing in different technologies. Auto-ID technology investment in ROL spare parts warehouse aims to increase company profits and benefits. Based on [14] said that in the case of mutually exclusive projects, the project with the larger NPV should be accepted. By using a discount rate of 10% and an average ROL demand for critical spare parts of 66,654 units with a standard deviation of 80776%, the total benefit value is Rp. 1,462,764,326.60 for RFID investment and Rp 269,305,292.29 for barcode investment. With the same discount rate of 10%, NPV RFID (3,659,844,069.75) > NPV barcode (913,723,450.85, but the percentage value of IRR RFID (29%) < IRR barcode (55%). Based on the IRR value, barcode technology investment can provide a higher level of certainty of profits compared to RFID investments with a faster payback time compared to RFID investments.

3.2 Monte Carlo Simulation Analysis

The actual demand for ROL spare parts (Q) is a variable with a random value; this is due to the level of uncertainty in demand for ROL spare parts. The company cannot ensure the number of requests for

ROL spare parts in the next period because end users can request ROL spare parts at any time when there is a problem with the machine and repairs or maintenance must be carried out.

By implementing the input data into the model, the results of the cost-benefit of Auto-ID investment are determined. The average amount of yearly spare part ROL demand of the company is 60,954 units with a standard deviation of 913%. The NPV and IRR value of Auto-ID investment in five years horizon is positive. The distribution of the NPV of RFID investment has the mean of Rp 3,660,399.54 and standard deviation of Rp 4,487,890.00 shown in Fig. 2.

Managers of industry manufacture can use this research study to perform a better analysis of possible Auto-ID providers and then choose the best solution provider for their organization. As shown in Fig.3 the output of alternative Auto-ID provider is determined. Now, based on Fig.4, we can rank the solution providers from best to worst as PT P, PT Z, PT Y, and PT X. On average, Provider P is the better alternative.

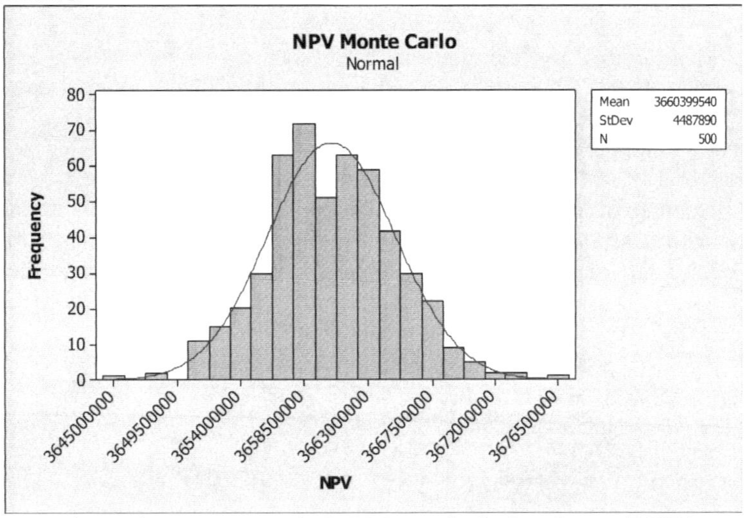

Figure 3. *Histogram of* NPV *Monte Carlo Simulation* (RFID)

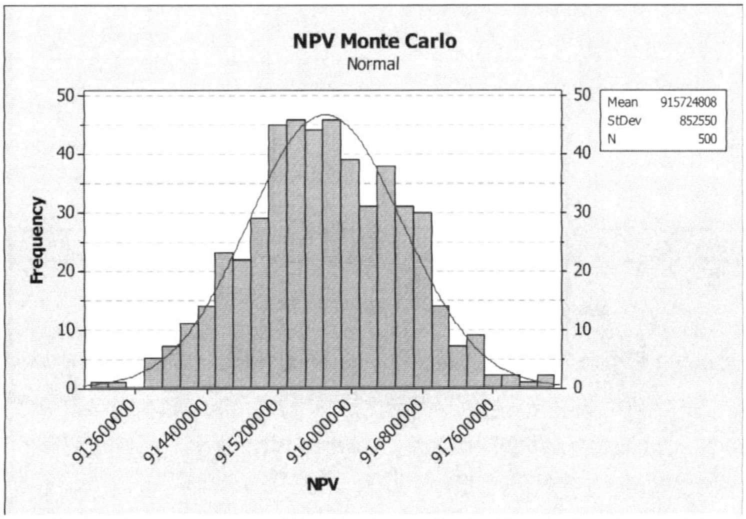

Figure 4. Histogram of NPV Monte Carlo Simulation (Barcode)

Based on the simulation results, the total cost saving for RFID is between Rp. 66,750,000.00 - Rp. 72,000,000.00, and the total benefit for RFID is between Rp. 1,453,500,000.00 and Rp. 1,458,750,000.00. Meanwhile, the total cost saving for barcodes is between Rp. 12,150,000.00 and Rp. 13,050,000.00, and the total benefit for barcodes is between Rp. 268.050,000.00 and Rp. 269,100,000.00. The NPV distribution of RFID technology investment has an average value of Rp. 3,660,399.54 and a standard deviation of Rp. 4,487,890.00 with an interval between Rp. 3,645,000,000.00 and Rp. 3,676,500,000.00, and the NPV of barcode investment has an average value of Rp. 915,724,808.00 and a standard deviation of Rp. 852,550.00 with an interval between Rp. 913,600,000.00 and IDR 917,600,000.00.

Based on the simulation results obtained, it can be concluded that the NPV of Auto-ID technology investment provides a positive value with a high degree of certainty. Therefore, company managers can decide to invest in RFID or barcode technology.

3.3 Auto-ID Supplier Analysis

After doing the calculations, the percentage value of the weights for each Auto-ID supplier is obtained. Figure 5.8 shows that supplier PT P has the highest weight value of 44.2% compared to the other three alternative suppliers. The next global weight ranking in order is PT Z, PT Y, and PT X, with 26.5%, 16.4%, and 12.9% weights. Supplier PT P has a value that is in accordance with what the company wants. With the advantage of being able to provide cost-effectiveness to consumers, PT P's suppliers are able to increase the company's profit according to the conditions desired by the company. In addition, PT P also provides customized tag services that can be adapted to the conditions and wishes of the company. Thus, supplier PT P is the best supplier that can be chosen by company managers.

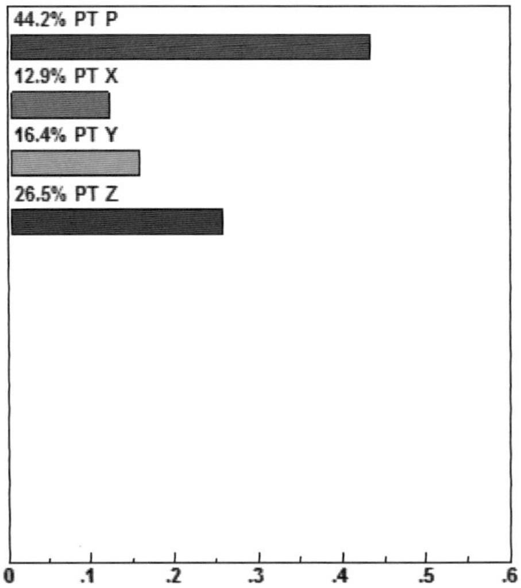

Figure 5. The output of alternative Auto-ID provider

4. Conclusion

The cost and benefit analysis is used to evaluate the economic analysis of Auto-ID investments within a corporation in this study. Hardware, middleware, software, and service charges are all included in the cost of deploying Auto-ID. The key advantages are cost savings, increased productivity, and easier maintenance. The goal of this study is to develop an integrated model that takes into account the increased productivity and maintenance that Auto-ID brings. The Auto-ID solution provider choice model is designed to assist managers of an organization or corporation in efficiently evaluating possible Auto-ID solution providers and then selecting the best one for their specific situation. The greater the cost reduction and the greater the increase in productivity and maintenance, the greater the total benefit obtained. Likewise, with the value of NPV and IRR. However, even though the production costs incurred are greater, the company will get big profits too, this is because production runs smoothly. Based on AHP weight calculation, the optimal suppliers is PT C with a weight of 39.3%, then PT A with a weight of 27.2%, PT D with a weight of 19.7%, and the last priority was PT B Indonesia with a weight of 13.8%. Company managers can make decisions to choose suppliers based on the criteria and according to the different interests. The criteria will have an effect on reducing production costs, increasing productivity, and user satisfaction.

5. References

[1] K. Finkenzeller, *RFID Handbook, Fundamentals and Applications in Contactless Smart Cards, RFID and near-field communication (third edition)*. 2010.

[2] U. Ramanathan, R; Ramanathan, "Adoption of RFID Technologies in UK Logistics: Moderating Roles of Size, barcode experience and Government Support. Application," *Expert Syst. Appl.*, vol. 41, no. 1, pp. 230–236, 2014.

[3] E. Schuster, "Auto-ID Technology: Creating an Intelligent Infrastructure for Business," *Bus. Intel. Rep.*, vol. 5786, no. c, pp. 1–38, 2005, [Online]. Available: http://web.mit.edu/edmund_w/www/CutterAutoIDReportV2.pdf.

[4] P. J. Sweeney II, "RFID for dummies," 2005.

[5] A. Üstündağ and E. Çevikcan, "Return on investment analysis for evaluation of RFID implementation on cargo operations," *2007 1st Annu. RFID Eurasia*, no. May 2019, 2007, doi: 10.1109/RFIDEURASIA.2007.4368145.

[6] A. Baysan, S ;Ustundag, "The Cost -Benefit Models for RFID Investments," 2013, [Online]. Available: http://doi.org/10.1007/978-1-4471-4345-1.

[7] L. Ivantysynova, "RFID in Manufacturing," *RFID Appl.*, pp. 321–324, 2007, doi: 10.1002/9780470168226.ch15.

[8] B. Srivastava, *Radio Frequency ID Technology: The Next Revolution in SCM*. 2004.

[9] K. H. Kurniawati, D; Yuliando, H; Widodo, "Kriteria Pemilihan Pemasok Menggunakan Anlaytical Network Process," 2013.

[10] udin Ciptomulyono, *Paradigma Pengambilan Keputusan Multikriteria dalam Perspektif Pengembangan Projek dan Industri yang Berwawasan Lingkungan*. 2010.

[11] A. Ustundag, M. S. Kılınç, and E. Cevikcan, "Fuzzy rule-based system for the economic analysis of RFID investments," *Expert Syst. Appl.*, vol. 37, no. 7, pp. 5300–5306, 2010, doi: 10.1016/j.eswa.2010.01.009.

[12] M. I. Dachyar, "Analisis Kelayakan Investasi Dan Risiko Proyek Pembangunan Pltu Indramayu Pt Pln (Persero) Analisis Kelayakan Investasi Dan Risiko Proyek Pembangunan Pltu Indramayu PT PLN (PERSERO)," 2012.

[13] K. Sari and M. Carlo, "Selection of RFID solution provider A fuzzy multi-criteria decision model with," 2014, doi: 10.1108/03684921311323680.

[14] S. R. Jory and N. O. Madichie, "Net Present Value Analysis and the Wealth Creation Process : A Case Illustration Abdelhafid Benamraoui Devkumar Roshan Boojihawon," *Account. Educ. J.*, vol. XXVI, pp. 85–99, 2016.

Design and Development of Water Distribution Monitoring System in Regional Drinking Water Companies (PDAM) Based On Internet Of Things

Efrizon[1], M. Irmansyah[2], Era Madona[3], Anggara. N[4] and Yultrisna[5]

[1,2,3,4] Electrical Engineering Polytechnic State of Padang

Jl. Kampus Limau Manis, Kec. Pauh, Kota Padang, Sumatera Barat, Indonesia

Email: efrizon@pnp.ac.id

Abstract. The purpose of this study is to create a prepaid PDAM clean water distribution system using a microcontroller based on the Internet of Things (IoT). The hardware used to realize the system consists of ultrasonic sensors, water flow sensors, relays, LCD buzzers and Arduino. ESP 8266 01 for delivery to the Thingspeak app. From the test results obtained HC-SR04 ultrasonic sensor reading error occurs when the water level is low and too high, the maximum measurable water level is 95%. When calculating the comparison between the water discharge that is read by the sensor and that measured by the measuring cup, the results are always not the same. The error when testing the water flow sensor at the water level is less than 49% this is influenced by the speed of the water fired by the pump, where the pump will be under low pressure when the water level is below that value. The system can monitor data readings from the water flow sensor using the ESP8266 monitored on the thinkspeak web server using a smartphone. Overall the tool can function well.

1. Introduction

Water is an essential compound for all forms of life. Humans are never separated from the use of water every day, both for consumption and for other daily uses. Clean water has become one of the basic needs for humans. Guaranteed water used by humans also affects health. Currently, the government distributes clean water through the Regional Drinking Water Company (PDAM). The distribution of PDAM water to the community is currently done on a postpaid basis. Where the community first uses PDAM water, then pays it once a month. The PDAM calculates the volume of water distributed using a water meter. The water meters currently in use require periodic calibration so that they are accurate in calculating the volume of water used by residents.

From the PDAM water distribution system, problems often occur including clean water from PDAM sometimes not flowing or dead, and frequent damage to residents' water meters so that the increasing PDAM water bill costs are not in accordance with the use of water, calibrating water meters regularly so that the level of problems with the cause of the damage to the water meter is reduced, residents often forget that it is time to pay for PDAM water and result in fines because they have passed the payment due date. This makes consumers who use PDAM water feel disappointed and complain about it. Based on these problems, this study designed a monitoring tool for PDAM water usage that enters residents' homes so that it can be monitored directly (online) and can be monitored remotely using the internet network. The development of information technology has given birth to technological innovations such as the Internet of Things (IoT) [1]. Currently, IoT has been applied to agriculture [2], health [3], parking monitoring [4] and industry[5], IoT technology in basically used as control [6] and monitoring [7].

Research related to water distribution has been done previously. The study conducted by [8] proposed a web-based real-time water monitoring system. The MPX5700AP sensor is used to measure water pressure. The GSM/GPRS module functions for communication with the server via the GSM/GPRS network, the results of the study show that the system can measure water pressure and display graphs on the web. A study conducted by [9] proposed a water monitoring system using an IoT-based model. The prototype is made using an ultrasonic sensor to detect the water level, the GSM module will send a message when there is activity on the water pump. A study conducted by [10] realized an IoT-based Smart Water Level Controlling System [WLCS] to measure water levels and control the filling of water tanks. Similar research was also carried out by [11] to realize a tool that can monitor water levels automatically. The ultrasonic sensor HC-SR04 will detect the water level. If the water reservoir is full or reaches the 100% level, the sensor will detect the water level and the water supply pump will automatically turn off.

In this study, a system was created to monitor the volume of water quota with ESP8266 using Thingspeak. The system made is a prototype with simulation testing. The purpose of this study is to create a prepaid PDAM clean water distribution system using a microcontroller based on the Internet of Things (IoT).

2. Research Method

The design of this system consists of hardware design and software design. Figure 1 is a block diagram of a water distribution monitoring system design at a Regional Drinking Water Company (PDAM) based on the Internet of Things (IoT).

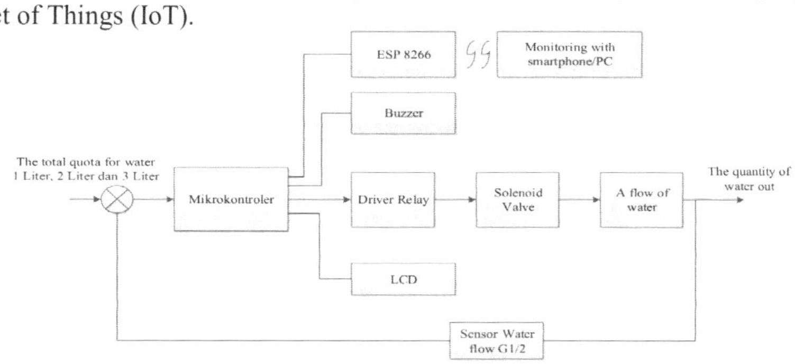

Figure 1. System Block Diagram

From Figure 1, it can be explained that the input of clean water to residents' homes is carried out using a personal computer (PC) through a web page that is connected to a LAN (local area network). If the data has been inputted into the microcontroller then the water quota can be used. by pressing the push button on the house meter so that the pump and solenoid valve are active and water can flow into people's homes. The data from the push button is received by the microcontroller to activate the pump and solenoid valve. When the water quota is up, the solenoid valve will automatically close. When water flows into people's homes, the water will go through the water flow sensor. By using this water flow sensor, the volume of water entering the residents' homes is calculated whose sensor reading data is sent to the microcontroller and then displayed on the LCD. Monitoring via a smartphone on this device uses an ESP8266 with a Thinkspeak web server. By using Thinkspeak the amount of water that has been used can be continuously monitored as long as you have internet access. The reading data from the Water Flow sensor is displayed on Thinkspeak in the form of a graph. If the calculation has reached the quota from the purchase of water, the selonoid valve closes and turns on the buzzer at the residents' houses as a warning for water consumption per 1 liter and when the water quota has run out.

2.1 Hardware Design

To control the 12Vdc solenoid valve, a switching circuit is used which is connected to pin 6 and pin 5. When the pin on the Arduino Mega 2560 is HIGH, the solenoid valve will open (normally open) and when the pin on the Arduino Mega 2560 is LOW, the solenoid valve will be closed (normally closed). In this circuit using an LED connected to a switching transistor as an indication of whether the solenoid valve is active or not. To control the ON/OFF conditions on the 220Vac water pump, a relay driver circuit is used which is connected to pins 7 and 8. This circuit also uses an LED which functions as an indicator when the water pump is ON or OFF as shown in Figure 2.

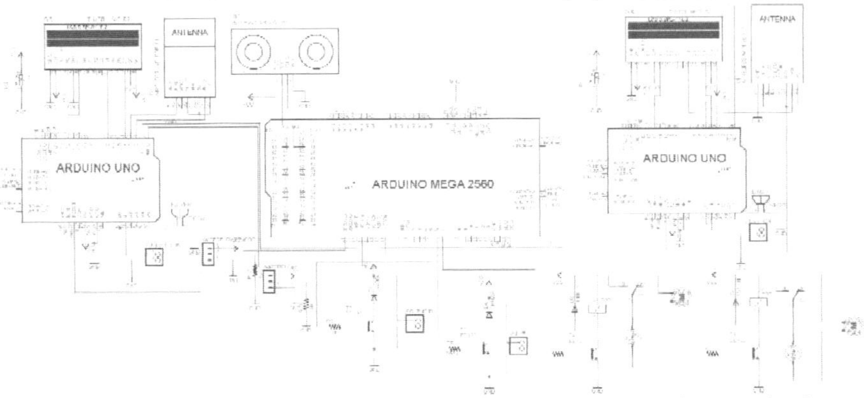

Figure 2. Overall Series of IoT-Based PDAM Water Distribution Monitoring

The Water Flow G1/2 sensor is used to calculate the amount of water that has flowed in the residents' homes connected to pins 20 and 21 by calculating the speed of water emitted by the water pump. ultrasonic sensor HC-SR04, the trigger pin is connected to pin 48 while the echo pin is connected to pin 46 of the microcontroller, it is used for reading water available in the main gallon. The data read is the water level in the main gallon whose reading results are displayed on a web page on a PC. The data that has been read by the G1/2 water flow sensor stored on the arduino mega 2560 is sent to the arduino uno R3 using serial communication using the TX pin and the Rx pin to be displayed on the arduino uno LCD and monitored with the thinkspeak account using the ESP8266.

This circuit 5Vdc buzzer connected to pin 6 arduino uno. When the pin on the Arduino Uno R3 is HIGH, the buzzer is on and when pin 6 is LOW, the buzzer will turn off. The series of push buttons on this tool are used to activate the water quota that has been inputted. When the push button is pressed after the water is input, the water can flow to the customer connected to pin 13. The ESP8266 circuit is a circuit made for communication with internet monitoring data that has been read by the sensor. In this circuit using a 3.3Vdc voltage which is connected to the 3.3V pin on the Arduino Uno. When this Arduino has a HIGH logic, the ESP8266 01 will be active and can send sensor data that is connected to the internet network. The mechanical design of the tool can be seen in Figure 3.

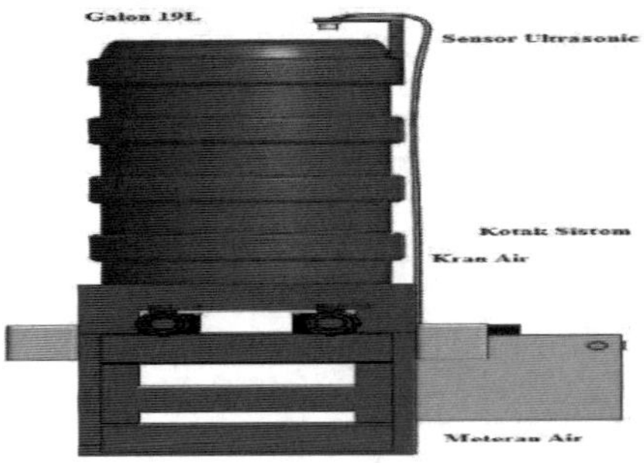

Figure 3. Mechanical Design of IoT-Based PDAM Water Distribution Monitoring

2.2 Software Design

The software design process starts from initialization, then the sensor will read the water level in the filling tank, if the water level is 90% then the water will be distributed and will display a web page for filling the water quota. After the water is distributed, the customer will press the push button. If the push button is HIGH then the solenoid valve is ON and the water flow sensor reading will be displayed on the LCD and the data is sent to the smartphone via ESP8266. If the water quota is the same as the water flow sensor reading, the buzzer is active and the solenoid valve is OFF, as shown in Figure 4.

3. Experiment Result

In this section, the system testing that has been created is carried out. The test was carried out in two stages. The first stage is hardware testing and software t.esting. The prototype of the tool can be seen in Figure 5.

3.1 Hardware Testing

***3.1.1.** HC-SR04 Ultrasonic Sensor Testing with Arduino Mega 2560*

Tests carried out on the ultrasonic sensor HC-SR04 to determine the performance of the sensor when used to transmit water level data in large gallons in the form of a centimeter (cm) distance which will be calculated into the water level level on the microcontroller. The test carried out on this sensor is to measure the water level which is read by the ultrasonic sensor HC-SR04 and the sensor data is compared with the ruler and the data measured by the sensor will be converted into a water level level in the form of percent, the data can be seen in table 1. the data obtained in testing the ultrasonic sensor HC-SR04 can be made in a comparison graph between the water level and the water level read by the sensor, figure 6 graphs of the comparison of the data.

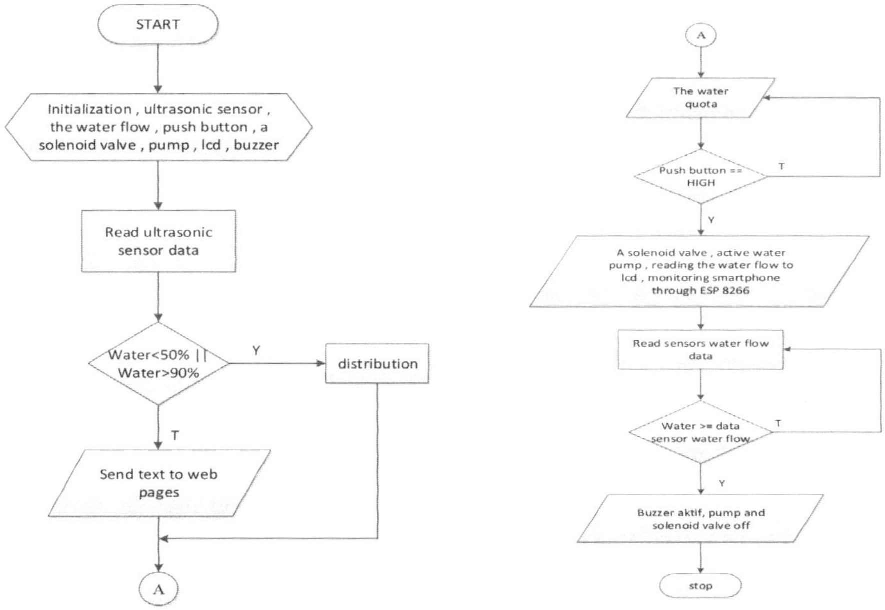

Figure 4. IoT-Based PDAM Water Distribution Monitoring Flowchart

Figure 5. Prototype of IoT-Based PDAM Water Distribution Monitoring

Table 1. Tests on the Ultrasonic Sensor HC-SR04

The distance condition sensor read ultrasonic (cm)	distance actually (cm)	Error measurement distance (%)	water level (%)
2	1	-100	95
2	2,5	20	95
4	4,5	11,11	95
6	6	0	84
8	8	0	84
10	10	0	73
12	12	0	67
14	14	0	62
18	18	0	50

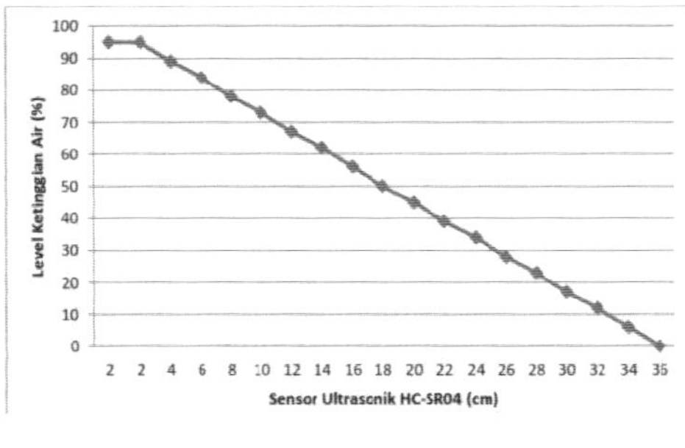

Figure 6. Water Level Chart

Ultrasonic sensors are sensors that work on the principle of ultrasonic waves. To get a distance measurement on the ultrasonic sensor, the trigger pin is first triggered so that it can emit ultrasonic waves which will be reflected by the water object and received by the echo pin on the ultrasonic sensor. The time difference between transmitting ultrasonic waves until it is received back by the ultrasonic receiver is a pulse that will be converted by Arduino to produce data from the amount of water capacity. The formula used in the program to convert the pulses captured by the echo is cm = duration * 0.034/2. Duration is the time span of pulse transmission from transmitter to receiver, and 0.034 is the speed of sound in microseconds. The error data contained in the table of experimental results is caused by the reflected plane of the signal from the transmitter to the receiver being different from the other data. The water storage tube is not cylindrical so that when the transmitter signal the field the reflection obtained is not from the surface of the water but from a concave wall. So based on the workings of the ultrasonic sensor where the time difference between the transmitter signal and the receiver signal becomes shorter. When the length of time is converted to distance, the value becomes smaller than the value of the measuring instrument. The shorter time causes the existing distance data to be small. Errors do not occur when the reading is in the middle of the tube, this proves the curvature at the bottom and top that affects the reading.

When the sensor reading remains 2 cm even though the distance obtained on the measuring instrument is already smaller than 2 cm, this is due to the ability of the ultrasonic sensor based on the data sheet which explains that the minimum distance of the ultrasonic sensor to read is 2 cm, after that the sensor cannot read anymore. The error value obtained based on the bottom reading of the tube proves that the more concave a field is, the faster the signal bounces then over time the concave plane returns to linear, this can be seen based on the smaller error value. The largest error value is when the sensor's minimum capability is so that the error becomes 100.

The water level is the reciprocal of the distance measured by the sensor, the smaller the sensor distance, the greater the water level, and vice versa. This value indicates that if the water approaches the ultrasonic sensor which is above the gallon, then the distance of the measured water is getting closer to the sensor or the water in gallons is higher. Therefore, the water level is getting bigger. The water level cannot reach 100 percent due to the sensor performance. Increase the water level level with a different sensor distance, where the sensor distance range is 2 cm while the water level increase range is 5 and 6 cm.

3.1.2. Testing the Water flow Sensor Circuit

Water flow sensor testing is done to determine how the performance of the water flow sensor in measuring water flow. To find out the time needed to reach the input quota, use a stopwatch. The quotas distributed to households are 1 liter, 2 liter, and 3 liter. The water flow sensor when the data pin is fed by water has a voltage of about 3.5 to 3.9 V but when it is not flowing water the voltage measured at the data pin is 0.08 V. So this sensor sends a voltage to the Arduino pin used by the sensor. The test results for the 1 liter quota can be seen in table 2.

Table 2. Water Flow Sensor Test Results

Input (Liter)	Water level	Time (s)	Measuring glass (mL)	Water Discharge Measured By Sensor (mL)	Error (%)
1 Liter	95	12,66	1000	1034	3,29
	90	12,55	970	994	2,41
	87	12,51	1000	1022	2,15
	82	12,45	1000	1025	2,44
	76	13,06	1050	1067	1,59
	71	12,72	1000	1043	4,12
	65	13,06	1000	1058	5,48
	60	13,26	950	1014	6,31
	52	13,01	1000	1041	3,94
	49	12,5	950	981	3,16
	25	14,59	960	993	3,32
	19	13,98	940	971	3,19
	9	13,89	925	970	4,64

This water flow sensor generates a frequency when the hall effect detects the magnetic field from the rotor, which generates a frequency pulse. The frequency pulse that will be calculated in the program to calculate the speed of the water flow, from the flow velocity can be used to calculate the water discharge. Based on the data that has been obtained, the resulting water discharge will be influenced by the level of the water level. The height of the water level is the pressure for the product of the rate at which water is fired at the pump. Based on table 2, it can be seen that when the water level starts to get low, the resulting water discharge will decrease from the quota, and the time required exceeds the average of 1 second. The formula used to obtain this flow rate is as follows:

flowRate = ((1000.0 / (millis() - oldTime)) * pulseCount) / calibrationFactor;

Since the water flow calculation is iterative, therefore by calculating the number of milliseconds that have elapsed since the last execution and using it to develop the sensor output results. The program formula also applies calibration to develop results based on the number of waves per second per unit of measure (liters/minute in this case) coming from the sensor, this calibration aims to get the number of pulses per second for the size per liter of water according to the speed of the water. This calibration is done by entering a value to get the output of the water discharge according to the readings from the program calculations. After getting the flow rate in liters per minute, this flow rate will be converted into milliliters per second using the formula:

flowMilliLitres = (flowRate / 60) * 1000;

Based on this formula, divide the flow rate by 60 seconds and multiply by 1000 mL. After getting the flow rate data in mL per second, this data is always added to get the water flow sensor reading data, using the program instructions:

totalMilliLitres += flowMilliLitres;

From these data, it has been obtained the water discharge from the speed of the water. So the water discharge counter sensor used in this tool will not always fit from the data, because the addition of total MilliLitres of water to achieve the desired data will not always be the same.

3.2 Software Testing

This test aims to determine the performance of ethernet shield and esp8266 on this prepaid PDAM tool. The Ethernet shield functions as a web server. Network programming using HTML programmed in arduino. Testing the Arduino web server is done by connecting the Arduino web server to a personal computer using a UTP cable. Then the IP address of the computer is configured so that it can connect to the arduino web server. The results of the quota input test and monitoring using a smartphone can be seen in Figure 7.

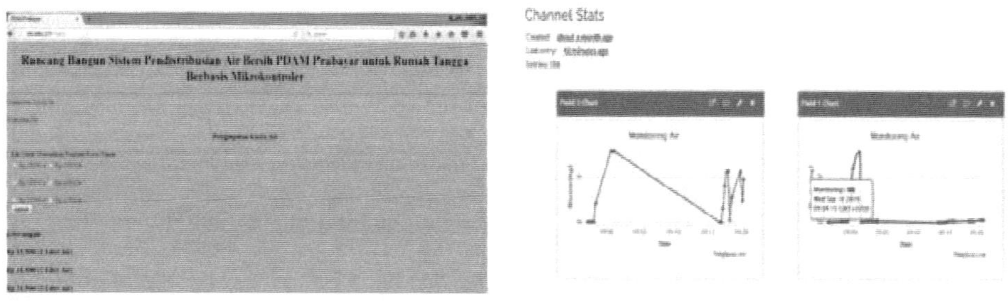

Figure 7. The results of the quota input test and monitoring using a smartphone

This prototype system uses a computer network to connect to devices that function to monitor water levels and transmit data via PC devices. If the system is connected to the internet network, the information and data can be accessed from anywhere as long as the device used to control it is also connected to the internet network, but this system is only based on a LAN network. Arduino functions as a webserver, where HTML programs are located on Arduino which does not require a data base. That is, it only displays the sensor data that we want to display on the web page according to the html program created, namely with the client.print(sensor data) program instructions. But to send data using the program instruction if(Readstring.indexOf(the page to be read by Arduino)) after that the action that will be carried out by Arduino, for example inputting 1 liter of water quota and the action performed on the program by providing 1000 data, then The data will be processed by Arduino according to the program that has been made.

4. Conclusion

Based on the design of the prepaid PDAM system that has been made, it can be concluded that (1) HC-SR04 ultrasonic sensor reading error occurs when the water level is low and too high, because the minimum reading of the HC-SR04 ultrasonic sensor is 2 cm. So the maximum measurable water level is 95%. (2) When calculating the comparison between the water discharge read by the sensor and that measured by the measuring cup, the results are always not the same. This error is influenced by the speed of the water fired by the pump, where the pump will be under low pressure when the water level is less than 49%. (3) The system can monitor data reading from the water flow sensor using the ESP8266 monitored on the thinkspeak web server using a smartphone.

References

[1] Miraz, M. H., Ali, M., Excell, P. S., & Picking, R. (2015). A review on Internet of Things (IoT), Internet of Everything (IoE) and Internet of Nano Things (IoNT). *2015 Internet Technologies and Applications, ITA 2015 - Proceedings of the 6th International Conference*, 219–224. https://doi.org/10.1109/ITechA.2015.7317398

[2] Farooq, M. S., Riaz, S., Abid, A., Abid, K., & Naeem, M. A. (2019). A Survey on the Role of IoT in Agriculture for the Implementation of Smart Farming. *IEEE Access, 7*, 156237–156271. https://doi.org/10.1109/ACCESS.2019.2949703

[3] Islam, S. M. R., Kwak, D., Kabir, M. H., Hossain, M., & Kwak, K. S. (2015). The internet of things for health care: A comprehensive survey. *IEEE Access*, *3*, 678–708. https://doi.org/10.1109/ACCESS.2015.2437951

[4] Irmansyah, M., Efrizon, E., Madona, E., & Putra, R. (2018). Perancangan Sistim Monitoring Parkir Mobil Berbasis Web. *Elektron : Jurnal Ilmiah*, *10*(2), 11–14. https://doi.org/10.30630/eji.10.2.72

[5] Christoulakis, F., & Thramboulidis, K. (2016). IoT-based integration of IEC 61131 industrial automation systems: The case of UML4IoT. *IEEE International Symposium on Industrial Electronics*, *2016-November*, 322–327. https://doi.org/10.1109/ISIE.2016.7744911

[6] Ali, M. A., Miry, A. H., & Salman, T. M. (2020). IoT Based Water Tank Level Control System Using PLC. *Proceedings of the 2020 International Conference on Computer Science and Software Engineering, CSASE 2020*, 7–12. https://doi.org/10.1109/CSASE48920.2020.9142067

[7] Prathibha, S. R., Hongal, A., & Jyothi, M. P. (2017). IOT Based Monitoring System in Smart Agriculture. *Proceedings - 2017 International Conference on Recent Advances in Electronics and Communication Technology, ICRAECT 2017*, 81–84. https://doi.org/10.1109/ICRAECT.2017.52

[8] Natividad, J. G., & Palaoag, T. D. (2019). IoT based model for monitoring and controlling water distribution. *IOP Conference Series: Materials Science and Engineering*, *482*(1). https://doi.org/10.1088/1757-899X/482/1/012045

[9] Shankar, S., & Dakshayini, M. (2018). IoT-Mobile Enabled Smart Water Level Controlling System to Regulate Water Wastage. *2018 International Conference on Advances in Computing, Communications and Informatics, ICACCI 2018*, 2045–2048. https://doi.org/10.1109/ICACCI.2018.8554373

[10] Salam, A. E. U., Muh, Tola, Selintung, M., & Maricar, F. (2014). Web based real time water pressure monitoring system. *International Conference on Electrical Engineering, Computer Science and Informatics (EECSI)*, *1*(August), 223–227. https://doi.org/10.11591/eecsi.1.406

[11] Wahyuni, R., Sentana, J. T., Muhardi, M., & Irawan, Y. (2021). Water Level Control Monitoring Based On Arduino Uno R3 Atmega 238p Using Lm016l LCD at STMIK Hang Tuah Pekanbaru. *Journal of Robotics and Control (JRC)*, *2*(4), 265–269. https://doi.org/10.18196/jrc.2489

Developing an engaging whiteboard animation video for vitamins

S Purwaningsih[1] and A A Anggraeni[2]

[1,2]Culinary Science Study Program, Faculty of Engineering, Universitas Negeri Yogyakarta, Indonesia

Email: andian_ari@uny.ac.id

Abstract. Whiteboard animation videos are engaging learning media for Generation Z. This study aimed to develop and assess the feasibility of a whiteboard animation video for vitamin in the Nutrition Science subject for class X culinary vocational school. This study was research and development using a 4D approach (define, design, develop, disseminate). The define stage analyzed the curriculum, material, student character, learning media, and school facilities. The design stage included the writing and assessing the material, the writing and assessing of storyboards, and production team selection. The develop stage was making a video and assessing the feasibility. Animation images were created using Paint Tool Sai. Image conversion from JPEG to SVG was performed with Inkscape. The dubber's voice was recorded with Universal Audio using the Apollo Twin audio interface and Avantone CV12 microphone. The audio was edited in Pro Tools 2020.9. To produce a whiteboard animation video, animation images, captions, dubber's voice, and back sound were combined with VideoScribe. Videos used mpg format. In order to maintain video duration below 10 minutes, the video was divided into two parts. The assessment of video feasibility was carried out by one media expert and two content experts. The assessment of feasibility at the disseminate stage was carried out to 30 users. Based on the feasibility assessment, this video was very suitable to be applied as a learning medium.

1. Introduction

Digitalization occurs in many areas of people's lives. Teaching and learning in schools must be integrated with digital technology. Teachers and students are required to master information and communication technology (ICT). Students can browse learning resources on the internet, so teachers are not the only source of learning [1].

The current vocational high school students are Generation Z, which was born between 1995-2010 [2]. Characteristics of Generation Z are, having digital literacy skills, mastering new media, and ICT. This generation grew up in the rapid development of digital technology [1]. Generation Z is often called homo globalists because technological developments cause them to grow up in the same culture [3]. They are very dependent on technology, fond of wireless technology, must always be connected to the internet, prefer diversity, and have good skills in social interaction in the digital world [1]. Generation Z students prefer flexible learning, which is not limited by location and time. However, they are less independent, so they need designed instruction, feedback, and regular evaluation [1]. Therefore, the application of ICT-based learning needs to be adapted to the characteristics of Generation Z, one of which is in the form of digital learning media.

Learning media that vocational students prefer today is the video [4]. Video is an audio-visual media that can convey material more effectively because video stimulates the sense of hearing and sight [5].

Video shows learning objects concretely, can achieve cognitive and psychomotor skills, motivates to learn, reduces boredom, increases memory retention, and is attractive to Generation Z and easy to distribute. The use of animation in videos is also preferred by Generation Z [6].

Whiteboard animation video is a popular learning media these days. Whiteboard animation video shows the process of drawing objects, and accompanied by narrator explanation. This animation is considered engaging. The most prominent feature of whiteboard animation is the simplicity of the images and short captions [7]. Whiteboard animation can be applied at any level of education.

Nutrition Science is a theoretical and practical subject in culinary vocational high school. Nutrition science is studying how to design and process the food needed by the body. Competencies in nutrition science include analyzing vitamins needed by the body [8]. Learning videos about vitamins are widely available on YouTube. However, those videos have many drawbacks. The material presented in the video is incomplete. Recommended daily value (RDV) for vitamins and a complete explanation of vitamin B complex have not been discussed in those current videos. The audio quality is also not good; the sound is noisy and the sound volume is unstable. Some videos do not have any audio. Besides, some videos use Korean-Pop songs as backsound and/or the back sound volume is too loud. Vocational school students, currently generation Z, favour learning videos with clear audio quality [6]. The visual appearance of those videos is also not attractive because it only displays text and a few images. The average length of those videos is 10-20 minutes, so it does not match the standard duration of a learning video which must be under 15 minutes [9]. Up to today, videos for vitamins are still rarely developed using whiteboard animation applications.

Therefore, this research would develop a whiteboard animation video for vitamin using whiteboard animation. This video discussed the description of vitamins, the function of vitamins, types of water-soluble and fat-soluble vitamins, food sources of vitamins, the body need for vitamins, and the effects of vitamin deficiency and hypervitaminosis. This study aimed to develop an animated whiteboard video for vitamins and assess the video feasibility.

2. Method

2.1 Research Method

This study used a research and development model using a 4D development model (define, design, develop, and disseminate). Figure 1 shows the development steps in detail.

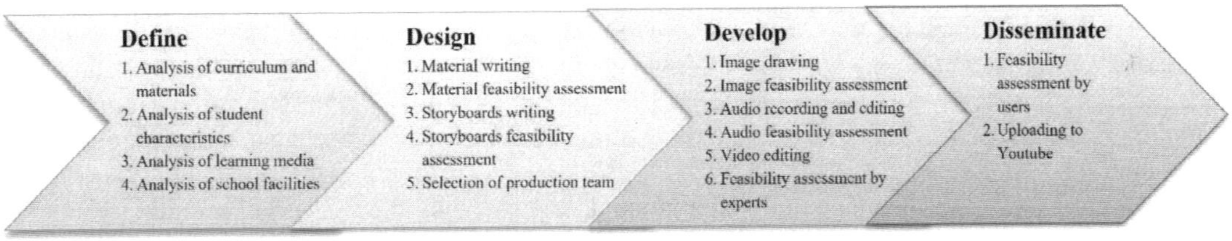

Figure 1. 4D model for whiteboard animation video development

2.2 Feasibility Assessment

The research instrument was composed of closed and open questionnaires of video feasibility for content expert, media expert, and user. The feasibility instrument for content expert covered aspects of learning, content, benefits, and usage. The instrument of media expert feasibility included aspects of media, benefits, and usage. Whereas the instrument for user feasibility consisted of aspects of learning, media, content, benefits, and usage [10]. The instrument was assessed by a construct validity test. The feasibility assessment for the video was carried out by two content experts, one media expert, and 30 users from class X culinary students of SMK N 6 Yogyakarta.

2.3 Data Analysis

The feasibility assessment applied a 4-Likert-scale. Scale meant as follow: 1 very unsuitable, 2 unsuitable, 3 suitable, and 4 very suitable. Data were descriptively analyzed according to formula (1).

$$\text{Feasibility percentage} = \frac{\text{Total score}}{\text{Maximum score}} x\ 100\ \% \qquad (1)$$

The feasibility percentage was transformed by categorizing data from interval-scale to ordinal-scale, as shown in Table 1 [11].

Table 1. Feasibility category

Feasibility Percentage	Feasibility Category
>80%	Very suitable
66%-80%	Suitable
56%-65%	Unsuitable
<56%	Very unsuitable

3. Result and Discussion

Whiteboard animation video for vitamin was developed with a 4D approach as follows.

3.1 Define

The curriculum, material, student character, learning media, and school facilities were analyzed at the define stage. Observations were made at SMK N 6 Yogyakarta. Based on curriculum and materials analysis, SMK N 6 Yogyakarta used student-centered learning and ICT in teaching and learning. However, this learning design has not been implemented optimally. The Nutrition Science teacher said that students still had difficulties in understanding the vitamin material. The results of student character analysis showed that student's character and learning style followed the trait of generation Z. Students preferred learning that was individual, interesting, and visual-based and used educational technology. However, learning media at SMK N 6 Yogyakarta was limited to verbal media such as presentation media, handouts, and books. SMK N 6 Yogyakarta had wifi internet facilities and generators that could support learning with ICT. Based on the analysis results at the define stage, this study would develop an animated whiteboard video for vitamins in Nutrition Science.

3.2 Design

The video was divided into two parts: 1) fat-soluble vitamins and 2) water-soluble vitamins. The first step in the design stage was material writing. The video material was adapted from a reference book for a culinary vocational school in nutrition science. Spoken language is different from written language. Therefore, complicated and complex written language must be transformed into a spoken language with a familiar and simple vocabulary and style [12]. A content expert then assessed the feasibility of the material. The material was then revised according to the content expert's feedback. The revision process of the material was carried out eight times.

The material was then developed into a storyboard. The storyboard consisted of a visual section and a narration section that the dubber would read. The storyboard was arranged in a table consisting of several columns: scene number, duration, visual source, and narration. The final visual source would be animated images. Designing animated images was carried out in several stages: drawing design, sketching, sketch feasibility assessments, sketching revisions, image coloring, colored image feasibility assessments, and image coloring revisions. Figure 2 shows the storyboard development flow in the visual section. Meanwhile, the narration writing process went from narration writing, assessing the feasibility of the narration, and revising the narration. Content expert and media expert carry out the feasibility assessment of the storyboard.

The production team consists of a scriptwriter, illustrator, image format editor, dubber, audio editor, and animator. The production team was selected by considering the background, experience, and expertise [13]. The election of illustrators was carried out by selecting a colored image portfolio. The selection was made based on the image's attractiveness and the coloring technique's suitability with the Scribe software. Scribe is not able to support very delicate color gradients and more suitable for bold colors. This whiteboard animation video used colored images. The dubber selection was made based on the absence of a strong regional accent and the intonation suitability.

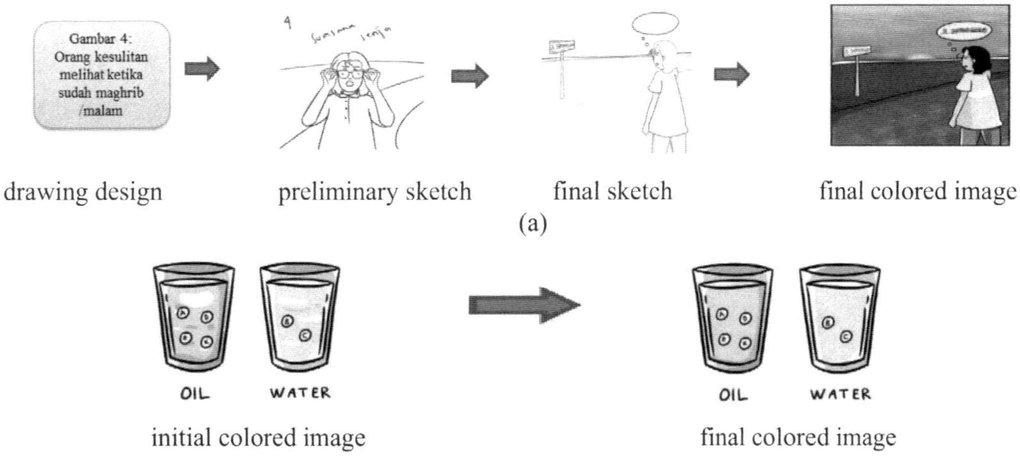

Figure 2. The development flow of the storyboard in the visual section includes: (a) the process from drawing design until it became the final-colored image and (b) the color revision in the image.

3.3 Develop

The development of animated videos was carried out with VideoScribe software. First of all, the animated image was drawn using the Paint Tool Sai software with the JPEG image format. The input image on VideoScibe needs the SVG format; therefore, the image format was converted from JPEG to SVG using Inkscape software. The dubber's voice was recorded with the Asus Zenfone m1 cellphone without editing, making the sound unclear and unstable. Based on media expert advice, the dubber's voice was then re-recorded. The dubber voice was then recorded with Universal Audio using the Apollo Twin audio interface and Avantone CV12 microphone. The audio was recorded in a soundproof recording studio. The audio file was then edited with the Pro Tools 2020.9 software. The content expert then suggested re-recording the dubber's voice in several scenes because of material substitutions or additions.

After the animated images and audio recordings were declared suitable, combining images, caption, audio, and back sound was carried out using the VideoScribe software. Captions in the form of keywords were added to the scene, to help students understand the subject [14]. Videos used mpg format. The duration of video 1 and 2 was 7 minutes 24 seconds and 10 minutes 19 seconds. Figures 3 and 4 show captures from the video of fat-soluble vitamins and water-soluble vitamins.

Figure 3(a) shows video thumbnails on YouTube. Thumbnails used attractive images that represent the content of the video. The title of the video was short and easy to understand. The video's title was written in black Comic Sans font, which was easy to read and gave the impression that the video material was not difficult. The main characteristic of whiteboard animation videos is the display of hand movements while drawing or writing. Most of this video's visual appearances were shown in Figure 3(b), where the hand movement was drawing a black and white sketch, then coloring the sketch image. Alternatively, hand gestures were shown while writing the caption. The final images were shown for the last 5 seconds of each scene. Figures 3(c), 3(d), 3(e), 3(f), 4(a), 4(b), 4(c), 4(d), and 4(e) show the visual appearance of the final images. The duration of each scene was determined by the length of the dubber's narration. Most whiteboard animation videos were developed with black and white sketches. However, this study developed a whiteboard animation video with colored images to make it more attractive to Generation Z students. The video was ended with a mind-mapping of all the material, as shown in Figure 4(f).

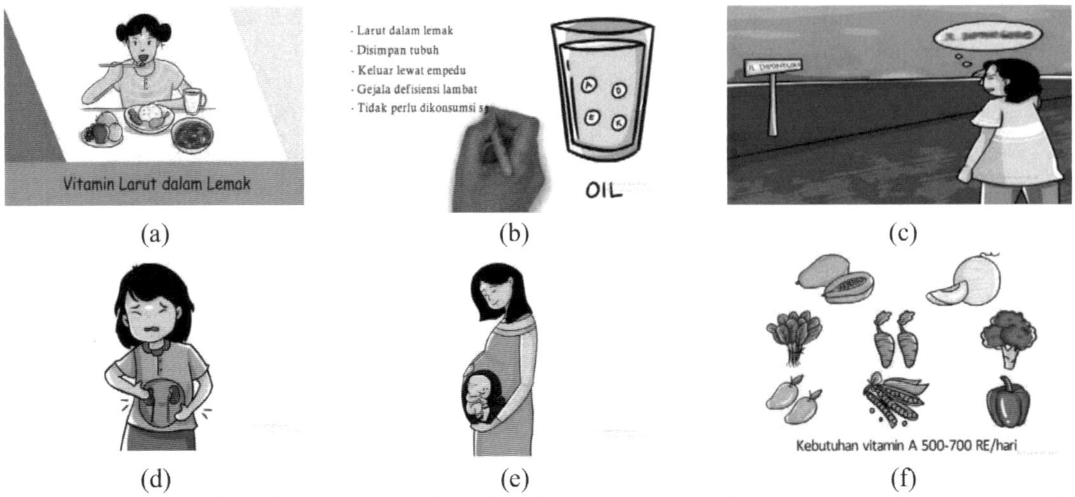

Figure 3. Video of fat-soluble vitamins, including (a) video thumbnails on Youtube, (b) properties of fat-soluble vitamins, (c) the effect of vitamin A deficiency, (d) the effect of vitamin D hypervitaminosis, (e) function of vitamin E, and (f) vegetable sources and RDV of vitamin A.

In order to help the students to understand the material, text captions using Calibri font were added to several scenes, as shown in Figures 3(b), 3(f), 4(a), and 4(d). Some of the images were designed to visualize the human internal organs. Figure 3(d) shows the kidneys, figure 3(e) shows the fetus, and Figure 4(b) shows the lungs. The human animations were drawn in several distances: long-distance images (Figures 3(c) and 4(e)), medium-distance images (Figures 3(d), 3(e), and 4(b)), and close-up (Figure 4(c)). The focus material needed to be conveyed would determine the distance of taking a human object. A long-distance image showed the uncomfortable feeling experienced by humans. A medium-distance image depicted the pain in the internal organs. In contrast, the visible organs, such as eyes, would be drawn as close-up images.

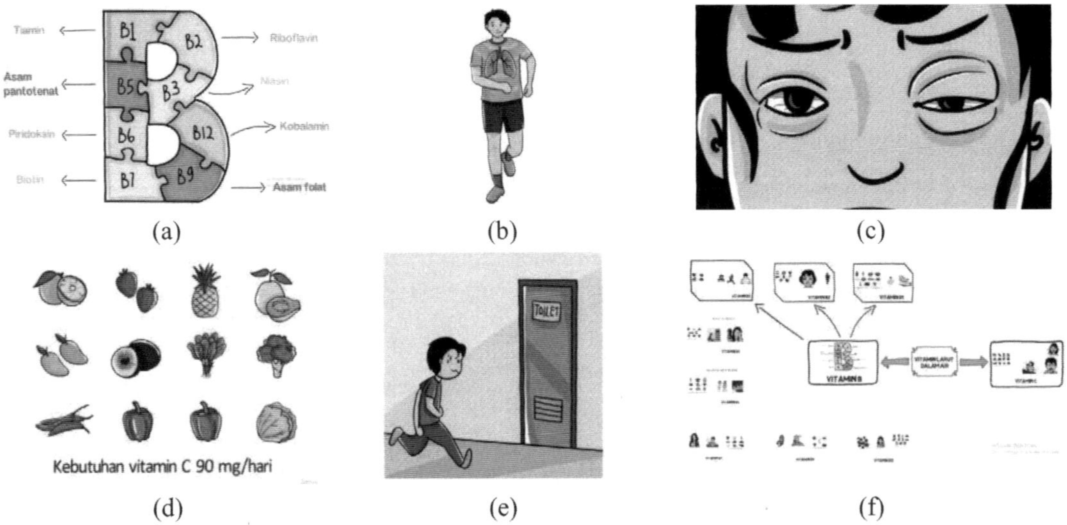

Figure 4. Video of water-soluble vitamins, including (a) types of vitamin B complex, (b) function of vitamin B_2, (c) the effect of vitamin B_2 deficiency, (d) sources and RDV of vitamin C, (e) the effect of vitamin B_5 hypervitaminosis, and (f) mind-mapping of video material.

Figures 3(f) and 4(d) show food sources of vitamins. Because these foodstuffs had different colors, the background was white to contrast with the main object. Some images used color backgrounds, such as Figures 3(c) and 4(e). Figure 4(a) shows a combination of animation images, text captions, and the narrator's voice in explaining the types of vitamins B complex. The vitamin B text font color was black so that it contrasted with the puzzled color that used soft colors. A contrasting color between the font and the background would make the text easier to read. The text color of vitamin B_1 terminology (e.g. thiamine, niacin, and others) were selected as the same color as the puzzle color to facilitate association. The association will reduce students' cognitive load when viewing videos.

The feasibility of animation videos was assessed by content experts twice and media expert once. The feasibility assessment results are presented in Table 2.

Table 2. Feasibility assessment of whiteboard animation video by experts

Aspect	Content expert	Media expert	Category
Learning	95.8 %	-	Very suitable
Content	88.8 %	-	Very suitable
Media	-	95.0 %	Very suitable
Benefit	85.0 %	100.0 %	Very suitable
Usage	92.5 %	100.0 %	Very suitable
Total	89.9 %	96.7 %	Very suitable

The first content expert, a lecturer in Nutrition Science at Universitas Negeri Yogyakarta, suggested several revisions. RDV of vitamins D, E, K, B_1, B_2, B_3, B_5, B_6, B_9, B_{12}, and C needed to be revised following the Minister of Health of the Republic of Indonesia's regulation number 28 of the year 2019 concerning the RDV for Indonesian. It also could refer to the regulation of the Head of the Food and Drug Supervisory Agency of the Republic of Indonesia number 9 of the year 2006 concerning nutrition labels. The first content expert also asked for additional explanations about 1) the effects of hypervitaminosis and deficiency, 2) hypervitaminosis of vitamin E and K, and 3) references at the end of the video. The video was then improved according to those suggestions. However, the second content expert, a teacher of Nutrition Science at SMK N 6 Yogyakarta, stated that the material in the video was coherent and very suitable. The characteristic of learning videos is to convey messages effectively [1].

Media experts provided suggestions for video improvement. The video of fat-soluble vitamins should be added with the caption "vitamin function" in the opening scene at time 0:36 and narrating the functions and properties of vitamins that were explained started at time 0:19. The media expert suggested adding captions to all vitamin B terminology. When the dubber mentions the phrase "vitamin B_1", the caption "thiamin" should be shown with an arrow pointing to B1 text. These revisions should be applied to other vitamin B complexes, as shown in Figure 5. The dubber's explanation can be clarified by adding captions to help the students to understand the subject [15].

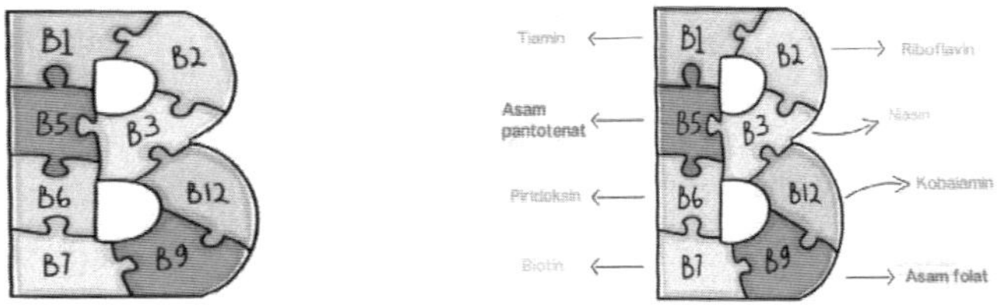

(a) (b)

Figure 5. Revision process based on suggestions from media expert: (a) before revision, and (b) after revision.

3.4 *Disseminate*

According to the advice from content experts and media experts, the video was revised and had a final duration of 7 minutes 21 seconds and 8 minutes 57 seconds, respectively, for fat-soluble vitamins and water-soluble vitamins. Users then carried out the video feasibility assessment. Table 3 shows the feasibility assessment result.

Table 3. Feasibility assessment of the whiteboard animation video by users

Aspect	Feasibility	Category
Learning	89.9 %	Very suitable
Content	91.3 %	Very suitable
Media	86.0 %	Very suitable
Benefit	90.8 %	Very suitable
Usage	89.3 %	Very suitable
Total	89.7 %	Very suitable

Users' media aspect assessment was 91.3%, which was lower than that of media experts of 95.5%. The users were Generation Z, while media experts were Generation X. Generation Z has a higher standard of audio-visual quality on video than the previous generation [4]. They might equate the quality standards of video clips or films with instructional videos. However, those differences were not a fundamental problem because the content quality in learning videos was more important than media quality [13]. One user scored 2 (unsuitable) about the video duration because the video was considered long. The video duration of 10-15 minutes was selected because it suited the characteristics of current students [9]. The duration of the vitamin videos has followed these standards, so revisions to the video duration were not carried out.

This whiteboard animation video was declared suitable for use as a learning medium after a feasibility examination by content experts, media experts, and users. This animation video made it easier for teachers to deliver learning materials and contributed to variety to learning media. This video was able to arouse students' interest in learning. The video was then distributed on YouTube, especially on the Boga UNY channel with the address https://youtu.be/Xe5Dzki9OaI and https://youtu.be/hbAsWnHok-Q, because YouTube has more exciting features and better quality [16].

4. Conclusion

Whiteboard animation videos for vitamins were divided into two parts, namely fat-soluble vitamins and water-soluble vitamins. The first video consisted of the definition, function, characteristic, and classification of fat-soluble vitamins and water-soluble vitamins and an explanation of vitamins A, D, E, and K, which included benefits, functions, and effects of deficiency/hypervitaminosis, sources, and RDV. The second video explained the types of vitamin B complexes, the benefits, functions, effect of deficiency/hypervitaminosis, sources, and RDV of vitamin B complexes and C. This animation used color images with bold colors and without gradations. Each scene began with an animation of hand movements when drawing a sketch and coloring the sketch. The final image appeared in the scene in the last 5 seconds of each scene. This whiteboard animation video combined animated color images, text captions, dubber voices, and back sounds to improve student understanding and maintain student engagement. The feasibility assessment of the video by content experts, media expert, and users showed a score of 89.9%, 96.7%, and 89.7%, so this whiteboard animation video was very suitable for use as a learning medium for Nutrition Science in culinary vocational schools.

5. Acknowledgement

We want to thank media expert, content experts, and users who have assessed the feasibility of this video. We are also grateful for the suggestion from reviewers to improve the writing of this article.

6. References

[1] Chicca J and Shellenbarger T, 2018 Connecting with Generation Z: Approaches in Nursing Education *Teach. Learn. Nurs.* **13**, 3 p. 180–184.

[2] Seemiller C and Grace M, 2017 Generation Z: Educating and Engaging the Next Generation of Students *About Campus* **22**, 3 p. 21–26.

[3] Törőcsik M Szűcs K and Kehl D, 2014 How Generations Think: Research on Generation Z *Acta Univ. Sapientiae, Commun.* **1** p. 23–45.
[4] Anggraeni A A and Surya R A, 2021 Developing a Demonstration Video on Making Dry Banana Preferred by Generation Z in *Journal of Physics: Conference Series* **1737** p. 012037.
[5] McKenny K, 2011 Using an online video to teach nursing skills *Teach. Learn. Nurs.* **6**, 4 p. 172–175.
[6] Rinawati W Ghassani P M and Anggraeni A A, 2021 The development of an engaging demonstration video for making shredded chicken in *Journal of Physics: Conference Series* **1810** p. 012043.
[7] Türkay S, 2016 The effects of whiteboard animations on retention and subjective experiences when learning advanced physics topics *Comput. Educ.* **98** p. 102–114.
[8] Miharti T, 2013 *Ilmu Gizi 1* 1st ed. Kementerian Pendidikan dan Kebudayaan, Direktorat Pembinaan SMK.
[9] Chauvet P *et al.*, 2020 What Is a Good Teaching Video? Results of an Online International Survey *J. Minim. Invasive Gynecol.* **27**, 3 p. 738–747.
[10] Anggraeni A A Sumaryana S Rinawati W and Murniati D E, 2021 The Development of Video on Pickled Fruit Production for Flipped Classroom *Int. J. Interact. Mob. Technol.* **15**, 19 p. 118–135.
[11] Mulyatiningsih E, 2011 *Riset Terapan Bidang Pendidikan dan Teknik* 1st ed. UNY Press.
[12] Alsaawi A, 2019 Spoken and Written Language as Medium of Communication: A Self-reflection *Int. J. Appl. Linguist. English Lit.* **8**, 2 p. 194.
[13] Hansch A Hillers L McConachie K Newman C Schildhauer T and Schmidt P, 2015 Video and Online Learning: Critical Reflections and Findings from the Field *HIIG Discuss. Pap. Ser.* **2**.
[14] Mayer R E, 2014 Incorporating motivation into multimedia learning *Learn. Instr.* **29** p. 171–173.
[15] Anggraeni A A Nurtriyanti D Rinawati W and Murniati D E, 2020 Developing video for agriculture product processing course in the subject of garlic snack production in *Journal of Physics: Conference Series* **1446** p. 012065.
[16] Jackman W M, 2019 YouTube usage in the university classroom: An argument for its pedagogical benefits *Int. J. Emerg. Technol. Learn.* **14**, 9 p. 157–165.

Microcontroller Based Portable Incubator Monitoring Tool with Short Message Service (SMS) notifications

Muhammad Irmansyah[1], Efrizon[2], Anggara Nasution[3] and Era Madona[4]

[1,2,3,4] Teknik Elektro, Politeknik Negeri Padang

Jl. Kampus Limau Manis, Kec. Pauh, Kota Padang, Sumatera Barat, Indonesia

Email: irmansyah@pnp.ac.id

Abstract. The aim of this research was applied a microcontroller, temperature sensor, weight sensor, heart rate sensor and GSM module to monitoring and notification of the condition of premature babies in portable incubators. The hardware used consists of a DS18B20 sensor, Load Cell, Pulse Heart Rate Sensor, Buzzer, LCD and SIM800L Module. The results showed the Pulse sensor and DS18B20 sensor could measure and detect the baby's heart rate and baby temperature. The result was on the LCD with an average error of 4.354% for heartrate and 1.437% for temperature. The loadcell sensor can detect weight with an error of 2.16%. The duration of sending SMS to Smartphone is 8s for each delivery. SMS was sent if the baby weak and critical condition.

1. Introduction

Globally 1.1 million children passed away in the first month of life in 2013. Almost 3000 newborn deaths every day, all child deaths under the age of 5-years around 47 %, up from 40% in 1990. Nearly 965,000 of all under-five child deaths are among newborn infants, babies in their first 28 days of life or the neonatal period. 125,000 child death before 1st birthday, and child mortality under 5 [1]. The babies with abnormal condition after they were born would help survive in new environmental using the incubator while in the neonatal intensive care unit (NICU). Premature babies, also known as preemies, put in an incubator with temperature in the incubator is controlled to keep baby's body temperature where it should be. After premature baby was moved to the incubator, he or she may follow a number of tests such as measuring the amount of oxygen in child's blood or oxygen saturation monitoring, collected through a needle or a heel stick inserted into a vein and Monitoring of baby's vital signs, sensors may be taped to the baby's body to monitor blood pressure, heart rate, breathing and temperature[2][3] and body weight [4]. It has been many Infant Incubator Project, one of these the infant incubator made by Prof. Dr. Ir. Raldi Artono Koestoer [5]. An infant-incubator using natural circulation and natural convection system. Profitting the Buoyanci force due to the difference of temperature between upper-side and lower-side, hot-air flow to the cabin where the baby's sleeping, It flows by itself without any force neither a fan or blower. Smallest power of 20-40 Watt will be sufficient to make a convenient limited environment for the newborn baby. And due to the small energy supply, the incubator doesn't need to be equiped by the electronic control system. In dry season of Indonesian climate, less power of 20 W can be used. But in the middle of rainy season due to lower average ambient temperature, the power of bulb heater should be increased to 40 Watts. Many similar studies have also been conducted

[6-12]. in general, the research conducted is more on monitoring [8] [10] and controlling temperature and humidity[6][7][11], whereas in addition to the temperature of the incubator the condition of the weight and heart rate are also important parameters to see the condition of premature babies health [4]

We designed and developed a monitoring system for measuring body temperature, heart rate and premature baby weight, all measurement data is displayed on the LCD and stored on the database server. The condition of the baby inside the incubator can be monitored by the parents and the medical team on the smartphone by utilizing the internet facilities through an Android-based website and mobile application in real time. The Websites will display baby identity and result of all parameter data measurement which known as medical record. Incubators made using the Grashof method based on research [5].

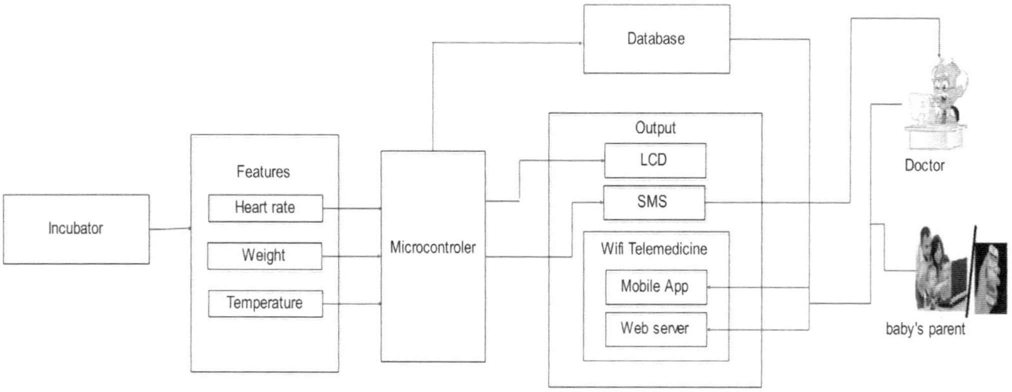

Figure 1. The proposed infant incubator monitoring system

On this study we only discussed three parameters of the system we made, namely heart rate, body temperature and baby body weight. Whole data is displayed on the LCD. The tool will send an SMS notification to the doctor if the baby is in a weakening or critical condition and the weight of a premature baby is normal. By this tool, the baby put in the incubator can be monitored in detail and deeply, while the doctor and the parents of the baby are easier to monitor.

2. **Research Method**

2.1. Block diagram System

This section discuss about the design and production of the system and how the system works not only in hardware and but also software. Figure 2 is a block diagram of the tool to be created.

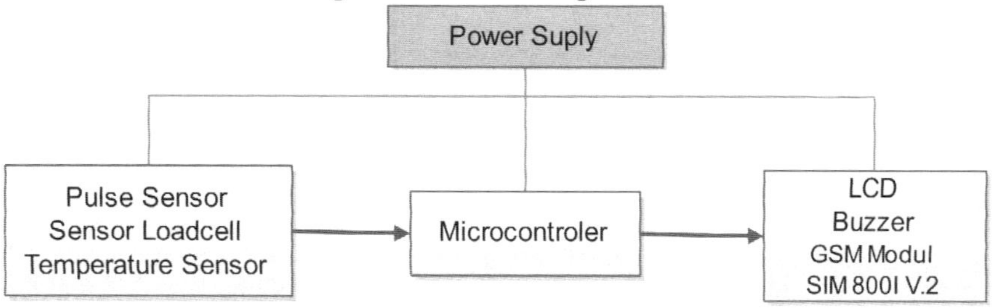

Figure 2.Tool Design Diagram Block

This tool has 3 measurement parameters, namely heart rate, body temperature and weight of the baby, with a device controller was a microcontroller. Measuring heart rate in infants is done by placing the heart sensor on the baby's wrist. Temperature sensor serves to measure the baby's body temperature placed in the armpit, where the normal body temperature is measured at 36OC-37.5OC per minute. The weight sensor is placed as a baby's sleeping pad which will weigh the baby's weight regularly. The microcontroller functions as a signal / data processor from the input variable. The signal / data is then processed and will produce output variables and are displayed on the LCD. if abnormal condition detected the buzzer will sound, the doctors will receive SMS, notice that condition of the baby is abnormal through GSM modules SIM800L V.2.

2.2. Circuit design

Figure 3 shows the electronic circuit of the device. This system consists of pulse sensors, DS18B20 sensors and Load Cells. Light was detected by pulse sensors. While this sensor was placed on skin's surface, organs and tissues (skin, bones, muscles, blood) would absorb or reflected the light of sensor but if they are thin enough some light would pass through body tissues. If the light intensity amount regarding the pulse sensor remains, then the signal value would the middle of the 10-bit ADC range be around 512. The ADC value depend on the light intensity. The signal produced by the sensor a wave called photoplethysmogram (PPG) as seen in Figure 4. PPG in the medical world is used to measure the respiratory rate [13] and heart rate[14].

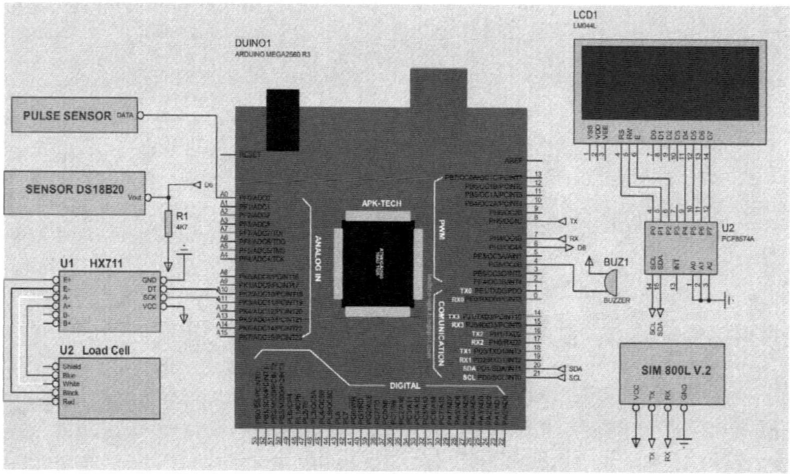

Figure 3. Electronic circuit monitoring system for premature infants

Blood would throughout the body while the heart pumps, every pulse that occurs is accompanied by the appearance of pulse waves like shock waves travel through arteries to the capillary layer of the hand (fingers), where the pulse sensor is installed. Blood speed flows slower than pulse waves. The number of heart beats per minute (BPM = beat per minute) is determined by sensor trough dividing 60000 (in milliseconds), the average value of ten IBI (inter beat intervals) that have been passed. IBI is the time difference between one point and the next point with the point value is 50% of the value of P (peak) minus T (valley) when the graph occurs extreme increase. Table 1 indicate the limits of a baby's heart rate from 1 to 60 days.

Table 1. Range BPM of Baby

Baby Age	BPM (Beat Per Minute)	Explanation
1-2 days	123-159/Minute	The activity of baby or child would influence the heart rate. For example, the heart rate
3-6 days	129-166/Minute	
1-3 weeks	107-182/Minute	

		can reach 180 times/minute while crying or pain. This heart rate would increase if child has fever or dehydration
1-2 months	121-179/Minute	

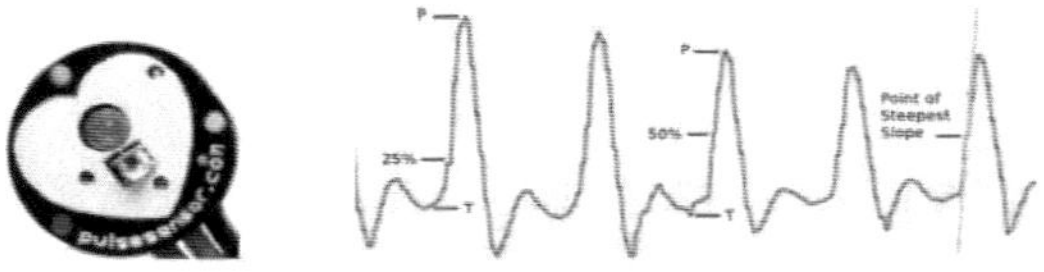

Figure 4. Pulse Sensor

Pulse sensor has 3 different purpose pins. Pin 1 was a data pin connected to pin A0 on the microcontroller. Pin 2 was Vcc resource and Pin 3 was Ground. To measure the baby's body temperature, the waterproof DS18B20 temperature sensor is used. Pin 1 is connected to Vcc, pin 2 is connected to ground and finally pin 3 is connected to pin D6 on the microcontroller. The three pins use a pin header that can be directly connected to the microcontroller. Load cell sensors are used to measure body weight. The load cell was put as a heavy sensor with 4 different function pins. The load cell used is equipped with an amplifier module, the 24X HX711 ADC module. The working principle of the loadcell sensor and hx711 module is when the baby put in, at that time the baby will put pressure or load on the loadcell sensor. The changing of force to the sensor would effect to the resistance and a voltage, so the value of the load also changes. This change in load value is used to inform the maximum value has been reached or not. Pin 1 connected to a + 5V source, pin 2 for data directly to pin A10 on the Atmega 2560 microcontroller, pin 3 as a clock which is also directly connected to the A11 pin microcontroller, while pin 4 is used for ground. To display the sensor readout value on the LCD used I2C LCD, this module is controlled serially synchronously with the I2C / IIC (inter integrated circuit) protocol or TWI (Two Wire interface) with addresses of 0x27 and 0x37. If the baby's condition is abnormal, the buzzer will sound as an indicator for the baby's parents. The doctors would receive an SMS if the condition of patient was abnormal through the SIM800L V.2 GSM module. The condition of an abnormal baby can be seen in table 2.

Table 2. Abnormal baby condition

No	Age (Year)	Heart rate (BPM)	Body Temp. (^{0}C)	Cond.
1	1-2 days	<123	< 35	weakened
	3-6 days	< 129	< 35	
	1-3 weeks	< 107	< 35	
	1-2 months	< 121	< 35	
2	1-2 days	> 159	> 37,5	weakened
	3-6 days	> 166	> 37,5	
	1-3 weeks	> 182	> 37,5	
	1-2 months	> 179	> 37,5	
3	1-2 days	> 159	< 35	critical
	3-6 days	> 166	< 35	

1-3 weeks	> 182	< 35
1-2 months	> 179	< 35

The monitoring tool algorithm can be seen in Figure 5. Data from the heart rate sensor, body temperature and body weight are then processed on the Arduino, and the sensor value will be displayed on the LCD. There are five conditions. The first condition of the value of the sensor data indicates under normal circumstances. The second to four conditions indicate the condition of the baby is weak and critical. If in this condition the buzzer on the device will be active and alert text will be sent to the doctor / health worker. Condition five is the condition of reading the baby's weight with a limit of 5000 grams, if the baby's weight has 2500gram, the LCD show "the baby comes out of the incubator and the sms will be sent to the doctor / health worker

Note : sign * can be seen in table 3

Table 3. Description of the condition of the baby on the Flowchart

Condition	Heart rate	Body Temperature	Weight
1*	1-2 days : 123-159/Minute 3-6 days : 129-166/Minute 1-3 weeks : 107-182/Minute 1-2 months : 121-179/Minute	35-37.5	
2*	1-2 days < 123/Minute 3-6 days < 129/Minute 1-3 weeks < 107/Minute 1-2 months < 121/Minute	< 35	
3*	1-2 days > 159/Minute 3-6 days > 166/Minute 1-3 weeks > 182/Minute 1-2 months > 179/Minute	> 37.5	
4*	1-2 days > 159/Minute 3-6 days > 166/Minute 1-3 weeks > 182/Minute 1-2 months > 179/Minute	< 35	
5*			2500 gram

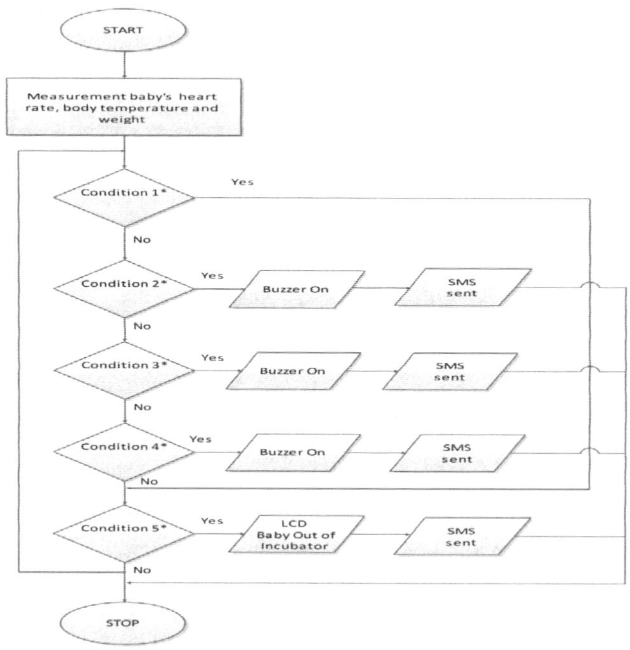

Figure 5. Flowchart monitoring tool

3. Experiment Results

This section explain the results of testing of the tools that have been made will be shown. First, testing of the tool as a whole is done. The results of the third reading of the sensors displayed on the LCD are then compared with manual measuring devices. Furthermore, testing system alerts, buzzers and SMS will be active according to the conditions that have been determined. Figure 6 is a picture of the tool as a whole.

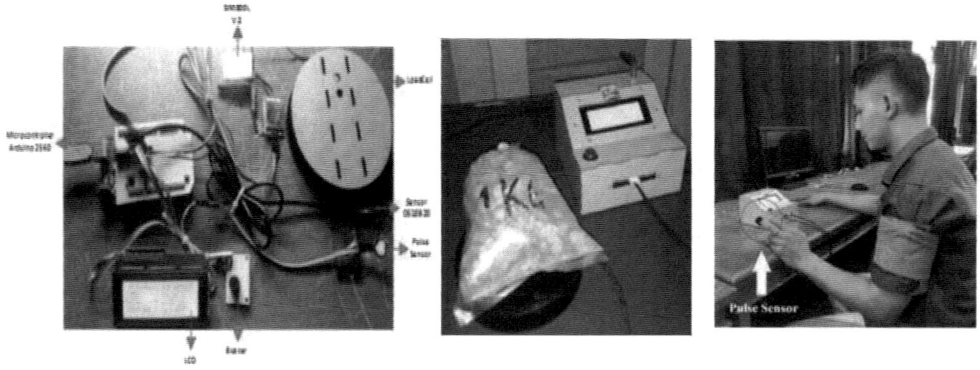

Figure 6. Testing Device for monitoring heart rate, body temperature, weight of premature babies

3.1 The Baby Heart Rate and Body Temperature Series (circuits)

This series of heart rate and body temperature consists of pulse sensors, microcontrollers, buzzers and LCDs. The reading of the BPM value (Bit Per Minute) was periodically and is in the range 50 / minute - 180 / minute, if the range of readings detected is above the specified range, the buzzer would on as a warning sign and the LCD show the BPM values. The number of samples of the people tested were 10 babies and children of different ages. Testing is carried out for 60 seconds / person. Tests and measurements were made by compared the measurements results between using a Pulse Oximeter and the results of circuit in this research. Pulse Oximeter used for premature babies or patients in special conditions to measure heart rate (HR = heart rate). The heart rate measure by place a pulse sensor at the tip of the index finger.

Tabel 4. Test results and measurements of the baby's heart rate and body temperature

Testing (person number)	Age (month)	Measurement result using Pulse Oximeter (BPM)	Measurement circuit result using Pulse Sensor on LCD (BPM)	Error (%)	Measurement result using Thermometer Digital (°C)	Measurement result using Sensor DS18B20 on LCD (°C)	Error (%)	Buzzer
1	4	108	112	3.70	36.9	36.3	1.62	off
2	8	177	181	2.25	38.1	37.7	1.04	on
3	13	111	117	5.40	37.3	36.8	1.34	off
4	14	101	106	4.95	37.4	36.8	1.60	off
5	17	113	118	4.42	37.4	36.9	1.33	off
6	21	103	107	3.88	37.4	36.9	1.33	off
7	24	89	93	4.49	37.6	37	1.59	off
8	35	98	103	5.10	37.6	37	1.59	off
9	40	94	98	4.25	37.3	36.8	1.34	off

10	50	98	103	5.10	37.6	37	1.59	off
Error average in measurement				4.354			1.437	

The average error can be seen in Table 4, this shows the results of testing and measuring the heart rate and temperature running well. DS18B20 pulse sensors could measure and detect heart rate, infant body temperature and display on LCD with an average error of 4.354% and -1.437%. Table 4 shows, testing was carried out with qualifying infants 4 - 50 months old. In the range of babies from 4-17 months the heart rate measured using pulse Oxymeter 103-108 BPM, while in the series 107-112 BPM. In the infant range of 21-50 months a heart rate measured using pulse Oxymeter 89-98 BPM, while in the 93-107 BPM series. Table 4 also shows the active buzzer when the heart rate is >179 BPM and body temperature is above 37.5 ° C in 8 months old babies.

3.2 Baby Weight Measuring Series

The weight measuring circuit or series consists of loadcell sensors, HX711 amplifiers, microcontrollers, and LCD. Load cell sensor and amplifier HX711 to retrieve baby's weight data, process data by microcontroller, and LCD show the result of measurement the baby weight data. It shows information about the normal condition of baby's weight so the baby can be removed. For loadcell testing by giving different loads were 500 grams, 1000 grams, 1500 grams, 2000 grams, 2500 grams and 3000 grams. The load cell output voltage was too small, so the HX711 module was used as an amplifier as well as an analog data converter from load cell to digital data before this sensor connected to microcontroller. Load cell programming using a library of HX711 modules on Arduino and the digital output will be converted by an Arduino microcontroller through the HX711 library to a heavy scale. The test results and measurements of the weight measuring circuit using loadcell sensors show in Table 5.

Tabel 5. Result of weight measure circuit

Load Weight (gram)	Test	Result of Testing (gram)	Avarage Weight of measure (gram)	Error (%)	LCD Indicator
500	1	515	512,8	2,5	_
	2	513			
	3	512			
	4	513			
	5	511			
1000	1	1028	1027	2,7	_
	2	1028			
	3	1027			
	4	1025			
	5	1027			
1500	1	1532	1530	2	_
	2	1530			
	3	1527			
	4	1528			
	5	1535			
2000	1	2045	2041	2,1	_
	2	2038			
	3	2042			
	4	2041			
	5	2039			
2500	1	2539	2542,8	1,7	baby out of the incubator
	2	2542			
	3	2540			
	4	2545			

	5	2548			
3000	1	3065	3060,8	2	baby out of the incubator
	2	3060			
	3	3058			
	4	3060			
	5	3061			
	average error on measuring			2,16%	

Testing is done by compared between the load measured data and the loadcell sensor with reference to the real weight of the load as shows in table 5. The measurement result of loadcell sensor data and the circuit works well. Loadcell sensor could detect the weight measured by the pressure or the weight measured through the LCD. When measuring with a weight of 500 grams, error is 2.5% of the actual load weight, 2.7% for 1000 grams, 2% for 1500 grams, 2.1% for 2500 grams, 2500 gram was 1.7%, and 2% for 3000 grams. So that the average error of the measurement of the weight of the load is 2.16%. In Table 5 we can also see that when testing and measuring the weight of the load is 500, 1000 and 2000 grams and LCDhave no information. While the weight were 2500 and 3000 grams the LCD have information "baby out of the incubator" will appear. This inform the baby's weight has on normal weight condition means the baby can be put out the incubator.

3.3 GSM Modem Series (SIM800L)

Testing of the SIM800L GSM Module is done to find out that this GSM modem can work properly. To be able to communicate between networks with a microcontroller module must be set when and to what number the short text message will be sent and the contents of a short text message that want to send. It send the conditions, namely the weakening and critical state. The duration of sending messages to a smartphone is around 2 seconds. The time depends on what we want by setting the delay. For warnings of weakening conditions and critical SMS sending takes place every 1 second and continues repeatedly before undesirable conditions occur in the patient. After the sensor reads the patient's condition the form of message / SMS that will be received to the medical staff's smartphone. The form of the signal sent by Arduino to the GSM SIM800L module. It is done to see the signal sent from the Arduino TX pin to the SIM800L GSM module. This test is done by sending the command "AT + CMGF" which is used by Arduino as the SMS sender command to the SIM800L module. The output voltage on SIM800l is 5 V, with the duration of sending SMS to Smartphone that is for 8s per shipment. SMS is sent if the patient is weak and critical can be seen in Figure 7.

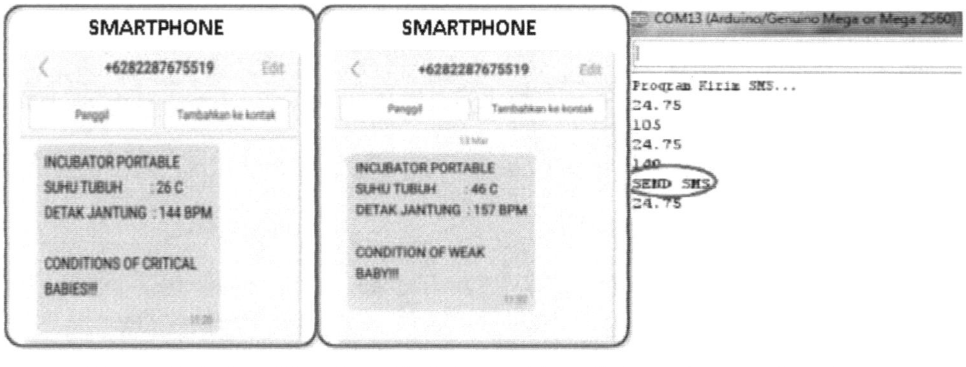

Figure 7. SMS notification of baby's condition if it is not in normal condition

4. Conclusion

Testing results shown the whole series can work well: DS18B20 pulse sensors and sensors detects heart rate and body temperature measured, the buzzer active and sent text as a danger sign if the baby's heart rate and body temperature are read above the specified range, loadcell sensor detects the weight measured, and displays inform the baby can be taken from inside incubator if the baby has weight more then 2500 grams. So, the tool can be implemented for babies in the incubator.

References

[1] Li Liu, PhD Shefali Oza, MSc Daniel Hogan, PhD Jamie Perin, PhD Prof Igor Rudan, MD Prof Joy E Lawn, MD. Global, regional, and national causes of child mortality in 2000–13, with projections to inform post-2015 priorities: an updated systematic analysis. The Lancet, Volume 385, ISSUE 9966, P430-440, January 31, 2015.
[2] D'Souza, S W, Janakova, H, Minors, D,Suri, R Waterhouse, JAppleton, G Ramesh, C Sims, D G Chiswick, M L. Blood pressure, heart rate, and skin temperature in preterm infants: associations with periventricular haemorrhage. Archives of Disease in Childhood Fetal & Neonatal Edition Volume 72. Issue 3.1995.
[3] P. B. J. Deacon, Core Curriculum Neonatal Intensive Care Nursing. (W.B Saunders Co., Philidelphia, 1993).
[4] Pensée Wu, Martha Gulati, Chun Shing Kwok, Chun Wai Wong, Aditya Narain, Shaughn O'Brien, Carolyn A. Chew-Graham, Ganga Verma, Umesh T. Kadam, Mamas A. Mamas. Preterm Delivery and Future Risk of Maternal Cardiovascular Disease: A Systematic Review and Meta-Analysis. Journal of the American Heart Association. 2018;7:e007809.
[5] Raldi A. Koestoer, (2016). Unpatented grashof-incubator as a part of community-engagement in mechanical
engineering university of Indonesia.
[6] Velagic, J.,Osmic, N.,Lutvica, K.,Kadic, N. Incubator System Identification and Temperature Control with PLC & HMI. 52nd International Symposium ELMAR-2010.
[7] Wang, Ruilan. The design of temperature and humidity control system in multi incubators based on single-chip microcomputer. Intelligence, Management Science and Electronic Commerce, AIMSEC 2011 - Proceedings
[8] Sahoo, Saikat,Champaty, Biswajeet, Pal, Kunal, Ray, Sirsendu S., Tibarewala, D. N. Wireless transmission of alarm signals from baby incubators to neonatal nursing station. 1st International Conference on Automation, Control, Energy and Systems - 2014, ACES 2014
[9] Yuliant, Alvien, Kowanda, Anacostiana, Salahuddin, Nur Sultan. Rancang Aplikasi Pemantau Suhu Dan Kelembaban Pada Inkubator Bayi Berbasis Internet. Seminar Nasional Aplikasi Teknologi Informasi. 2015
[10] Asmidar, Nor, Binti, Natasha, Fudzi, Mohd, Othman, Nurmiza Binti. Development of Infant Incubator for Clinic in the Rural Area of Malaysia. 2016 IEEE EMBS Conference on Biomedical Engineering and Sciences (IECBES)
[11] Shaib, M, Rashid, M, Hamawy, L, Arnout, M, Majzoub, I El, Zaylaa, A J. Advanced Portable Preterm Baby Incubator. 2017 Fourth International Conference on Advances in Biomedical Engineering (ICABME) 2017. ISBN 9781538616420.
[12] Muhammad Irmansyah, Era Madona, Anggara Nasution. Design And Application Of Portable Heart Rate And Weight Measuring Tool For Premature Baby With Microcontroller Base. International Journal of GEOMATE, Sept., 2019 Vol.17, Issue 61, pp. 195 -201
[13] Zainudin, N. Ahmad Mansor, W. Lee, Khuan Y. Sani, N. H Mohd Mahrim, S. A. Respiratory rate of photoplethysmogram signal from anaesthetic patients. ISCAIE 2015 - 2015 IEEE Symposium on Computer Applications and Industrial Electronics
[14] Ave, Arrozaq Fauzan, Hamdan Adhitya, S. Rhandy Zakaria, Hasballah. Early detection of cardiovascular disease with photoplethysmogram(PPG) sensor. Proceedings - 5th International Conference on Electrical Engineering and Informatics: Bridging the Knowledge between Academic, Industry, and Community, ICEEI 2015.

Journal of Physics: Conference Series **2111** (2021) 012027 doi:10.1088/1742-6596/2111/1/012027

Development of Training Kit Basic Aerial Robotics Using RC Plane Robot for Robotics Course

F N Aziz[1] and Mashoedah[2]

[1,2] Department of Electronics and Informatics Engineering Education, Postgraduate Program, Universitas Negeri Yogyakarta, Indonesia

Email: fardiansyahnur.2021@student.uny.ac.id

Abstract. RC Plane Robot and basic aerial robotics are technologies that are widely used, so training kits are needed to learn them. This study aims to: (1) Design and build, (2) Test performance and (3) Determine the feasibility level of training kit basic aerial robotics using RC plane robot. The study used the Research and Development method with the following procedures: (1) Potential and problem, (2) Data collection, (3) Design, (4) Validation, (5) Revision and (6) Trial. The object of the research is the RC Plane Robot training kit and the practicum jobsheet. The research subjects were students of the Electronic Engineering Education Study Program, Faculty of Engineering, Universitas Negeri Yogyakarta. The results of the research are in the form of hardware and practicum jobsheets. The hardware part consists of a robot plane, a work area (Box) and a remote control. The level of feasibility of training kits by material experts gets a percentage of 91%. By media experts get a percentage of 89.5%. By respondents (students) get a percentage of 86%. Based on these results, the training kit basic aerial robotics using RC plane robot is very feasible to be applied to the Robotics course.

1. Introduction

The development of the role of robots in the industrial world is currently increasing along with the times. Robot is a mechanical device that can do work like humans or behave like humans [1]. In the industrial sector, the technology used is developing so fast. Because of this, RC Plane Robots are widely used in the Unmaned Aerial Vehicle (UAV) industry to activate human activities in carrying out various tasks/missions. The application of the RC Plane Robot is used to find victims when there is a disaster, make maps for geographical purposes and be able to observe many areas or extensive agriculture from aerial monitoring and so on [2]. The development from time to time that is rapidly accelerating makes the function or use of UAV technology increasingly widespread, from its initial function only being used to assist military defense [3].

In order to support the operation of the RC Plane Robot, of course it is very necessary to know and understand the various uses of technology and the characteristics of the RC Plane Robot used. Based on this, it further strengthens that it is necessary to study various applications of the use of technology, especially in the industrial world that develops from time to time. Therefore, HR (Human Resources) to support this is needed. The way to improve competence and learn about the development of industrial technology that is always evolving is with a training kit [4].

In the robotics course, the Department of Electronics and Informatics Engineering Education, Universitas Negeri Yogyakarta teaches several topics of discussion regarding the field of robotics.

Based on observations, information data was obtained, namely the absence of a training kit that supports the basic material for the application of aerial robotics systems that are applied especially to airplanes. This causes students not to be able to know the application and basic material of aerial robotics applied to aircraft, thus causing the lack of achievement of aerial robotics competence. These competencies are 1) understand the servo configuration on 3-axis arm-base and elevons motion control, 2) understand how to control the speed of brushless DC motor, 3) understand how to communicate data wirelessly using NRF24L01, 4) understand the basic motion of plane robot flying wing. These four competencies packaged in single training kit as a RC Plane Robot.

Based on description above, the several problems can be identified are 1) lack of supporting training kit devices of competence to understand about servo configuration, understand how to control brushless DC motor, understand how to communicate data using NRF24L01, and understand the basic motion of plane robot, 2) the four aerial robotics competence on above has not yet been achieved, 3) there is no supporting training kit, causing students not to be able to know the application and basic material of aerial robotics applied to plane robot. This study aims to 1) Design and build, 2) Test performance and, 3) Determine the feasibility level of training kit basic aerial robotics using RC Plane Robot.

Aspects that include psychomotor, cognitive, and affective aspects can be applied by the use of assistive devices/training kits during the learning process [5]. The training kit has the same function or role as the learning media. Some of the roles of using learning media during the practicum process include 1) facilitate learning, 2) provide real experience, 3) learning becomes less boring and more fun, 4) activate all the senses of students as well as, and 5) attract the attention of students [6].

RC Plane Robot is an Unmaned Aerial Vehicle (UAV) type robot that can fly without a pilot inside it [7]. RC Plane Robot is an unmanned aircraft, in which the only method of take-off and landing is horizontal (HTOL) [8]. This research uses flying wing type aircraft. Flying wing is a fixed-wing aircraft with no definite fuselage [9]. The main control to move the RC Plane flying wing type is the thrust and elevons that provide aerodynamic strength so that the aircraft can follow the desired flight path. Thrust is a mandatory input for every movement [10]. Every movement of the RC Plane Robot has its own term. Pitch is a term for the movement up or down the nose of the aircraft. Roll is a term for the movement of tilted to the right or tilted to the left of the aircraft. In addition there is also a movement to turn right or turn left, this movement can be called yaw [11].

2. Methods

The research is carried out using a development model by means of R&D or Research and Development. The stages of the development procedure carried out using six steps, including 1) Potential and problem, 2) Data collection, 3) Product design, 4) Product validation, 5) Product revision and 6) Product trial [12]. The research method used to produce a particular product and used to assess the effectiveness of the product is called R&D or Research and Development [12]. All these stages are shown in the image below:

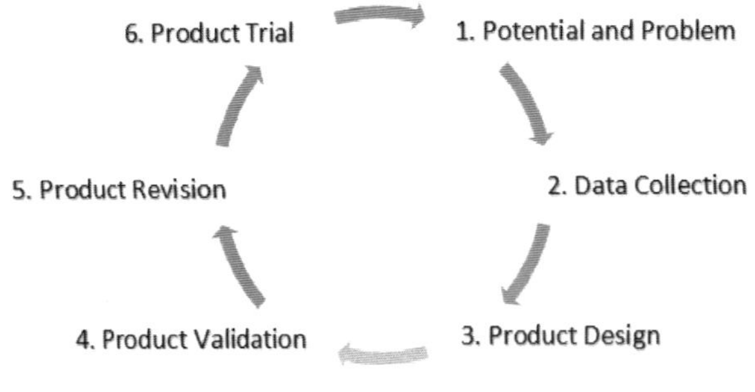

Figure 1. R&D model [12]

2.1. Time, location, data collection and analysis

The implementation process of the research was carried out in the month of May 2021. This research was conducted at the Department of Electronics and Informatics Engineering Education, Universitas Negeri Yogyakarta. The object under research is the Basic Aerial Robotics Training Kit using an RC Plane Robot Based on an Arduino Microcontroller and supported by a practicum jobsheet. The research subjects are material expert lecturers and media expert lecturers with the role of evaluator. In addition, there are also students of the Electronic Engineering Education Study Program who have taken robotics courses.

The process of collecting research data through measurement can use a tool called an instrument [13]. Questionnaires, graded scales, tests, interview guides, observation guidelines, and checklists are types of instruments [14]. In this research, data collection was carried out by observation and distributing questionnaires with consist of 4 choice answer using a Likert scale to material expert lecturers, media expert lecturers and students who in this case acted as respondents. Due to the Covid-19 pandemic, the observation and data collection was carried out online using a google form.

The data analysis technique in this study is a quantitative descriptive analysis technique, which serves to determine the feasibility level of the training kit. The use of this analytical technique can analyze and present the data obtained so that it becomes communicative and meaningful [15]. Determination of the feasibility level of the training kit is done by weighting the answer scores from the questionnaire filled out by lecturers and users. The weighting of the scores is described in the following table:

Table 1. Questionnaire score criteria

Criteria	Score
Totally Agree	4
Agree	3
Disagree	2
Totally Disagree	1

In order to produce the average value of the feasibility of the training kit, the following formula is used:

$$\overline{X} = \frac{\sum \text{rater score}}{\sum \text{number of respondents}} \tag{1}$$

How to get the percentage of the average calculation results in order to determine the feasibility of the training kit is determined by the formula below:

$$\text{Feasibility (\%)} = \frac{\text{observed score}}{\text{maximum score}} \text{ x } 100\% \tag{2}$$

To find out the level of feasibility based on the percentage of feasibility obtained using a rating scale benchmark. The rating scale benchmark serves to convert quantitative data to qualitative data [16]. The feasibility percentage category shown in the following table:

Table 2. Feasibility percentage category

Feasibility (%)	Category
>75 - 100 %	Very Feasible
>50 - 75 %	Feasible
>25 - 50 %	Unfeasible
0 – 25 %	Very Unfeasible

3. Results and discussion

3.1. Training kit design

This training kit is designed for learning about the movement of flying wing aircraft. Therefore, there is a 3 Axis arm which functions to simulate the movement of pitch, yaw and roll. This 3 Axis arm is placed just below the plane robot. We get the 3 Axis arm design in Figure 3 based on the principle of movement on the plane robot flying wing type in Figure 2.

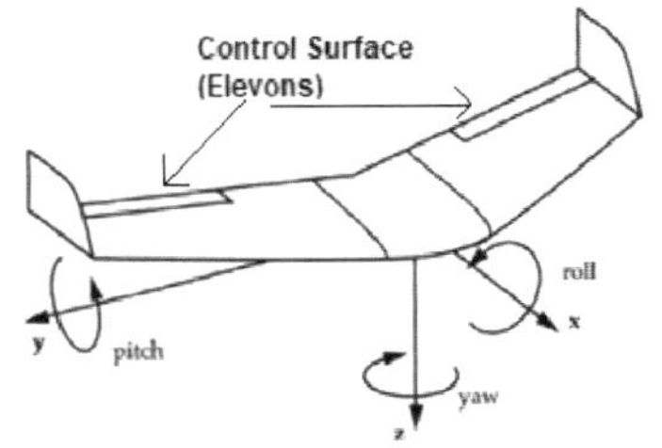

Figure 2. Principle of movement plane robot flying wing type

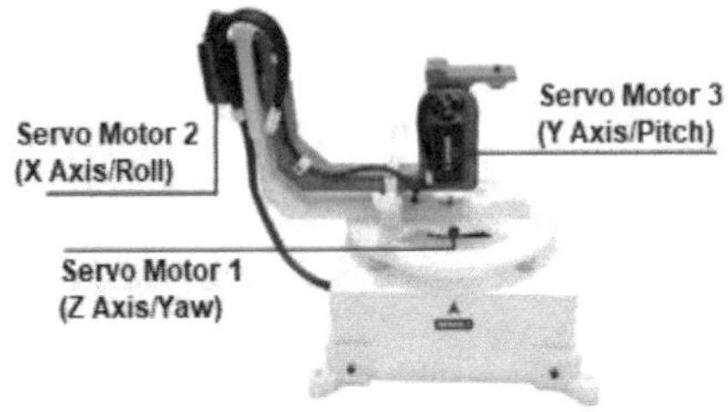

Figure 3. 3 Axis arm to simulate the movement

Plane robot and 3 axis arm are controlled wirelessly via remote control. On the remote control there are two joysticks, where the left joystick functions to adjust the speed of the brushless motor (throttle). The joystick on the right is used to control the movement of the aircraft. This training kit has a design in diagram block like following below:

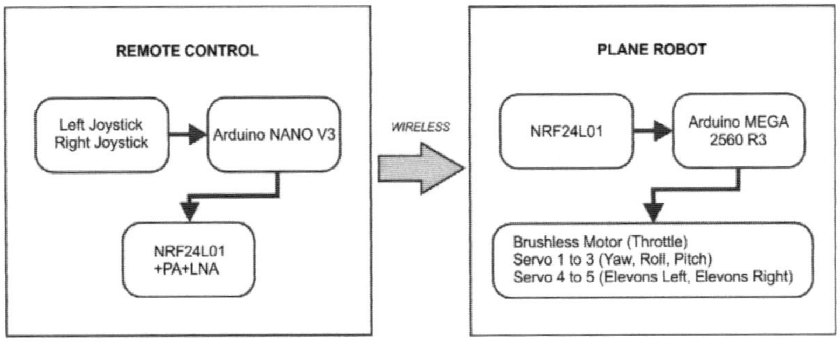

Figure 4. RC plane robot block diagram

The working principle is, the pilot sends a signal from the remote control (transmitter) to the receiver on the plane, then the receiver gives orders to the servo and brushless motor on the plane to move according to the pilot's wishes. The principle for controlling aircraft movement is identical to the control on conventional aircraft, namely using a propeller to push the plane so that it moves forward, and using the control surface (elevons) to maneuver. Because of training kit is designed for learning about the movement of flying wing aircraft, then the speed of brushless motor are limited only 15%, it's for safety.

Figure 5. Remote controller

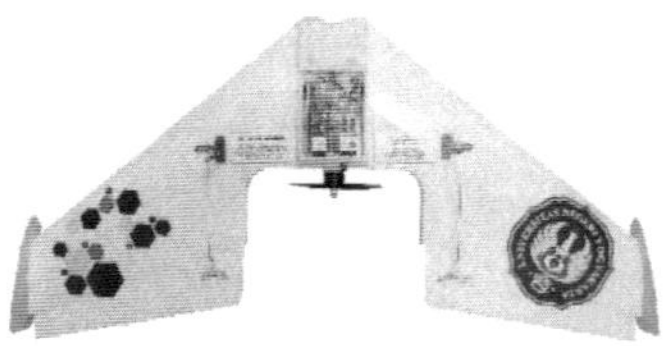

Figure 6. Plane robot

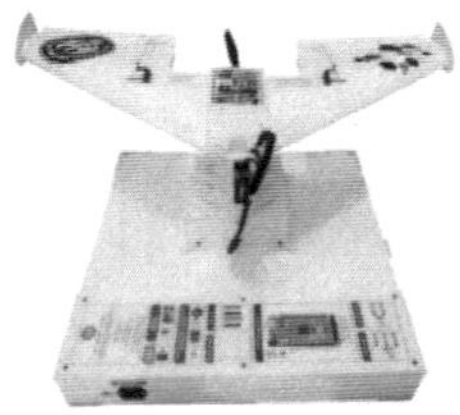

Figure 7. Plane robot with 3 axis arm movement simulators

A complete design training kit in the form of a robot serves to provide students with real application experience of aerial robotics on RC plane robots. In the development of the RC Plane robot training kit, it is also equipped with a jobsheet. The practicum jobsheet can be used as a guide for students as well as helping lecturers to teach aerial robotics material on the RC Plane Robot.

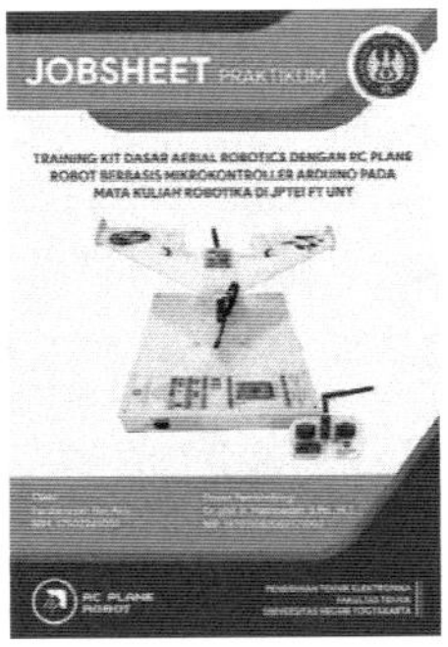

Figure 8. Practicum jobsheet

3.2. Training kit performance
The performance of the training kit was tested by testing the angle of each arm axis (Roll, pitch, yaw), Tx-Rx communication and correlation between joystick position with robot movement.

3.2.1. 3 Axis arm testing. The 3 Axis arm testing is aims to determine angle movement that produced by servo motor that can be found in each arm axis. Testing is done by giving the PWM value for the default position. The default angle position is 90 degree for each axis. To measure the result of angle movement, we use bow. The testing result of 3 Axis arm is shown in the following table:

Table 3. Testing result of 3 axis arm

Servo	Axis	Angle Result Each Axis (degree)			Error (%)
		Trial 1	Trial 2	Trial 3	
1	Z (Yaw)	96	98	94	6,67
2	X (Roll)	83	93	85	3,33
3	Y (Pitch)	90	89	93	0,74

Based on the result above, there are contains error in measurement of x, y and z axis. This error caused by design mechanic less precise and contains friction especially for x and z axis. Besides that, it caused by offset servo value and manual measurement using a bow resulting error.

3.2.2. Tx and Rx communication. The Tx and Rx communication testing is aims to determines the performance of NRF24L01 in sending analog data from the joystick on the remote controller and receiving that data on the plane robot. The testing result of Tx and Rx communication is shown in the table 4.

Table 4. Testing result of Tx and Rx communication

Trial	Transmitter			Receiver			Error (%)	Delay (ms)
	CH 1	CH 2	CH 3	CH 1	CH 2	CH 3		
1	76	124	121	76	124	121	0	1
2	122	50	220	122	50	220	0	3
3	250	180	75	250	180	75	0	1

Based on the result that shown in table 4, it shows that the NRF24L01 module, both transmitter and receiver, is able to send and receive data properly in each channel without contains error. No noticeable latency. Testing from anywhere in the practicum room with 1-10 meters distance (with thick walls and lots of wifi networks around) results in 1-3ms lags with nothing data loss.

3.2.3. Correlation between joystick position with robot movement. Changes in the position of each joystick on the remote controller will cause movement effects on the 3 Axis arm simulation, control surface (elevons) and position of the plane robot. That correlation is shown in the following table:

Table 5. Correlation between joystick with robot movement result

Input		Output		
Left Joystick	Right Joystick	Elevons Left	Elevons Right	Robot Movement
Move Up	-	-	-	More throttle (forward)
Move Down	-	-	-	Less throttle (silent)
-	Move Up	Down	Down	Nose down
-	Move Down	Up	Up	Nose up
-	Move Left	Up	Down	Tilt to the left
-	Move Right	Down	Up	Tilt to the right
-	Move Left & down	Up (Flexible)	Down (Flexible)	Turn to left
-	Move Right & down	Down (Flexible)	Up (Flexible)	Turn to right

3.3. Training kit feasibility level

The feasibility level of training kit product is carried out by material experts, media experts and users (students). Feasibility level is described below:

3.3.1. Based on material expert. The assessment carried out covers the quality aspect of the material and its expediency aspect. The quality aspect of the material contains 6 parameters which include 16 assessments. For expediency aspect contains 2 parameters that include 4 assessments. Maximum score from all aspects is 80 and minimum score is 20.

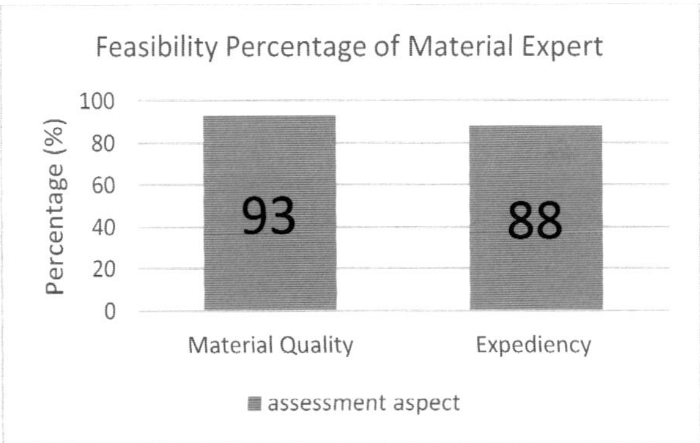

Figure 9. Feasibility percentage graphics of material expert

The result of the assessment by one material expert on graphic figure 9, feasibility of material quality aspect get percentage score 93%, feasibility of expediency aspect get percentage score 88% and total of average score is 91%. Besides that, material expert giving feedback to apply other robotics topics for next studies on this training kits, example: stability system, graphic user interface, or etc. to get completely good material.

3.3.2. Based on media expert. The assessment carried out covers the quality aspect of the display, technical quality aspect and its expediency aspect. The quality aspect of the display contains 4 parameters which include 9 assessments. For technical quality aspect contains 3 parameters that include 7 assessments. For expediency aspect contains 4 parameters that include 8 assessments. Maximum score from all aspects is 96 and minimum score is 24.

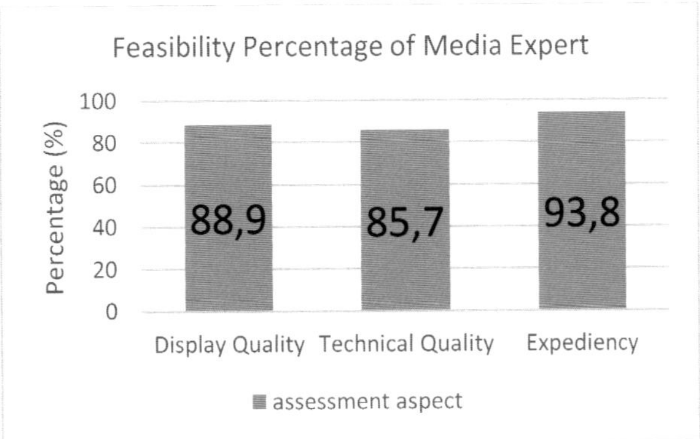

Figure 10. Feasibility percentage graphics of media expert

The result of the assessment by one media expert on graphic figure 10, feasibility of display quality aspect get percentage score 88,9%, feasibility of technical quality aspect get percentage score 85,7%, feasibility of expediency aspect get percentage score 93,8% and total of average score is 89,5%.

3.3.3. Based on users (students). The assessment carried out covers the quality aspect of the display, technical quality aspect, material quality aspect and its expediency aspect. The quality aspect of the display contains 4 parameters which include 9 assessments. For technical quality aspect contains 3 parameters that include 6 assessments. For material quality aspect contains 3 parameters that include 6 assessments. For expediency aspect contains 4 parameters that include 8 assessments. Maximum score from all aspects is 116 and minimum score is 29.

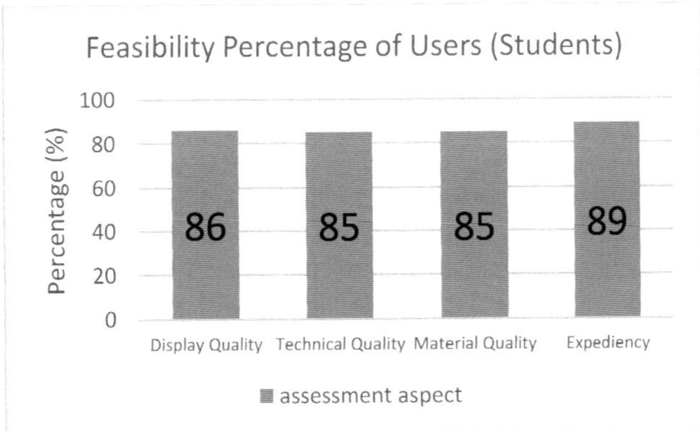

Figure 11. Feasibility percentage graphics of users (students)

The result of the assessment by users (students) on graphic figure 11, feasibility of display quality aspect get percentage score 86%, feasibility of technical quality aspect get percentage score 85%, feasibility of material quality aspect get percentage score 85%, feasibility of expediency aspect get percentage score 89% and total of average score is 86%.

4. Conclusion

From the research and discussions it can be concluded that, 1) the design of the training kit basic aerial robotics can be well developed by using the RC Plane Robot in the form of tools and practicum jobsheet, 2) in the performance test of the training kit, each axis on the 3 axis arm can simulate various basic movements of a flying wing type robot aircraft. Each axis can move with an error percentage x of 3,33%, y of 0,74% and z of 6,67%. The control surface and brushless motor can be controlled using a remote control wirelessly without contains error, 3) the feasibility level of the training kit based on the assessment of material expert lecturers gets an average score of 91% (very feasible), based on media expert lecturers gets an average score of 89.5% (very feasible) and based on users (students) gets an average score of 86% (very feasible). Based on these results, the training kit basic aerial robotics using RC plane robot is very feasible to be applied to the Robotics course. As a future improvement, to provide a complete topic on the study of robotics, using a stability system or Graphic User Interface (GUI) can get good material. In addition, by using the GUI, the system will be communicative.

References

[1] Suyadhi T D S 2010 *Buku Pintar Robotika Bagaimana Merancang dan Membuat Robot Sendiri*, Yogyakarta: ANDI Yogyakarta

[2] Sugeng S, Putra R A, Muslim R F and Septianto Y 2019 Unmanned Aerial Vehicle (UAV) for Mapping Plantation Area. *Telekontran*. **7**(1), pp. 79-89

[3] Barton J D 2012 Fundamental of Small Unmanned Aircraft Flight. *John Hopkins APL Technical Digest.* **31**(2), pp. 132-149

[4] Mashoedah 2015 Kajian Penggunaan Media Pembelajaran dalam Pelatihan Peningkatan Kompetensi Profesional Guru. *Jurnal Electronics, Informatics, and Vocational Education (ELINVO)*. **1**(1), pp. 17-25

[5] Anderson R H 1987 *Pemilihan dan Pengembangan Media untuk Pembelajaran*. (Translated by Yusufhadi Miarso, dkk). Jakarta: Rajawali, pp. 183-186

[6] Asnawir and Usman M B 2002 *Media Pembelajaran*, Jakarta: Ciputat Pers, p. 24

[7] Patel K, Parmar Y, Parmar V, Mistry H and Tandel M 2019 A Review Article on Design and Development of Radio Controlled Airplane. *International Journal of Advances in Engineering and Management (IJAEM)*. **2**(7), pp. 41-46

[8] Muchowski J, Szumski M and Krzysiak A 2018 Aerodynamic Concept of the UAV in the Gyrodyne Configuration. *Transactions on Aerospace Research*. **1**, pp. 49-66

[9] Hamada A A, Sultan A and Abdelrahman M M 2018 Design, Build and Fly a Flying Wing. *Athens Journal of Technology and Engineering*. **5**(3), pp. 223-250

[10] Warsi F A, Desa H, Ahmed S F, Joyo M K, Tanveer M H, Kamarudin H and Razlan Z M 2014 Yaw, Pitch and Roll controller design for fixed-wing UAV under uncertainty and perturbed condition. *IEEE 10th International Colloquium on Signal Processing and its Applications*, pp. 151-156

[11] Wiratama C 2016 *Control Surface Pada Pesawat Terbang*. from: https://www.aeroengineering.co.id/2016/01/control-surface-pada-pesawat-terbang/

[12] Sugiyono 2015 *Metode Penelitian & Pengembangan Research and Development*. Bandung: Alfabeta, p. 409

[13] Purwanto 2010 *Instrumen Penelitian Sosial dan Pendidikan: Pengembangan dan Pemanfaatan*, Yogyakarta: Pustaka Pelajar

[14] Arikunto S 2013 *Dasar-Dasar Evaluasi Pendidikan*, Jakarta: Bumi Aksara,.

[15] Purwanto E A and Sulistyastuti D R 2017 *Metode Penelitian Kuantitatif Untuk Administrasi Publik Dan Masalah-Masalah Sosial*, Yogyakarta: ANDI Yogyakarta

[16] Sugiyono 2009 *Metode Penelitian Pendidikan (Pendekatan Kuantitatif, Kualitatif dan R&D)*. Bandung: Alfabeta, p. 171

Improving online learning interactivity with 3D virtual classroom models

R Wardani[1*], H Jati[1], Nurkhamid[2], Y Indrihapsari[2], P Setialana[2], A M Budiyanto[1] and M N Ardiansyah[1]

[1]Electronics Engineering Education, UNY, Yogyakarta
[2]Information Technology, UNY, Yogyakarta

Email: ratna@uny.ac.id

Abstract. The world of education is experiencing a very significant impact of the Covid-19 pandemic. Online learning is a solution in overcoming learning limitations due to social distancing. Researchers developed a 3d virtual classroom using a Minecraft model to increase online learning interactivity. The 3 dimensional virtual classroom model design that has been developed has received respondents' acceptance of 90.00% in the "Very Eligible" category. Testing the 3 virtual classroom features feature got a percentage of 90.83% with the "Very Eligible" category to produce a sense of presence. In the interactivity test, character interaction is 89.17% in the "Very Eligible" category. The effectiveness of online learning continuity using the 3D virtual classroom model tested gets 89.17% with the "Very Eligible" category. In addition, the test uses an average score of SUS (System Usability Scale) of 73.67, so it can be concluded that the 3D virtual classroom model is "acceptable" as an online learning medium.

1. Introduction

The United Nations stated that one of the sectors affected by the Covid-19 pandemic is the world of education. Based on this, the Indonesian government made several efforts to respond to the spread of Covid-19. One of them is issuing the Circular Letter of the Ministry of Education and Culture (Kemendikbud) of the Directorate of Higher Education No. 1 of 2020 regarding preventing the spread of Covid-19 in the world of education. The Ministry of Education and Culture provides instructions that learning can be carried out remotely so that students, including students, are advised to study from their respective homes.

Along with this, technological developments in the Covid-19 pandemic have increased, especially in education. Technological transformation occurs in teaching and learning activities and can be a change from conventional to modern methods. Students are directed to indirect learning by utilizing online or as an educational solution during the Covid-19 pandemic. Online learning aims to provide quality learning services in a massive and open network to reach more and more enthusiasts of the learning space [1]. Online learning is the right solution in overcoming learning limitations during the pandemic, but it is not without its weaknesses. One of the weaknesses of online learning today is that it still has a low level of interaction and communication during the learning process. In everyday life, communication has a critical role as a medium for forming and developing soft skills in an individual and relationships or social contacts.

Several researchers have developed various methods to overcome this, such as emphasizes the implementation aspect of the 3D-based virtual classroom system to interact with objects in the virtual classroom to carry out learning activities [2]. The researcher testing the implementation of 3D virtual classes was only carried out based on learning outcomes, namely grades and acceptance tests using TAM. The test did not show the level of student interactivity after the implementation. Another research [3], aims to implement a real-time interactive virtual classroom model that provides real-time interactive classes with a teacher and student collaboration in one portal. This study is less detailed to provide details of the features that must be developed and provide a teaching and learning system only [4]. The research aims were to describe the effectiveness of virtual reality-based 3-dimensional (3D) learning media in increasing fifth-grade elementary school students' interest and learning outcomes. However, this study has not discussed what features can impact students' interest in learning compared to other learning media.

Another research aims to develop a 3D Virtual Reality Application for Introduction to Indonesian Early Humans for class X [5]. The goal is to find out the response of class X students to implementing the 3D Virtual Reality Application Recognizing the Early Man of the Indonesian nation, but it is unfortunate that students only interacted with objects as an introduction to ancient humans so that there was no direct interactivity with the teacher in the virtual world. Another study [6] entitled "Design and Implementation of 3D Virtual Campus Online Interaction Based on Unity 3D" aims to design a virtual campus roaming system at Shanxi Datong University. It is a 3D virtual campus system with powerful sight, hearing, and interaction. In this study, users interact with objects in a virtual campus, but unfortunately, the interactivity is still limited between humans and computers, so it has not been able to bring humans to humans like students and lecturers.

The researcher answers the problem by developing and use Virtual Reality (VR) technology in learning. A virtual classroom is a place for the online learning process that allows students and educators to communicate synchronously through audio, video, text, interactive whiteboards, exchange opinions, and other similar features as if they were realistically interacting in the classroom [7]. One type of virtual classroom that can be developed is the 3D Virtual Classroom which provides a realistic three-dimensional environment, offers an exciting and interactive experience. Integrating VR technology through 3D Virtual Classroom into the education sector can invite students to interact in the learning process.

In this study, the researcher uses Minecraft as a 3D classroom model. Researchers learned that Minecraft is not simply a video game that allows youth to build and create virtual worlds. Minecraft has now become an educational tool used as a vehicle for teaching critical content. The student who plays this game can take control and be active learners, thus enhancing their motivation for learning. It also acts as a supplement in today's classrooms-a popular learning activity in content areas such as science, math, history, engineering, architecture, and computer coding. It is expected to increase the level of interactivity and the active role of students. This research aims (1) to develop the 3D Virtual Classroom design that can improve interest in online learning and (2) to analyze the development of 3D Virtual Classroom models that can increase student interactivity in online learning.

2. Literature Review

2.1 Virtual Worlds in Education

Virtual worlds are places that exist entirely in a networked environment where people coexist, communicate, and interact through their avatars. The world is a dynamic and interactive environment that supports various social, entertainment, educational, and productive activities loosely based on physical activities. A common metaphor for the design of this world is the concept of place. Most designs use the concept of the place to design the surrounding environment and objects in the virtual world. A virtual world is designed for a specific purpose and can support various online activities.

The place design in the virtual world refers to people's experience and knowledge of architectural design in the physical world. Metaphors of places and references to architectural design provide a consistent basis that is familiar to people who visit the virtual world and designers of the virtual world. These metaphors and references can help facilitate interaction between users of the virtual world and the environment that has been designed for each other. Effectively, the physical world inspired the design of the virtual world but nowadays, beginning to see the opposite effect, that the design of the virtual world inspires the design of places in the physical world. The characteristics of the virtual world that are dynamic, adaptive, and interactive are very relevant to contemporary place designs and feel like the times ahead in the physical world. This characteristic makes the virtual world an ideal testing ground for developing and refining design areas such as intelligent environments, multi-premises computing, distributed computing broadly, and interaction design [8].

The virtual world is an application that has been around for more than 20 years, presenting different characteristics and functions such as Multi-User Dungeons (MUDs) and Object-Oriented Multi-User Dungeons (MOOs) so that it continues to grow, leading to the development of a 3-dimensional Virtual World. Virtual world three-dimensional or 3D Virtual Worlds (VWs) are essential for education today because they provide realistic three-dimensional environments, offer engaging, interactive, and immersive experiences, and create new opportunities related to learning and teaching. This evolution in the three-dimensional virtual world was made possible due to several factors, such as the evolution of web technologies and standards that allow interoperability between different web applications and broad broadband access. Historically, the most critical design consideration for three-dimensional virtual worlds has been their ability to constantly "move" between the "real" and "virtual" worlds, providing opportunities for experiences to simulating "real" world experiences. Otherwise leads to the hypothesis that experiences presented in a three-dimensional virtual world can feel like "real" as those presented in the "real" world.

Three-dimensional virtual worlds offer unique learning and teaching opportunities as they present a rich, engaging, immersive, motivating, and highly interactive environment because they can: (1) recreate a sense of presence, (2) be felt directly, (3) quickly adapt, (4) offer the possibility to simulate the "real" world, (5) offer the possibility to create new experiences that might be difficult to represent in the "real" world, (6) can be offered for experimentation, and (7) enable synchronous communication and collaboration. It can be considered helpful for teachers who wish to apply a three-dimensional virtual world in educational practice [9].

It can be concluded that the use of this three-dimensional virtual world can facilitate teachers to understand foreign concepts presented in 3D virtual worlds by exploring the similarities and differences between a "traditional" class and a three-dimensional virtual classroom or a 3D virtual classroom. This concept allows the teachers to design simple educational activities by transferring their previous experiences and ideas using objects presented in 3D virtual classrooms.

2.2 3D virtual Classroom

A virtual or virtual classroom is a place for learning by using computers and the internet as media. Synchronous online learning or virtual classrooms are used to apply the principle of combining technological aspects to support distance learning that requires support and guidance, feeling a social presence, share responsibilities, and get motivation from educators. Virtual classrooms allow learners and educators to communicate synchronously via audio, video, text chat, interactive whiteboards, app sharing, brainstorming, and other similar features as if they were interacting face-to-face in the classroom.

A 3D virtual classroom is a virtual classroom model designed and developed in three dimensions regarding components and how to use them. In the learning process using the online method, space a three-dimensional virtual classroom can provide the essence of learning as it is done in an actual physical classroom.

2.3 Minecratf Model

Minecraft is the first sandbox game published by Mojang that can be played in multiplayer and developed on a Java basis. Minecraft carries the concept of placing blocks where players are asked to collect objects to create the environment around them as they please. The Minecraft feature gives players a challenge to complete their world in their way. With the multiplayer experience, players can coordinate to create amazing feats such as replicas of famous buildings or places [10].

Minecraft in education can be interpreted as creating virtual classroom models with room designs like real classes. By creating a virtual class that looks real, players can feel the natural aspects of implementing learning in a virtual classroom. Some tangible aspects can be felt. Namely, the habits carried out in the classroom, such as entering a learning room through a door, looking for a seat, interacting, or conversing with other friends in the room. Minecraft allows the user to modify game rules, alter game content, redesind textures and give players new abilities-summarised with the Mod, one can do anything to the game.

Figure 1. Minecraft Model

3. Research Methods

3.1 Waterfall Methods

In this research, the researcher using the waterfall method with 4D model specifications. The 4D model is one of the research and development methods. 4D models are used to develop learning tools. The 4D model was developed by S. Thiagarajan, et al [11]. The 4D model consists of 4 main stages, namely (1) Define, (2) Design (Design), (3) Develop (Development), and (4) Disseminate (Spread).

3.2 Materials dan Instrument of the Study

The study used various tools in hardware and software such as operating systems, software libraries, development kits, and database management systems. It used oculus rift, Minecraft software ver. 1.17.0, license Minecraft and Discord software. Minimum specification computer to run Minecraft software: CPU at least: Intel i5 4950, GPU Nvidia GeForce GTX1060, RAM 8 GB, and Hardisk SSD 1TB. Data collection was carried out in online lectures because there was no face-to-face offline class due to the Covid-19 pandemic.

3.3 Data Collection Techniques and Instruments

3.3.1 Development Questionnaire Method

Researchers prepared several questions to use as data material and relevant sources in this study. This questionnaire contains ten questions with four answer options, namely disagree, disagree, agree, and strongly agree. Questions regarding the design of virtual classrooms, entrance features, seating selection features, initial preparation process before learning, features attendance, interaction with classmates, interaction with lecturers, learning interest using virtual classrooms, focus level during learning, and avatar activities.

3.3.2 *Usability Questionnaire Method*

In this method, the questionnaire used as a usability testing technique in evaluating the 3D Virtual Classroom model developed is the System Usability Scale (SUS). The use of the SUS questionnaire in this study means that this study will use ten questions with five answer choices are available to evaluate the model that has been developed.

3.4 Data analysis technique

Analysis of the results to be carried out using SUS according to the rules the calculation of the score in SUS is as follows:

3.4.1 Each question with an odd number, the score of each question obtained from the respondent's score will be reduced by 1.

3.4.2 Every question with an even number, the final score is obtained from the result of the value of 5 is reduced by the score of the answers obtained from the respondents.

3.4.3 The final SUS score is obtained from the sum of the scores for each question then multiplied by 2.5.

The rules for calculating the score only apply to one respondent. The next step calculation is to find the average score of each respondent's SUS score, add up all scores, and divide by the number of respondents. Here is the formula to calculate the SUS score:

$$\bar{x} = \frac{\sum x}{n}, \text{ where} \tag{1}$$

$\bar{x}$ = mean score

$\sum x$ = total SUS score

n = number of respondents

4 Result and Discussion

4.1 Proof of Concept: 3D Virtual Classroom

3D Virtual Classroom conceptual model developed in this research is stated through a formal specification. In the formal specification, the scope of the Virtual Classroom 3D conceptual model can identify the physical artifacts represented in the virtual model domain.

4.1.1 Vi (Virtual Classroom) represents the real object of the model domain being discussed. In this case, the object in the model domain is a virtual classroom. A Virtual Classroom (Vi) can involve several Actor (Ai), (Aj), (Ak), … (An), and each Actor can perform several actions or activities or Operations (Oi) related to Virtual Classroom (Vi), which are limited in Context (Ci);

4.1.2 Ci (Context) is a boundary of the classroom that binds (binding) the scope of interaction that occurs between subjects (in the future referred to as Actor) through the actions or activities (in the future referred to as Operations) that they perform. The context here is in the form of courses or subjects;

4.1.3 Ai (Actor) is the subject/user involved in a Virtual Classroom (Vi). Actors in the domain of this model are students and educators;

4.1.4 Oi (Operation) represents any activity, action or operations that the Actor (Ai) can do in the Context (Ci). Operations in models in this domain are in the form of activities that occur in the learning process.

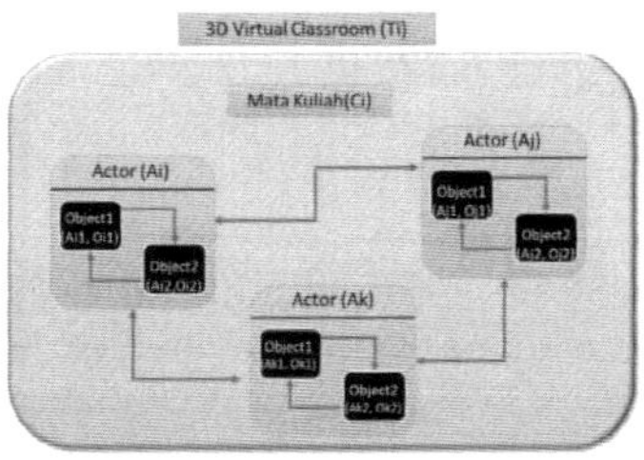

Figure 2. 3D Virtual Classroom Conceptual Model

The 3D Virtual Classroom conceptual model can be expressed through the following schema:

A Virtual Classroom (Vi) is a collection of objects that interact with each other in a Context (Ci) that represents a learning activity, expressed by:

$$\{Ci\} \in Vi \tag{2}$$

Context Ci is a conceptual boundary that allows activities in the learning process (presentations, discussions, assignments, and other learning activities) to be carried out in a contextual context. A Context consists of zero or more Actor A and zero or more Operation O. Actor is an object that represents participants in one or more learning processes. *Operation* is an activity that an actor can do in a learning process. Based on this specification, it can be defined the relationship between the Actor and the Operation performed. A Context will be related to a topic.

$$Ci : \{\{Ai\} \wedge \{Oi\}\} \tag{3}$$

Based on this relation, the researcher can define several operations or activities that are implemented in the *Context* and its components:

$$op : \{present; askQ; respQ; discussM; assignA; assignmentA; null\} \quad (4)$$

Each operation or activity can be described as follows:
Present operations or activities are activities (Ai) that represent the scenario of the educator explaining the material/topic (M) in a course session (*Context* Ci). This activity can be expressed as:

$$present(Ai; M; Cj) \quad (5)$$

AskQ operations or activities are activities (Ai) that represent the scenario of the educator asking the material/topic (M) to students or vice versa, students asking the material/topic (M) to the educator (Aj). This activity can be expressed as:

$$askQ(Ai; Aj; M) \quad (6)$$

Operations or *respQ* activities are activities (Ai) that represent the scenario of educators or students answering questions about the material/topic (M)

$$respQ(Ai; Aj; M) \quad (7)$$

Operation or *discuss M* activity is an activity (Ai) that represents group discussion scenarios between students about the topic (M) in a particular course or *Context* (Ci) session. A discussion is usually carried out in certain groups, for this operation will require determining the members in the discussion groups and topics for each group. The scheme for modeling discussion scenarios in class can be stated as follows:

$$discuss(M): Ci \quad (8)$$

$$assignA(Ai\,..\,An; M)$$

4.2 3D Virtual Classroom Instructional Design
Several concepts are used to represent the instructional aspects that will be implemented in the 3D Virtual Classroom. These instructional aspects can be explained as follows:
4.2.1 *Sense of Presence*, educators, and students are represented in the form of Avatars in order to participate in learning activities. The avatar of each actor (educator and student) is recognized through the institutional account (e.g. institutional email given to each educator and student). This account is the authorization to be able to join the virtual class.
4.2.2 *Virtual Immediateness*. The tools provided in the 3D Virtual Classroom provide a feature for feedback on the interaction between educators and students in the virtual classroom.
4.2.3 *Adaptability*. Tools can be manipulated and configured by educators as needed when carrying out learning activities.
4.2.4 *Real World Simulation*. The tools provided in 3D Virtual Classroom represent or imitate the appearance and functionality of traditional/actual classes.
4.2.5 *New Experience*. The tools provided in the 3D Virtual Classroom provide educators with additional functionality that is impossible to represent in a real classroom.
4.2.6 *Experiment*. Configuration and modification of tools made by educators can be returned to the initial settings if desired.
4.2.7 *Synchronous communication and collaborations*. Tools provide features for interactive communication in the learning process and carry out collaborative activities using certain features.

Tabel 1. *3D Virtual Classroom Instructional Design*

Design		Implementation	
Step	Learning Activities	Implementasi Konsep	Virtual Space (Tools and Functionality)
Introduction	▪ Class Opening ▪ Students wait in the waiting room	*Sense of Presence*: *Avatar* represents *the actor (Ai)*. *Avatars* must be able to represent	Functionality: ▪ Class meeting feature that allows an educator's Actor (Ai) / Avatar to

	▪ Greet students ▪ Presence	educators and students. The identity of each *Actor (Ai)* is represented by the account used.	manage and allow students to enter virtual classrooms (access control to limit participants who can enter classes according to KRS Tools: ▪ Avatar, virtual lobby,skin editor, button, password
	Delivering apperception	*Real-World Simulations:* The tools contained in the 3D Virtual Classroom can represent the appearance and functionality as in the "traditional" learning process (real class)	Functionality: ▪ Movement availability allows Avatar to carry out learning activities in class such as talking, walking, approaching students, and others in offline classes. Tools: ▪ Keyboard, voice chat
Presentation/core section	Delivering learning materials through presentations	*Real-World Simulations:* The tools contained in the 3D Virtual Classroom can represent the appearance and functionality as in the "traditional" learning process (real class) *Adaptability:* The tools available in the 3D Virtual Classroom must be configurable and manipulated by the educator. *Experiment:* configuration and modification of tools made by educators can be returned to the initial settings if desired	Functionality: ▪ Virtual whiteboard feature that allows an educator's Avatar to manage and display his presentation ▪ Screen sharing feature (using OHP) to present material ▪ Web browser features to manage the presentation of web-based content ▪ The desktop features are integrated with web browsers so students can browse the content they need Tools: ▪ Screen writing, screen capture, internal browser, voice chat
	Displaying simulations that represent activities such as demonstrations, tutorials, and experiments.	*New Experience:* tools used to enable simulations on activities that might be difficult to do in a real classroom.	Functionality: ▪ Doing several learning methods application Tools: ▪ 3D Model
	Interact with students through question and answer activities. The teacher provides discussion topics and divides students into several groups to carry out brainstorming and group discussions.	*Immediatness*: All the tools provided in 3D Virtual Classroom provide responsive feedback both textually and graphically. *Synchronous communication and collaborations*: Special tools are available for conducting discussions/chats, presentations	Functionality: ▪ Enabling students for an interactive learning atmosphere. Tools: ▪ Voice chat Functionality: ▪ The feature is to conduct discussions in small groups, and it is also possible to share screens for presentations in discussions. ▪ Synchronous communication Tools: ▪ Voice chat, breakout room, screen capture. screen writing
	Students can present the discussion result.	*Real World Simulation*: The tools contained in the 3D Virtual Classroom can represent the appearance and functionality as in the "traditional" learning process (real class)	Functionality: ▪ Screen sharing feature (using OHP) to present material ▪ Communication Tools: ▪ Screen writing, screen capture
	Educators provide feedback on the discussion results	*Immediatness*: All the tools provided in 3D Virtual Classroom provide	Functionality:

		responsive feedback both textually and graphically.	▪ Submission of feedback for each presentation and delivery of assignments Tools: ▪ Screen writing, voice chat
Closing and Follow-up	Gives reinforcement	*Immediatness*: All the tools provided in 3D Virtual Classroom provide responsive feedback both textually and graphically.	Functionality: ▪ Submission of confirmation and reinforcement to students Tools: ▪ Screen writing,voice chat
	Making conclusions	*Immediatness*: All the tools provided in 3D Virtual Classroom provide responsive feedback both textually and graphically.	Functionality: ▪ Submission of learning summaries at the end of the lecture session. Tools: ▪ Screen writing, voice chat
	Giving assignments/quizzes	*Real World Simulation*: The tools contained in the 3D Virtual Classroom can represent the appearance and functionality as in the "traditional" learning process (real class)	▪ Quiz feature that allows students to participate in quizzes ▪ Assignment features that allow students to Tools: ▪ Screen capture, screen writing, screen sharing, LMS

4.3 Diagram Sequensial

Figure 3 describes how the entities in the system interact, including the messages used during the interaction. There are two users and three objects, namely: activity component, content component, and interaction component. First, user one will enter the activity component screen to register. From the activity component screen, the admin will authenticate and confirm. After the user gets confirmation, the user can log in, the admin authenticates and confirms again. After that, the user can see the class model and change the class model on the content component screen. The user can continue the activity with the joint class. User activities are continued with general activities in actual classes such as talking, discussing, sharing computer screens, making 3d models, which are carried out on the interaction component screen. User one can also create and follow breakout rooms with other users.

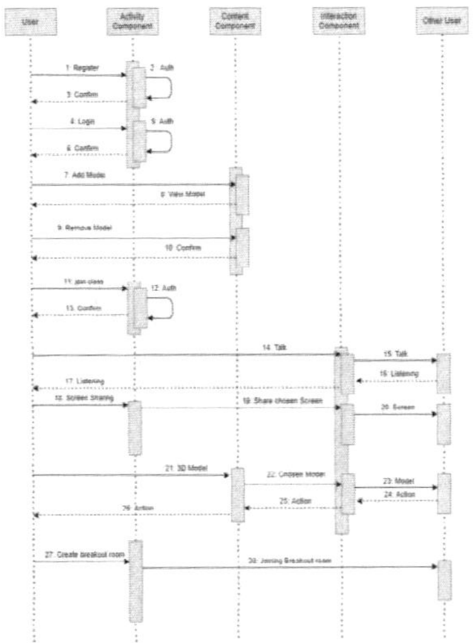

Figure 3. Diagram Sequensial

4.4 3D Virtual Classroom Features

Some of the features developed in the virtual 3D model of this classroom are as follows:

4.4.1 Custom Avatars. Adjustments can be made to the user's wishes when using a 3 (three) dimensional avatar. Besides being able to eliminating boredom, this realistic representation can also give the impression of "not alone" for users in a virtual space.
4.4.2 Real-Time Voice Chat. Real-time audio and video communication is the main feature of this application. Audio and video communication can be implemented in real-time.
4.4.3 Notes. During learning, it is not uncommon for students to need personal records. This record cannot be logged on other devices because the user will be in one virtual place, making note-taking in the application very significant.
4.4.4 Custom Map. One virtual laboratory cannot be used repeatedly by users. In addition to the boredom factor, not every learning can be done in the same virtual lab, so a virtual lab adapts to user needs.
4.4.5 Custom Media. Each lesson requires a different model. Like virtual labs, developer provision of demonstrations becomes ineffective. In addition, there may be errors in the display. It is necessary to display a display that can be entered independently by the user.

Figure 4. 3D Virtual Classroom Design

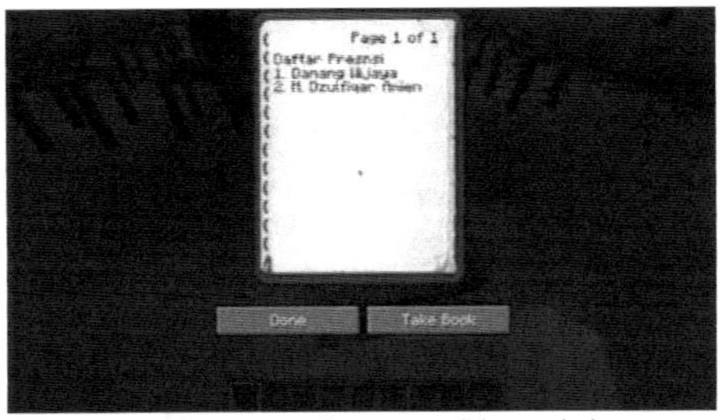

Figure 5. System view to add participant

Figure 6. Point of Student View

At this stage, the 3D virtual classroom model development and usability test using the System Usability Scale (SUS) was also carried out to 30 students of Yogyakarta State University.

5 Conclussion

This research is devoted to increasing interactivity during online learning by developing a 3D virtual classroom model using Minecraft as an online learning platform. The design of a 3 (three) dimensional virtual classroom model using Minecraft can be accepted by students of the Department of Electronics and Informatics Engineering Education, FT UNY in online learning, getting a percentage of 90.00% in the "Very Eligible" category. Several classroom feature tests were added to the 3D virtual classroom model, namely the entrance feature before entering the building or classroom, the seat or table position selection feature, the attendance record feature, and the avatar customization feature, getting a percentage of 90.83% with the "Very Eligible" category to produce a sense of presence. The test results on the aspects of character interaction get 89.17% with the "Very Eligible" category and successfully represent actions or interactivity such as in classrooms when learning offline. Based on testing the effectiveness level of online learning using a 3D virtual classroom model that has been tested, the percentage is 89.17% with the "Very Eligible" category. By testing using the System Usability Scale

(SUS), the average score of SUS is 73.67, so it can be concluded that the 3D virtual classroom model is "acceptable" as an online learning medium.
This study did not examine the effectiveness or improvement of student learning outcomes by using a 3D virtual classroom model. So that in the future, research will be focused on the effectiveness or improvement of student learning outcomes by using a 3D virtual classroom model.

References

[1] Sourial, N. *et al.* (2018) 'Daring to draw causal claims from non-randomized studies of primary care interventions', *Family Practice*, 35(5), pp. 639–643
[2] Rizal, A., Adam, R. I. and Susilawati, S. (2018) 'Sistem Kelas Virtual dan Pengelolaan Pembelajaran Berbasis 3-Dimensional Virtual World', *Jurnal Edukasi dan Penelitian Informatika (JEPIN)*, 4(2), p. 132
[3] Rion, Z. K. and Hasan, M. (2015) 'An Implementation of Virtual Classroom and Performance Analysis of Teaching-Learning Outcome', 15(7)
[4] Dewi, R. K. (2020) 'Pemanfaatan Media 3 Dimensi Berbasis Virtual Reality Untuk Meningkatkan Minat Dan Hasil Belajar Ipa Siswa Kelas V Sd', *Jurnal Pendidikan*, 21(1), pp. 28–37
[5] Wardhana, N. J., Arthana, I. K. R. and Divayana, D. G. H. (2019) 'Pengembangan Aplikasi Virtual Reality 3D Pengenalan Manusia Purba Bangsa Indonesia untuk SMA Kelas X (Studi Kasus SMA Negeri 2 Singaraja)', *Jurnal Nasional Pendidikan Teknik Informatika,* 8(2), p. 109
[6] Li, W. (2019) 'Design and Implementation of 3D Virtual Campus Online Interaction Based on Untiy3D', *2019 IEEE 3rd Information Technology, Networking, Electronic and Automation Control Conference (ITNEC)*, (Itnec), pp. 2066–2070
[7] Al-Nuaim, H. A. (2012) 'The use of virtual classrooms in e-learning: A case study in King Abdulaziz University, Saudi Arabia', *E-Learning and Digital Media*, 9(2), pp. 211–222
[8] Gu, N. and Maher, M. Lou (2014) *Designing Adaptive Virtual Worlds*
[9] Kallonis, P. and Sampson, D. G. (2010) 'Implementing a 3D Virtual Classroom Simulation for teacher's Continuing Professional Development', *International Conference on Computers in Education,* pp. 36–44
[10] Ekaputra, G., Lim, C. and Eng, K. I. (2015) 'Minecraft: A Game as an Education and Scientific Learning Tool', *Information Systems International Conference (ISICO)*, pp. 2–4.
[11] Thiagarajan, S., Semmel, D. S & Semmel, M. I. (1974) 'Instructional Development for Training Teachers of Expectional Children', Minneapolis, *Minnesota: Leadership Training Institute/Special Education, University of Minnesota*

Development of Mobile Augmented Reality Based Media for An Electrical Measurement Instrument

Y M Nursita[1] and S Hadi[2]

[1,2]Electrical Engineering Education Department, Engineering Faculty, Yogyakarta State University, Yogyakarta, Indonesia

E-mail: [1]yushanafi.mursid2016@student.uny.ac.id , [2]samsul.hd@gmail.com

Abstract. The research aims to develop mobile learning media with augmented reality for electrical measurement instruments. The learner can use this application to improve their skills and knowledge about using electrical measurement instruments correctly. One of the essential skills for electricians is using voltmeter, ammeter, and ohmmeter. The result of measuring they can do some analysis about an issue or troubleshooting the electrical field. From the development research, was produced learning media application product was named ARAVO. ARAVO is an abbreviation of augmented reality of Ammeter, Voltmeter, and Ohmmeter. ARAVO helps learners and even lecturers to simulate the use of electrical measuring instruments by combining virtual objects such as multimeter with the real world. Thus will provide a more visible visualization of how to use electrical measuring instruments before they practice directly with actual measuring instruments. ARAVO is a mobile application that can run on the smartphone platform mobile Android. Into the development of this application must go through several stages before it is ready for use.

1. Introduction

Electrical power installation engineering is one of the expertise programs held at vocational high schools. The purpose of this expertise program refers to the Law on the National Education System number 20 of 2003 article 15 states that vocational education is secondary education that prepares learners to continue their studies in the vocational field or to enter the world of work directly. Specifically, the Electrical Power Installation Engineering Skills Competency is that learners are equipped with skills, knowledge, and competencies in the electricity sector in a professional skills.

The Indonesian National Competency Standards as work competency standards in a field are used as reference competency standards in the development of the Electrical Power Installation Engineering Skills Competency curriculum. These competencies are built on the foundation of normative, adaptive, and productive basic competencies. One of the competencies that must be mastered by prospective professionals in the electrical field is the use of electrical measurement instruments.

In vocational high schools, electric measurement instruments are the competency of expertise in electrical power installation engineering. In this competency, learners learn about the introduction of electric measurement instruments, work principles, measurement procedures, and analysis of measuring electrical quantities results. This competency is used as is a primary reference in developing mobile learning augmented reality-based media. The augmented reality technology has been chosen as mobile learning media because of several advantages such as Augmented Reality (AR) is a technology that can change the way people learn and educate. Aside from that, the widespread adoption of mobile devices has sparked a surge of interest in combining the advantages of mobile learning and augmented reality applications [1]. Visual augmentation, in which computer graphics are combined with real-world imagery, is the most popular application of AR. AR blends graphics and video using a handheld computer such as a smartphone or tablet. This is referred to as "handheld camera see-through" [2]. Augmented Reality (AR) promises to be used in educational methods in colleges [3]. Augmented Reality (AR) as mixed and enhanced reality has interesting features for educational purposes; its potential and possibilities can be further expanded when AR systems are designed by connecting different types of technologies. [4].

Augmented reality as a learning medium can also overcome various obstacles in the teaching and learning process in vocational high school. Numerous studies have shown that Augmented Reality (AR) technologies have a beneficial impact on student motivation, learning benefits, communication, engagement, learning behaviors, and enjoyment in all levels of education. In addition, little research has been done on the benefits of AR applications in VET in terms of their impact on addressing various special educational needs, such as learning for disabilities. [5]. During the next few decades, multimedia will cause significant changes in the teaching process, particularly as smart learners discover that they can go beyond the limitations of traditional teaching methods. There is a change in learning from the transmitting or passive-learner model to the experiential or active-learner model [6 In vocational schools, a healthy learning environment learning process is needed to increase students' interest in learning, reduce stress on students' cognitive abilities, and improve students' ability to visualize the material being taught. these things are the impact of the development of augmented reality learning media. Furthermore, the teaching and learning process is more engaging because of the lecturer's teaching style. Furthermore, the teaching and learning process is more engaging due to the lecturer's style of combining theory and practice at the same time. The use of augmented reality in education not only aids in the creation of a healthy teaching and learning atmosphere, but it also improves learners' abstract visualization, which is beneficial when performing experiments [7].

2. Design of Mobile Augmented Reality Based Electrical Measurement Instruments

2.1 Development Instruments Introduction

Unity3D is a cross-platform game development framework with a fully integrated professional game engine that includes rendering, physics engine, script engine, lighting mapping, and scene management functionality. Unity3D is primarily used to build mobile games or computer games, and create immersive 3D and 2D environments such as structural visualization, training simulations, and medical visualization. A complete Unity3D software consists of several scenes, each containing a large number of game object and their actions, which are managed by scripts (such as JavaScript, C #, and others). The camera presents and controls what we see in the image [8].

As a mobile application developer platform, Unity does not work alone. There are several development instruments that can be used to create an augmented reality-based mobile application, one of which is the Vuforia SDK. Vuforia is a software development kit developed by Qualcomm with augmented reality capabilities that allow developers to create applications based on augmented reality technology. It recognizes and tracks planar images and 3D objects in real time using computer vision technology. When viewing virtual objects such as 3D models and other media through a mobile device camera, this image registration feature allows developers to position and align them with real things. Virtual objects appear alive and blended into the environmental background. [9].

2.2 The Concept of Learning Media

A study produced a mobile application based on Augmented Reality for electrical measurement instruments in a vocational high school called ARAVO. The apps became a learning media that compiled using AR-based technology. A technology that overlays virtual objects (augmented components) into the physical world is known as augmented reality (AR). These virtual objects then tend to share the same space as real-world objects [1]. Augmented Reality (AR) is a kind of Virtual Environment (VE) or Virtual Reality (VR), as it's also known. Virtual reality systems completely immerse the user in a synthetic environment, and the user does not perceive the physical world around them. On the other hand, this AR technology takes digital or computer-generated content such as 3D objects, images, audio, or video and adds it to the real world. [10]. This technology is a method of viewing the real world (either directly or through a camera system that creates a visual representation of the real world) and enhancing it visually with computer-generated inputs such as still images, audio, or video, differs from virtual reality in that it enhances a real-world or current scene rather than doing something completely different [11].

The ARAVO is Augmented Reality from Ammeter, Ohmmeter, and Voltmeter. The interface and learning features inside ARAVO are designed to resemble the actual form of using electrical measurement instruments in virtual objects. Mobile AR systems can strengthen cooperative learning and ubiquitous scenes with the help of virtual objects in real environments. Everything in real life can be used as an aid in learning augmented reality to achieve comfort, interactivity, situational awareness, accessibility, and personalization. The ARAVO had five main menus, namely the Home, Competency Standard (SKKD), Play AR, material, and information menu. To understand the work system of the Learning Media of mobile augmented reality-based electric measurement instruments, figure 1 is presented below:

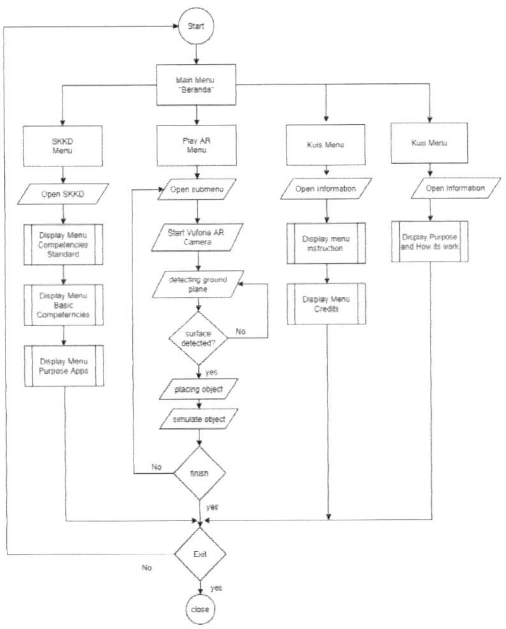

Figure 1. Flowchart of Mobile Augmented Reality-Based Media Electrical Measurement Instruments

The Main menu consisted of the being this application purpose made and instructions on how to use the learning media for electrical measurement instruments. The Information menu also can be found here. The Information menu consisted of a detailed explanation of each button and function in apps. The credits menu shows sources of every part and component 3D model, icon, and pictures that exist inside apps. The competency standard menu is a list of Competencies standards, and Basic Competencies is part of the curriculum vocational high school major of electrical engineering. This menu helps a user understand that the application is part of the competency standard in vocational high school.

The Play AR is the main menu for a Learning Media for electrical measurement instruments with augmented-based technology. There five sub-menus called Resistance measuring with Ohmmeter, DC electric current measuring with Amperemeter, DC voltage measuring with Voltmeter, AC voltage measuring with Voltmeter, and extras.

2.3 Media Design Display

The design of the user interface is one of the most important things while developing a mobile application.

Figure 2. Main Menu Display

Figure 2 shows the initial view and face of the learning media for electric measuring instruments based on augmented reality. There are 5 main menus displayed, namely homepage, competency, AR play, materials, and information. The workflow for each menu is described in Figure 1 above.

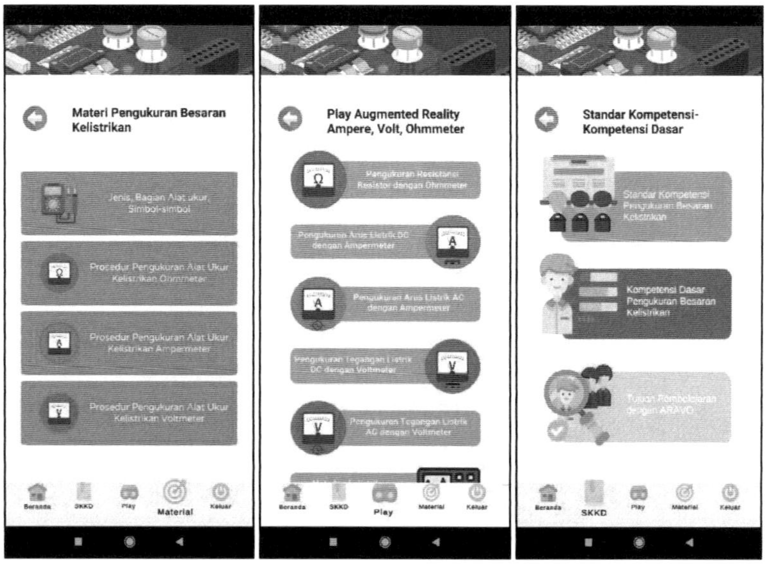

Figure 3 Sub-main menu display

Figure 3 shows the content in each sub-main menu. Every sub-menus was directly connected to the previously programmed AR-based technology in Unity3D. Vuforia AR platform used in this program. The augmented reality method is not marker detection but ground plane detection. The ground plane detection method combines a camera and sensor in smartphones to detect flat surfaces in the environment such as tabletop or floor. After that, an AR program will run and put the 3D model of the electric measurement instrument top on it. The main 3D model virtually for electrical measurement instruments is the multimeter.

A Multimeter is a measuring instrument that can measure various electrical properties like voltage, current and electric resistance. A user can use this menu to simulate how to operate measurement instruments correctly. The Multimeter has simulation can see in smartphones with camera-assisted. As an example measuring DC electric voltage with a voltmeter. As an example measuring DC electric voltage with a voltmeter can see how real is, the voltmeter in a virtual world can use to measure the DC electric voltage as in the physical world perspective.

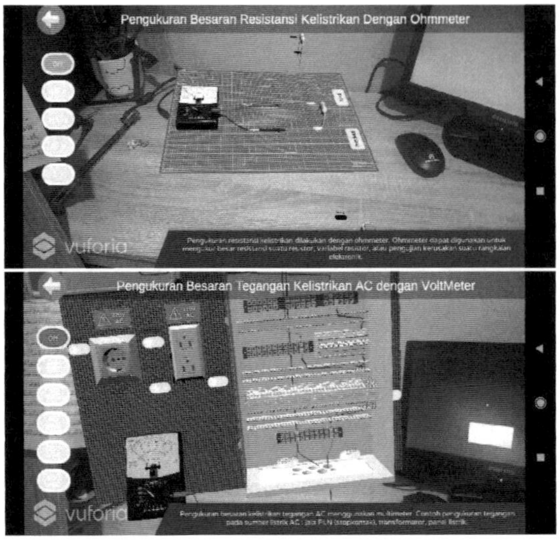

Figure 4. Display of Multimeter and electric panel based on augmented reality

Figure 4 shows a simulation of using a multimeter to measure voltage with augmented reality technology. The Vuforia Ground Plane allows you to place digital content on flat surfaces in your environment, such as floors and tables. Supports detection and tracking of flat surfaces and also allows you to position content in the air using anchor points [12]. The detection technique does not use markers like AR applications in general but uses the ground plane detection method. The virtual model can access by pointing the camera at a flat area such as a tabletop or floor. The multimeter acts as a 3d model is displayed on the table. The model is visible with the help of a smartphone camera. Each material is equipped with a virtual button that functions to simulate measurements of various electrical quantities

The multimeter has a selector knob to determine its function and measuring limits. For example, users take measurements of electrical voltage quantities, users define their own measuring limits. The user interface (UI) button on the left side of the smartphone screen is used to control the selector knobs. The multimeter probe cable is moved by pressing a virtual button pin that is visible through the cellphone camera so that the application will pinpoint the target point for measuring voltage. In addition to the virtual button pins, there are virtual previous and next buttons, used to replace the components to be measured.

3. Results

From this development research, was produced a learning media application product named ARAVO electrical measuring instrument. ARAVO is an abbreviation of Augmented-Reality an amperemeter, a voltmeter, and an ohmmeter. This application runs on smartphones that use the Android mobile platform. This research is still limited to the functional test and must through expert judgment to assess the feasibility of the product. ARAVO requires an expert assessment to ensure that the displayed content is worthy of being presented. Two categories of feasibility were assessed, the feasibility of media and material.

Based on the research, the results of the media feasibility assessment by media experts on the visual dimension obtained an average score of 24 and a feasibility percentage of 85.7%. With the eligibility criteria "very eligible". The software dimension assessment score obtained an average score of 33 and a feasibility percentage of 91.7% with the eligibility criteria "very eligible". The usability dimension assessment score obtained an average score of 28.5 and a feasibility percentage of 89.1% with the eligibility criteria "very eligible". The portability dimension assessment score obtained an average score of 32 and a feasibility percentage of 88.9% with the "very feasible" eligibility criteria. Figure 5 will show the result of media assessment.

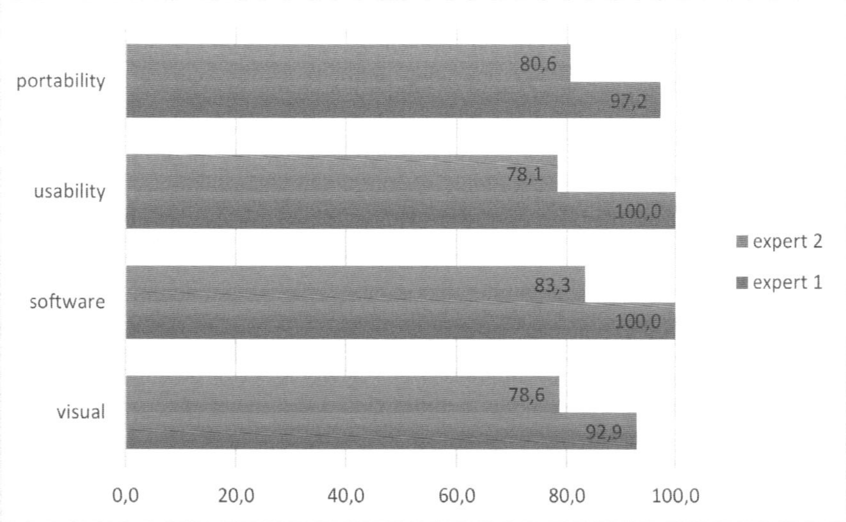

Figure 5. The result of media assessment.

The results of the material feasibility assessment by material experts on the material feasibility dimension obtained an average score of 42.5 and a feasibility percentage of 88.5%. With the eligibility criteria "very eligible". The learning dimension assessment score obtained an average score of 21,5 and a feasibility percentage of 89.6% with the "very eligible" eligibility criteria. The functional dimension assessment score obtained an average score of 13.5 and a feasibility percentage of 84.4% with the "very eligible" eligibility criteria. Figure 6 will show the material assessment.

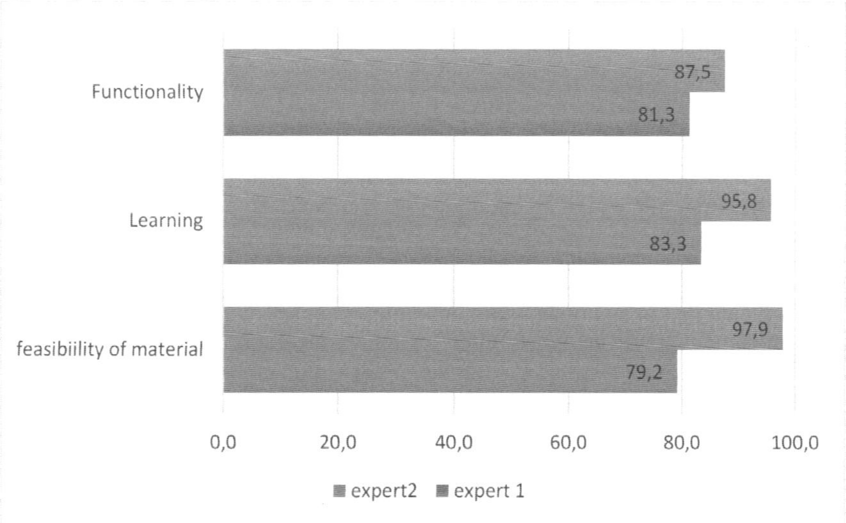

Figure 6. The result of material assessment.

The functional test is installing the application on several android devices from Android version API Level 19 till Android version API level 30. The results obtained are that the ARAVO application can run well on the Android platform and above.

4. Conclusion
The goal of creating augmented reality-based mobile applications is to make learning with smartphones more engaging. This application is a way to overcome the limitations of practical instruments, particularly the competence of electrical measuring instruments. In order to master this competency, the student must first master the understanding and skills of electrical measuring instruments. This electric measuring instrument's augmented reality learning media is predicted to boost a lecturer's efficiency in delivering electrical measuring instrument content that is full of practical learning. Students can digitally replicate the measurement of numerous electrical values such as voltage, resistance, and electric current using a smartphone, just as they exercise with true measuring equipment instruments.

5. References
[1] Nincarean, D., Ali, M. B., Abdul Hamid, N. D., & AR, M. M. (2013). Mobile Augmented Reality: the potential for education. *Procedia - Social and Behavioral Sciences 103* , 657 – 664
[2] Linowes, J., & Babilinski, K. (2017). *Augmented Reality for Developer: Build practical augmented reality applications with Unity,.* Birmingham: Packt Publishing.
[3] Akçayır, M., & Akçayir, G. (2016). Advantages and challenges associated with augmented reality for education: A Systemic review of the literature. *Educational Research Review.*
[4] Hsin, K. W., Silvia Wen, Y. L., Hsin, Y. C., & Jyh, C. L. (2013). Current status, opportunities and challenges of augmented reality in education. *Sciencedirect: Computer and Education Volume 62.*
[5] Acosta, J. B., Fabregat, R., Baldiris, S., & Graf, S. (2014). Augmented Reality Trends in Education: A Systematic Review of Research and Applications. *J. Educ. Technol. Soc.17 (4)*, 133–149.
[6] Vaughan, T. (2011). *Multimedia : Making It Work,eight edition.* McGraw-Hills.
[7] Ismail, M. E., Utami, P., Ismail, I. M., Khairudin, M., Amiruddin, M. H., Lastariwati, B., & Maneetien, N. (October 2018). The effect of an augmented reality teaching kit on visualization, cognitive load and teaching styles. *Jurnal Pendidikan Teknologi dan Kejuruan, Vol. 24, No. 2*, 178-184.
[8] Li, C. Y., & Tang, B. (2018). *Research on The Application of AR technology Based on Unity 3D in Education.* IOP Conf. Series: Journal of Physics: Conf. Series 1168 .
[9] Getting Started with Vuforia Engine in Unity: About Vuforia. Accessed April 30, 2021 https://library.vuforia.com/articles/Training/getting-started-with-vuforia-in-unity.html#about
[10] Kipper, G., & Rampolla, J. (2013). *Augmented Reality: An Emerging Technologies Guide to AR.* 225 Wyman Street, Waltham, MA 02451, USA: Elsvier.
[11] Mealy, P. (2018). *VIrtual and Augmented Reality for Dummies.* New Jersey: John Wiley & Sons, Inc.
[12] Ground Plane Vuforia. Accessed April 30, 2021 https://library.vuforia.com/features/environments/ground-plane-guide.html

Journal of Physics: Conference Series **2111** (2021) 012030 doi:10.1088/1742-6596/2111/1/012030

Utilization Extract, Transform, Load For Developing Data Warehouse In Education Using Pentaho Data Integration

A D Barahama[1] and R Wardani[2]

[1,2]Department of Electronics and Informatics Engineering Education, Post-Graduate Program, Universitas Negeri Yogyakarta Jalan Colombo 01, Daerah Istimewa Yogyakarta 55281, Indonesia.

Email: ameilianidianira.2019@student.uny.ac.id

Abstract. The utilization of data warehouses in various fields is an absolute necessity. A data warehouse is a database that contains large amounts of data that aims to help organizations, fields, and institutions specifically for decision making. Data warehouses can produce important information in the future. Loading data from various sources and processed through an ETL (Extract, Transform, Load) process that displays data consistently is the basis for creating a data warehouse architecture. The development of a data warehouse in education will provide significant benefits for the progress of education. Integration of data and processing results stored in the data warehouse can be the basis for evaluating better planning. Development of data warehouse adopt the multidimensional modelling method which consists of four stages: select the business process, declare the grain, select dimensions, and identify facts. This stage produces a data warehouse architecture and influences and contributes to the advanced information technology in education.

1. Introduction

Supporting the achievement of Indonesian education can be measured through 6 indicators: students, teachers, infrastructure, costs, curriculum, and involvement. Elements of society are interrelated and have their respective duties and roles as educators and teachers. The availability of educators must be adequate in each region, one of which is the implementation of technology information. The use of information technology in data integration is a vital need in education. The data integration process can produce information that can help users find appropriate and efficient solutions to be used as a basis for decision making. The availability of data will support success in the field of education. Extract-Transform-Load has the task and is responsible for integrating data into a data warehouse [1]. Process implementation data into information can be used as a basis for decision making and is helpful for knowledge system management in organizations to improve human resources and information technology. [2].

Data availability can be support by the existence of a Data Warehouse which aims to process, store and manage large amounts of data for analysis and decision-making needs. [3] The analytical process stored in the data warehouse will be helpful in the future as a support in obtaining efficient decision making [4] The data warehouse will provide sources of information that work directly in solving problems so that they find the solution [5]. This research will use the Pentaho Data Integration-kettle as a tool that makes it easier to overcome the diversity of data in education to produce a data analysis process for educators. Data integration through extracts, transformations, loading in the development of Data Warehouses in education can assist in conducting evaluations, planning and an accurate basis to support decision making in the future [6].

2. Methods

2.1. Preparing Data source

In developing a data warehouse, the basic thing needed is to prepare data sources. Data source will be processed with the stages of data cleaning, data selection so that the data is ready to be transformed into a data warehouse. Preparation of data sources will be used in development of data warehouse.

2.2. Design of data warehouse

The basis of data warehouse development is to have dimensional data modeling. According to Ralph Kimball and Joe Caserta in their book The Data Warehouse ETL Toolkit (practical techniques for extracting, cleaning, and delivering data), there are four stages in developing a Data Warehouse.[7]

2.2.1. Select the business process

In the first stage, the thing to do is identify the requirements that must be available, the data sources to be processed, understand what kind of modeling and requirement are needed.

2.2.2. Declare the grain

The second stage is the process of identifying the problems found. What are the right dimension and the right way to solve the problem.

2.2.3. Choose the dimension

This stage is the stage to choose the appropriate dimensions so that it can produce a fact table.

2.2.4. Identify the facts

This stage is the stage to choose the dimensions so that it can produce an appropriate fact table. This stage is the last step of dimensional data modeling, the stage to produce a fact table which will be the core of dimensional data modeling from data warehouse development.

2.3. Dimensional data modeling

Multidimensional modeling is an old technique for creating simple databases. Dimensional data modeling will produce several dimensions and fact tables, which will explain the flow of the data integration process into a data warehouse.

2.4. Extract, Transform, Load – ETL

ETL is the process of extracting data from various sources, transforming it and loading it into a data warehouse. The ETL process is an important step, the data stored in the data warehouse can be used as a source to obtain information that can be used as a basis for decision making or as a knowledge management system within the organization. In the ETL process, the main things to do are extract, clean, conform and deliver. There are important parts in the ETL process namely: XML data collection, using an independent DBMS, having a schema, normalized entity/relationship (E/R), and dimensional data modeling [7]. ETL plays an important role in the data interaction process because it can help determine the outcome of the problem so that it can get the appropriate retrieval [2]. ETL can support the Identification of relevant information from its source, perform information extraction, customization, and integration from various sources, perform cleaning and load it in the big data into a Data Warehouse. [8]. The purpose of the ETL process is described in figure 1:

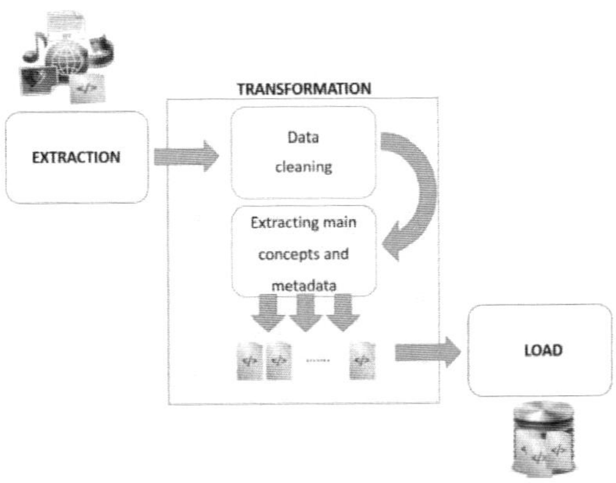

Figure 1. The proposed ETL tool [9]

Figure 1. Describes the integration process starting from extraction, transformation where the data is then cleaned, processed in meta data and loaded into the data warehouse.

2.5. *Result of data warehouse*

Experiment results from the initial stages of preparing data source, design of data warehouse, star schema until the completion of the extract, transform, load stages will produce a data warehouse that is ready to be used.

3. Result and discussion

3.1. *Data source*

In developing the data warehouse, data source comes from raw data in Microsoft Excel format containing educational data, namely teacher and educator data, then cleaning the selection data and adding data types until the data is ready to be integrated through the extract, transform, load process. The following illustrates the change of raw data into data that is ready to be integrated into the Extract, Transform, Load (ETL) process.

KODE GURU	NO GURU	NAMA GURU	KODE KUALIFIKASI	thn_lulus	JENIS KELAMIN	KODE SEKOLA	KODE STATUS SEKOLAH	NAMA_SEKOLAH
44123	99123456781	Margaretha Juliana Wattimena	18	1991	Perempuan	40102618	628123	SMAN 1 MANADO
44124	99123456782	Jodie Janie Albert Tulung	18	1987	Laki-Laki	40102618	628123	SMAN 1 MANADO
44125	99123456783	Boyke Mentang	18	1988	Laki-Laki	40102618	628123	SMAN 1 MANADO
44126	99123456784	Betsie Parengkuan	19	2014	Perempuan	40102618	628123	SMAN 1 MANADO
44127	99123456785	Vivi Zeila Kumaat	18	1988	Perempuan	40102618	628123	SMAN 1 MANADO
44128	99123456786	Dewi Norin Jans Lumintang	18	2013	Perempuar	40102618	628123	SMAN 1 MANADO
44129	99123456787	Dyterzet Kamrun	18	1991	Laki-Laki	40102621	628123	SMAN 3 MANADO
44130	99123456788	Junifke Antameng	18	1981	Perempuan	40102621	628123	SMAN 3 MANADO
44131	99123456789	Alex. Victor Octavianus Sumarauw	18	2014	Laki-Laki	40102621	628123	SMAN 3 MANADO
44132	99123456790	Sri Wahyuni	18	2012	Perempuan	40102747	628123	SMAN 6 MANADO
44133	99123456791	Julisi Risat Supit	18	2006	Laki-Laki	40102761	628123	SMAN 7 MANADO
44134	99123456792	Randi Kaseger	18	2010	Laki-Laki	40102761	628123	SMAN 7 MANADO
44135	99123456793	Naomi Sartje Coni Mogonta	18	1977	Perempuan	40102761	628123	SMAN 7 MANADO
44136	99123456794	Marlon Ferdy Wellem Rompas	18	1984	Laki-Laki	40102761	628123	SMAN 7 MANADO
44137	99123456795	Alce A. Seke	18	1984	Perempuan	40102620	628123	SMAN 2 MANADO
44138	99123456796	Trusje Penny Antou	18	2006	Perempuan	40102620	628123	SMAN 2 MANADO
44139	99123456797	Altje Kalalo	18	2005	Perempuan	40102762	628123	SMAN 8 MANADO
44140	99123456798	Femmij Tulangow	18	2011	Perempuan	40102620	628123	SMAN 2 MANADO
44141	99123456799	Novi Nixon Sonces Ratu	18	1983	Laki-Laki	40102619	628123	SMAN 10 MANADO
44142	99123456800	Marlin Bernadet Taariwuan, S.pd	18	2010	Perempuan	40102618	628123	SMAN 1 MANADO

Figure 2. Modified data source table

Figure 2. Describes the raw data table that has been processed and cleaned to prepare the data for the ETL process.

3.2. *Design of data warehouse*

The data warehouse produces from the completion of the multidimensional modeling stage. A data warehouse is a specific database that generates from various sources and information systems of organizations, fields of companies, and institution [12]. A data warehouse makes as a multidimensional database that contains large amounts of data that can be organized through several dimensions and fact tables and becomes the basis for an organization, company in the decision-making process[13]. A data warehouse is an organized collection of data that aims to provide the right solution so that it helps in the decision making [14]. According to William H., Inmon Data warehouse is a collection of subject-oriented, integrated, time-variant, and non-volatile data to support the decision-making. [15]. *The four main features of a data warehouse are:*

1. Subject Oriented**:** A data warehouse is organized around the main subject concentrating on the day-to-day operations and transaction processing of an organization, data Warehouse focuses on modeling and analyzing data for decision-making.
2. Data warehouses are built by integrating several heterogeneous sources, such as relational databases, files, and online transaction records.
3. Time Variant**:** Data is stored to provide information from a historical perspective (e.g., the last 5-10 years)
4. Nonvolatile: A data warehouse is always a physically separate data mart that is the transformation from the application data found in the operational environment. The data warehouse is not updated in real-time but is refreshed from the operating system regularly continually and incrementally to be integrated with previous data.

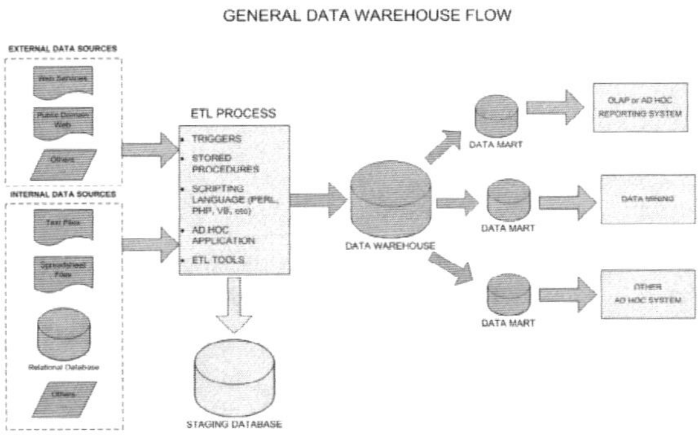

Figure 3. General Data Warehouse [12]

Figure 3. Describes the subject-oriented data warehouse architecture and comes from different sources integrated into a single unit through ETL (extract, transform, load) so that it can achieve the goal of producing a data warehouse.

3.2.1 Architectural design data warehouse for teacher in education.

Extraction, transformation, and loading (ETL) and creates an improved and consistent appearance materialized from a collection of data sources that will be used as the basis for the data warehouse architecture. [8] The construction of a data warehouse for educators in education source from the educational database, which is the process through Microsoft Excel and lustration of data will be processed using Pentaho Data Integration tools. The results of the extraction that depicted out produce an architecture that can explain as follows:

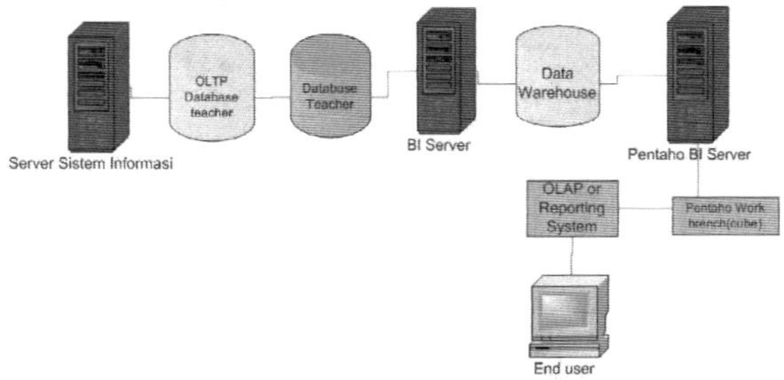

Figure 4. Physical Data Warehouse Architecture in education

Figure 4. Describes the data warehouse architecture for educators in education, data sources originated from information systems are then utilized and integrated through extraction transformation and loading to produce a special database in the form of a data warehouse which is a necessity to assist stakeholders in institutions education for decision making.

3.3 Dimensional data modeling

Extraction, transformation, and loading (ETL) results create dimensions that will be related to each other to produce a fact table, these dimensions are generated using Pentaho data interaction tools through excel input data sources and generate databases stored in SQL yog, and have a surrogate key that is a key containing an integer value.

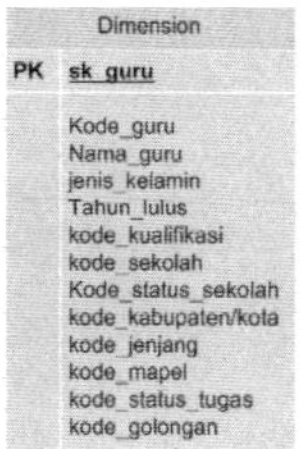

Figure 5. Dimension_teacher

Figure 5. Illustrates the teacher dimension as one of the important dimensions in dimensional data modeling. This dimension contains integer, date and character string values. An integer value in the form of a code will be an entity from another dimension. Sk_guru becomes the sourgate key on this dimension.

3.4 Extract-Transform-Load (ETL)

Extract-Transform-Load (ETL) process that has been used in computing for integration of data from multiple sources. This process refers to the development of a data warehouse that extracts data from outside sources, transforms it according to operational needs, and then loads it in a data mart or data warehouse[10]. The three phases of this process are described as follows:

1. The first process of this phase is the extraction of data available in formats such as xls, cvs and txt. the procedure of extracting data from different sources is done by reading the data and purifying the information received to be used as guidelines and metadata.[11]
2. The second phase is data cleaning carried out according to a predetermined scheme, involving normalization of data, deleting data, adding data, sorting, and grouping them according to need.
3. The third phase aims to load the data into a data mart or data warehouse that is ready for use. Composing the data into the objective database. ETL process move data starting with one

database then onto the next, to shape data marts and data warehouse furthermore to change over databases starting with one organization or institution [11]

Utilization of extract, transform, load in data integration to develop teacher data warehouse using Pentaho data integration aims to facilitate data processing can be used in the future for analysis, evaluation, and decision-making purposes. The following illustrates the integration of teacher data through the ETL process.

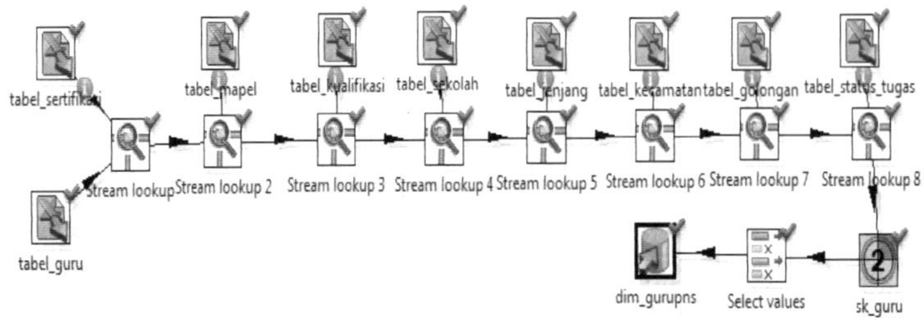

Figure 6. Output dimension teacher from ETL process Pentaho data integration tools.

Figure 6. Describes the ETL process from several tables that are determined to be processed to produce teacher dimensions, connected to the stream lookup, selected with select values, a database query is carried out which will produce the teacher dimensions. The results of the process carried out can be seen through the data preview shown in Figure 7.

First rows Last rows Off

	kode_kualifikasi	kualifikasi	nama_kualifikasi	sk_kualifikasi
1	11	SMP	Sekolah menengah pertama	1
2	12	SMA	Sekolah menengah atas	2
3	13	SMK	Sekolah menengah kejuruan	3
4	14	D1	Diploma tingkat satu	4
5	15	D2	Diploma tingkat dua	5
6	16	D3	Dipploma tingkat tiga	6
7	17	D4	Diploma tingkat empat	7
8	18	S1	Sarjana	8
9	19	S2	Master	9
10	20	S3	Doctor	10

Figure 7. Preview data qualification dimension

Figure 7. Preview data from one dimension that is closely related to the data for educators, a data contains the qualification dimension. Explains the resulting dimension table, in this case the data selected to be displayed is one of the dimensions related to the teacher dimension, namely the qualification dimension.

The final result of each dimension that has been generated is to combine them to produce a fact table. The fact table contains a collection of dimensions that will produce a fact table and display dimensional data modeling. The following is the result of data integration that combines the existing dimensions into a fact table:

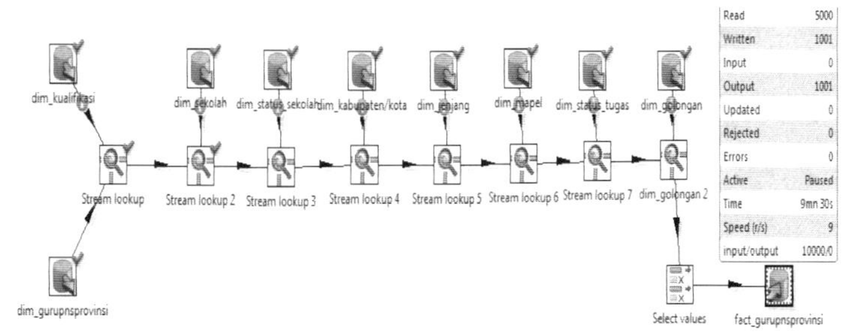

Figure 8. Fact_tabel (fact_guru)

Figure 8. Describes the dimensions associated with stream_lookup and performs metadata on the attributes generated from the dimensions using select values will generate fact_tables. will display data in the form of preview data which can be seen in figure 9.

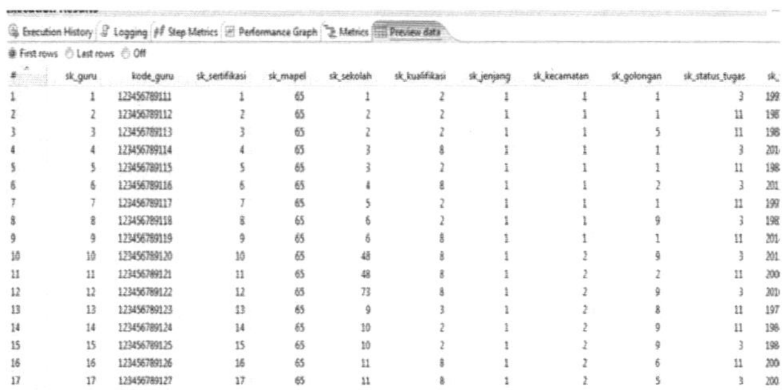
Execution History | Logging | Step Metrics | Performance Graph | Metrics | Preview data

First rows | Last rows | Off

#	sk_guru	kode_guru	sk_sertifikasi	sk_mapel	sk_sekolah	sk_kualifikasi	sk_jenjang	sk_kecamatan	sk_golongan	sk_status_tugas	sk_
1	1	123456789111	1	65	1	2	1	1	1	3	199
2	2	123456789112	2	65	2	2	1	1	1	11	198
3	3	123456789113	3	65	2	2	1	1	5	11	198
4	4	123456789114	4	65	3	8	1	1	1	3	201
5	5	123456789115	5	65	3	2	1	1	1	11	198
6	6	123456789116	6	65	4	8	1	1	2	3	201
7	7	123456789117	7	65	5	2	1	1	1	11	199
8	8	123456789118	8	65	6	2	1	1	9	3	198
9	9	123456789119	9	65	6	8	1	1	1	11	201
10	10	123456789120	10	65	48	8	1	2	9	3	201
11	11	123456789121	11	65	48	8	1	2	2	11	200
12	12	123456789122	12	65	73	8	1	2	9	3	201
13	13	123456789123	13	65	9	3	1	2	8	11	197
14	14	123456789124	14	65	10	2	1	2	9	11	198
15	15	123456789125	15	65	10	2	1	2	9	3	198
16	16	123456789126	16	65	11	8	1	2	6	11	200
17	17	123456789127	17	65	11	8	1	2	5	3	200

Figure 9. The result - transformasi fact_tabel

Figure 9. describes the results of the fact_table transformation of the dimensions that are used as a special database through the Extraction, transformation and load process with the Pentaho Data Integration-kettle tool that displays the fact_table results.

Based on the stages of research carried out for data warehouse development through extract, transform, load with Pentaho data integration tools, good results are obtained and explained in the data preview that has been described. The stages of research carried out from preparing data sources, making designs, and processing them with extract, transform, load stages that produce several dimensions and one fact table obtain the results that the use of the extract, transform, load in the development of teacher data warehouses in the field of education will help parties interested in conducting analysis, evaluation and assisting in decision making.

4. Conclusion

Thereby, the building of a data warehouse requires the execution of the Extraction-Transformation-Load (ETL) process. Integration of data from different sources forms a data warehouse that can be used as a core source in obtaining accurate information in making decisions for the advancement of education. The data warehouse offers a summary of operational data that is simple and easy for management to understand if needed. The process carried out in building a data warehouse for teachers and educators in the field of education using Pentaho data integration tools facilitates data processing and analysis. The source of data obtained from raw data is then processed following the stages of extract, transform, load to produce several dimensions and fact tables which are the key to the development of this data warehouse. Thus, the development of this data warehouse will make it easier for decision-makers to determine policies, facilitate the development of DSS and EIS applications, support a better analysis process for decision-makers for the advancement of education.

Acknowledgment
The authors thank Dr. Ratna Wardani M.T in Yogyakarta State University for advice in writing of this manuscript. This study was supported in part by Dinas Pendidikan Provinsi Sulawesi Utara.

References

[1] P. S. Diouf, A. Boly, and S. Ndiaye, "Variety of data in the ETL processes in the cloud: State of the art," *2018 IEEE Int. Conf. Innov. Res. Dev. ICIRD 2018*, no. May, pp. 1–5, 2018, doi: 10.1109/ICIRD.2018.8376308.
[2] P. S. Diouf, A. Boly, and S. Ndiaye, "Performance of the ETL processes in terms of volume and velocity in the cloud: State of the art," *4th IEEE Int. Conf. Eng. Technol. Appl. Sci. ICETAS 2017*, vol. 2018-Janua, pp. 1–5, 2018, doi: 10.1109/ICETAS.2017.8277875.
[3] O. Moscoso-Zea, J. Paredes-Gualtor, and S. Luján-Mora, "A Holistic View of Data Warehousing in Education," *IEEE Access*, vol. 6, pp. 64659–64673, 2018, doi: 10.1109/ACCESS.2018.2876753.
[4] G. Blaži, P. Poš, and D. Jakši, "Data Warehouse Architecture Classification," pp. 1491–1495, 2017.
[5] I. I. Kholod and M. S. Efimova, "Using ETL Tools for Developing a Virtual Data Warehouse," pp. 351–354, 2016.
[6] A. D. Barahama and R. Wardani, "Data analysis and data warehouse design based on Pentaho data integration (kettle) to support the determination of student learning achievement," *IOP Conf. Ser. Mater. Sci. Eng.*, vol. 1098, no. 5, p. 052089, 2021, doi: 10.1088/1757-899x/1098/5/052089.
[7] Ralph Kimball & Joe Caserta, *The Data Warehouse ETL Toolkit (Parctical Techniques for Extracting,Cleaning,Conforming, and Delivering Data*. 2004.
[8] S. Bergamaschi, F. Guerra, M. Orsini, C. Sartori, and M. Vincini, "A semantic approach to ETL technologies," *Data Knowl. Eng.*, vol. 70, no. 8, pp. 717–731, 2011, doi: 10.1016/j.datak.2011.03.003.
[9] S. Souissi and M. Benayed, "GENUS: An ETL tool treating the Big Data Variety," *Proc. IEEE/ACS Int. Conf. Comput. Syst. Appl. AICCSA*, vol. 0, 2016, doi: 10.1109/AICCSA.2016.7945615.
[10] J. Chakraborty, A. Padki, and S. K. Bansal, "Semantic ETL-State-of-the-Art and Open Research Challenges," *Proc. - IEEE 11th Int. Conf. Semant. Comput. ICSC 2017*, pp. 413–418, 2017, doi: 10.1109/ICSC.2017.94.
[11] L. Fu, "A Recommendation System Using OLAP Approach," pp. 622–625, 2016, doi: 10.1109/WI.2016.108.
[12] R. J. Salaki, J. Waworuntu, and I. R. H. T. Tangkawarow, "Extract transformation loading from OLTP to OLAP data using pentaho data integration," *IOP Conf. Ser. Mater. Sci. Eng.*, vol. 128, no. 1, 2016, doi: 10.1088/1757-899X/128/1/012020.
[13] G. Garani, A. V. Chernov, I. K. Savvas, and M. A. Butakova, "A Data Warehouse Approach for Business Intelligence," *Proc. - 2019 IEEE 28th Int. Conf. Enabling Technol. Infrastruct. Collab. Enterp. WETICE 2019*, pp. 70–75, 2019, doi: 10.1109/WETICE.2019.00022.
[14] A. Lamani, B. Erraha, M. Elkyal, and A. Sair, "Data mining techniques application for prediction in OLAP cube," *Int. J. Electr. Comput. Eng.*, vol. 9, no. 3, pp. 2094–2102, 2019, doi: 10.11591/ijece.v9i3.pp2094-2102.
[15] J. P. Jiawei Han, Micheline Kamber, *Data mining: Data mining concepts and techniques*, 3rd ed. San Fransisco, 2014.

Evaluating UI in Digital Skills Training: Human Centered-Design approach from UI Sanbercode Bootcamp

Fiducia[1] and Suprapto[2]

[1,2] Electronics and Informatics Engineering Education, Postgraduate, Universitas Negeri Yogyakarta, Daerah Istimewa Yogyakarta 55281, Indonesia

Email: [1]fiducia.2019@student.uny.ac.id; [2]suprapto@uny.ac.id

Abstract. Sanbercode is a digital skill training program, which can be studied by various classes of training. In the Covid-19 pandemic, much is felt: health, economics, to learning activities. These conditions require learning activities to be carried out at home. The purpose of this training is to enhance a person's digital abilities or digital skilss. Sanbercode training is done online the web that is easy to access, it is hoped can take place more interactive, real, and fun. The approach applied to UI involves human centered-design. Found eight issues on the features consistent with human centered-design and conducted heuristic evaluations by evaluators. The results show a decline in the number following improvement solutions.

1. Introduction

A Person needs knowledge as a provision for himself to know the wide world and have reliable abilities. Technological developments are apparent in today's life. In general, learning activities are carried out at schools, universities, seminars, and others that require face-to-face contact. However, these learning activities have changed since the COVID-19 pandemic. The COVID-19 pandemic is a global health crisis that affects all sectors [1], especially teaching and learning activities. Some countries have regulations for out-of-house reduction to avoid the spread of the virus. So, the education sector to several can be done online and at home. Learning activities are realized and can be affectively automated.

Indonesia's IT industry is moving to develop online learning. Sanbercode Bootcamp is a training program for people who want to increase their IT skills. There are programming training, data science, web app development, UI/UX design, graphic design, and other promising programs in IT. The learning system at Sanbercode Bootcamp uses web applications, zoom meetings, and discussion rooms with experienced trainers and other participants [2]. This digital skills training has a fun learning space, with material facilities, daily assignments, weekly quizzes, and final projects.

Human-centered design becomes an important reference to assess how successful a technology development and system design is. System designed using human-centered methods improve quality by: (1) increasing the productivity of user and the operational efficiency of organizations, (2) being easier to understand and use, thus reducing training and support cost, (3) increasing usability for people with a wider range of capabilities and thus increasing accessibility, (4) improving user experience, (5) reducing discomfort and stress, (6) providing a competitive advantage, (7) contributing towards sustainability objectives [3]. In this context, there are findings of deficiencies in the Sanbercode web such as the layout of the menu that has a lot of free space, the size of the menu select category is too small, and the display of the active page is not clear. Thus, identification of specific problems is carried out to be used as an evaluation based on the findings of the User [4]. The application of human centered-design is implemented properly to improve the quality of this web application. The most appropriate for analyzing tasks run on form sections of the website, it can judge the usability and function of the user interface. Ten rules in heuristic evaluation (1) Visibility of system status, (2) Match between system and the real world, (3) User control and freedom, (4) Consistency and standards, (5) Error prevention, (6) Recognition rather than recall, (7) Flexibility and efficiency of use, (8) Aesthetic and minimalist design, (9) Help and user recognize, diagnose, and recover from errors, (10) Help and documentation [5]. The evaluation carried out is expected to improve the appearance of the Sanbercode Bootcamp web application and be comfortable to use for this digital skill training.

2. Methods

This section contains an explanation of the methodology used in evaluating the User Interface in Sanbercode Bootcamp.

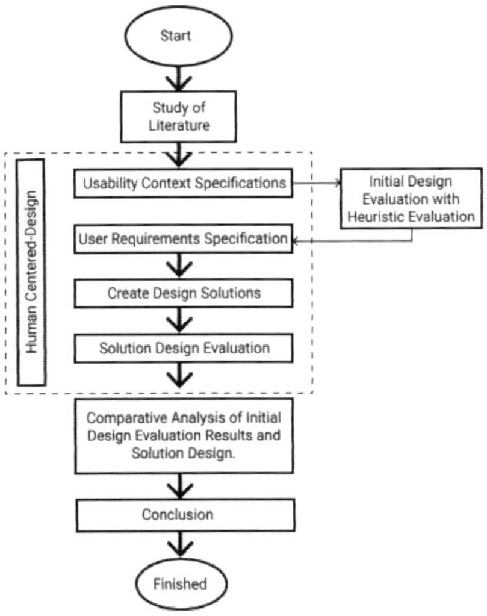

Figure 1. A Methodology of UI Evaluations

2.1. *Analysis of Literature Study*

The process of analyzing to find research needs such as a literature study, problem finding, design. The stage of collecting a deep knowledge so that the problem-finding process will be seen and strengthen the research carried out [6].

2.2. *Usability Context Specifications*

After the analysis has been found, the next step is to identify the usability context specification carried out by stakeholders and user. Usability context specification will found tge activities of the needs, characteristics, and environment of the system. Sanbercode web will be dissected carefully so that a deficiency can be found.

2.3. *Initial Design Evaluation with Heuristic Evaluation*

The third stage is an initial design heuristic evaluation which is an evaluation that will be carried out on this Sanbercode web. Two evaluators have UI background and experience. So that it can obtain heuristic findings that can overcome improvements.

2.4. *User Requirements Specifications*
The next step is done requirement specification, namely processing the data from the evaluation so that it is fulfilled user needs, so that developers can design appropriate repair. Repair design using Figma application. This facilitates communication by allowing requirements from users' language, we follow the "what you needs" approach[7].

2.5. *Create Design Solutions*
Solution design refers to the wireframe display and improved interface. Keep using the Figma app to make design improvements. Furthermore, the stakeholders are checking the solution and the suitability of the repair.

2.6. *Solution Design Evaluations*
The design solution evaluation stage refers to the heuristic evaluation principle. The evaluation aims to assess and provide suggestions contained in the initial design. So, the Sanbercode web is perfect than the previous design.

2.7. *Comparative Analysis of Initial Design Evaluation Results and Solution Design*
Re-evaluation is carried out at this stage to find a comparison between initial design evaluation and solution design.

2.8. *Conclusion*
The conclusion is the final stage used to answer the problems in this study. Furthermore, the suggestions can be used for further research if this research has shortcomings. Re-evaluation is carried out at this stage to find a comparison between initial design evaluation and solution design.

3. Analysis Results

3.1. *Usability Contexts Specification Results*
This process is interviewing the participants of the Sanbercode digital skills training. Users want clarity of system flow. The implementation hours are well listed, the material easily accessible, and career and affiliation registrations have been provided by the organizer.

3.2. *User Requirement Specification Results*
According to the heuristic evaluation, the rules for the category H1 (Visibility of System Status) found a problem where the affiliate program status was always active, the word "new" did not disappear. Look at the figure 2.

Figure 2 Affiliate Menu (NEW)

Figure 3 in work status, the assignment status has two confusing pieces of information, and it's related to the interface. The status area has followed and the quiz recap button. It doesn't look like a typical status the layout, and in one column.

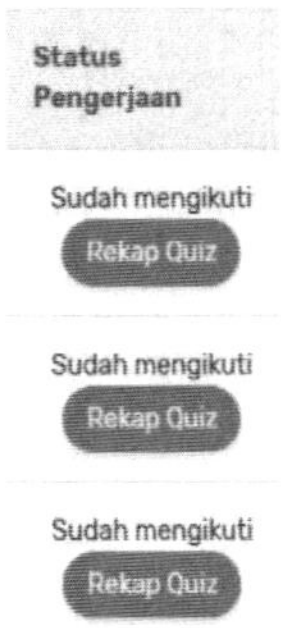

Figure 3 Work Status

Furthermore, in the H2 (Match between System and The Real World) category of inconsistent use of language, there are two languages, Indonesian and English. Look at the figure 4.

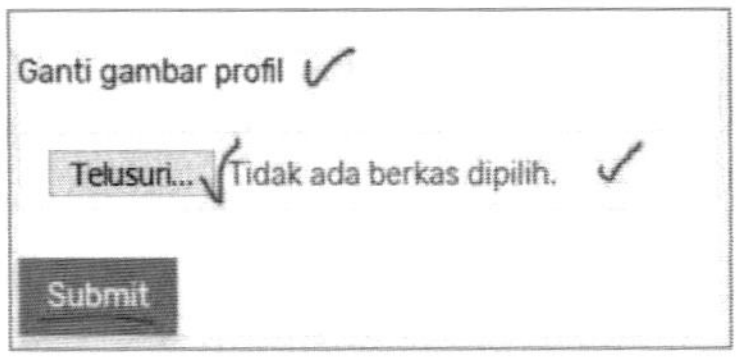

Figure 4 Language Clarity

In the H6 (Recognition Rather Than Recall) category, there is no difference in the color of the active and inactive buttons. Look at the figure 5.

Figure 5 The H6 (Recognition Rather Than Recall) category

In the H8 (Aesthetic and Minimalist Design) category, the display has a lot of free space, although the size of the text is good enough to be read by the user, there is a space still used. Example: the main Sanbercode web page (www.sanbercode.com).

3.3. Design Solutions

The design solution is made for improvements according to the user needs analysis, where some have yet to match.

Figure 6 Affiliate Menu View Solution

Figure 6. A Comparison is how of initial and new design prototypes for affiliate menu buttons. Previously when the button clicked, the word "new" was still listed even though clicked several times, then the solution was to delete the word "new" when one click and so on was no word "new". To avoid confusion of users on the information button that this information is new or not.

Figure 7 Quiz Recap Display Solution

Figure 7. The solution designed is for the location of the Quiz Recap buttons. Previously this button was located in one column with the work status section. Place the split button with other rows. So you can see the difference between the text and the button.

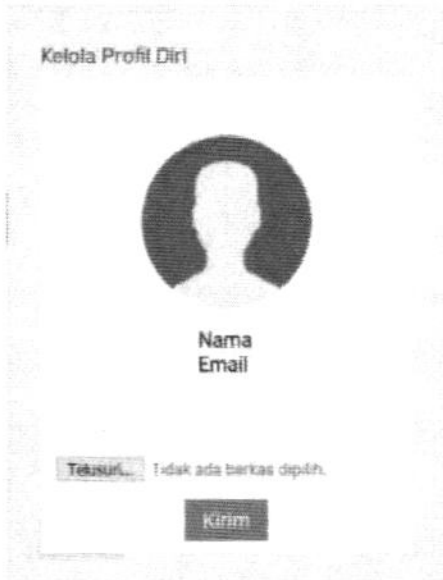

Figure 8 Profile Management Menu Solutions

Figure 8. Language improvement solutions to remain consistent according to heuristic guidelines, namely in the H2 (Match between System and The Real World) category. Other inconsistent words are also applied, this is better to use one language.

Figure 9 Active Menu Solution

Figure 9. The active tab is visible made it easier for users to know which part of the page is currently operating. The marked with a different color from the bars next to it.

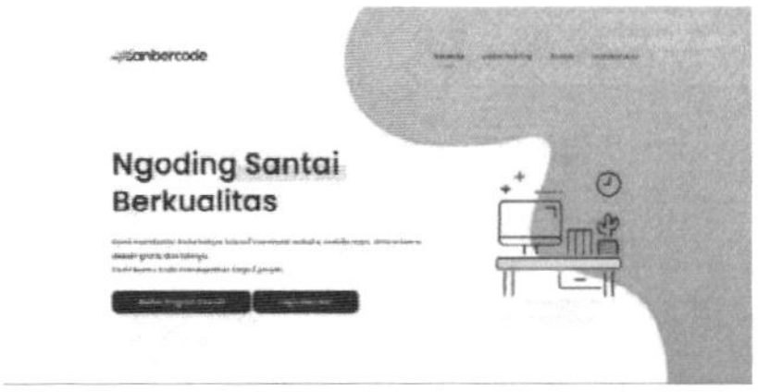

Figure 10 Main Menu Solution

Figure 10. The size of the main menu page section looks well, and there is not much free space. The size of the main button is reasonable than before, so users know that the buttons menu is significant and visible.

3.4. Comparative Analysis of Initial Design Evaluation Results and Solution Design

The following is the number of findings by two evaluators to assess the problems found on the Sanbercode web.

Table 1. Total of problems found in Initial Heuristic Evaluations.

Heuristic Evaluation	Evaluator 1	Evaluator 2	Total
H1	0	2	2
H2	1	1	2
H3	0	0	0
H4	0	0	0
H5	0	0	0
H6	1	1	2
H7	0	0	0
H8	1	1	2
H9	0	0	0
H10	0	0	0
Total	3	5	8

Based on Table 1. It can be seen that the average score entered in either category. There are 8 (eight) problems found by the two evaluators.

This study examines a small part of Web Sanbercode, namely the main menu in general. Human centered-design reference is even wider so that UI design can be even better, and heuristic evaluation is one of the right tests. The findings made, there are at least eight problems that exist on the web. Followed by a Human centered-design approach and heuristic evaluation. The categories found are:

- H1 (Visibility of System Status) shows the notification of the status of the "new" affiliate button that does not disappear when it has been more than one visit (click), and there is a placement of work status still joined to the quiz recap button. Then after being given a solution design, the evaluator assesses that there is 1 (one) problem that still arises, namely the placement of the colored new affiliate icon.
- There are two problems that arise by evaluators that have been assessed in H2 (Match between System and The Real World), the use of language that is more appropriate to use one language. And the placement of the writing is not parallel.
- H6 (Recognition Rather Than Recall) category has two problems. there is no difference in the color of the active and inactive buttons. So, Reduced design solutions with one fix.

- The interface is no problem found in the H8 category (Aesthetic and Minimalist Design) after the final evaluation by the evaluator.

Table 2. Number of Findings on Solution Design

Heuristic Evaluation	Evaluator 1	Evaluator 2	Total
H1	0	1	1
H2	0	0	0
H3	0	0	0
H4	0	0	0
H5	0	0	0
H6	1	0	1
H7	0	0	0
H8	0	0	0
H9	0	0	0
H10	0	0	0
Total	1	1	2

4. Conclusion

Sanbercode Bootcamp is the driver and organizer of Indonesia Mengoding which facilitates users to learn digital skills. Characteristics of the world of work and the qualifications of labor that the industry needs in the industrial era 4.0 are changing rapidly[8]. Participants are provided with knowledge for the future. Training programs ranging from design to programming have grade levels: Basic Class, Project Class, and Partner Class. Many of the boot camp participants who took part in this program still put forward a review of the results of the Sanbercode Bootcamp.

The design of the design solution resulted in improvement and reduction of the problem that was resolved after a heuristic evaluation according to human centered-design method was carried out. There are many aspects of the user experience that can be taken into consideration when designing interactive products. What matters most is utility, functionality, aesthetics, content, look and feel, as well as sensual and emotional appeal [9]. With this research, it is hoped that there will be improvements by the Sanbercode team so that they pay more attention to details to facilitate members, can pay attention to the user interface, and have a better user experience.

References

[1] W. Nur Hidayat, H. Suswanto, C. Wijaya Kristanto, A. Pramudya Wardhani, A. Hamdan, and R. Kartika Sari, "The effectiveness of interactive digital evaluation training for improving teacher skills in the covid-19 pandemic period," *4th Int. Conf. Vocat. Educ. Training, ICOVET 2020*, pp. 310–314, 2020, doi: 10.1109/ICOVET50258.2020.9230070.

[2] P. Berkualitas Syberindo Santai, "Sanbercode." https://sanbercode.com.

[3] S. SAITO and K. OGAWA, "Ergonomics of human-system interaction," *Japanese J. Ergon.*, vol. 30, no. 1, pp. 1–1, 1994, doi: 10.5100/jje.30.1.

[4] A. Dix, J. Finlay, G. D. Abowd, and R. Beale, *Human-Computer Interaction Ch. 9 Evaluation Techniques*. 2004.

[5] N. Jakob, "NN/g Nielsen Norman Group World Leaders in Research-Based User Experience." https://riyanthisianturi.com/10-usability-heuristics-jakob-nielsen.

[6] R. Irwandana, A. D. Herlambang, M. C. Saputra, D. A. Fatah, A. P. Novitasari, and H. Tolle, "Evaluasi dan Perancangan User Interface untuk Meningkatkan User Experience Menggunakan Metode Human-Centered Design dan Heuristic Evaluation Pada Aplikasi

Ezypos," *Rekayasa*, vol. 2, no. 2, pp. 130–143, 2018.

[7] K. S. Mukasa and H. Kaindl, "An integration of requirements and user interface specifications," *Proc. 16th IEEE Int. Requir. Eng. Conf. RE'08*, pp. 327–328, 2008, doi: 10.1109/RE.2008.55.

[8] E. Tome, "Tomé, E. (2007). Employability, skills and training in Portugal (1988-2000) evidence from official data. Journal of European Industrial Training, 31(5), 336–357. doi10.11080309059071075 6792.pdf." .

[9] M. Pratama, N. A. Setiawan, and S. Wibirama, "User interface design for android-based family genealogy social media," *Proc. - 2017 7th Int. Annu. Eng. Semin. Ina. 2017*, 2017, doi: 10.1109/INAES.2017.8068557.

Need Analysis of Interactive Multimedia Development for Virtual Learning in Pandemic

J F Lempas[1], S Soenarto[2] and A D Barahama[3]

[1,2]Instructional Technology Post-Graduate Program, Universitas Negeri Yogyakarta Jalan Colombo 01, Daerah Istimewa Yogyakarta 55281, Indonesia.
[3]Department of Electronics and Informatics Engineering Education, Post-Graduate Program, Universitas Negeri Yogyakarta Jalan Colombo 01, Daerah Istimewa Yogyakarta 55281, Indonesia.

Email: josianefianny.2017@student.uny.ac.id

Abstract. The pandemic that has not ended has an impact on changing the educational paradigm. Existing conditions demand to continue to use virtual learning. During the epidemic, this study will examine students' demand for interactive multimedia in virtual learning. The interviews reveal that the media that teachers provide in virtual learning during this pandemic has limits. Data was gathered through interviews and questionnaires given to teachers and students. The student requirements questionnaire reveals that interactive multimedia is the most effective instructional medium for pupils. This study is part of a larger investigation of the evolution of multimedia learning during a pandemic.

1. Introduction

Pandemic Covid-19, which wreaked havoc on Indonesia and the rest of the world, altered almost every element of life, including schooling. The government has implemented a number of strategies to break the cycle of the Covid-19 virus's transmission, one of which involves learning from one's own home. It has previously altered the way people learn. Many schools and institutions have been forced to close temporarily due of the Covid-19 pandemic. According to the researchers' judgment, it is improbable to revert to past learning in the near future. The application of social distancing at this stage will harm learning opportunities. Given this situation, it requires teachers to overcome this urgent situation[1].

Online learning is an educational model that is suitable for this pandemic period. Combining face-to-face learning with technology produces blended learning to increase students' learning potential[2]. Educators can now use digital media in their classrooms because to technological advancements. Therefore, teachers are selective in choosing learning media that students can understand concepts well without causing misconceptions.

The major outbreak of Corona Virus sickness may provide another another case for virtual learning. Unfortunately, necessary preparation and infrastructure, such as a decent internet connection, a laptop or iPad as a learning aid, teachers who have completed the training period, and pupils who have been trained to learn independently, were not provided. There are numerous advantages to studying online. In other words, virtual learning can be a lifesaver in a crisis. They contend that it is critical to evaluate software as it is being produced, by testing it on learners and making changes as needed [3]. The use of technology has recently increased rapidly, especially in the world of education. The state of this pandemic era requires learning to continue virtually.

Virtual Learning Environments are currently commonly used by many educational institutions to enhance the learning experience by combining several multimedia technologies. Virtual learning is regarded as a form of learning via technological mediation that fills the gap formed when teachers and students are isolated in location and time [4]. In addition, it allows real-time learning to take place between students and educators in two separate ways. In this example, the computer has a strong influence on the learning setting. The role of teachers and students changed because of the influence of media and technology. Teachers and books are no longer the only sources of information for students, but rather serve as a guide to the methods of knowledge.

Unfortunately, in fact, teachers' online learning processes still employ the lecture style. In lectures, the teacher is the only source of student learning, so students are not creative and are not allowed to learn more about the subjects they are studying. Supposedly, students are more active in developing lessons [5]. The teacher is only a companion in charge of planning and creating more learning resources to create a conductive learning environment.

The researchers are interested in conducting a need analysis study connected to interactive multimedia based on the analysis of numerous studies mentioned above. Developers construct learning based on the goals to be attained by educators and learners, depending on the results of the requirements analysis [6]. The goal of this research is to present something different from earlier research. As a result, having another alternate media that can be utilized as a tool to aid students in achieving their specific needs and requirements is critical. High school students were used as research subjects, and the descriptive qualitative approach was applied.

Based on the analysis of the problem it was found that the learning process that took place during the Covid-19 pandemic was not yet efficient because learning was still in a conventional form, even though the current situation required the learning process to take place online, and was more effective. As a result, interactive learning tools are required to increase student interest in learning. It can be supported by the use of interactive learning media as a learning aid, thus the first step is to assess the learning media requirements during the pandemic, which piques researchers' interest in doing this study.

2. Methods

This research is qualitative descriptive study conducted by senior high school students that used a survey method. There are 18 teachers and 90 students who participated in this study. The following describes the steps carried out in this study;

2.1. Data collection

In this study, to find out and measure what the needs of students are, data collection techniques are carried out in the form of interviews and questionnaires. To collect additional data sources, the researcher conducted interviews with a number of teachers.

2.1.1 Need analysis

It is vital to examine the media needs in order to determine what is required to support the learning process. This analysis includes an analysis of student needs and what media are used in learning.

1. Questionnaire.

Furthermore, data collection was carried out using a questionnaire. Questionnaires are disseminated online using a google-form comprising questions in order to gather information regarding the demands of learning material during online learning during the pandemic.

2. Interviews

To obtain information related to the needs analysis carried out, the researcher conducted interviews with several teachers regarding the media used in the learning process.

2.2. *Data processing*
Data processing is carried out to produce information from raw data into information that can be understood in this study using descriptive-qualitative analysis. In processing this data, the data that has been collected is processed from quantitative data into qualitative data.

The score is not transformed into interval data. On a five-point score, the quantitative data is converted to qualitative data. The qualitative data will be transformed according to the conversion table following the processes done after getting quantitative data from the student.

Table 1. Conversion of quantitative data

Interval of Data	Qualitative Information
5	Very Needed
4	Needed
3	Enough
2	Less Needed
1	Very Little Needed

After the data has been converted, it is processed by calculating the mean using the formula:

$$\bar{X} = \frac{\sum x}{N} \quad (1)$$

Information:

$\bar{X}$: Each component's average score
$\sum x$: Score in total
N : The number of indicators that have been evaluated is large

3. Result and discussion

3.1. *Data collecting*
In this study, descriptive quantitative approaches were used to analyze the needs of students of multimedia in virtual learning through a questionnaire. The questionnaire is based on numerous factors, including needs and attractiveness, that are adapt to the characteristics of students. Aspects of needs refer to the students' background in using multimedia. The attractiveness aspect refers to multimedia criteria which are important to the students in terms of appearance and content.

3.1.1 Needs analysis
1. Questionnaire.

Students with a score of 1 strongly disagree, disagree with a score of 2, agree with a score of 3, and strongly agree with a score of 4 with the statements provided are given four response columns. The table is an illustration of how a questionnaire was used in this study: students were provided a few statements as be asked to respond by putting a checkmark in the space provided.

Table 2. Need analysis of student [7]

No.	Statement	SD	D	A	SA
Aspect of need					
1.	Learning to use books is a lot more fun.				
2.	Aside from books, I need a new sort of learning media.				
3.	When there are only texts in the learning media, I am content.				
Aspect of need					
4.	When there is audio in the learning media, I am happy.				
5.	When there are videos in learning medium, I am happy.				
6.	When there are visuals in the learning material, I am happy.				
7.	When there is animation in studying medium, it makes me happy.				
8.	When learning medium includes text, audio, and video, I am happy.				
9	When learning media includes text, graphics, and animations, I am happy.				
10.	When learning media includes text, movies, and animations, I am happy.				

2. Interviews

The data obtained from interviews with teachers show in table 2. Results are shown in table of a student questionnaire using learning media during the pandemic. The findings of the questionnaire in relation to the planned media needs are shown in Table 3.

Table 3. The result of the interviews.

No.	The result of the interviews
1.	Educators deliver glued material from textbooks provided by the government.
2.	The method most often used when learning online is just lectures.
3.	The limitations of educators are using learning media in the form of links to access videos from YouTube and files sent via WhatsApp groups.
4.	Some students have difficulty understanding the content of the material.
5.	Teachers deliver structured assignments through WhatsApp groups, Google Classroom, and Zoom meetings.
6.	Teachers have difficulty accessing learning media.

Based on the results, the teacher is still obsessed on the concept of textbook material. In essence, the textbook suffices as a guide, and the remainder can be developed based on teacher creativity.

This element can be demonstrated by the media that teachers use to engage pupils in learning, such as PowerPoint presentations, YouTube videos, and television broadcasts about learning themes. Many of them struggle to comprehend the learning materials.

According to the findings of interviews with numerous educators, throughout the pandemic, most teachers just applied WhatsApp to deliver structured projects, while others used Google Classroom and Zoom Meeting to aid them in teaching. This data demonstrates that online learning at home is confined to completing prescribed homework from teachers. Because of rapid advancement of internet technology, educators now have the opportunity to use it as an innovative online learning medium [8] . Certainly, if learning is not diverse, students will become bored and burdened.

This condition can be caused by a variety of issues, including teachers' limits in establishing communicative instructional tools for students and a lack of effort in developing interactive learning, which makes learning a chore for students.

Many of them have difficulties comprehending the material. As a result, the learning that is done has no meaning. It is not optional for student learning outcomes to use the online learning paradigm that is being used. As a result, the learning that is done has no meaning.

Table 4. Data relate to learning media was given during the pandemic.

No.	Media	Amount %
1.	Textbooks for lesson.	35,8 %
2.	Ms. Office: Word & Power Point	29,4 %
3.	Link YouTube for video learning	24,2 %
4.	Television	10,6 %

Textbooks, Ms. Office, and YouTube video links are among the learning resources supplied during the pandemic, according to interview data. The importance of requirements analysis in the development of content and the creation of tools to aid instructors in generating material that fits the needs of target learners and learning needs is crucial [9]. In this case, the table shows is in the highest score (35.8%), than followed by Ms. Office (29.4%), link YouTube (24.2%), and television (10.6%). The results, that most of teachers delivered students literature in the form of presentation slides. Conventional learning does not allow for the presentation of interactive multimedia content. One way to connect the teacher as a source of student learning is through learning media. Video displays, animations, and simulations are some examples of interactive multimedia presentations [10]. Media is an educational instrument that may be utilized to help students learn more effectively and efficiently by acting as an intermediate in the learning process.

3.2. Data processing

The processes performed to acquire data for this survey by reviewing data from previous surveys. Establishing criteria for characteristics of necessity as well as aspects of learning media's attractiveness then, based on quantitative data gleaned from student evaluation results, criteria for components of need and factors of learning media appeal were developed.

After finding the average score for each component, the data is translated to qualitative data using Widoyoko's conversion.

Table 5.A slightly more complex table with a narrow caption

Range of scores	Average Score	Category
$X > \bar{X}i + 1,8$ sbi	X > 3,4	Very Needed
$\bar{X}i + 0,6$ sbi $< X \leq \bar{X}i + 1,8$ sbi	2,8 < X ≤ 3,4	Needed
$\bar{X}i - 0,6$ sbi $< X \leq \bar{X}i + 0,6$ sbi	2,2 < X ≤ 2,8	Enough
$\bar{X}i - 1,8$ sbi $< X \leq \bar{X}i - 0,6$ sbi	1,6 < X ≤ 2,2	Less Needed
$X < \bar{X}i - 1,8$ sbi	X < 1,6	Very Little Needed

Description:

X	: mean score
$\bar{X}i$	: ideal mean
sbi	: standard deviation

The conversion method is then applied using a five-point scale, with a high score of 4 and a lowest score of 1. The result of converting to a 5-point scale is as follows.

high	: 4
lowest	: 1
$\bar{X}i$	: $\frac{1}{2}$ (4+1)
	: 2,5
sbi	: $\frac{1}{6}$ (4-1)
	: 0,5

Table 6. Scale 1-5 conversion calculation

Scale	Calculation
5	Scale 5= $X > \bar{X}i$+ 1,8 sbi = $X >$ 2,5+ (1,8 x 0,5) = $X >$ 2,5+ 0,9 = $X >$ 3,4
4	Scale 4= $\bar{X}i$ + 0,6 sbi < $X \leq \bar{X}i$ + 1,8 sbi = 2,5+(0,6 x 0,5) < $X \leq$ 2,5+ (1,8 x 0,5) = 2,5+0,3< $X \leq$ 3,4 = 2,8 < $X \leq$ 3,4
3	Scale 3= $\bar{X}i$ - 0,6 sbi < $X \leq \bar{X}i$ + 0,6 sbi = 2,5-0,3< $X \leq$ 2,8 = 2,2< $X \leq$ 2,8 Scale 3= $\bar{X}i$ - 0,6 sbi < $X \leq \bar{X}i$ + 0,6 sbi = 2,5-0,3< $X \leq$ 2,8 = 2,2< $X \leq$ 2,8
2	Scale 2= $\bar{X}i$ – 1,8 sbi < $X \leq \bar{X}i$ - 0,6 sbi = 2,5 - (1,8 x 0,5) < $X \leq$ 2,2 = 2,5 – 0,9 < $X \leq$ 2,2 = 1,6 < $X \leq$ 2,2
1	Scale 1= $X \leq \bar{X}i$ - 1,8 sbi = $X \leq$ 2,5 – 0,9 = $X \leq$ 1,6

The 1-5 range conversion can be used to draw the following conclusions about student needs.

Table 7. The result of integrating quantitative and qualitative information

Quantitative Data	Range	Qualitative Data
5	$X >$ 3,4	Very Needed
4	2,8 < $X \leq$ 3,4	Needed
3	2,2 < $X \leq$ 2,8	Enough
2	1,6 < $X \leq$ 2,2	Less Needed
1	$X \leq$ 1,6	Very Little Needed

In one multimedia, text, graphics, audio, video, animation, and interaction are all mixed to make learning more enjoyable than previously. All learning types, including auditory, kinesthetic, and visual, can benefit from multimedia. Researchers discovered numerous constraints in the continuing learning process analysis of interactive learning multimedia needs was "required" to promote learning based on the findings of their research.

4. Conclusion

The absence of interactive media that can give understanding and are engaging in presenting content while learning is one of the barriers to online learning, according to research on this topic. This study serves as a foundation for future research to produce appropriate media to be offered to students during the epidemic such that the subsequent learning is enjoyable and of high quality.

The researcher thinks that the findings of this study can help and serve as a reference for teachers as they try to innovate in generating media that can boost student knowledge, so that it can be utilized as a reference and input for future scientific growth.

Acknowledgment

The authors acknowledge the assistance of Dr. Sunaryo Soenarto M.Pd of Universitas Negeri Yogyakarta in the preparation of this work. SMA Anthonius Manado, North Sulawesi, contributed to this research.

References

[1] S. Dhawan, "Online Learning: A Panacea in the Time of COVID-19 Crisis," *J. Educ. Technol. Syst.*, vol. 49, no. 1, pp. 5–22, 2020.
[2] M. H. Bahiraey, "Quality of Collaborative and Individual Learning in Virtual," *Proc. Second Int. Conf. E-Learning E-Teaching (ICELET 2010)*, no. Icelet, pp. 33–39, 2010.
[3] S. Tzur, A. Katz, and N. Davidovich, "Learning Supported by Technology: Effectiveness with Educational Software," *Eur. J. Educ. Res.*, vol. 10, no. 3, pp. 1137–1156, 2021.
[4] P. W. Stonebraker and J. E. Hazeltine, "Virtual learning effectiveness: An examination of the process," *Learn. Organ.*, vol. 11, no. 3, pp. 209–225, 2004.
[5] A.-N. Septiani, T. Rejekiningsih, Triyanto, and Rusnaini, "Development of Interactive Multimedia Learning Courseware to Strengthen Students' Character," *Eur. J. Educ. Res.*, vol. 9, no. 3, pp. 1267–1279, 2018.
[6] F. N. Biduri and Y. Rasyid, "THE ANALYSIS OF NEEDS ON LEARNING MATERIALS IN CONTEXT- BASED READING MANDARIN LANGUAGES AND CULTURE," *J. Educ. Teach. Learn.*, vol. 3, no. 1, pp. 9–16, 2018.
[7] J. F. Lempas and S. Soenarto, "Analysis of learning multimedia development needs for network infrastructure architecture," *IOP Conf. Ser. Mater. Sci. Eng.*, vol. 1098, no. 5, p. 052090, 2021.
[8] M. A. Gilbert, *Strengthening Your Social Media Marketing with Live Streaming Video*. 2019.
[9] S. Menggo, I. M. Suastra, M. Budiarsa, and N. N. Padmadewi, "Needs analysis of academic-English speaking material in promoting 21 st century skills," *Int. J. Instr.*, vol. 12, no. 2, pp. 739–754, 2019.
[10] S. W. Widyaningsih, I. Yusuf, Z. K. Prasetyo, and E. Istiyono, "Online Interactive Multimedia Oriented to HOTS through E-Learning on Physics Material about Electrical Circuit," vol. 9, no. 1, pp. 1–14, 2020.
[11] E. P. Widoyoko, *Evaluasi Program Pembelajaran*. Yogyakarta: Pustaka Pelajar, 2009.
[12] E. . Anisimova, "Digital Literacy of Future Preschool Teachers Ellina Sergeevna Anisimova 1," *J. Soc. Stud. Educ. Res.*, vol. 11, no. 1, pp. 230–253, 2020.

Item Selection on The Moodle-Based Computerized Adaptive Test

Haryanto[1], Y Neng-Shu[2], S Hadi[1], M Ali[1], AF Husna[1], ASJ Wardhana[1]

[1]Electrical Engineering Education Dept., Universitas Negeri Yogyakarta, Indonesia
[2]Science and Technology Dept., National Yunlin University

e-mail: haryanto@uny.ac.id

Abstract. In this industrial era, all areas of life have been entered. There are five central issues that support performance, namely numerical physical devices, production tools, programs, interfaces, and networks. The Modular Object-Oriented Dynamic Learning Environment (Moodle) in order to support Industrial technology has been equipped with adaptive test facilities. Adaptive Moodle can be used to organize a communicative and interactive test process because of the communication features (chat, messaging, or forum). In addition, Adaptive Moodle can be used to administer online tests. Adaptive tests are tests whose quiz presentations will adjust to the user's abilities. The results of the research that has been carried out regarding the selection of items on the Moodle-based computerized adaptive test (CAT) were obtained: (1) The Moodle adaptive test worked successfully in accordance with the research objectives, (2) Based on user responses, the selection of items on the Moodle adaptive test items had worked well for the exam, (3) The Moodle adaptive test can run according to its function, namely adaptive to the user's ability.

1. Introduction

In this industrial era, all areas of life have been entered. There are five central issues that support performance, namely numerical physical devices, production tools, programs, interfaces, and networks. [5]. Utilization of Industry 4.0 technology for testing purposes in various fields can be done by applying the Adaptive module in the Modular Object-Oriented Dynamic Learning Environment (Moodle). Adaptive Moodle can be used to organize a communicative and interactive test process because of the communication features (chat, messaging, or forum). In addition, Adaptive Moodle can be used to administer online tests [1]. Adaptive tests are tests whose quiz presentations will adjust to the user's abilities [2][4].

The development of the Moodle adaptive test program is to utilize at least four of the five industrial 4.0 technology centers. Four technologies that can be utilized for this include numerical physical devices, production tools, programs, interfaces, and networks. The four technologies have been integrated into one device called a smartphone, computer and Wide Fidelity. [5].

In this article the use of industrial technology is applied in exam management. With the help of Moodle which has the capability to be used for network-based exam management. Moodle was chosen because it is an open source and free tool. Moodle users to date have recorded 104,361 websites, 21,000,000 administrators, and 179,000,000 users, spread across 232 countries. The use of Moodle shows an increasing trend, due to the addition of new features including the computerized adaptive test library (CAT).

Moodle software has several advantages. First, the features are complete. Moodle can be used for communication (chat, messaging, or forums), document creation and administration, data tracking, and plugin extensibility capabilities. Second, ease of use. This is because Moodle can be adjusted flexibly according to the policies and needs of each user institution [1].

The superior features, ease of use, and nature of Moodle as open source should make us a reason to abandon the traditional test process management model. For that, we can develop and design a Moodle adaptive test process. With the Moodle Test, the presentation of quizzes can be more interesting because it utilizes multimedia. The test process does not need to be face-to-face, because users can work anywhere according to a predetermined schedule [2].

The Adaptive Moodle test has a quiz feature that can be used to measure the user's ability. This is a series of activities that aim to measure the level of mastery of a material. This activity is a process of measuring and evaluating abilities. For this purpose, a measuring instrument is needed, which in the test process is often referred to as a test or assignment tool [2][6].

The change in the way of testing in this industrial era is the use of internet-based test applications [5]. The ability to present test items with a level of difficulty according to the user's ability is an advantage of internet-based test applications. Such abilities are called adaptive tests. The test items in the question bank, which have been standardized according to content parameters and level of difficulty, are the main source for the Moodle adaptive test [4]. In its use, the measurement of the user's ability by the Moodle adaptive test can be carried out better and more accurately, although between users varies in the number of items obtained and the time of the test. With certain analysis through a program, the user's ability is estimated and the score is known immediately after the test is finished. All estimated scores are provided to the system administrator. These adaptive tests can be done well by the Adaptive Quiz module installed into Moodle [1]. This study was conducted to determine the ability of the Moodle adaptive test which has not been widely used. In detail, this research article discusses Moodle's adaptive test ability, item selection based on user responses, and the function of measuring user abilities.

Adaptive test-based on Moodle is different from computer-based testing (CBT). The difference between CBT and Adaptive tests-based on Moodle lies in the algorithm approach used to select a set of quizzes that are given to the user during the test process [2]. Adaptive tests-based on Moodle is software that contains content and/or quiz sequences that will automatically adapt to the abilities of each user's abilities. The term Adaptive Moodle-based test is commonly used to describe Computer Assisted Assessment (CAA) software [3]. In terms of adaptive, Item Response Theory (IRT) is used to estimate user capabilities. After that, based on these estimates, the user will be given the next quiz with the level of difficulty according to the results of the estimated ability [2][6].

Adaptive tests based on Moodle center on classifying users into one of a number of discrete categories, such as pass or fail. An adaptive test based on Moodle in which a set of If-then rules that are created are then used to select the next given quiz. In other words, the Adaptive test system based on Moodle will classify the user's abilities [3]. Based on the ability classification, the user will be taken to an advanced quiz according to the ability of each user.

There are four components of the Moodle-based Adaptive test, namely the item bank, item selection procedure, ability estimation and termination rules. Another component of the Moodle-based Adaptive test that is often considered is the content balance and control items that often appear (exposure control items) [8].

The concept of the algorithm used by the Adaptive test-based on Moodle is that if the user cannot answer correctly on a given quiz, the system will give a quiz that has a lower level of difficulty. Conversely, if the user can answer correctly, then the next quiz given is a quiz with a higher difficulty level. The chart of the Adaptive test process based on Moodle can be seen in Figure 1 [8].

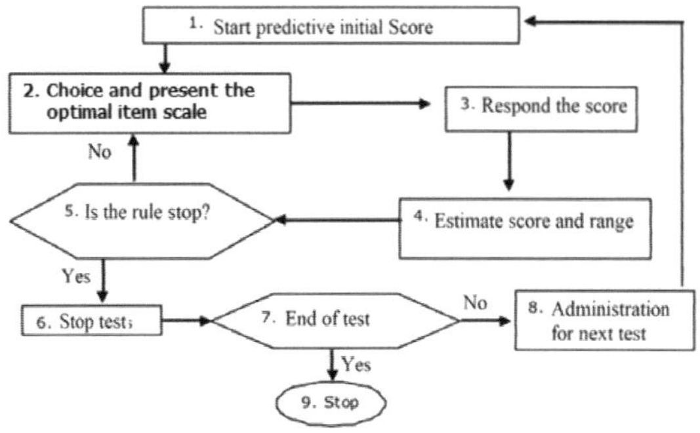

Figure 1 Adaptive test-based on Moodle process chart [8]

In the Adaptive test-based on Moodle, quiz item difficulty is estimated using the Rasch model or item response theory. The Rasch model was developed by a Danish mathematician named George Rasch. The Rasch model is mathematically written as follows:

$$P_i(\theta) = \frac{e^{D(\theta - b_i)}}{1 + e^{D(\theta - b_i)}} \quad (1)$$

P_i (θ) = probability that the user's ablity with the level of ability θ answered item i correctly, e = 2.7183; D = 1.7, and b_i = item difficulty level [2][6].

Moodle is a software package produced for internet-based managerial activities and websites using social construction principles. Instructor assignments will change from being a source of information to being an influencer. The instructor's role in the Moodle system includes: connecting with individual users to understand their needs and moderating discussions and activities that lead users to achieve capability goals [1].

Moodle is a description of the application of concepts and mechanisms of the testing process electronically by utilizing information technology. Moodle under the GNU license can be used freely as an open source product. The advantage of Moodle is that it can be installed on any computer and operating system as long as it can run PHP and supports SQL databases. Some of the features that Moodle has include task submission, discussion forums, download archives, ratings, chat, online calendar, news, online quizzes, and Wiki. By installing the Adaptive Quiz plugin, Moodle's online quiz capabilities are even better [1].

2. Method

This study uses a research and development approach. The development research model used is a scale construction process [4]. After that proceed with the installation of the Adaptive Quiz plugin in Moodle, user testing, data analysis and reporting [1][3].

The adaptive test-based on Moodle grade test tools in this study consisted of the test of knowledge aspects. The test tool to assess the knowledge aspect is made in the form of multiple choice which is equipped with a level of difficulty for each item. The test tools for aspects of knowledge is filled in by users, while the skills test tools are filled by the observer. MCQs are chosen in the assessment of aspects of knowledge because this model can cover a wide range of material, the test is objective so it is suitable for use in the adaptive test-based on Moodle.

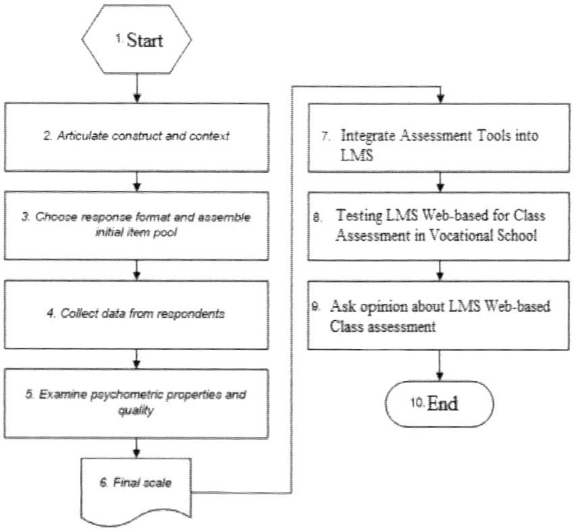

Figure 2 Research process [3][4]

There are two types of research population. First is the expertise competency of the users to be chosen. The second population is the name of the subject to be developed in the test tool. The sample of expertise competency and the name of the subjects in this study will be chosen purposively. The sample of users where the data will be collected will be determined randomly in Yunlin University, Taiwan (sample 1) and Yogyakarta (sample 2). Data collection techniques in this study used instruments in the form of multiple choice online tests to assess aspects of knowledge.

The quality of instruments for a number of quizez in the question bank will be analyzed based on item response theory. In addition, the item items will also be analyzed with the Rasch model to obtain the level of difficulty that will be used as the basis for item selection in CAT.

To determine eligibility, estimate user's ability, research data will be analyzed using descriptive statistics. Descriptive statistics will display data trends and data variability. Trending data will also be displayed in tabular or graphical form.

3. Results and Discussion

Based on the results of research and development, the result in the first stage is to install a Moodle-based adaptive test. The following is a display of the start page, which contains information in two languages, namely Indonesian and Taiwanese (System View in link address https://catpkln.com).

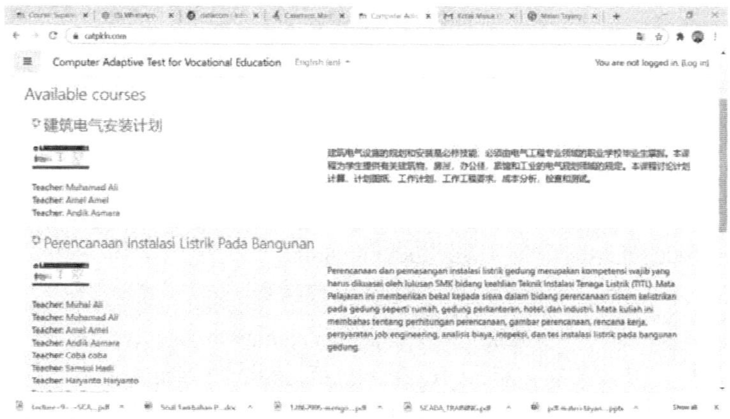

Figure 3 The start page is Indonesian and Taiwanese.

Before entering the system to start the test, the user first logs in. The purpose of the login system is for data security, because only registered users can enter the system. By logging in, the observer knows the number of users who took the test. The following illustrates how to login to the system and the pages of the system Figure 4.

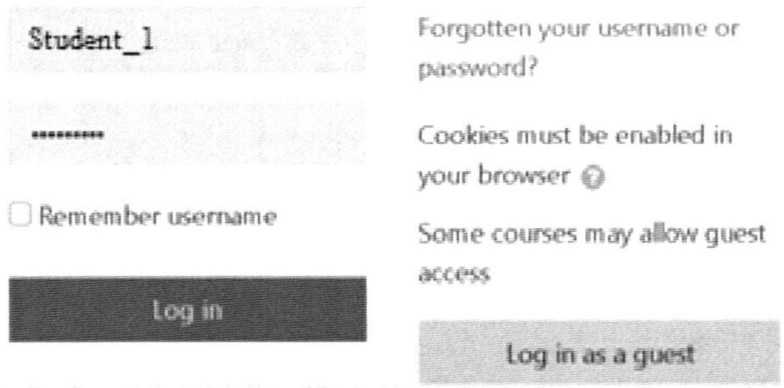

Figure 4 User login page

Based on the test results, the Maximum value data for the 1st sample is 37.78, for the 2nd sample, it is 69.30. The minimum value for the 1st sample is 20.00, while for the 2nd sample it is 23.00. The average value for the 1st sample is 26.59 for the 2nd sample, it is 49.26. The materials tested for both countries are the same. The language used in the exam questions for both countries has been translated into their respective languages. The test material tested has been selected for the generally accepted international Electrical Installation Engineering competency. From these results, there is a clear difference, that the competence of the 2nd sample is higher than that of the 1st sample.

Based on the results of observations of the data on the test items received by the user, it has varied according to the correct and incorrect responses of the user's answers. Users also said that the adaptive test model was good for the test. This shows that the test items given to students are different and adaptive according to the student's ability level (Figure 5).

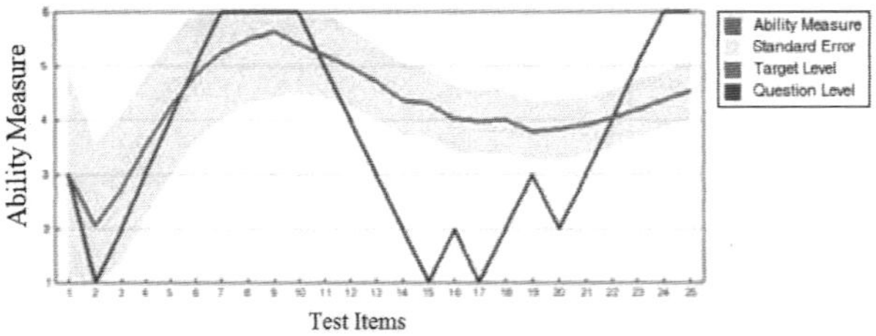

Figure 5 Result page

Figure 5 above shows that the selection of test items is indicated by a blue line. The user's abilities are indicated by a red line. While the standard error is described by a pattern of pink areas. Based on these results (blue line) it can be seen that the rise and fall of the line depends on the right and wrong of the user's response. The user's ability can be directly seen from the red line. The same thing was also obtained by other users immediately after the exam was over, namely getting different images according to the user's answers correctly or wrongly. This shows that the Moodle adaptive test is able to choose the test items given to the user, adaptive according to the user's ability level.

4. Conclusion

The user responses that were collected using a questionnaire generally stated that the Moodle-based adaptive test was good. This is because the test questions from the computer that are given to be done to the user are different and adaptive, according to the level of the user's ability.

The conclusions from the results obtained are: (1) The Moodle-based adaptive test has succeeded in working and functioning properly according to the user's response, (2) The Moodle-based adaptive test has been able to work on choosing the right questions according to the right and wrong user responses. (3) Moodle-based adaptive test is good enough to be used for the test in order to measure the ability of the user's cognitive level.

References

[1] Adaptive Quiz, diunduh dari https://moodle.org/plugins/mod_adaptivequiz pada tanggal 9 Februari 2020
[2] Haryanto, 2011. Pengembangan Computerized Adaptive Testing (CAT) Dengan Algoritma Logika Fuzzy. Jurnal Penelitian dan Evaluasi Pendidikan Tahun 15, Nomor 1, 2011. Hal. 47 – 70.
[3] Lilley, M. (2007). The Development and Application of Computer Adaptive Testing in a Higher Education Environment. Doctoral Dissertation. University of Hertfordshire.
[4] Magis, D., Yan, D. & von Davier, A.A. (2017). Computerized adaptive and multistage testing with R. Switzerland: Springer International Publishing.
[5] Magno, C. (2019). What happens to educational assessment in the 4th industrial revolution. Diambil dari https://www.pemea.org/single-post/2019/02/03/What-Happens-to-Educational-Assessment-in-the-4th-Industrial-Revolution pada tanggal 19 Juli 2019.
[6] Samsul Hadi, 2013. Pengembangan Computerized Adaptive Test Berbasis Web. Yogyakarta, Aswaja Pressindo.
[7] Statistics. Diambil dari https://stats.moodle.org/ pada tanggal 6 Februari 2020
[8] Wainer, H. (1990). Computerized Adaptive Testing: A Primer (2nd ed., pp. 161-186). Hillsdale, NJ: Lawrence Erlbaum Associates.

Development of IoT-Based Smart Home Electrical Protection System

[1] **Renaldo R.Y.**, [2] **Edy Supriyadi.**

[1, 2] Electrical Engineering Department Faculty of Engineering Universitas Negeri Yogyakarta, Indonesia

Email: [1] renaldorizqi.2018@student.uny.ac.id; [2]edy_supriyadi@uny.ac.id

Abstract: This study aims to: (1) develop a smart home electrical protection system, (2) determine the feasibility, and (3) user responses to the developed system. Development research was conducted using the ADDIE and Waterfall models. Product trials were conducted on 94 respondents from college students. Data collection techniques using a questionnaire. Data were analyzed using descriptive techniques. The results showed that: (1) there was an IoT-based electrical protection system that was feasible to be applied in smart homes. (2) the feasibility of the developed media is included in the "Very Appropriate" category according to the expert, and (3) user responses to the IoT-based electrical protection system are in the "Very Good" category.

Keywords: IoT, Protection System, Smart home

1. Introduction

Higher education is a level of education that aims to develop the potential of students to become human beings who believe and fear God Almighty and have a noble character, are healthy, knowledgeable, capable, creative, independent, skilled, competent, and cultured for the benefit of the nation. Universities are expected to produce graduates who master the branches of Science and/or Technology to fulfil the national interest and increase the nation's competitiveness. [1]

Learning or learning activities are carried out to develop student's cognitive, affective, and psychomotor aspects. Learning is expected to bring students to mastery of competence as a whole as expected. Several factors influence the quality of a learning activity. The competence of educators, characteristics of students, learning atmosphere, available learning facilities and resources, as well as school or college conditions affect the quality of learning. The implementation of monitoring and evaluation is believed to be able to improve the quality of learning.[2] All of these factors influence each other. If one of the factors is inadequate, then the quality of the learning activities is less than optimal.[3]

The development of information and communication technology, sensors, and relays supports the development of IoT so that it has great potential to be utilized in more effective teaching and learning activities. IoT has the characteristics of increasing interaction and response between objects. These characteristics are the capital to make a positive contribution to the learning process, especially in increasing the interaction between learning participants and learning media as well as among fellow learning participants. Learning by implementing IoT is starting to become a special concern in the scope of education.[4]

One way to take advantage of technological advances to meet the needs of practice learning of protection systems is to create IoT-based learning media. Learning media can eliminate the possibility of work accidents in the laboratory. Spending funds on laboratory equipment will also become more efficient because laboratory equipment and tools will be replaced with IoT with the same functions as the original state. Users will also know the use of these tools according to procedures directly without direct contact with these tools using IoT technology. IoT can support the implementation of conventional practicum on protection system practice learning and experiments can be carried out at home because they are not limited by time and place. In this study, the development of learning media for electric power protection systems will be carried out by utilizing IoT-based android technology so that learning media can be in accordance with technological developments so that there is an increase in student competence because they can learn mobile anywhere and anytime.

Based on the conditions that have been described, it is necessary to develop learning media for the practice of IoT-based electric power protection systems in the Electrical Power Protection System Practice course in the Electrical Engineering Study Program.

2. Related works

2.1. Computer-Based Learning Media

Computer-based learning technology is a way of delivering material using a device sourced from the microprocessor.[5] Computer-based technology displays information to students through displays on the monitor screen. Computer application programs for learning or known as Computer assisted instructional (cia), Computer based instructional (cbi), or Computer managed instructional (cmi), instruction assisted learning (ial), instructional application of Computer (iac) which in Indonesian means Computer-assisted learning.

Characteristics of Computer technology include the following: 1) besides being used linearly, it can be used randomly, 2) it can be used with participants according to the wishes of the students in addition to the way as designed by the developer, 3) ideas are usually expressed abstractly using ~~using~~ words, symbols, and graphics, 4) cognitive science principles are applied during development, and 5) learning can be learner-centred with a high level of interactivity. Computer technology can be used as a medium that allows someone to learn independently in understanding a concept. It is possible because Computer technology has the ability to: (a) prepare and manipulate alphanumeric data; (b) properly perform some operations in an appropriate manner; and (c) combining text, colour, motion (animation), sound and video, and contains an "intelligence" that can present an interactive process.[5]

Learning media is appropriate if it meets the following principles: (a) the use of media is following learning needs, (b) the media is not only for entertainment but also for learning, (c) the media has easy and flexible access, (d) the media is an expression of the creative self, (e) the media triggers critical thinking, (f) the media reflects the targets and objectives of the results produced, (g) the media must be connected to real life, and (h) the media produced must be of high quality and have relevance to the learning material.[6] Assessment of mobile learning development is based on several components, namely, learning content, technology skills, design skills, and media suitability. The development of ICT-based teaching materials, namely the substance of the material, learning design, display, and use of software.

An application requires a criterion to measure and evaluate its feasibility. However, the criteria for developing learning applications also refer to the development of teaching materials. ISO 9126 is one of the standards used to test the feasibility of software. ISO 9126 states that six characteristics can be used as an indicator to measure the feasibility of a software. The six characteristics are functionality, reliability, usability, efficiency, maintainability, and portability.

2.2. Internet of Things (IoT)

IoT is all activities whose actors interact with each other and are carried out by utilizing the internet. In its use, the Internet of Things is found in various activities, for example: the number of online transportation, e-commerce, online ticket reservations, live streaming, e-learning and others even tools to assist in certain fields such as remote temperature sensors. GPS tracking, and so on that use the internet or network as a medium to do so. With the many benefits of the Internet of Things, it makes everything easier, in the field of IoT education it is very necessary to carry out all activities using a system and organized and proper filing system.

Internet of Things concepts and ideas have begun to be applied in the world of education. One of them is integrating interactive English learning with sound and visual sensors to capture students' pronunciation skills. According to Wang, the internet has the characteristics of motivating and delights teachers and enables them to teach students according to their individual talents. Teachers can choose teaching materials according to the characteristics of students. In contrast, students themselves can learn according to their speed and ability, so they do not have to be confined by a single learning program that must be followed by all learning participants.

The ESP 8266 is a complete chip that includes a processor, memory and access to General Purpose input-output (GPIO). It causes the ESP8266 to directly replace the Arduino and coupled with its ability to support direct WiFi connections. The Internet of Things (IoT) is growing along with the development of microcontrollers, more and more and more various modules based on Ethernet and WiFi, starting from Wiznet, Ethernet shield to the newest WiFi module known as ESP8266. Several types of ESP8266 that can be found in the market, but the easiest to find in Indonesia is the ESP type 01,07, and 12 with the same function. The difference lies in the GPIO pin provided. Here are some types of ESP8266.[7]

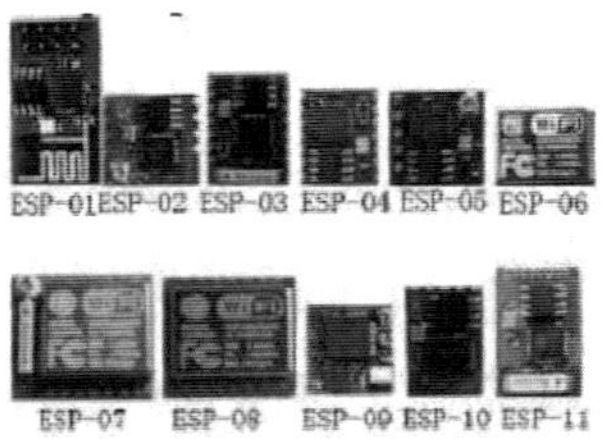

Figure 1. ESP Family

ESP-8266 is a component with a working voltage of 3.3V so if additional devices are needed, it is necessary to pay attention to the working voltage of the installed board. but it would be better if a separate level switch was made for communication and voltage sources in this WiFi module. This WiFi module is equipped with a Microcontroller and GPIO so that many firmwares have been developed to make it easier to use. The firmware is used so that this WiFi module can work standalone (Arafat, 2016). ESP8266 is a microcontroller that has a WIFI connection facility. Because it is a microcontroller, this ESP8266 module has a processor and memory, which can be integrated with sensors and actuators via the GPIO pin. This module has features such as supporting the IEEE 802.11 b/g/n standard, can be used for WiFi direct (P2P), AccessPoint soft-AP, has 81 Mb RAM and 1 Mb Flash memory, speeds up to 160 MHz, and output power of 19.5 dBm.

2.3. Electric Power Protection System

The protection system is one of the most vital things in the world of electricity. The protection system is used to maintain the continuity of electricity distribution and to secure both electrical network equipment and electrical loads or consumers due to disturbances. The generator or transmission protection system must work according to the conditions, including reacting quickly in the event of a disturbance, being selective, sensitive/sensitive to disturbance, reliable, stability and economical.[8] If these conditions are not met, it will affect performance. The reliability and ability of an electric power system in serving consumers are highly dependent on the protection system used. Therefore, in designing an electric power system, it is necessary to consider the conditions of disturbance conditions that may occur in the power system.[9]

2.3.1. Overvoltage and Voltage Drop

Overvoltage disturbances in the power grid can be caused by lightning strikes to overhead transmission lines, circuit breaker switching operations and faults. Power transformers may be exposed to such transient overvoltages during operation. Transient overvoltages with steep wavefronts impact the insulating dielectric stresses of the first few windings of the winding or in the case of resonant voltages built up locally within the windings. The amount and amplitude of the voltage stress on the isolation depends on various parameters such as the lightning strike in the considered area, as it determines how often the transformer is stressed by the lightning overvoltage. Since the amplitude of the overvoltage at the transformer terminals is usually unknown, an online overvoltage transient recorder is used. with the ability to analyze and store transients in real-time.[10]

The value of the voltage variation shall be limited to within ±10% of the nominal supply voltage (Un). The transmission system is not interconnected with the power supply in the network, or for special long-distance network equipment, the variation in the magnitude of the voltage should not exceed + 10% / - 15% of Un. At least 95% of the time interval of 10 minutes, the average rms value of the supply voltage must be above the lower limit of 90% Un. So, over a time interval of 10 minutes, the average rms value of the supply voltage should not be higher than +10% or lower than -15% of Un. Also the actual power consumption required by individual network users is not completely predictable, in terms of quantity and contemporaneity.

Distribution of electric power from a source to a load in an installation, there will be a voltage difference between the voltage on the source side and the voltage on the load side. Where the voltage on the source side is greater than the voltage on the load side. This is due to the voltage drop in the installation system. The voltage drop between the consumer terminal and any point of the installation shall not exceed 5% of the rated voltage at the consumer terminal. The voltage drop between the consumer terminal and any point of the installation shall not exceed 4% of the rated voltage at the consumer terminal. [11]

2.3.2. Overcurrent Fault

Overcurrent faults must be protected immediately, Humans or livestock must be protected from injury, and property must be protected. protected from damage due to excessive temperature or electromechanical stress due to overcurrent which may arise in the conductors. Fault current limiting shall be chosen to limit the instantaneous fault current to a value within the limits of the equipment being protected. In selecting a fault current limiter, the following two factors should be considered: a) the prospective short-circuit current of the supply system b) the rated value and characteristics of the equipment c) the rated value and characteristics of the protective equipment associated with the equipment to which it is connected.[11]

2.3.3. Leakage Current Fault

Disturbances that cause leakage currents in a conductor material object can be caused by the presence of live electrical parts connected to the ground. Faults on the live part that is connected to ground can occur directly because the conductor object touches the active electric phase. Other disturbances can occur if or touch parts of active electrical equipment that experience insulation failure so that the object gets voltage. Touch voltage which results in leakage current to the ground through objects that touch the voltage equipment.[12] Fault in the live part connected to ground occurs indirectly, for example because the object is close to the active voltage source where the object is induced by an electromagnetic field (field discharge), this process is known as step voltage disturbance. (voltage between the legs of the object) V which can generate current. Leak flows through between the legs of the object to the ground.[12]

Table 1. Leakage Current Standards

Equipment	*Type*	*Max. Leakage Current*
Double Insulated	*All*	*25mA*
Grounded	*Handheld*	*75 mA*
	Movable	*3.5 mA*
	Stationary	*3.5 mA*

3. Methods

Research conducted to develop IoT-based learning media on electrical protection systems is Research and Development (R&D). Development research is research to develop a new product or improve an existing product with a certain process or development step. The results of the research can be used to increase the productivity of education, the quality of education, and as a means to achieve educational goals. The research method is shown in Figure 2.

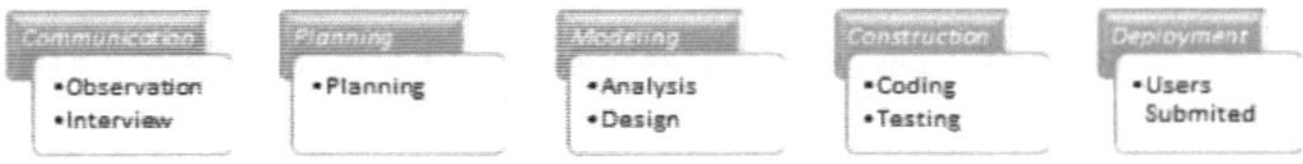

Figure 2. Research Methods

This research is a media development research as a learning tool for students of Electrical Engineering Education. The development research model approach used is the waterfall model for software development and ADDIE for the development of learning materials.

In this research on the development of IoT-based learning media, data collection will be carried out in the Electrical Engineering Education Study Program starting in the 2019/2020 academic year.

The research subjects were Lecturers of the Department of Electrical Engineering Education as instrument validators, material experts, and media experts. Students of the Electrical Engineering Education Study Program who are currently taking and have taken electrical power protection system courses as users.

3.1. Procedure

The development procedure of this research is in accordance with the waterfall and ADDIE models which include:

a. Communication (Communication)

Communication is carried out to find out the description of the practical activities of the electrical protection system and the needs that need to be met in the learning process. Communication is expected to make IoT-based learning media feasible and can be used properly.

b. Planning (Planning)

Planning is done by planning the manufacture of learning media using Arduino Software (IDE) as a device used to do programming. The developed media is supported by a prototype of a real condition that can be used as an object to be controlled by Android

c. Modeling (Design)

The design of learning media is adjusted to the results of observations made at the communication stage. The design is based on the analysis of laboratory requirements and the practice of electrical protection systems. The analysis is carried out according to the results of interviews conducted and the needs suitable for use in learning the practice of electrical protection systems. The design images are shown in Figure 3 until Figure 6.

Figure 3. Building Design

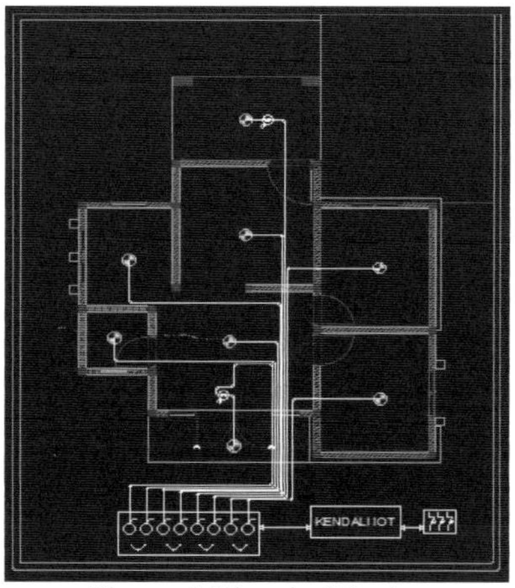

Figure 4. Electrical Design

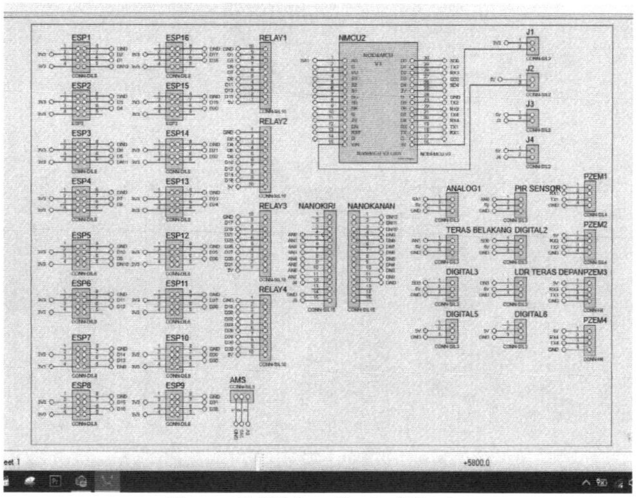

Figure 5. Wiring Control Design

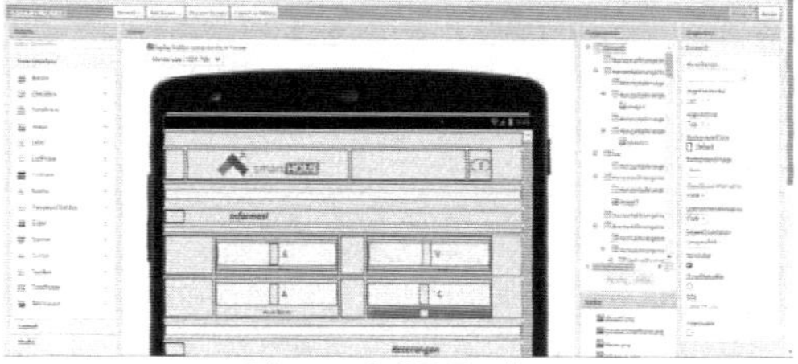

Figure 6. UI Design

d. Construction (Construction)

Construction is in the form of coding from learning media developed in the IDE as a programming algorithm used in learning media. Making the form of media adapted to the laboratory used for learning the practice of electrical protection systems. The UI Construction is shown in Figure 7.

Figure 7. UI Construction

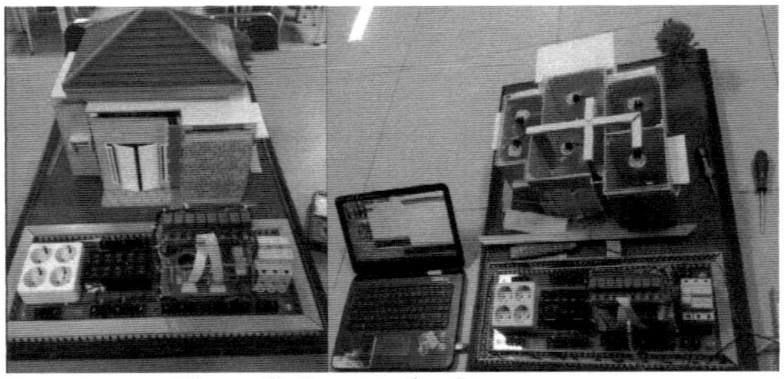

Figure 8. Construction Process

e. Deployment

Submission of learning media to the teaching team of electrical protection system courses to be given to students for practical learning activities of electrical protection systems. The use of students and lecturers is expected to help learn electrical protection systems and problems with the use of media that are not used properly. The use of electrical protection system learning media is expected to help students and lecturers in carrying out electrical protection system practice activities.

4. Result and Discussions

The product resulting from this development research is an Internet of Things (IoT)-based learning media. The resulting learning media is used for learning activities of electric power protection systems, especially electrical protection systems in Smart Buildings in learning held at the Electrical Engineering Education Study Program. Research on the development of learning media products applies the ADDIE development model for material development and Waterfall for software development.

4.1. Media Performance

Test The performance test of the IoT-based electric power protection system learning media was obtained through the black box test. The black box test is done by looking at the results of any given function or event by not looking at the given program code. Based on the results of the black box test, obtained a score of 34 on the function aspect, with the "Very Eligible" category and a score of 4 on the performance aspect, with the "Very Eligible" category. The learning performance of the IoT-based electric power protection system from the function aspect gets a percentage of 100%, the performance aspect gets a percentage of 100%. Thus the total score of the black box test gets a percentage of 100%, with the performance category "Very Good". The assessment score can be seen in table 2.

Table 2. Black box test

Respondent	Aspect			
	Function		Performance	
	(%)	Category	(%)	Category
Respondent 1	100	Very Feasible	100	Very Feasible
Respondent 2	100	Very Feasible	100	Very Feasible
Average	100	Very Feasible	100	Very Feasible

4.2. Feasibility Test

4.2.1. Content Feasibility

The feasibility of learning media materials for IoT-based electric power protection systems includes three aspects, content, learning, and efficiency. Aspects of the content of the material include several indicators, learning achievement, learning material, depth of material, and current. The learning aspect includes several indicators, namely learning strategies, learning motivation, and opportunities to learn. The efficiency aspect includes several indicators, namely time and resources. The items in the material expert questionnaire instrument are the application of several indicators on aspects of material content, learning, and efficiency. The material expert questionnaire that has been prepared is then assessed by the material expert and all indicators have been met.

The feasibility of the application program in terms of assessment by material experts includes aspects of material content, learning aspects, and efficiency aspects. All aspects of the material feasibility assessment received a score of 93.5 and were categorized as "Very Feasible". The assessment score can be seen in table 3.

Tabel 3. Content Feasibility

Expert	Aspect		
	Content	Learning	*efficiency*
Expert 1	52	20	22
Expert 2	50	19	24
Average	51	19,5	23

The results of the feasibility assessment of the IoT-based electric power protection system learning media by material experts can be seen in Figure 9.

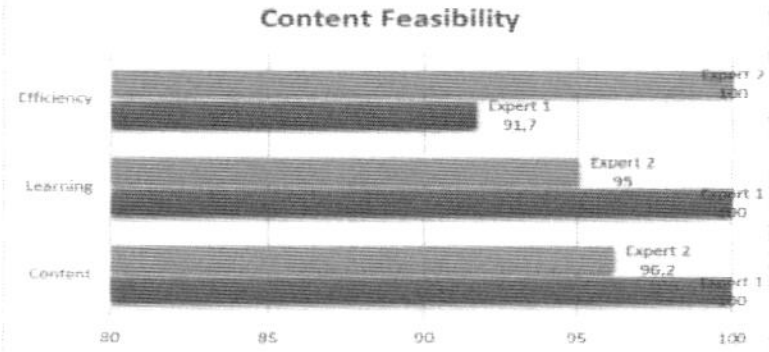

Figure 9. Expert Judgement

4.2.2. Media Feasibility

The feasibility of the IoT-based electrical protection system media includes four aspects, namely visual, software, usability, and portability. The visual aspect includes several indicators, display design, color composition, and legibility. the software aspect includes several indicators, namely application programs, documentation, and interactive. usability aspects include several indicators, operation, usability, and attractiveness. portability aspects include two indicators, namely installation and adaptation. The items in the media expert questionnaire are the implementation of the indicators on the visual, software, usability, and portability aspects. The media expert questionnaire that has been compiled is then assessed by media experts and all indicators have been met.

The feasibility of the application program is reviewed from an assessment by media experts covering visual aspects, software aspects, usability aspects, and portability aspects. All aspects of the media feasibility assessment received an average score of 108.5 and were categorized as "Very Feasible". The assessment score can be seen in table 4.

Table 4. Expert Judgement

Expert	Aspect			
	Display	Software	*Usability*	*Portability*
Expert 1	40	27	28	16
Expert 2	37	26	28	15
Average	38,5	26,5	28	15,5

The results of the feasibility assessment of the IoT-based electric power protection system learning media by media experts can be seen in Figure 10.

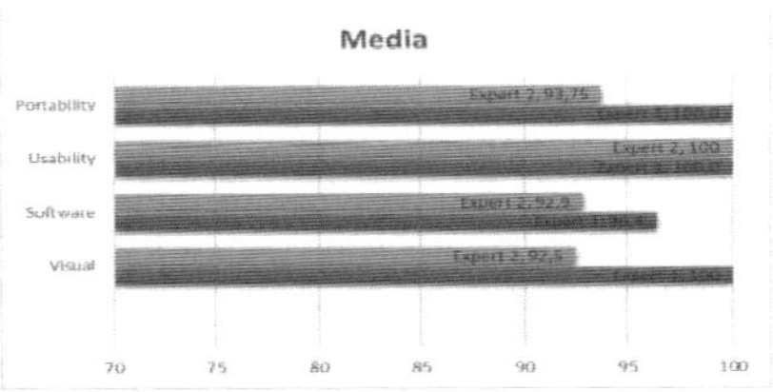

Gambar 10. Expert Judgement

4.3. User Feedback

The user response test was carried out to find out the responses of students using the IoT-based electric power protection system learning media product as a learning medium for the practice of electric power protection systems. Products developed must get good responses from users. The user response test was carried out by ninety-four people, including 31 Electrical Engineering students and 63 undergraduate Electrical Engineering Education students, Faculty of Engineering, Yogyakarta State University, who had taken electrical power protection system courses. There are five aspects to the user's response test, including learning, visual, software, usability, and portability aspects. responses by users using an instrument in the form of a questionnaire with a total of 31 items covering, 3 items of learning aspects, 10 items of visual aspects, 7 items of software aspects, 7 items of usability aspects, and 4 items of portability aspects. The assessment score uses a Likert scale of 1-4. The assessment score can be seen in table 5.

Tabel 5. User Feedback

Respondent	Aspect				
	Learning	Display	Software	*Usability*	*Portability*
Average	10,9	33,7	23,8	24,6	13,8
Standard Deviation	1,5	5	3,5	3,5	2

Minimum	3	10	7	7	4
Maximum	12	40	28	28	16

The results of the assessment of the IoT-based electric power protection system learning media by users can be seen in Figure 11.

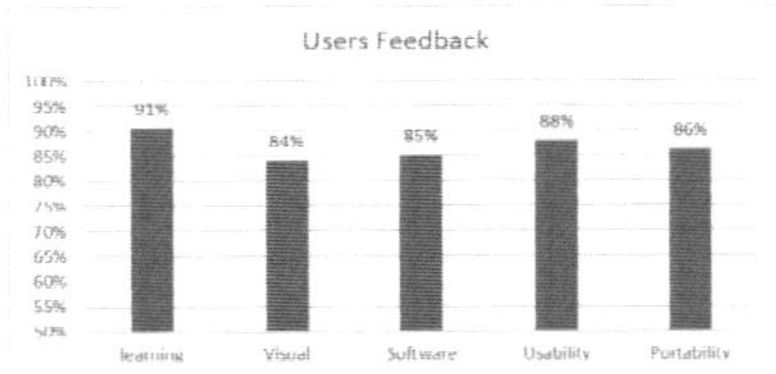

Figure 11. User Feedback

5. Conclusion

Based on the problem formulation, research objectives, research questions, and study of the results of research and development of learning media for IoT-based electric power protection systems, the conclusions are as follows. An IoT-based electric power protection system learning media has been developed that can simulate the testing of electrical protection devices with the correct procedure. The performance of the IoT-based electric power protection system learning media is known through the black box test, obtained a score of 38 so that it can be categorized as "Very Good". The feasibility of the application program is reviewed from an assessment by a material expert covering aspects of material content, learning aspects, and efficiency aspects. All aspects of the material feasibility assessment received a score of 93.5 and were categorized as "Very Feasible". The feasibility of the application program is reviewed from an assessment by media experts covering visual aspects, software aspects, usability aspects, and portability aspects. All aspects of the media feasibility assessment received an average score of 108.5 and were categorized as "Very Feasible".

User responses by students include learning aspects, visual aspects, software aspects, usability aspects, and portability aspects. All aspects of user response by students got an average score of 106.7 and were categorized as "Very Good".

References

[1] Republik indonesia. (2012). *Undang-Undang Nomor 12, Tahun 2012, tentang Pendidikan Tinggi.*

[2] Supriyadi, E., Indro, Y. H., Priyanto, E., & Surwi, F. 2020. Students' Evaluation on Teaching in Vocational and Technical Schools. *International Journal of Instruction*, 13(2), 621-636. https://doi.org/10.29333/iji.2020.13242a

[3] Nurrita, Teni. (2018). *Pengembangan Media Pembelajaran untuk Meningkatkan Hasil Belajar Siswa.* Jurnal Misykat vol 3 (1). 171-187.

[4] Bakri, Muhammad Amin. 2016. *STUDI AWAL IMPLEMENTASI INTERNET OF THINGS PADA BIDANG PENDIDIKAN.* Journal of Electrical and Electronics, Vol 4 (1). 18-23.

[5] Warsita, Bambang. (2008). *Teknologi Pembelajaran: Landasan &Aplikasinya,* Jakarta: Rineka.

[6] Zhang, M., Zhang, Z., Chang, Y., Aziz, E., & Chassapis, C. (2018). Recent Developments in Game-Based Virtual Reality Educational Laboratories Using the Microsoft Kinect. *International Journal of Emerging Technologies in Learning. Vol. 13(1),* 138-159.

[7] Arafat. (2016). Sistem Pengamanan Pintu Rumah Berbasis *Internet Of Things (IoT) dengan ESP8266.* Jurnal Technologia Vol 7 (4). 262-268.

[8] Firman, Abidin dan Irwan Nimara. 2012. Proteksi Petir Terhadap Sistem Tenaga Listrik.

[9] Mery, Seksio Kardila, Yulisman, Budi Santosa. 2013. *Sistem Proteksi Pembangkit Jenis.*

[10] B.Filipović-Grčić, B. Franc, I. Uglešić, I. Pavić, S. Keitoue, I. Murat, I. Ivanković, "Monitoring of transient overvoltages on the power transformers and shunt reactors – field experience in the Croatian power transmission system", Procedia Engineering, vol. 202, pp. 29-42, 2017.

[11] BSN, "Persyaratan Umum Instalasi Listrik 2011 (PUIL 2011)," SNI 0225:2011/Amd 1:2013.

[12] Syukriyadin. Sistem Proteksi Arus Bocor Menggunakan Earth Leakage Circuit Breaker Berbasis Arduino. 2016. Jurnal Rekayasa Elektrika Vol 12.111-118.

IoT Enabled Ventilator Monitoring System for Covid-19 Patients

Mashoedah[1], Umi Rochayati[1], Indra Hidayatulloh[1], Arya Sony[1], Ferda Ernawan[2] Fardiansyah[1], Apri Nuryanto[1]

[1]Universitas Negeri Yogyakarta Indonesia
[2] University Malaysia Pahang, Malaysia)

E-mail: mashoedah@uny.ac.id

Abstract. The use of the IoT protocol on medical equipment is expected to provide protection for medical personnel in dealing with Covid-19 patients, especially when medical personnel are monitoring and setting up an equipment. This study aims to (1) produce a monitoring and control system for a breathing apparatus (Ventilator) based on the Internet of Thing (IoT), (2) test the ventilator control function, (3) test the data transmission function with the IoT protocol. The method used is Define, Design, Develop, and Disseminate (4D). Data collection is done through (1) Testing and Observation (2) Limited field test. This research produces a control and monitoring system for mechanical ventilators. The mechanical ventilator consists of a gripper motion mechanism driven by a dc motor. The movement of the gripper creates pressure and releases pressure on the ambu bag. The depth of pressure exerted by the gripper is measured as the volume and pressure of the air delivered to the lungs. The rate of pressure exerted is measured as the velocity of air flowing into the lungs.
Keywords: Ventilator, IoT, Covid19

1. Introduction

We currently face a disease pandemic that known as Coronavirus Disease 2019 (Covid 19). Covid 19 is transmitted through droplets and spread very quickly in humans. This virus attacks the human airways. If infected with this virus, a person will experience a high fever and difficulty breathing. One of the medical equipment that is needed to treat corona patients is a ventilator. A ventilator is a device used by medical personnel for patients who experience respiratory distress. The need for ventilators oday is very high globally. This is because the number of corona patients has increased very sharply while the number of ventilators is very limited[1].

The ventilator serves to exhale gas (in this case oxygen) into the patient's lungs and helps strengthen the work of the respiratory muscles[2]. The working phase of the ventilator is shown in the Figure 1[3].

\

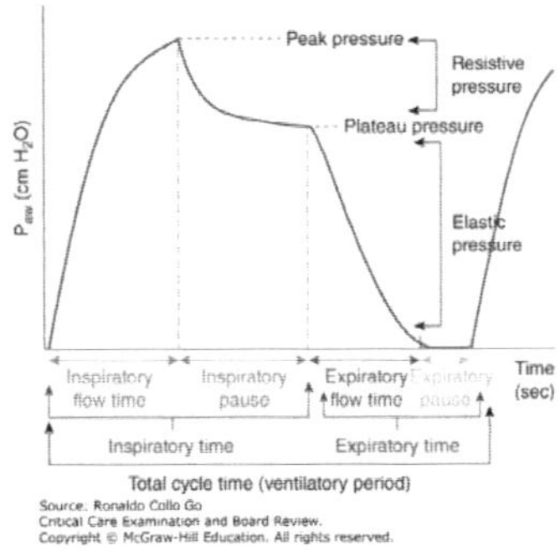

Figure 1. Expiratory and Inspiratory Phases of the Ventilator

The role of the ventilator is vital and is needed in the handling of covid19 patients. However, with a limited amount, not all patients can be treated using this equipment.As reported by several print or electronic media, transmission of covid19 virus is through droplets (small droplets) that come out of the patient's body and transmission can also be through the air. Covid19 transmission mechanism like this requires us to maintain a physical distance (physical distancing) of at least 1-2 meters when talking with others.

Medical staff (nurses, doctors and other health workers) are people who are at high risk of **contracting** this virus. This is caused by the frequency of interaction between medical personnel and patients that very high. To prevent and reduce the possibility of medical personnel contracting by the viruses from patients treated, medical personnel need to be equipped with Personal Protection Equipment (PPE). The use of PPE does not always guarantee that medical personnel are free from contracting this virus. This is evidenced by the high number of medical staff who died due to **contracting the covid19 virus**. Then, it is necessary to look for ways that can prevent medical personnel from contracting covid19. One of the ways is to reduce interaction and maintain distance between medical personnel and patients treated.

1.1. Medical Internet of Thing

Internet of Things (IoT) is a concept that aims to expand the function of the internet network. The functions and benefits of internet networks such as sharing data, remote control, and so on can connect objects in the real world, so that objects in the real world can communicate with each other. In general, the IoT architecture can be divided into three layers, the first is the application layer, the second is network layer and the third is the sensing layer. One form of application layer is service, which includes health service and emergency service.

IoT technology provides convenience features for remote control. The control network can use Bluetooth, WiFi, or the internet. With IoT technology, it is possible to implement a remote monitoring and control system. From the problems outlined above, this study aims to create an IoT-based ventilator monitoring and control system, which is expected to help prevent the transmission of the Covid19 virus and deal with Covi19 patients who have difficulty breathing.

One of the fastest-growing Internet of Things application is in medicine. Medical IoT (MIoT) is a term that refers to the application of IoT to medicine [4][5][2]. The rapid development of technology in the area of IoT has enabled for pervasive monitoring of health[6][7]. IoT has filled the gap in the integration of wearable sensors, mobile phones, smart devices and cloud platform so that remote monitoring of individual's health has been possible to perform [8][9][10].

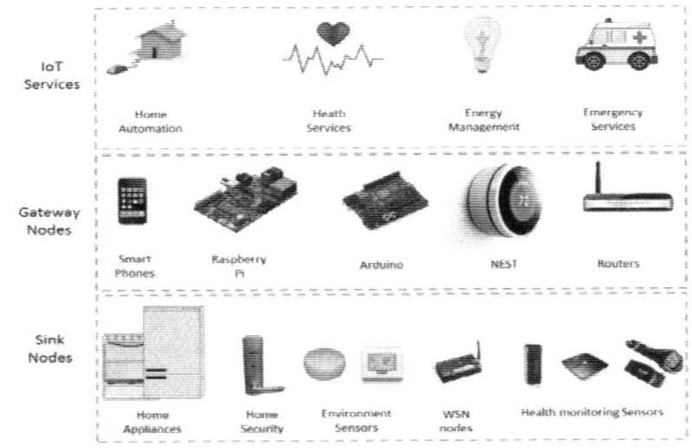

Figure 2. An Architecture for Open IoT Services

As proof of how important MIoT is, during the pandemic of Coronavirus 2019 (Covid-19) in China, IoT application called nCapp has been proposed to be used by chinese experts for Covid-19 intelligence diagnosis, treatment assistant program and online monitoring system[11] According to the the consensus results, nCapp has obtained warm welcomed from specialists and even managers at all levels of hospitals, local community development corporations, and public health centers.

nCapp has helped diagnosis process of Covid-19 patients and classified the severity level into mild, moderate, severe and critical pneumonia. For patients with pneumonia who have respiratory failure and require mechanical ventilation, they are categorized critical pneumonia. Furthermore, the patients will be monitored online and the system will send an alarm to remind the medical staff to take corrective measures when the data exceed the normal value or emergency occurs. However, it is likely that the proposed system didn't considered to monitor the ventilator it self whereas it is important to make sure the ventilator working well and the setting is suitable with the patient condition.

1.2. An Alternative Ventilator Monitoring System .

The function of installing Ventilator equipment for patients is to improve oxygenation, help eliminate carbon dioxide, and help the respiratory muscles work.

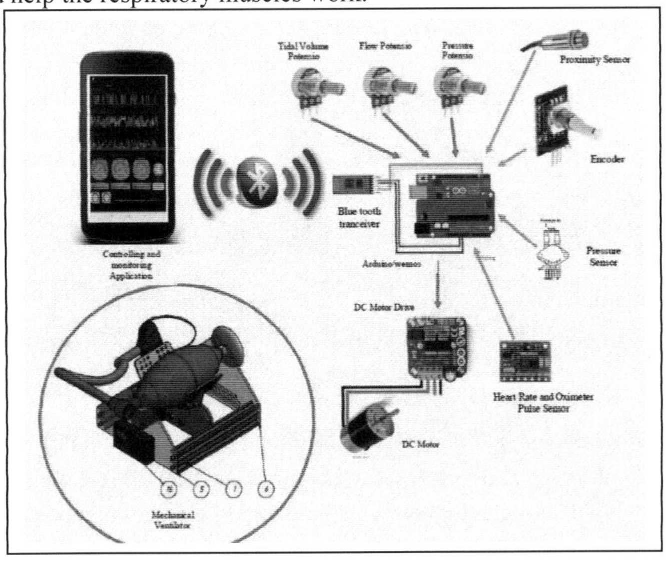

Figure 3. The Ventilator Monitoring System Component Diagram

The ventilator control system uses pwm controlling system, with the main control chip using the WEMOS LOLIN32 ESP32 chip. The use of the ESP32 chip aims to control IoT (Internet Of Thing)-based ventilators with bluetooth communication. So that in the process of monitoring and controlling the ventilator for COVID-19 patients, it will be possible to do it at a certain distance (Physical Distancing). In addition to the IoT-based monitoring and controlling function, the ventilator can also be adjusted with several control buttons provided near the system. The setting button consists of an inspiration button, an expiration button, and a pause button. Refer to the figure 3, some of the sensors used in this system are lung pressure sensor, blood oxygen levels sensor (oximeter), air flow sensor (flow sensor), current sensor, proximity sensor, and angular position sensor (encoder).

2. Method

The research was carried out in the Industrial Electronics laboratory, Faculty of Engineering, Yogyakarta State University. The object under study is the IoT Enabled Ventilator Monitoring System for Covid-19 Patients. A product that can be used by medical personnel or doctors to monitor patient data without having to approach the patient and the ventilator unit.

The research was conducted through 4 stages, starting from 1) Define, 2) Design, 3) Develop, and 4) Disseminate, as shown in Figure 4 below:

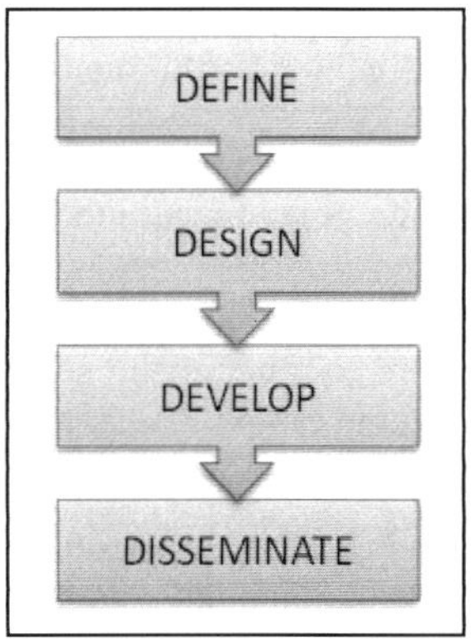

Figure 4. Research Method

The Define stage conducting needs analysis, and conducting literature study. Design Phase Designing ventilator hardware and software, determining programming algorithms, compiling programs, and conducting Simulations. Develop Phase validating the product by experts, product revision based on expert suggestions, conducting limited trials, and doing revision of test results products. at the Disseminate stage, implementation and product diffusion have been carried out.Result and Discussion.

The iot enabled ventilator monitoring system for covid-19 patients is a device to facilitate the transport of oxygen and carbon dioxide between the atmosphere and the alveoli for the purpose of improving lung gas exchange which can be monitored for measurement parameters via smart devices with a dedicated IoT network (Internet of Thing). for Covid-19 patients.

Medical staff need to be equipped with Personal Protective Equipment (PPE). So it is necessary to find ways that can prevent medical personnel from contracting COVID-19, one of which is to reduce interactions and maintain a distance between medical personnel and patients being treated.

IoT technology provides convenience features for remote control. In the implementation of the tool using bluetooth. with the consideration that monitoring and control devices do not need to have a very long distance, but enough distance is maintained between medical personnel and patients as a condition for meeting health protocols.

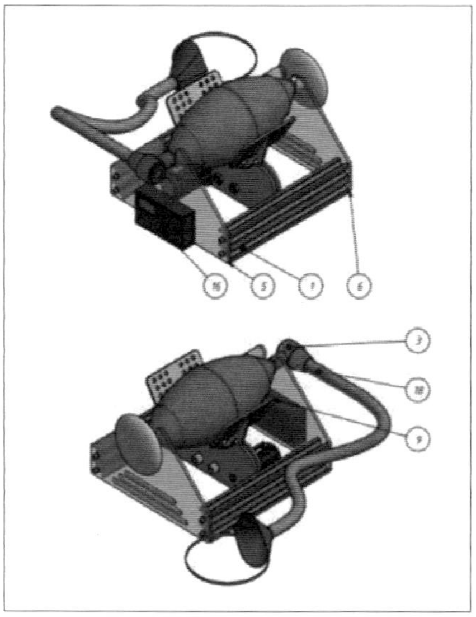

Figure 5. The Design of Mechanical Ventilator

The designed ventilator has several parts consisting of: mechanical hardware consisting of a gripper mechanism which is driven by a DC motor. The gripper functions to apply pressure to the ambu bag to provide inspiratory and expiratory pressure. The mechanical design uses a Gripper model that can apply pressure to an ambu bag.

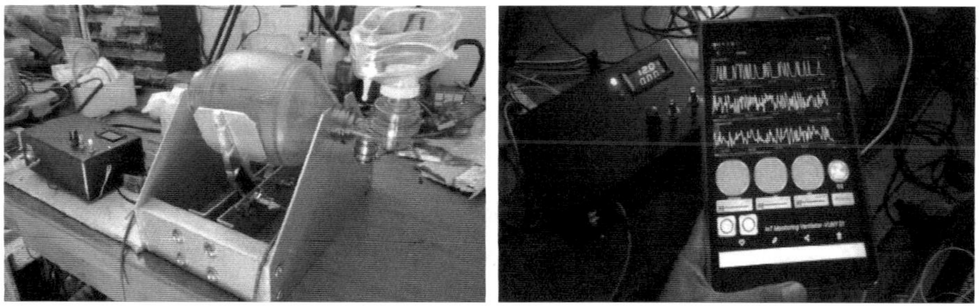

Figure 6. Prototype of Ventilator Monitoring System

The prototype IoT Enabled Ventilator Monitoring System has been functionally tested at a lab scale. The motion function of the gripper ambu bag has worked as expected, the motion control system of the gripper is obtained from the DC motor control circuit. Speed control using PWM programmed on arduino/wemos esp32.

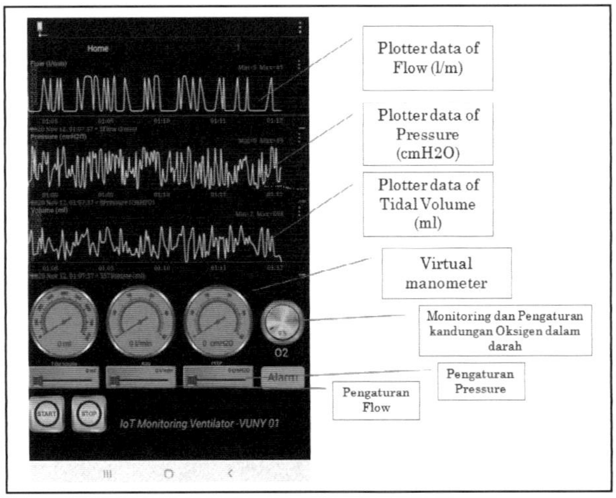

Figure 7. User Interface of Mechanical Ventilator Using Virtuino

At the application level this product uses Virtuino. Virtuino is an HMI platform for IoT servers, Arduino ESP and so on. Virtuino is installed on Android to monitor sensors or control the ventilator, with the aim of being the client side of the smartphone. The measurement parameters and ventilator settings are displayed on the smartphone screen. Besides that, monitoring and control are also provided on the hardware side of this ventilator device.

3. Conclusion

This research produces a control and monitoring system for a mechanical ventilator. The mechanical ventilator consists of a gripper motion mechanism driven by a dc motor. The motion of the gripper creates pressure and releases pressure on an ambu bag. The depth of pressure exerted by the gripper is measured as the volume and pressure of air delivered to the lungs. The rate of pressure exerted is measured as the velocity of air flowing into the lungs.

The control and monitoring system uses Arduino components as the main control. The bluetooth-based android application chosen is Virtuino, because it has a user-friendly interface and can work efficiently with Arduino to control and monitor variables on the ventilator via smart devices. Arduino is programmed to interact with the Virtuino application. The bluetooth module, alarm, air pressure sensor, blood oxygen level sensor (SpO2) and speed sensor will be connected directly to Arduino.

Acknowledgments

Thanks to Yogyakarta State University for providing funding support for the implementation of this research. We also express our gratitude to the Faculty of Engineering, Yogyakarta State University for providing support for all research activities, all colleagues, research partners and technicians who have provided time for the implementation of this research.

4. References

[1] J. S.-G. and N. K. Sarah Kliff, Adam Satariano, "There Aren't Enough Ventilators to Cope With the Coronavirus," *The New York Times*, 2020. [Online]. Available: https://www.nytimes.com/2020/03/18/business/coronavirus-ventilator-shortage.html.

[2] Y. Song, J. Jiang, X. Wang, D. Yang, and C. Bai, "Prospect and application of Internet of Things technology for prevention of SARIs," *Clin. eHealth*, vol. 3, pp. 1–4, 2020, doi: 10.1016/j.ceh.2020.02.001.

[3] Drzezo, "Mechanical Ventilation," 2018. [Online]. Available: https://aneskey.com/mechanical-ventilation-12/.

[4] C. Bai and J. Zhao, *Medical Internet of Things*. Beijing: Science Press, 2016.

[5] C. Bai, *Practical medical Internet of Things*. Beijing: People's Medical Publishing House Co., LTD, 2014.

[6] G. Elhayatmy, N. Dey, and A. S. Ashour, "Internet of Things Based Wireless Body Area Network in Healthcare," pp. 3–20, 2018, doi: 10.1007/978-3-319-60435-0_1.

[7] M. Haghi, K. Thurow, and R. Stoll, "Wearable devices in medical internet of things: Scientific research and commercially available devices," *Healthc. Inform. Res.*, vol. 23, no. 1, pp. 4–15, 2017, doi: 10.4258/hir.2017.23.1.4.

[8] U. Anliker *et al.*, "AMON: A wearable multiparameter medical monitoring and alert system," *IEEE Trans. Inf. Technol. Biomed.*, vol. 8, no. 4, pp. 415–427, 2004, doi: 10.1109/TITB.2004.837888.

[9] N. Dey, A. S. Ashour, F. Shi, S. J. Fong, and R. S. Sherratt, "Ref11 in Wsn Was 31," *IEEE Trans. Consum. Electron.*, vol. 63, no. 4, pp. 442–449, 2017.

[10] L. Cai *et al.*, "OSA patient monitoring system based on the Internet of Things framework," *2019 4th Int. Conf. Smart Sustain. Technol. Split. 2019*, pp. 0–3, 2019, doi: 10.23919/SpliTech.2019.8783103.

[11] L. Bai *et al.*, "Chinese experts' consensus on the Internet of Things-aided diagnosis and treatment of coronavirus disease 2019 (COVID-19)," *Clin. eHealth*, vol. 3, pp. 7–15, 2020, doi: 10.1016/j.ceh.2020.03.001.

Performance of Frequency Regulation on Power in the Development of Variable Speed Drive Training Kit as a Learning Media for Motor Control Practices

[1]T H T Maryadi, [2]A S J Wardhana, [3]E S Damarwan, [4]T Sukisno

[1234] Department of Electrical Engineering Education, Faculty of Engineering, UNY

E-mail: totokheru@uny.ac.id

Abstract. This study develops a variable speed drive training kit for electric motor control practice media in vocational education. This is research and development with the ADDIE model. The stages of the ADDIE model carried out are Analyze, Design, Development, Implementation, and Evaluation. A Variable Speed Drive (VSD) system consists of two panels, 1) a Variable Speed Drive panel and 2) a control and metering panel. The VSD Training kit module unit using Altivar 71 is a learning medium about electric motor control systems. The VSD Training kit can work well based on trials. The test results of the training kit variable speed drive function test show that at a frequency of 20 Hz, the magnitude of the current in the R, S, and T phases get the lowest value.

Keywords: Motor control, training kit, variable speed drive

1. Introduction

Vocational education is one of the formal educations to focus more on increasing the ability or competence to be ready to work in specific majors or fields. One of the fields in Vocational Education is Electricity. A quality learning process is needed to prepare graduates with reliable electrical competence to compete in the world of work. The learning process in vocational education is divided into three, namely theory, practice, and field. One of the practical electrical courses given is the Control System Practice Course. In the learning process, students will practice several types of control systems, working principles and components needed in the control system. In the industrial revolution 4.0, as it is today, the control system is an essential part. In the field of electricity, electric motor control competence is one of the mandatory things that students must be possessed.

The learning process of Control System Practice indeed cannot be separated from adequate learning facilities and infrastructure. According to the Government Regulation of the Republic of Indonesia No. 19 of 2005 concerning National Education Standards, every academic unit must have facilities, furniture, educational equipment, educational media, books and other learning resources, consumables, and other equipment. Needed to support an orderly and continuous learning process [1]. The learning process of the control system practice can be said to be of high quality. One of the indicators is the achievement of the learning objectives of the course.

To achieve the learning objectives, a media that can improve the ability of students in the learning process is needed [2]. Learning media can be used to stimulate the thoughts, feelings, and interests and willingness of students so that the learning process occurs to achieve learning objectives effectively [3]. Based on the above opinion, it can be concluded that to achieve the learning objectives of the control system practice in Vocational Education. It is necessary to have learning media that can stimulate the thoughts, interests and feelings of students to improve their abilities.

Based on the results of observations and interviews with lecturers in control systems courses, data is obtained if: (1) there is no learning media in the form of a training kit that makes it easier for students to assemble control systems as to electric motor speed control. (2) The need to increase the student's motivation in the learning process of electric motor control. Increased motivation can be through media that is interesting and easily understood by students. In the practice of control systems, especially motor control competencies, there is not yet available learning media in the form of an integrated training kit, making it difficult for students in the process of assembling and retrieving measurement data according to the worksheet. That is very unfortunate because competence regarding the control of electric motors such as variable speed drives is very much needed in the era of the industrial revolution 4.0 as it is today.

In the industrial world, by using a variable speed drive, electric motors can be controlled for their performance and speed as needed. Variable speed drives (VSDs) have gained more popularity in recent years, as VSDs not only improve motor load performance but can also provide frequency support in the event of a circuit fault [4]. The development of a Variable Speed Drive Training kit for Electric Motor Control Competency Practice Media in Vocational Education is the title of the research offered to improve the quality and competence of students in learning control system practices in Vocational Education.

2. Main Theory

Learning media is a tool that can help the teaching and learning process and serves to clarify the meaning of the message conveyed so that learning objectives can be achieved properly and perfectly. Learning media can be understood as anything that can share and distribute messages from sources in a planned manner to create a conducive learning environment where the recipient can carry out the learning process efficiently and effectively [5]. Learning media is anything that can be used to channel messages from the sender to the recipient to stimulate the thoughts, feelings, and interests and willingness of students in such a way that the learning process occurs to achieve learning objectives effectively [3]. Based on the description above, it can be concluded that the notion of learning media is anything that is used to clarify the meaning of the message conveyed to stimulate the thoughts, feelings, attention, and interests and willingness of students so that learning objectives are achieved effectively and efficiently.

2.1. Learning Media Training kit

Training kit is a learning media that is included in the form of mock-ups. The Training kit is included in the original object learning media. Learning media of original objects that are not modified are actual objects as they are, without changes, except they are only moved from their original places [5]. These objects have the following characteristics: (1) can be used; (2) in standard size; and (3) known by their real names. Based on this explanation, it can be concluded that the trainer-kit is a learning media consisting of simplification of original objects according to the subject matter that serves to simulate learning activities to increase the effectiveness of learning.

2.2. Variable Speed Drive

VSD or variable frequency drive is a device used to control the speed of an electric motor (AC) by controlling the frequency of the electric power supplied to the motor [6]. VSD can control motor speed (including rotation direction) and operation with sinusoidal current and unitary power factor on the network side [7]. The advantages of VSD compared to other induction motor speed controllers are energy-saving, process optimization, smoother machine operation, and reduced maintenance [8].

The VSD training module unit using Altivar 71 is a learning medium about the Variable Speed Drive system. The entire system in the components and circuits used in this trainingis also adapted to the existing Variable Speed Drive system. However, so that this trainingis suitable and worthy as a learning medium, this trainingwill be designed with a modular concept. Each component is made separately with an attractive design, equipped with the name of the component, the appropriate symbol and various information about the component. Through these designs, it will be easy to use, maintain and can facilitate understanding the concept of the system and installation of the Variable Speed Drive module unit. Training kits can be installed and assembled repeatedly. A VSD system consists of two panels, namely the Variable Speed Drive panel itself and the control and metering panels. The picture of the Altivar 71 module circuit with the ATV71HU40NHZ series used is shown in Figure 1 below [9].

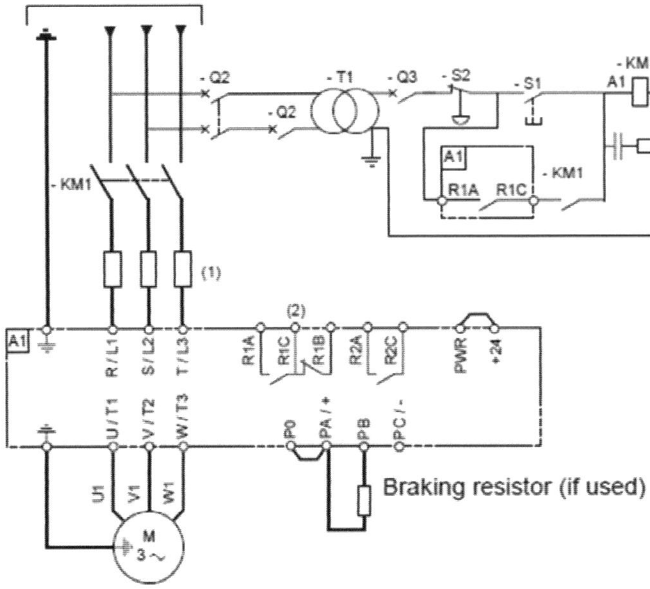

Figure 1. Altivar 71 Circuit.

3. Methods

This research is included in research and development (R&D) research. The purpose of this development research is to develop a Variable Speed Drive training kit as a tool or learning media in vocational education. The development model used in this research is the ADDIE approach. The ADDIE approach includes: Analyze (analyze), design (design), develop (develop), implementation (apply), and evaluate (evaluate) [10]. The implementation chart of the ADDIE model is shown in Figure 2 below.

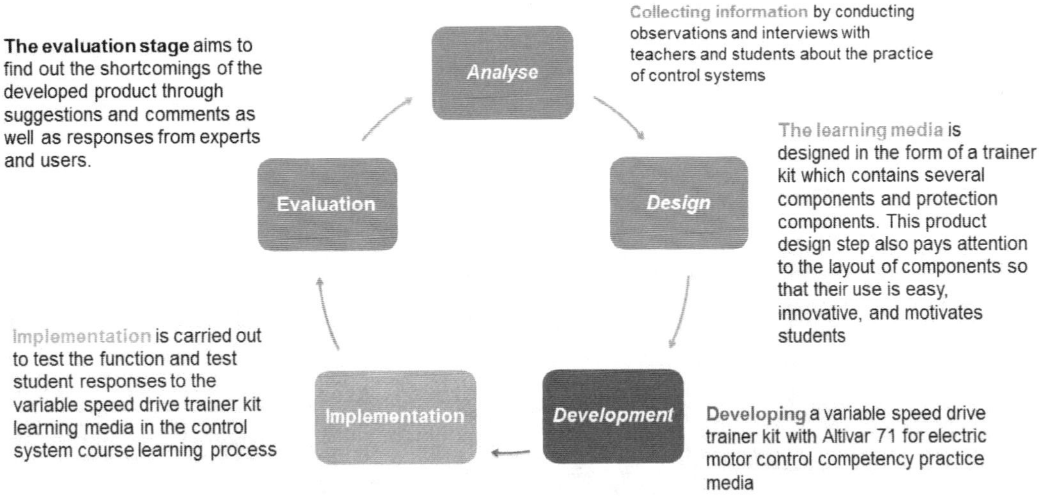

Figure 2. Implementation of the ADDIE Model

The analysis activity identifies the problem of implementing basic motor control learning on electric motor control competencies applied to vocational education. They are looking for differences between learning objectives and the reality of the learning process that has been implemented. It is analyzing the urgency of whether or not the development of learning media is necessary and adjusted to various aspects of supporting the variable speed drive training kit. The results of the analysis show that: (1) there is no learning media in the form of a training kit that makes it easier for students to assemble a control system, especially related to electric motor speed control. (2) The need to increase the motivation of students in the learning process of electric motor control. Increased motivation can be through media that is interesting and easily understood by students. In the practice of control systems, especially motor control competencies, there is not yet available learning media in the form of an integrated training kit, making it difficult for students in the process of assembling and retrieving measurement data. That is very unfortunate because competence regarding the control of electric motors such as variable speed drives is very much needed in the era of the industrial revolution 4.0 as it is today.

The following stage designs. It is an activity to design learning media products that will be developed starting from the design of the devices needed in the variable speed drive training kit. Determine the layout of each component of the variable speed drive training kit. The design of the trainingframe is shown in Figure 3 below.

Figure 3. Trainingframe

The third stage of development is the stage used to realize the concepts that have been prepared in the design stage into products that are ready to be implemented. At this stage, the researcher developed a variable speed drive training kit with Altivar 71 to practice electric motor control competence. The design and development is carried out by implementing the training kit design into the original training kit form. The image of making the training kit is shown in Figure 4 below.

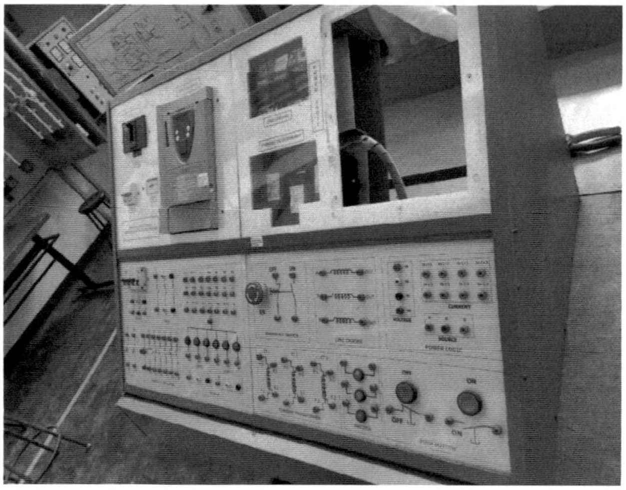

Figure 4. Installing Components on the Trainingboard.

The next stage is implementation, which is the stage to apply the product that has been developed in actual conditions. Evaluations are the last stage. After implementing in real conditions, the developed variable speed drive training kit was evaluated to provide feedback on the following performance.

4. Results and Discussion

In the development stage, after installing the variable speed drive components to the trainingbox and checking and adjusting the variable speed drive wiring, the next step is finishing the variable speed drive training kit. The final results of the variable speed drive training kit design are shown in Figure 5 below.

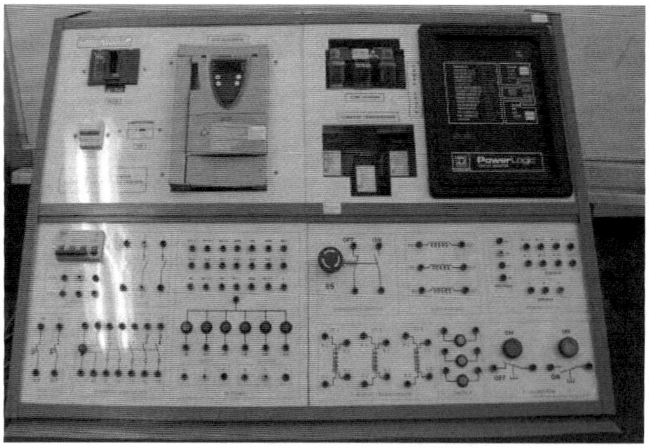

Figure 5. Training Kit Variable Speed Drive

After the variable speed drive training kit has been assembled, the next stage is a functional test. Functional testing was conducted to determine the performance of the variable speed drive training kit. The load used is a 3-phase induction motor 0.75 kW, 220/380 V, 1450 RPM, 50 Hz. Based on the test results, it is known the amount of power, Irms, and Vrms. The results of the experiment related to the effect of frequency on power on the variable speed drive training kit are shown in Table 1 below.

Table 1. Effect of Frequency on Power

F	**Power (kW)**		
	P1	**P2**	**P3**
1	0.034	0.006	0.004
5	0.099	0.026	0.004
10	0.007	0.001	0.006
15	0.019	0.003	0.003
20	0.515	3.889	1.734
25	0.018	0.011	0.002
30	0.001	0.005	0.001
35	0.002	0.004	0.002
40	0.035	0.031	0.038
45	0.222	0.225	0.230
50	0.028	0.020	0.057
55	0.192	0.159	0.196
60	0.007	0.007	0.003

Based on Table 1, it can be explained that at a frequency of 20 Hz, the required electrical power consumption is the highest compared to other frequencies. Variable Speed Drive (VSD) is used to overcome energy inefficiency in operating a 3-phase induction motor [11].

The next test stage is the effect of frequency on Vrms. In detail, the results of the frequency influence test on Vrms are shown in Table 2 below

Table 2. Effect of Frequency on Vrms

F	Vrms (Volt)		
	VR	VS	VT
1	229.9	223.6	224.5
5	230.5	225.2	224.2
10	229.9	223.5	224.5
15	229.8	223.8	224.6
20	231.4	227.4	226.5
25	231.6	228.9	227.1
30	232.2	229.7	227.5
35	231.4	229.1	227
40	230.5	227.6	227.9
45	230.7	227.6	228.5
50	230.7	227.3	226.7
55	230.1	226.7	226.9
60	228.9	225.4	225.7

Based on Table 2, it can be explained that the magnitude of Vrms VR, VS, VT at a frequency of 1 to 60 Hz ranges from 223 V to 230 V. This means that the voltage is still in accordance with the standard.

The results of the next trial tested the effect of frequency on Irms. In detail, the results of the test of the effect of frequency on Irms are shown in Figure 6 below.

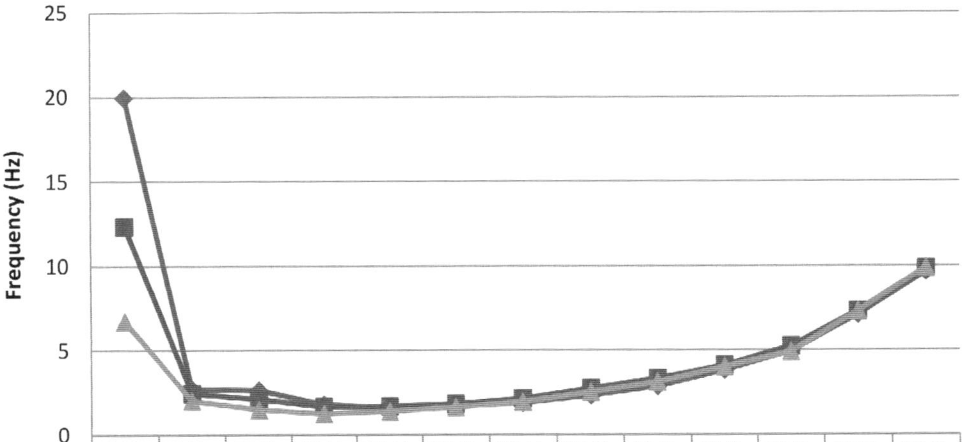

	1 Hz	5 Hz	10 Hz	15 Hz	20 Hz	25 Hz	30 Hz	35 Hz	40 Hz	45 Hz	50 Hz	55 Hz	60 Hz
IR	19,95	2,72	2,65	1,79	1,6	1,71	1,88	2,34	2,87	3,82	4,94	7,09	9,64
IS	12,34	2,45	2,11	1,67	1,7	1,83	2,15	2,75	3,33	4,11	5,24	7,32	9,86
IT	6,7	2,02	1,51	1,25	1,4	1,64	1,96	2,54	3,12	4,02	4,93	7,3	9,89

Figure 6. Effect of Frequency (Hz) on Irms (A)

Based on Figure 6, it can be seen that the magnitude of the current at a frequency of 1 Hz is the most significant current. That is because at that the motor starting current condition. Furthermore, at a frequency of 15-20 Hz, the current in each phase is at its lowest condition. After that, the current gradually increased. Based on the test results, it can be concluded that the variable speed drive training kit can work properly. In all load operation cases, the use of the VSD SW pump has a positive effect on reducing the total fuel consumption [12]. The existence of a variable speed drive can assist in operating the motor load.

5. Conclusion

Development of a VSD training kit using the ADDIE model. The stages of development carried out are Analyze, Design, Development, Implementation, and Evaluation. The VSD training kit consists of 2 panels, a VSD panel and a control and metering panel. The VSD training kit is designed using the Altivar 71 series of modules with the ATV71HU40NHZ series. The VSD Training kit can work well based on trials. The effect of frequency on Vrms of each phase, the voltage is between 223 V to 230 V. The training kit VSD function test show that at a frequency of 20 Hz, the magnitude of the current in the R, S, and T phases get the lowest value.

References

[1] Government Regulation of the Republic of Indonesia No. 19 of 2005 concerning National Education Standards, CHAPTER VII concerning Facilities and Infrastructure.

[2] Kartikasari. Pengaruh Media Pembelaajaran Berbasis Multimedia Terhadap motivasi dan Hasil Belajar Materi System Pencernaan Manusia. *Jurnal Dinamika Penelitian (63)*. 2016. 63-68

[3] Sukiman. *Pengembangan Media Pembelajaran*. Yogyakarta: PEDAGOGIA (2012)
[4] Wang, S., Ma, Y., Short, T., Tolbert, L. M., & Wang, F.. Power emulator of variable speed drive with grid frequency support in multi-converter based power grid emulation system. *In 2020 IEEE Energy Conversion Congress and Exposition* (ECCE) (2020, October) (pp. 1694-1701). IEEE.
[5] Munadi, Yudhi. *Media Pembelajaran: Sebuah Pendekatan Baru*. Jakarta: Referensi Melinda. 2013.
[6] Atmam, Abrar Tanjung, & Zulfahri. Analisis Penggunaan Energi Listrik Motor Induksi Tiga Phasa Menggunakan Variable Speed Drive (VSD). *Jurnal Sain, Energi, Teknologi & Industri*, Vol. 2 No. 2, Juni 2018, pp. 52 – 59
[7] Machado, L., Sousa, T. J., Pedrosa, D., Monteiro, V., Pinto, J. G., & Afonso, J. L.. A three-phase bidirectional variable speed drive: an experimental validation for a three-phase induction motor. In *International Conference on Sustainable Energy for Smart Cities* (pp. 47-57). (2019, December). Springer, Cham.
[8] Gomgom, I. Effendi, Penerapan Variable Frequency Drive Pada Motor Fuel Screw Feeder Untuk Bahan Bakar Pada Sistem Boiler, *J. Desiminasi Teknol.* 2 (2014) 50–59.
[9] Manual Book DIA2ED2050104EN Altivar 71.
[10] Robert Maribe Branch. *Instructional Design: The ADDIE Approach*. USA. Springer Science Business Media. 2009.
[11] Sa'ban, M. A. F., Wijanarko, Y., & Pratama, D. A.. Analisis Efisiensi Energi dengan Implementasi Variable Speed Drive Pada Raw Mill Fan Pabrik Baturaja 1 PT Semen Baturaja (Persero) Tbk. *In Electro National Conference (ENACO) Politeknik Negeri Sriwijaya* .2021, June. (Vol. 1, No. 1 Juni, pp. 33-48).
[12] Pariotis, E. G., Zannis, T. C., & Katsanis, J. S.. An integrated approach for the assessment of central cooling retrofit using variable speed drive pump in marine applications. *Journal of Marine Science and Engineering*, 2019. *7*(8), 253.

Development of Android-based Mobile Application "Cyber Gen" for Genetic Counselling Implementation among Thalassemia Patients

Henri Setiawan[1], Doni Setiawan[1], Suhanda[1], Acep Hidayatul Mustopa[1]

[1]STIKes Muhammadiyah Ciamis

Corresponding author: henrisetiawan@stikesmucis.ac.id

Abstract. Thalassemia is a genetic disease that causes various physical and psychosocial problems. Patients will experience psychosocial and emotional disorders in the form of feelings of depression, anxiety and other problems. Nurses as counselors need to provide genetic counseling interventions as a strategy to overcome these problems through information technology media such as android-based mobile applications. Android-based information technology is the most common and easy-to-use medium for nurses to provide health services such as genetic counseling. This research method used Research and Development design which aimed to design, produce and test the validity of the products that have been produced. The research phase used a qualitative approach to find out the problems and needs of the android application for the implementation of genetic counseling. Furthermore, the system development stage used the FAST (Framework for Application of Systems Technology) method which consists of preliminary investigation, problem analysis, requirements analysis, decision analysis, system design and construction. Testing of the Android-based Cyber Gene mobile application for the implementation of genetic counseling in thalassemia patients was carried out by testing all existing menus using secondary data. The result of this research is the Android-based Mobile Application "Cyber Gene" for Genetic Counseling Implementation among Thalassemia Patients. Nurses, thalassemia patients, and their caregivers can use this application for the comprehensive genetic counseling intervention process with instant messaging through the Live Consultation menu based on the predetermined procedure. Testing the system installed on android version 1.1 with the result that all menus and submenus on the system can be accessed and work well. Therefore, the Cyber Gene application can be used for the implementation of genetic counseling in thalassemia patients, both in hospitals and other health facilities.

1. Introduction

Thalassemia is a hereditary disease that occurs due to a lack of red blood cells in the human body so that haemoglobin production is reduced [1]. Although thalassemia is a genetic disease, this case is not difficult to find in Indonesia and even the world. Through its official website, the World Health Organization (WHO) states that 7% of the total world population are carriers of thalassemia traits. In addition, there are 300-400 thousand new births of thalassemia per year. Not much different, the case of thalassemia in Indonesia continues to increase until it ranks 5th in the world. In Indonesia, thalassemia is the most common genetic disorder and is the most common among hemolytic anemia groups [2]. The prevalence of thalassemia major in Indonesia based on UKK Hematology data from the Indonesian Pediatrician Association reached 9,121 people. Based on data from Yayasan Thalassemia

Indonesia/Persatuan Orangtua Penderita Thalassemia Indonesia (YTI / POPTI) it is known that people with thalassemia in Indonesia have increased from 4,896 thalassemia sufferers in 2012 to 9,028 people with thalassemia in 2018 [3].

Thalassemia results from changes in alpha or beta globin synthesis. Beta-thalassemia is caused by impaired production of beta globin chains, reduced globulin supply reduces the production of haemoglobin tetramers, thereby causing hypochromia and microcytosis [4]. The globin gene is part of a group of genes located on chromosome 11. Decreased production of the globin gene product, either -1 globin or -2 globin (the -globin gene is present in duplicate located on chromosome 16 to be exact at 16p13.3 either Both HBA1 and HBA2), produce a relative beta chain excess, which results in a less stable chain, resulting in a clinical disease known as alpha thalassemia [1].

Gene polymorphisms have an impact on intensive care in the long term [3]. Besides being able to cause severe clinical manifestations, polymorphisms can be passed on to the next generation with inheritance patterns that follow Mendelian laws, including autosomal recessive, autosomal dominant, X-linked recessive and X-linked dominant [5]. Thalassemia is inherited in an autosomal recessive manner that carries a risk of more than 25%. Clinical manifestations vary widely from asymptomatic to severe due to an excess of unstable alpha globin chains [6]. Beta thalassemia minor (BTM) is the term used for heterozygotes who have inherited a single gene leading to reduced beta globin production [7].

Thalassemia, not only causes physical problems, but also affects the patient's psychological condition. Physical changes or disorders that may arise include weakness, pale face, difficulty sleeping, and enlarged lymph nodes. So that it will have an impact on the emotional state of the client. Clients will experience psychosocial and emotional disorders in the form of feelings of depression, anxiety and other problems [8]. To overcome this problem, nurses need to provide nursing care with the main intervention of genetic counseling which has been proven to reduce depression and improve quality of life [9]. However, the times have opened up opportunities for the use of information technology as a medium for the delivery of genetic counseling interventions such as websites and mobile applications [10].

Research conducted by the PEW Research Center, (2018) describes that one of the most widely accessed media is a mobile device or cell-phone, therefore efforts to increase knowledge through an Android-based cell-phone application are quite effective in providing knowledge and also reducing the psychosocial impact on thalassemia patients [11]. One of the efforts to reduce the psychosocial impact of thalassemia children is to provide genetic counseling health services based on the Cyber Gen application.

This Cyber Gen application offers an intervention which includes thalassemia material, recurrent risk calculation, carrier screening test and decision making, clients can also consult at any time via the application. So, with this it can affect the client's psychosocial status for the better. With increasing information about diseases, especially genetic diseases experienced, the average individual feels calm because they already have certainty about the conditions experienced. In some developed countries such as Australia, the United States and the Netherlands, genetic counseling is carried out as a permanent procedure to treat psychological problems in parents who have children with genetic disorders, including thalassemia [12].

2. Materials and Method

This study used Research and Development (R and D) design, as a scientific way to design, produce and test the validity of the products that have been produced. The research phase used a qualitative approach to find out the problems and needs of the android application for the implementation of genetic counseling. Furthermore, the system development stage used the FAST (Framework for Application of Systems Technology) method [13]. At the research stage, the data collection method used was documentation, namely the data collection method carried out through reference books on Thalassemia which includes definitions, etiology, gene factors, inheritance patterns, diagnostic screening, medication and therapy, and adaptation. At this stage, an analysis of the need for the system to be developed was carried out, including the menus, functions, reports and the required interface. Meanwhile, the system development stage was the system development stage that will be carried out by researchers in full using the FAST (Framework for the Application of System Thinking) method depicted in figure 1.

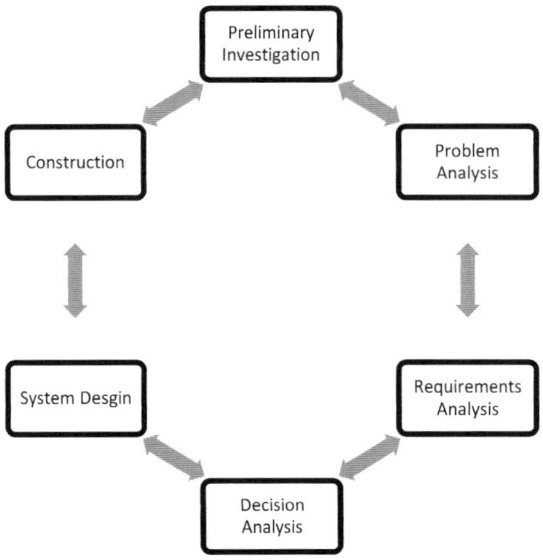

Figure 1. Framework for the Application of System Thinking

2.1. Preliminary Investigation
At this stage, researchers looked for references from reading books and scientific articles from online journals about thalassemia and research in similar fields as reference materials [14], [15]. The data and information become the basis for compiling information that will be displayed on the system.

2.2. Problem Analysis
At this stage the researchers studied and analyzed the problems found through the anamnesis of the thalassemia patient and the nurse on duty [16]. Some of the main problems that arise are the absence of genetic counseling services, patient knowledge about the disease is not comprehensive, genetic counseling has not become a standard procedure for implementing nursing care, and the knowledge and competence of nurses is not adequate.

2.3. Requirements Analysis
Needs analysis was done by means of a literature study about the needs and information that needs to be conveyed in genetic counseling [17]. From the problems that have been collected, it could be concluded that there was a need to create a mobile application that can be installed on the cell-phones of patients and caregivers as clients, and nurses and doctors as counselors.

2.4. Decision Analysis
At this stage the researcher carried out the process of identifying alternative systems in this case determining the system application design [13].

2.5. System Design
At this stage the researcher designed the application, created the necessary database and made menus and display interfaces on Android. Researchers divide the menu into 3 submenus which include basic knowledge, social support and genetic counseling. The design of the developed application can be seen in figure 2.

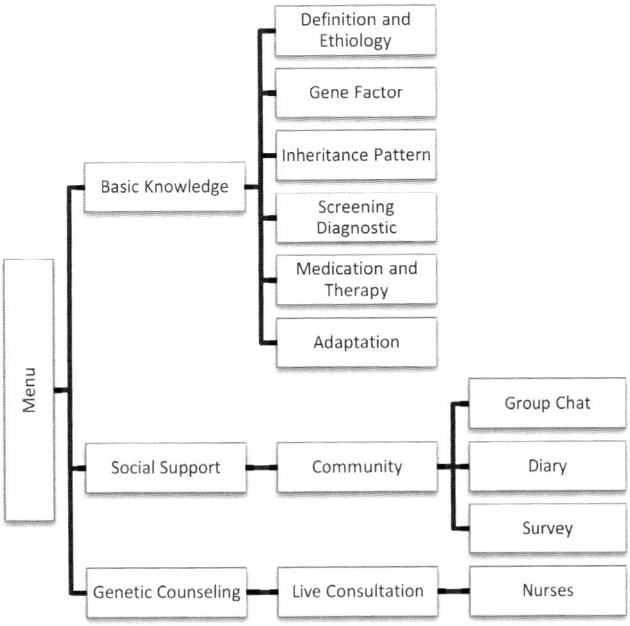

Figure 2. System Design

2.6. Construction

At this stage, the researcher translated the results of the design into a computer program using the MIT App Inventor application. After this stage, the next step is to test the implementation of the system that was built. At this stage, a functional test of the completed mobile application is carried out. The technique used in the system function test uses a closed list of questions.

3. Result and Discussion

The developed application in this study is called Cyber Gen which was tested by being installed on an android device version 1.1. Figure 3 is a display of the Cyber Gen application shortcut and register form. After installing the program, users can use this application by registering first. The registration form includes the user's identity, education and occupation. It is intended that each user can be classified into 2 groups, namely doctors and nurses as counselors, and patients and their caregivers as clients.

Six of the nine menus provided are references to thalassemia disease as basic knowledge that can be read by users or clients, both thalassemia patients and their caregivers. This basic knowledge includes definitions, causes or etiology, clinical manifestations, gene factors, inheritance patterns, screening diagnostics, medication and therapy, and adaptation. All information provided is the essence of various references to strengthen user knowledge about thalassemia.

Meanwhile Survey, diary and Group chat are facilities provided as a means to strengthen social support. In the survey menu, counselors (both doctors and nurses) can send questionnaires or questionnaires that measure the user's physical and psychosocial status. Diary submenu, provided to facilitate clients in expressing their feelings and experiences. The chatgroup submenu can be used as a means of interaction between users so that there is a fairly strong interpersonal relationship.

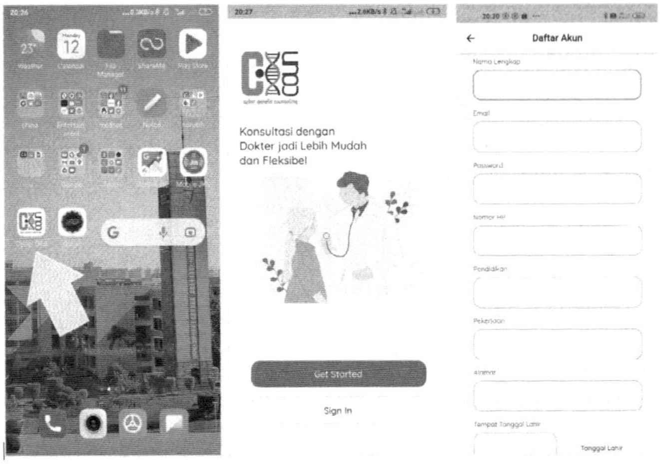

Figure 3. Application Shortcut and Register Form

In this section, users could register by filling in all available data fields including user name, email address, password, cellphone number, education, occupation, address and date of birth. This registration process is needed to make it easier for counselors to identify user data (patients and caregivers).

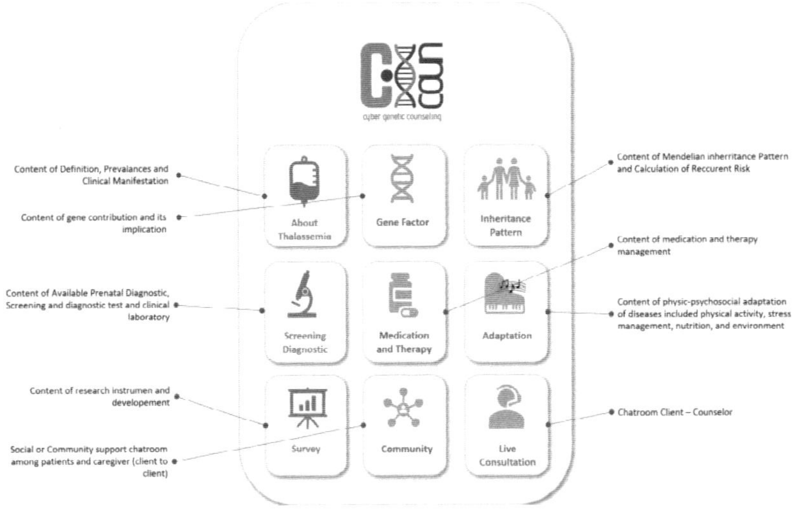

Figure 4. Application Menus

The front view of the application menu, shown in figure 4, is a menu of choices that can be used by patients as basic knowledge of Thalassemia which aims to increase user knowledge level. The available content includes "about thalassemia" which contains the definition, prevalence and epidemiology of thalassemia. While the "gene factor" menu contains the involvement of genes that influence the occurrence of thalassemia disease. In the "Inheritance pattern" menu, users can read content about the inheritance pattern of thalassemia based on Mendel's law and the calculation of recurrent risk that may appear in the next generation.

The "diagnostic screening" menu was provided to explain of the appropriate diagnostic tests to determine the diagnosis of thalassemia disease. This serves to strengthen the suspicion of carriers in groups prone to thalassemia, such as parents of the patient and other family members in the same pedigree. As for the "medication and therapy" menu, it provides information on therapeutic management that needs to be done in thalassemia patients. While the "adaptation" menu contains information about the future of thalassemia patients related to school, marriage, work to planning a pregnancy.

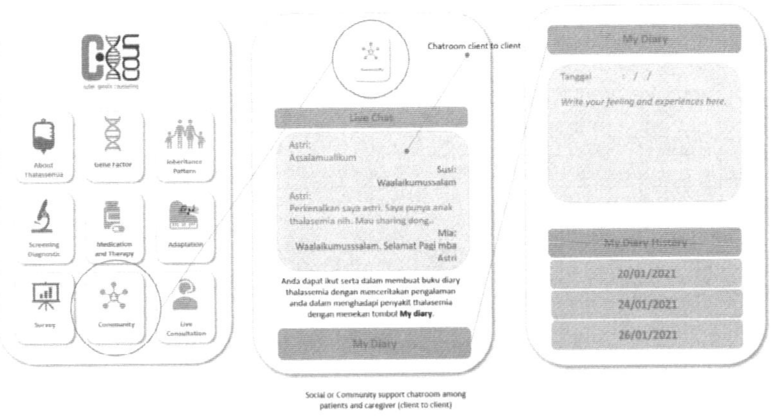

Figure 5. Social support menu

The "Survey" menu in figure 5 contains questions to measure the user's psychosocial status. While the "Community" menu is one of the facilities provided to strengthen social support between users (client-counselor, client-client, and caregiver-caregiver).

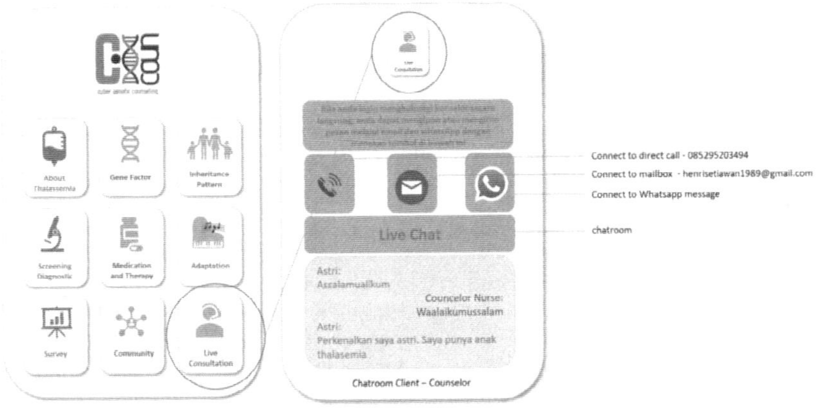

Figure 6. Live Consultation for genetic counseling implementation

Live consultation is the main interaction that can be used as a medium for the implementation of genetic counseling. All users, both clients and counselors can send messages directly and privately to discuss matters relating to the client's condition. Figure 4-6 are the main menu display, while the audio-visual guide manual for using the application has been uploaded to the YouTube which can be accessed by all users at the following link: https://youtu.be/jylsM5XBXBI

Testing of the Android-based Cyber Gene mobile application for the implementation of genetic counseling in thalassemia patients was carried out by testing all existing menus using secondary data. As a result, the whole system works fine. The three main menus (Basic Knowledge, Social Support, and Genetic Counseling Chat Room) and their derivative submenus work well. Table 1 shows the results of testing the menu on the application.

Table 1. System Test Results on All Menus and Submenus

No	Menu/Submenu	Result
1	Register Form/Login	Accessible and working well
2	Menu Utama	Accessible and working well
3	Basic Knowledge	Accessible and working well
4	About Thalassemia	Accessible and working well
5	Gene Factor	Accessible and working well
6	Inheritance Pattern	Accessible and working well
7	Screening Diagnostic	Accessible and working well
8	Medication and Therapy	Accessible and working well
9	Adaptation	Accessible and working well
10	Social Support	Accessible and working well
11	Survey	Can Save Data
12	Group Chat Room	Interactive
13	My Diary	Can Save Data
14	Live Consultation	Can Perform Inter-user Interaction

The test results of the developed system function show that all the features on the system can function and work well. All submenus on the basic knowledge feature can be accessed and read by the user as a source of scientific information to increase user knowledge [18]. Similar applications developed in Indonesia, such as Alodokter, provide basic knowledge about a disease. Although the variety of diseases described in the application is greater, the information provided does not focus on one disease in detail and in depth. That is why, Alodokter provides consulting services about a disease for users who do not fully understand it [19].

The Cyber Gene application is not only aimed at increasing the knowledge of thalassemia patients about the disease they are suffering from, but can adapt to all the implications of the gene problems they are experiencing. For this reason, Cyber Gen focuses more on the implementation of genetic counseling which is one of the nursing interventions for thalassemia sufferers [20]. That is why Cyber Gen adds another feature in the social support or community support menu. The forms of social support provided are chat groups and diaries. Social support refers to the psychological and material resources provided by a social network to help individuals cope with stress [1], [21]. Such social support may come in different forms, and might involve helping a person with various daily tasks when they are ill or offering financial assistance when they are in need, giving advice to a friend when they are facing a difficult situation and providing caring, empathy, and concern for loved ones in need.

The research, published in the September issue of APA's Journal of Experimental Psychology: General (JEP: General) (Vol. 130, No. 3), indicates that expressive writing reduces intrusive and avoidant thoughts about negative events and improves working memory. These improvements, researchers believe, may in turn free up our cognitive resources for other mental activities, including our ability to cope more effectively with stress [22], [23]. This is the difference between the Cyber Gen application and other similar applications. However, the application developed in this study is not integrated with the web, only using a mobile device. So, this is a weakness in this study, and an opportunity for other researchers to continue.

4. Conclusion

The Cyber Gen application for the implementation of genetic counseling in thalassemia patients can run well and testing on menus and sub menus works well. Nurses, thalassemia patients, and their caregivers can use this application for the comprehensive genetic counseling intervention process with instant messaging through the Live Consultation menu based on the predetermined procedure. The resulting database can be accessed by the user. Application construction has not been integrated with the website, so it needs to be developed in further research. The potential for abuse of this application can be controlled with a neat and orderly mechanism. Counselor and Client use applications based on face to face appointments made during routine therapy at the hospital. Thus, all application users are personal registered at the hospital.

Acknowledgments
We would like to thank the Ministry of Education and Culture of the Republic of Indonesia and the National Research and Innovation Agency for funding this research.

Conflict of Interest
Author declare there is no conflict of interest.

References
[1] A. Maheri *et al.*, "Depression, Anxiety, and Perceived Social Support among Adults with Beta-Thalassemia Major: Cross-Sectional Study.," *Korean J. Fam. Med.*, vol. 39, no. 2, pp. 101–107, Mar. 2018.
[2] M. M. Djer, S. L. Anggriawan, D. Gatot, P. Amalia, S. Sastroasmoro, and P. Widjaja, "Correlation between T2* cardiovascular magnetic resonance with left ventricular function and mass in adolescent and adult major thalassemia patients with iron overload," *Acta Med Indones*, vol. 45, no. 4, pp. 295–301, 2013.
[3] H. Setiawan, A. Ediati, and T. I. Winarni, "Genetic Counseling to Reduce the Level of Depression in Parents of Children with Thalassemia Major," *2nd Int. Conf. Sport Sci. Heal. Physiscal Educ.*, no. Icsshpe, pp. 102–106, 2017.
[4] C. Huda and A. Indrayati, "Molecular Detection of Exon 3 of Beta Globin Gene from Thalassemia Beta Major Patients in RSUD DR. Soeroto Ngawi using Polymerase Chain Reaction – Single Strand Conformation Polimorfism Method," *Farm. Indones.*, vol. 16, no. 1, pp. 24–33, 2019.
[5] F. Danjou, F. Anni, and R. Galanello, "Beta-thalassemia : from genotype to phenotype," *hematologica*, vol. 96, no. 11, pp. 1573–1575, 2011.
[6] E. A. Rachmilewitz, "Pathophysiology and treatment of patients with beta-thalassemia – an update [version 1 ; referees : 2 approved] Eitan Fibach Referee Status :," *PubMade*, vol. 6, no. 0, pp. 1–12, 2017.
[7] Galanello and Raffaella, "Beta-thalassemia," *orphanet J. pare Dis.*, vol. 5, no. 11, pp. 1–15, 2010.
[8] F. Mohamadia, M. Bagheri, M. S. Hashemi, and H. K. Sani, "The Effects of Cognitive Behavioral Therapy on Depression and Anxiety among Patients with Thalassemia : a Randomized Controlled Trial," *J. Caring Sci.*, vol. 7, no. 4, pp. 219–224, 2018.
[9] H. Setiawan, E. Roslianti, and A. Firmansyah, "Theory Development of Genetic Counseling among Patient with Genetic Diseases," *Int. J. Nurs. Sci. Heal. Serv.*, vol. 3, no. 6, pp. 709–715, 2020.
[10] H. Setiawan, R. N. Khaerunnisa, H. Ariyanto, and F. A. Firdaus, "Telenursing meningkatkan kualitas hidup pasien dengan penyakit kronis," vol. 3, no. 2, pp. 95–104, 2020.
[11] Monica Anderson and Jingjing Jiang, "Teens, Social Media & Technology 2018," *pew Res. Cent.*, vol. 31, pp. 1–10, 2018.
[12] C. I. Voils *et al.*, "Does Type 2 Diabetes Genetic Testing and Counseling Reduce Modifiable Risk Factors ? A Randomized Controlled Trial of Veterans," *JGIM*, pp. 1591–1598, 2015.
[13] M. Azmi, Y. Sonatha, E. Asri, Rasyidah, and D. S. Putra, "Implementing fast method on the development of object-oriented cooperative information systems," *Int. J. Informatics Vis.*, vol. 2, no. 4–2, pp. 366–371, 2018.
[14] A. Ahmar and A. Rahman, "Development of teaching material using an Android," *Glob. J. Eng. Educ.*, vol. 19, no. 1, 2017.
[15] B. Panda, S. C. Paul, L. J. Hui, Y. W. D. Tay, and M. J. Tan, "Additive manufacturing of geopolymer for sustainable built environment," *J. Clean. Prod.*, vol. 167, pp. 281–288, 2017.
[16] A. O. Sari and E. Nuari, "Rancang Bangun Sistem Informasi Persediaan Barang Berbasis Web

Dengan Metode Fast (Framework For The Applications)," *J. PILAR Nusa Mandiri*, vol. 13, no. 2, pp. 261–266, 2017.

[17] R. I. Vogel, K. Niendorf, H. Lee, S. Petzel, H. Y. Lee, and M. A. Geller, "A qualitative study of barriers to genetic counseling and potential for mobile technology education among women with ovarian cancer," *Hered. Cancer Clin. Pract.*, vol. 16, no. 1, pp. 1–7, 2018.

[18] K. T. Martono and O. D. Nurhayati, "Implementation of android based mobile Learning application as a flexible learning Media," *Int. J. Comput. Sci. Issues*, vol. 11, no. 3, p. 168, 2014.

[19] T. Fidowaty, E. Engwin, A. A. Nosprosanti, and M. Aisyah, "Development of E-Health Information Technology During The Covid-19," *Int. J. Educ. Inf. Technol. Others*, vol. 4, no. 2, pp. 282–286, 2021.

[20] P. Hudson, K. Zajo, C. A. Gerhardt, J. Stanek, and E. Varga, "Defining the role of a genetic counselor within pediatric hematology and oncology comprehensive care teams: Perspectives of the provider team and patients," *J. Genet. Couns.*, vol. 28, no. 6, pp. 1139–1147, 2019.

[21] H. Setiawan, R. Nantia Khaerunnisa, H. Ariyanto, A. Fitriani, F. Anisa Firdaus, and D. Nugraha, "Yoga Meningkatkan Kualitas Hidup Pada Pasien Kanker: Literature Review," *J. Holist. Nurs. Sci.*, vol. 8, no. 1, pp. 75–88, 2021.

[22] J. M. Smyth, J. A. Johnson, B. J. Auer, E. Lehman, G. Talamo, and C. N. Sciamanna, "Online positive affect journaling in the improvement of mental distress and well-being in general medical patients with elevated anxiety symptoms: A preliminary randomized controlled trial," *JMIR Ment. Heal.*, vol. 5, no. 4, p. e11290, 2018.

[23] A. Firmansyah, H. Setiawan, and H. Ariyanto, "Studi Kasus Implementasi Evidence-Based Nursing: Water Tepid Sponge Bath Untuk Menurunkan Demam Pasien Tifoid," *Viva Med. J. Kesehatan, Kebidanan dan Keperawatan*, vol. 14, no. 02, pp. 174–181, 2021.

Redesign The Electricity System of PT Barata Indonesia To Reduce Annual Power Loss

[1]Muhamad Ali, [2]Zaenal Arifin

[1] Electrical Engineering Department Faculty of Engineering, Universitas Negeri Yogyakarta, Indonesia
[2] Electrical Engineering Staff of Power Generation Division PT. Barata Indonesia

Email: [1]muhal@uny.ac.id, [2]zaenal.arifin@barata.id

Abstract. PT Barata is one of Indonesia's leading turbine component manufacturers, requiring a large amount of electricity for production. Along with the increase in production and factory expansion, the need for electricity continues to increase so that it continues to add power to reach 3780 KVA. One of the weaknesses of the electric power system at PT Barata is using six (6) 20kV/400 V, 630 kVA transformer units. This system produces a no-load or core loss of 8150 Watts. In one year, the power loss reached 71.394 kWh at the cost of IDR 118,514,040.00. Power loss is permanent and lasts all the time during the use of the Transformer. This article aims to redesign the electrical system at PT Barata using a single 4000 kVA transformer. The redesign of the electric power system at PT Barata was carried out by 1. Observing and measuring the electric power system at PT Barata, 2. calculating the resulting losses, 3. conducting studies and designs, 4 comparing the ratio of the total power loss of the electric power system currently with a new design, 5. Propose a redesign to management for implementation. Calculation and analysis of the new design with a 4000 kVA transformer obtained a core loss of 4450 Watt at the cost of Rp. 64,905.030.00, resulting in a savings of IDR. 53,965,980.00 per year. The new design will save 45% in operating costs compared to the previous six transformer units.

Keywords: electrical technician, adaptive, test, competency

1. Introduction

PT Barata is one of the leading Turbine Component factories in Indonesia, previously known as PT Siemens Indonesia Cilegon Factory. Along with the development of world demand for generator components, the company gradually increased its production capacity. This condition forced the company to increase its investment in the form of large-scale production machines. The consequence of growing production machines is the addition of electrical power with the appropriate capacity. Initially, the electricity demand plan did not pay attention to the need for investment in devices in the future, which must be supplied with electrical energy with adequate transformer infrastructure. The existing condition is that every new machine investment must install a new transformer. Until now, there were 6 (six) distribution transformer units (6x630 kVA) powerhouses operating.

As an electrical power supply, six transformer units impact operational costs due to transformer core losses of 71.394 kWh per year. For that, the company has to pay for electricity of IDR 118,514,040.00 per year. Core losses are permanent and will last throughout the life of the transformer [1], [2]. The cost of this power loss is invisible because it has been counted as an integral part of the electricity bill [2], [3].

Currently, the turbine component factory of PT Barata Indonesia receives electrical energy from PT KDL with a capacity of 3558 kVA. The company's energy user list notes that if all loads operate, the power required is 2819 kVA. The factory only consumes a maximum of 79% of the existing capacity. With the use of a 4000 kVA power transformer with the same load, the Transformer only bears an electrical burden of 70% of its total capacity. In reality, it has never happened that all electrical loads operate simultaneously.

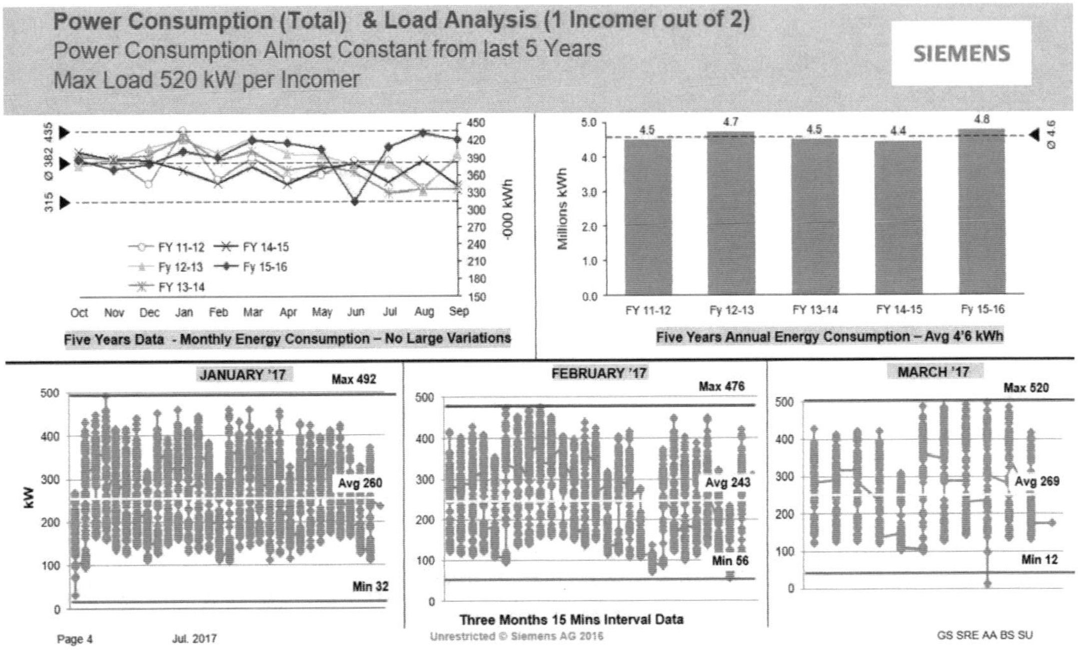

Figure 1. Loading Profile of PT Barata Indonesia

Until now, there has not been found a transformer that can work without loss. In general, the transformer core material is made of 0.27 mm Grain Oriented Silicon Steel M4, designed so that iron losses are low. The losses in the Transformer are divided into two parts, namely losses when the Transformer works without load (P-Fe) and losses when the Transformer works with loading (P-Cu). No-load losses arise because the primary circuit is given a voltage, and the secondary circuit is open [2], [4].

One of the efforts to reduce power loss is to choose the type of Transformer with the lowest possible core loss characteristics. Transformer losses relate to core and copper losses directly proportional to loading [4, 6]. Based on the technical specification data of the Transformer, the greater the power capacity of the Transformer, the greater the power consumption [5]. The total capacity of six 630 kVA transformer units is equal to 3780kVA, equivalent to a single transformer produced by a 4000 kVA manufacturer.

The amount of power consumption in using a 6-unit transformer is still more significant than a transformer with equivalent power. The 630 kVA transformer losses core PFe = 1300 Watt per Transformer, the total core loss is 6 x 1300 Watt or 7800 Watt, while the 3500 kVA transformer losses 3700 Watt, and the 4000 kVA transformer loses its core 4450 Watt.

In order not to deviate from the subject matter that has been determined, it is necessary to convey that what is meant by power loss in this article is core loss, and then we will limit the problems and objectives as follows:

- Identify the amount of power loss in the operation of a 6x630 kVA power transformer with a voltage of 20 kV/0.4 kV.
- Propose a redesign of the powerhouse electrical system using a single transformer to minimize power losses.

2. Method

The solution to the problem of how to minimize annual power losses in the powerhouse system of PT Barata Indonesia is to redesign the powerhouse system, namely by replacing the operation of six 6 x 630 kVA transformer units with a single transformer with an equivalent power of 4000 kVA, with the following stages.

2.1. Analyze and study the current (existing) condition of the Power House

There are six distribution transformers installed in the powerhouse of PT Barata Indonesia (Persero) (6x630 kVA) with a total installed capacity of 3780 kVA. The primary source of medium voltage is supplied through two feeders with two meters. The location occupies three powerhouse rooms, Power House 1, Powerhouse 2, and Powerhouse 3. Each Transformer is distributed through a low voltage panel (LVMDB), and each Transformer is equipped with an LVMDB panel so that in the powerhouse, six units, LVMDB 1 to LVMDB, are installed. LVMDB 6.

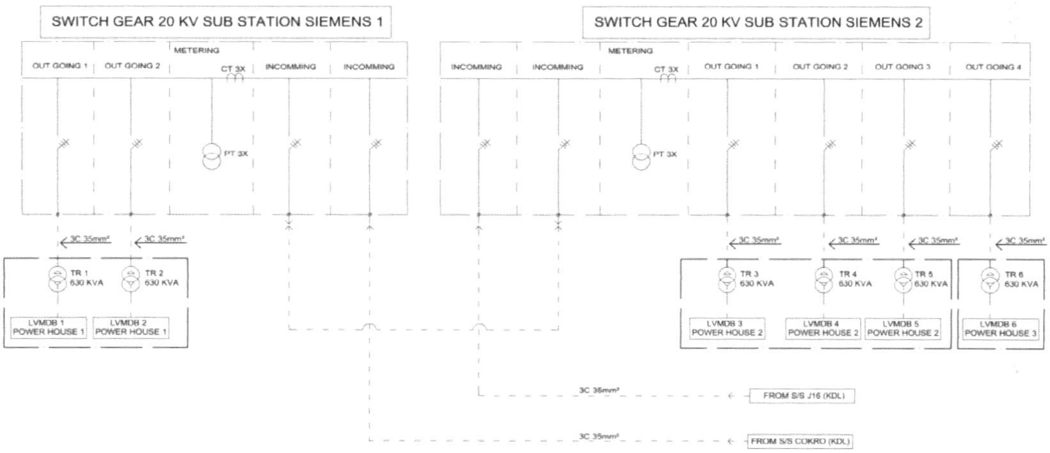

Figure 2. Single Line Diagram of Existing Powerhouse

Technical specifications of dry type transformer with Cast-Resin insulation Geafol brand with a power of 630 kVA. Core power dissipation (core loss) 1650-Watt, copper loss 6800-Watt, and a total of 8450-Watt at full load [7], [12].

Table 1. Dry Type Transformer Technical Specification Data

Rated power	Rated primary voltage1	Rated secondary voltage	Insulation level HV	Insulation level LV	Impedance voltage	No-load losses	Load losses at 120 °C	Noise level	Total weight
kVA	Ur HV kV	Ur LV kV	kV	kV	uzr %	Po W	Pk120 W	LWA dB	kg
630	20	0.4	50/ 95	3/–	6	1650	6800	70	1750

The technical specifications for transformers 2 to 6 are identical, transformer type Oil Immersed Transformer, oil-cooled transformer Trafindo with a power of 630 kVA, core power dissipation (core loss) 1300-Watt, copper loss 6500-Watt and a total of 7800-Watt at full load [8], [11], [12].

Table 2. Oil Immersed Transformer Technical Specification Data

Capacity	**Dimension**			**No Load Losses**	**Load Losses (70°)**	**Total Losses**	**The Impedance at (70°)**	**Exciting Current**	**Oil Volume**	**Total Weight**
kVA	Length	Width	Height	Watt	Watt	Watt	%	%	Liters	kgs
630	1680	960	1365	1300	6500	7800	4	1.8	510	2150

2.2. System design using 4000 kVA Transformer

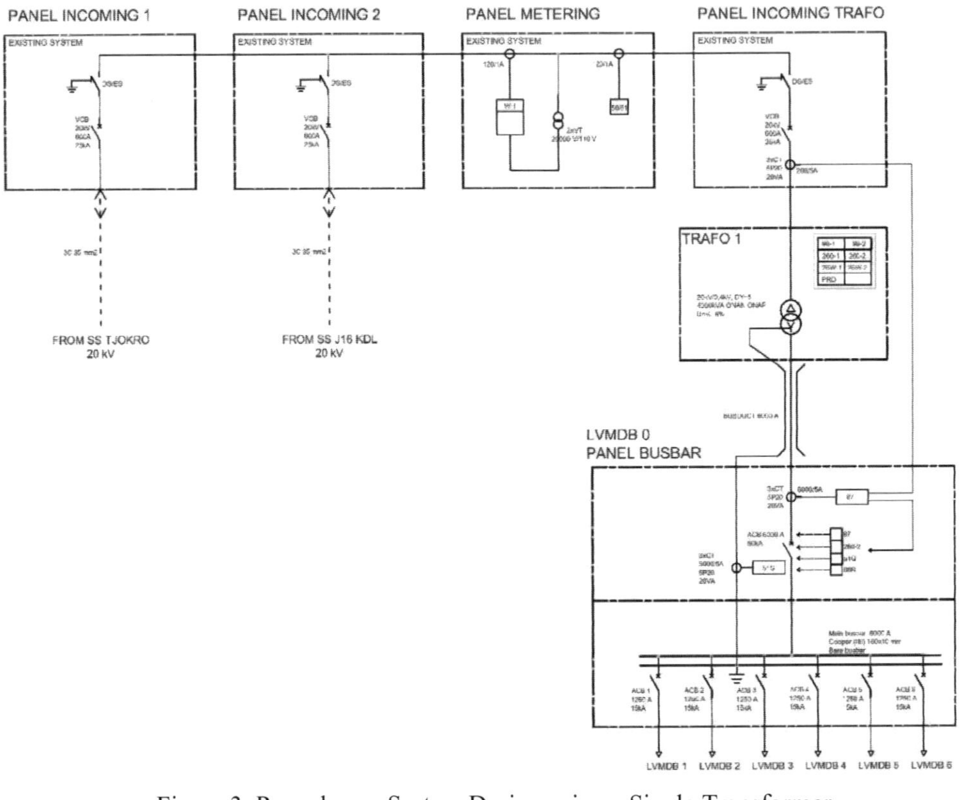

Figure 3. Powerhouse System Design using a Single Transformer

The design of PT Barata's electric power system with a single 4000 kVA transformer requires changes to the layout and low voltage distribution system on the secondary side of the Transformer so that not all existing systems have to be changed. LVMDB 1 to LVMDB 6 and all MDBs do not require changes. All LVMDB panels are repositioned. On the medium voltage side, it is still supplied from two feeders and a consumption metering system (energy metering) into one metering system. Changes were only made to the transformer control system, which used six units to become one unit. The addition of one LVMDB 0 busbar panel unit is needed because it functions as a low voltage distribution control panel to serve each LVMDB

3. Results and Discussion

3.1. Technical Analysis

The calculation and analysis results compare the power loss due to annual core loss from a six-unit 630 kVA transformer, with an equivalent power transformer of 3500 kVA and 4000 kVA.

Table 3. Comparison of Loss Calculation Results on Three Transformer Systems

Transformers Operation	Core Losses (kWH)	Power Losses Cost	Saving
6 Unit x 630 kVA	71,394.00	IDR 118,514,040	0%
1 Unit x 4000 kVA	38,982.00	IDR 64,710,120	45%

The results of the redesign of the supply system from medium voltage to low voltage of 400 Volts and their distribution in a single line diagram of the redesign are as follows.

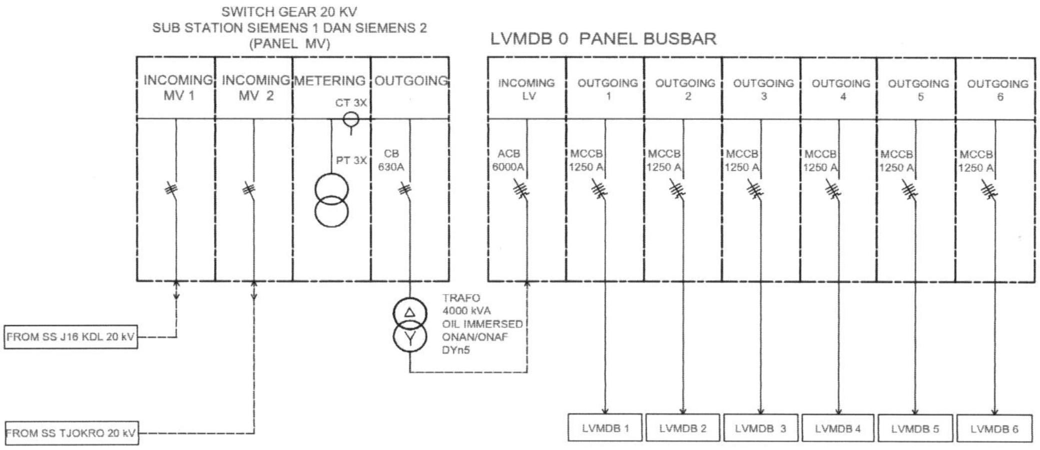

Figure 4. Redesign of Single Line Diagram Power House

The redesign of the powerhouse electrical system does not change the existing system as a whole. The low-voltage distribution panels (LVMDB 1 to LVMDB 6) and the entire MDB require no alterations. This panel is only being relocated according to the layout plan. The medium voltage panel is supplied by two feeders, a change to the discharging system to a single meter will limit power consumption to 3550 kVA (replacing two 2520 kVA and 1038 kVA metering). The transformer control system underwent a change from initially using six breaker units in two MV cubicles (Switchgear 20 kV) to one MV cubicle unit. The redesign requires adding one LVMDB busbar panel unit that functions as a 400-volt low voltage main divider panel output from the Transformer to be distributed to each LVMDB 1 to LVMDB 6. The topology of the powerhouse electrical system using a single 4000 kVA transformer is described in the following.

The technical advantage of using a single 4000 kVA transformer is that the fault localization system can be carried out with only one branch circuit breaker termination. Transformer maintenance work becomes lighter with fewer equipment requirements. The only drawback is that there is no backup. The new system only relies on one transformer unit as the primary source of electrical energy so that in the event of a transformer failure, blackouts in all load locations will occur. For that, we need a reliable protection system and limiting the Transformer's operation to 80% of its nominal power so that the Transformer works in a safe loading area. The operating limit of PT Barata Indonesia's turbine component generator transformer is 80%. The operating conditions of 80% of the transformers can meet the electrical power needs of all the equipment used. The company only uses less than 50% of the existing capacity based on the loading profile data. The Transformer only works at 0.5 x 80% of the nominal power or operates at 40% of the transformer power. These operating conditions can increase the lifetime of the Transformer.

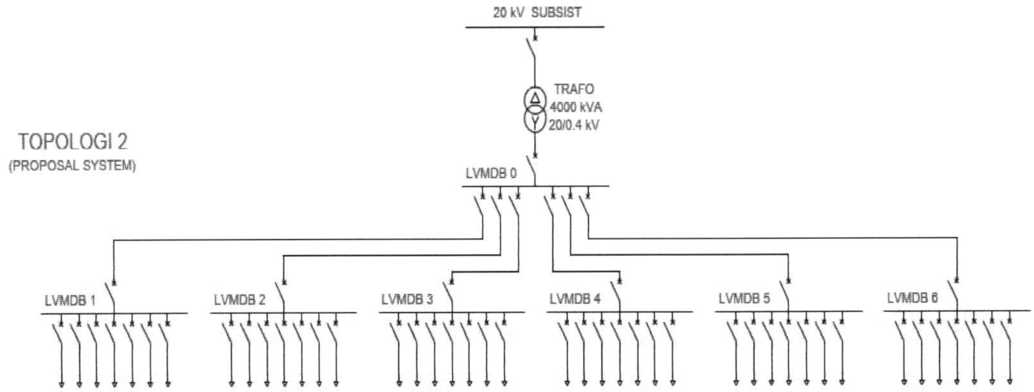

Figure 5. Topology of the Redesigned Powerhouse System

The technical constraint is that the redesign of the powerhouse uses a 4000 kVA transformer when the maximum loading on the low voltage side (secondary) will flow 5774 Amperes, so that the distribution line must use a busbar or bus duct with a capacity of 6000 A. The use of a busbar or bus duct requires that the busbar panel (LVMDB 0) be in a position adjacent to the Transformer. This condition can be overcome by placing the LVMDB 0 busbar panel close to the Transformer's low voltage (secondary) side so that the bus duct installation is more efficient than the old design. The redesigned powerhouse layout can be seen in Figure 6.

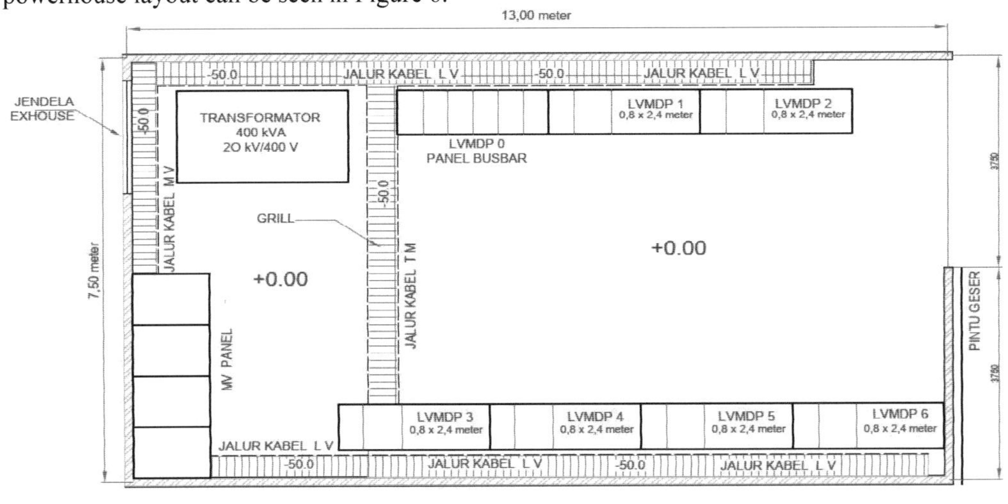

Figure 6. Layout Plan of the Redesigned Power House

Two alternatives are using a single transformer in the redesign of the powerhouse of PT Barata Indonesia, namely the use of a single 3500 kVA or 4000 kVA transformer. The Calculation of losses and savings is as follows.

Table 4. Oil Immersed Transformer Technical Specification Data

Capacity	Dimension			No Load Losses	Load Losses (70°)	Total Losses	The Impedance at (70°)	Exciting Current	Oil Volume	Total Weight
kVA	Length	Width	Height	Watt	Watt	Watt	%	%	Liters	kgs
630	1680	960	1365	1300	6500	7800	4	1.8	510	2150
4000	2730	1635	2530	4450	44000	48450	7.5	1.8	2830	9240

The technical advantage of using a single 4000 kVA transformer is that the fault localization system can be carried out with only one branch circuit breaker termination—less maintenance of transformers and equipment. The only drawback is that there is no backup. This system only relies on one transformer unit as a source.

The electric power system must have reliable protection to overcome disturbances and avoid damage [10]. In addition, the loading on the Transformer needs to be limited to a maximum of 80% of its nominal power [10]. Transformer operations of 80% can still meet the needs of all electrical equipment in the factory. For the last two years (2019-2020), the electricity demand has been less than 50% of the existing capacity. Thus, the Transformer only works at 0.5 x 80% of its nominal power or 40% of the transformer power. The low load condition of the Transformer is very beneficial for the life of the Transformer because the Transformer does not experience excessive heat.

4. Cost Estimation

The financing structure of the design powerhouse uses a single 4000 kVA transformer in outline. It can be described as follows in table 5.

Tabel 5. Cost Struktur Redesign System Electricity Powerhouse

No.	type of work	Qty	Unit	Material (IDR)	Equipment (IDR)	Service (IDR)	Sub Total (IDR)
1.	Procurement	1	Unit	954.314.000	-	-	954.314.000
a.	Procurement of transformer 4000 kVA	1	Unit	954.314.000	-	-	954.314.000
b.	Protection	1	Unit	500.000.000	-	-	500.000.000
c.	Busbar panel, ACB, and Instrumentation	1	Unit	250.000.000	-	-	250.000.000
d.	Busduct/Busbar	1	Unit	50.000.000			50.000.000
2.	Dismantling	1	Unit	-	68.000.000		68.000.000
3.	Powerhouse repair	1	Unit	31.000.000			31.000.000
4.	Installation work	1	Unit		64.000.000		64.000.000
5.	Testing, Commissioning, and operation worthy certificate (SLO)	1	Unit			55.906.000	
			Total				1.973.220.800

5. Return Investment (ROI) and Break-Even Point (BEP)

An investment needs to be analyzed based on ROI and BEP to determine its feasibility. An investment needs to be analyzed based on ROI and BEP to determine its feasibility. An investment is worth funding if the ROI value is positive, providing profits and a fast BEP.

$$\text{Return on investment (ROI)} = \frac{(\text{Ni+PROFIT})-\text{Ni}}{\text{Ni}} \text{ x } 100 = \frac{53.97 \; \textit{Milliom}}{1.97 \text{ Billion}} \text{x } 100\% = 3\%$$

$$\text{Payback Periode} = \frac{\text{Investation}}{\text{Savings}} = \frac{1.97 \text{ Billion}}{53.97 \text{ Million}} = 36{,}5 \text{ years}$$

ROI of 3% means the return on investment has a gain of 3% of the investment value for a year from power loss savings compared to the operation of the previous six transformer units. Meanwhile, the payback period of 36.5 years still seems too long. This Calculation is only carried out on the efficiency aspect of transformer core losses. The BEP value is much shorter if we consider maintenance savings, service utilization, and production loss costs due to power failures.

6. Conclusion

One of the actions that can increase the efficiency of PT Barata Indonesia powerhouse is to redesign it to replace 6 630 kVA transformer with a single transformer with an enormous power equivalent to 4000 kVA. The redesign of the powerhouse electrical system uses a single 4000 kVA transformer, making power losses due to transformer core losses decreased from 71.394 kWh to 38.982 kWh and savings of IDR 53,965,980.00 per year, 45% lower than the use of the previous 6-unit Transformer.

References

[1] Juan C, Olivares-Galvan, Pavlos S, Giorgilakis & Rodrigo, 2009, A Review of Transformer Losses, Electric Power Components and Systems, Volume 37, 2009 - Issue 9

[2] Adriano Galindo Leal; JosÉ Antonio Jardini; Luiz Carlos Magrini, 2009, Distribution Transformer Losses Evaluation: A New Analytical Methodology and Artificial Neural Network Approach, IEEE Transactions on Power Systems publication information

[3] Vladimir I. Biryulin; Alexey N. Gorlov; Oleg M. Larin; Daria V. Kudelina, 2016, Calculation of power losses in the transformer substation, 2016 13th International Scientific-Technical Conference on Actual Problems of Electronic Instrument Engineering (APEIE)

[4] Quan You Yue, Can-bing Li; Yijia Cao, Yuqing He, Binglin Cai, Qiuwei Wu, Bin Zhou, 2018, Comprehensive Power Losses Model for Electronic Power Transformer, IEEE, VOLUME 6, 2018

[5] A. A. A. El-Ela, R. A. El-Sehiemy, and A. S. Abbas, "Optimal Placement and Sizing of Distributed Generation and Capacitor Banks in Distribution Systems Using Water Cycle Algorithm," pp. 1–8, 2018.

[6] Abdul Kadir. (2010). Transformator. Jakarta. Penerbit Universitas Indonesia (UI-Press).

[7] Catalog TV1. (2013). GEAFOL Cast-Resin Transformers 100 to 16,000 kVA. URL:https://www.simosynergy.com/download/Catalog_TV1_GEAFOL_Castresin_Transformers.pdf

[8] Centrado Prima Int'l, PT. (Januari 2010). Product Catalog Centrado Distribution&Power Transformers. URL: https://centrado.co.id/images/download/brosur-trafo-distribusi-centrado-indonesia.pdf

[9] http://www.ee.ui.ac.id/online/semtafull/20110711143745-sm7044-tp4-BensonMarn-Jurnalp.pdf

[10] Sanjaya, P.I.G.A., Arjana, D.I.G., & Rina, I.W. (Juli 2017). Analisis Setting Pengaman Transformator Daya di Gardu Induk Nusa Dua Jika Terjadi Gangguan Hubung Singkat Satu Phasa Tanah. Jurnal Spektrum. 4(1), 28-35. https://ojs.unud.ac.id/index.php/spektrum/article/view/31509

[11] Trafoindo Prima Perkasa, PT. (2010), Catalog tp Trafindo Instrument transformer up to Um 24 kV for Indoor and outdoor Installation.

[12] Trafoindo Prima Perkasa, PT. (Juni 2021). Product Catalog Oil Immersed Transformer. URL. https://www.trafoindonesia.com/pdf/all-product-catalog/Trafoindo%20catalogue%20oil%20immersed%20transformers.pdf.

Mobile Augmented Reality Application for Component Identification

Y. I. Hatmojo[1], T. H. T. Maryadi[2], R. Badarudin[3], B. Indrawati[4]

[1,2,3,4] Dept. of Electrical Engineering Education, Faculty of Engineering, Universitas Negeri Yogyakarta, Indonesia

E-mail: [1]yuwono_indro76@uny.ac.id; [2]totokheru@uny.ac.id; [3]rohjai.badarudin@uny.ac.id; [4]bettyindrawati.2017@student.uny.ac.id

Abstract. This article discusses the development of augmented reality (AR) based object identification applications. The purpose of this study is to improve the distribution station component identification workbook. Through the implementation of the developed application, it is assumed that it will be easier to understand the functionality of the components of the distributing station. The augmented reality application was developed for the android platform. The result showed the application of Distributing Station-AR has been developed by applying a marker-based tracking method. The application test has successfully displayed the three-dimension components of distributing station using a marker on the component identification workbook of distributing station. Product performance is known to be very good and product validation is also very good, including material validation, media validation, and first user responses. For further development, this application is expected to add animation of the working principle of AR-based components.

1. Introduction

Augmented Reality (AR) is a technology that combines two and/or three-dimensional objects into a virtual environment in real time [1][2]. AR is a technology that can display a virtual object contained in a virtual environment to the real world [3]. The AR concept combines computer vision, image processing, and computer graphics to form a new paradigm of interaction between humans and computers [4][5]. AR is classified into two types, namely Geographic location-based AR and Computer vision-based AR. Computer vision-based AR is classified into two types, namely Marker-based AR and Marker-less AR [6]. In this paper, we apply the concept of Marker-based AR as the type of AR technology adopted. Marker Based Tracking will distinguish between existing markers and objects that are outside the marker or other objects in the real world. Position and orientation are considered in markers, where multiple positions and content will result in multiple markers. The implementation of many markers is very visible when capturing marker data on 3D objects. The marker will be divided into black border and white background as an illustration at the implementation stage [7].

AR applications can be an alternative interactive learning media. In the era of distance learning or online learning, it is necessary to renew learning media as a source of learning for students. Learning materials are complex, abstract, and difficult to understand, with AR technology to make it easier for students to understand the material [8][9]. Currently, smartphones are the most widely used devices by the community, including students. Distance learning requires every student to have a smartphone to participate in distance learning, so this condition is very supportive in the application of AR-based learning media. The purpose of this study is to introduce AR applications as interactive learning media. This application complements the learner's workbook for machine component identification with additional information displayed in the virtual world. This application was developed for the Android platform and the first version has been tested to material validation, media validation, and first user validation.

2. Methods

This study is development research by adopting the ADDIE development model [10]. ADDIE stands for Analysis, Design, Development, Implementation, and Evaluation. Figure 1 is a model of the development stages.

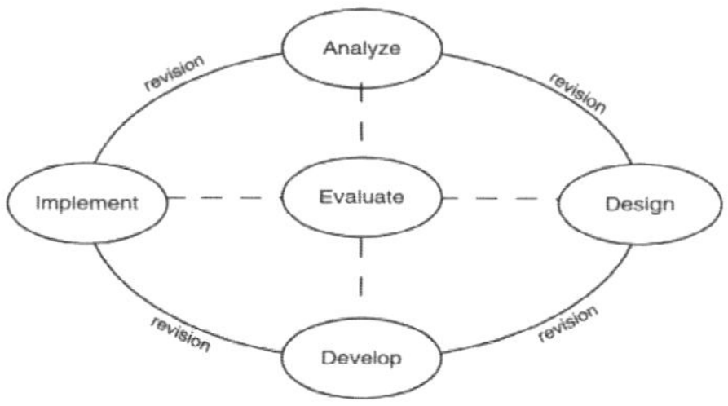

Figure 1. The ADDIE model [10]

The analysis phase aims to determine the current conditions and conditions that should be expected, to set goals and determine priorities for action. During the covid-19 pandemic, in the online learning process problems were found, namely students had difficulty understanding machine components caused by the limitations of the media used to visualize component images online. So that in the analysis stage it has been determined to identify AR-based components. The design phase aims to establish the expected performance and the appropriate testing strategy. Software requirements, user interface, and flow charts are also developed at this stage. Software requirements are described separately in the sub-chapters of this paper. Storyboards are used to describe the initial visualization of the user interface on the screen. The user interface is shown in Figure 2 up to Figure 7. Flowcharts are used to simplify the development process. The flow chart is shown in Figure 8. The development stage aims to produce and validate the selected learning product. The application developed is an application with AR technology that can project a two-dimensional image marker into three-dimensional form. A two-dimensional image is projected from the workbook. The implementation stage is to prepare a learning environment by involving students. This study discusses the steps of product development to the step of testing the performance of the product. The implementation stage has not been carried out in this study. The evaluation stage is to assess product quality and evaluate the learning process, both before and after implementation. The first product evaluation is done by black box test. Black box test is performed to assess product performance. Then continued validation by experts in the field of software engineering and experts in the field of pneumatic machine component materials, as well as responses from potential users to the developed AR products.

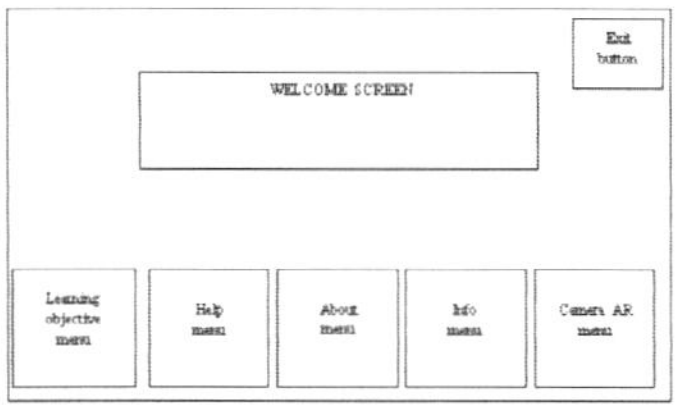

Figure 2. User interface of main menu page

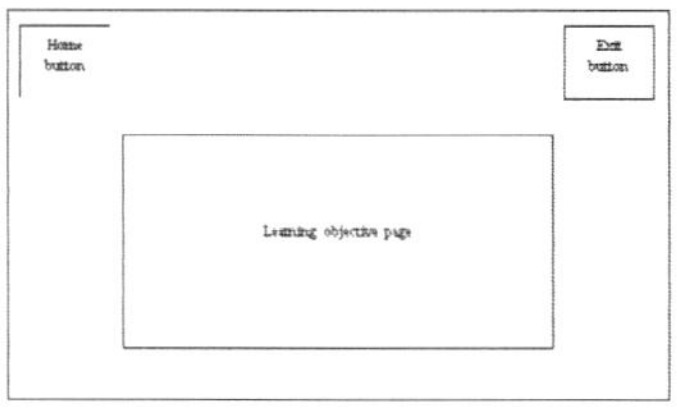

Figure 3. User interface of learning objective page

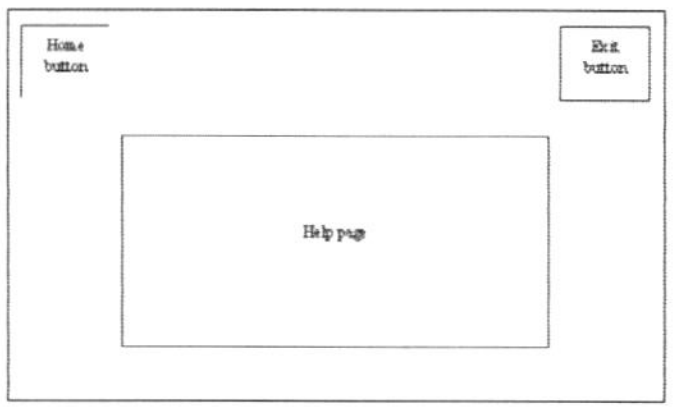

Figure 4. User interface of help page

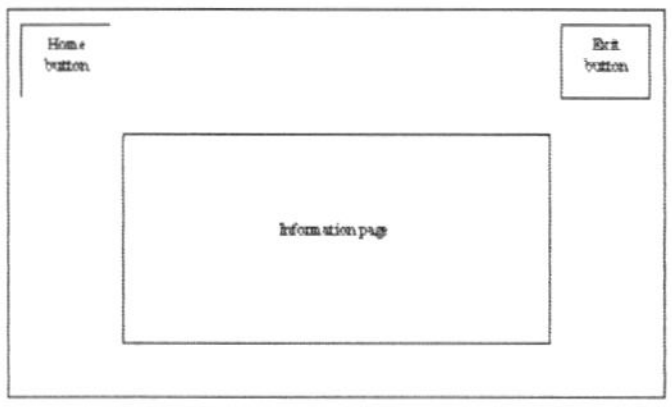

Figure 5. User interface of information page

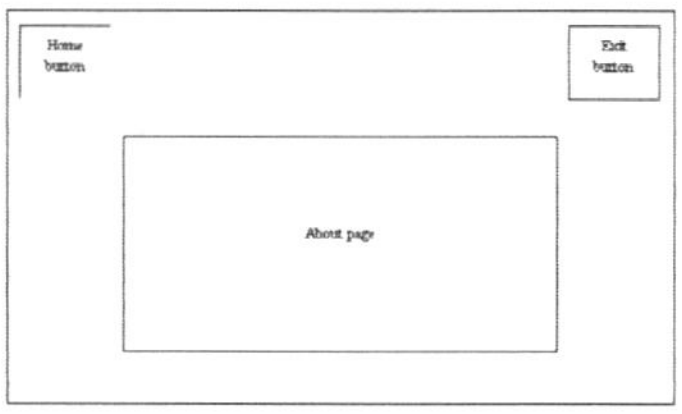

Figure 6. User interface of about the application page

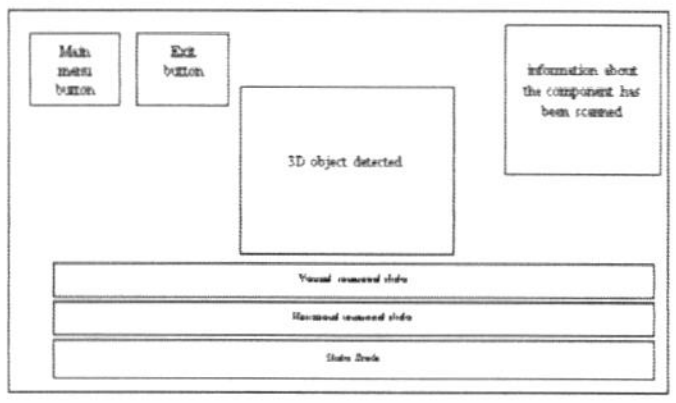

Figure 7. User interface of AR marker scanner camera page

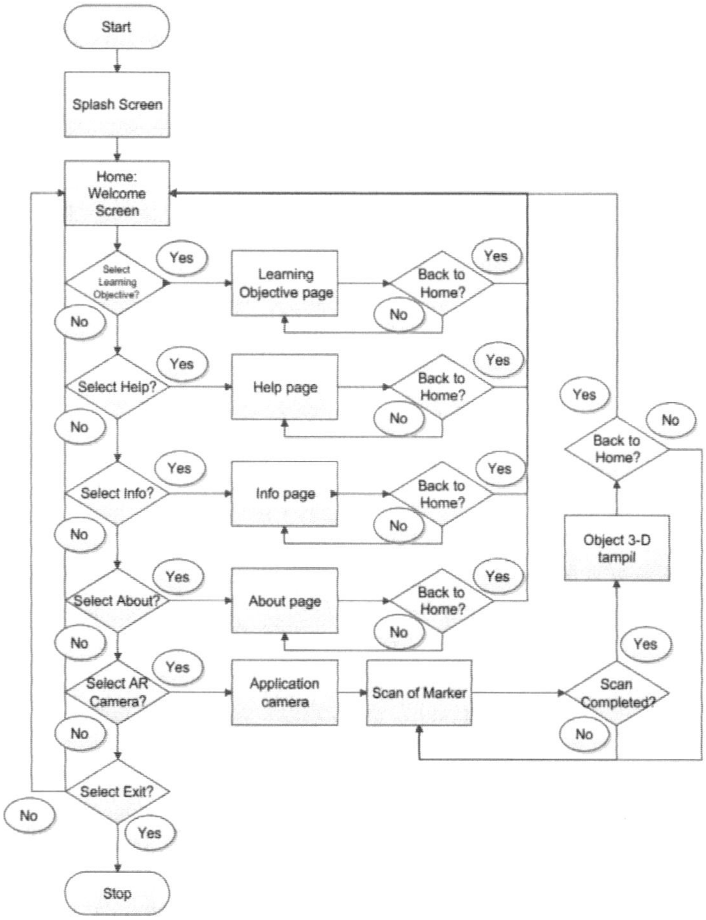

Figure 8. The flowchart of application

3. Technologies and Tools

AR technology is the latest trend and continues to grow in shaping human-computer interaction behavior. There are several platforms used to develop AR technology.

3.1. Unity

Unity is software that provides various tools in it to make it easier to make games (game engines) through objects that are already available or by adding scripts that are used to adjust the game as desired. Unity 3D is a software that is usually used to create multi-platform games that have high-optimized graphics performance. Applications created using Unity 3D can be run on Windows, Nintendo Wii, Mac, PS3, Xbox 360, Pad, Android platforms, and iPhone operating systems [11].

3.2. Visual Studio

Visual studio is software that can be used to develop applications with various programming languages [12]. Microsoft Visual Studio is integrated development software that can be used to develop applications in native code or managed code. Visual Studio can also be used to develop Silverlight applications, windows mobile applications (running on .NET Compact Framework). Microsoft Visual Studio is supported by several programming languages such as C++, Visual C#, and Visual Basic. Visual Studio can also support other programming languages, such as M, python, and ruby, all of which are included in additional packages that are separate from visual studio [11].

3.3. Vuforia

Vuforia is software used to support AR creation. Vuforia which focuses on image recognition (image recognition). The features and capabilities provided by Vuforia are very diverse, so that it can help AR application developers to realize their thoughts without any technical limitations [13].

3.4. Blender

Blender is software that can be used to create 3-dimensional objects in the form of models and animations. Blender can also be used as software used to create games (game engine) for free with a file size of less than 100 MB and is available for various operating systems such as Windows, Mac OSX, Linux, and Free BSD. Blender is a software for three-dimensional modeling based on Windows. In the world of architecture, blender has been widely used to create three-dimensional modeling which is used as a presentation medium to be presented to customers because the results of three-dimensional shapes resemble real shapes [11].

3.5. Android SDK

The Android SDK is a toolkit that functions as an emulator when the application is completed. In the manufacture of AR technology, the Android SDK is not only used as an emulator, but also at the deployment stage in the form of an "*.apk" application. If the Android SDK is not installed, the application that has been developed cannot be deployed in the form of an application [14].

3.6. Canva

Canva is a comprehensive website that can be used for designing covering a myriad of graphic design concepts. The basic features are free, very suitable for the development of info graphic designs, presentation slides, certificates, posters, banners, greeting cards, and so on [15].

4. Finding and discussions

This section will be divided into three topics, including component identification workbooks, Distributing Station-AR applications, and product validation results. The AR application is devoted to the identification of distribution station components. Students install AR applications on smartphones, then with guidance on workbooks, students identify components on the distribution station machine.

The workbook is equipped with the steps of the work that must be done, as well as the parts that must be scanned with the AR application to see in detail in three-dimensional form.

4.1. The Component Identification Workbook

Assembling, programming, and testing industrial machines requires an understanding of the machine. Work principles, electrical wiring, pneumatic wiring, component names, function of components, etc., must be mastered by an engineer. Distributing station is one of the replicas of industrial machines that can be disassembled for assembling, programming, and testing training. Learning about distributing station, already equipped with a workbook. The distributing station workbook contains instructions regarding the identification of component names, program descriptions, and program evaluation. This workbook is equipped with AR technology that can be scanned with a smartphone to display three-dimensional shapes. The images presented in the workbook are two-dimensional images, so they can only be observed from one side. With AR technology, objects displayed can be observed from various sides. The identification of the distributing station component is presented in the form of an image accompanied by an explanation of the component name. Figure 9 shows a page snippet from the distribution station component identification workbook.

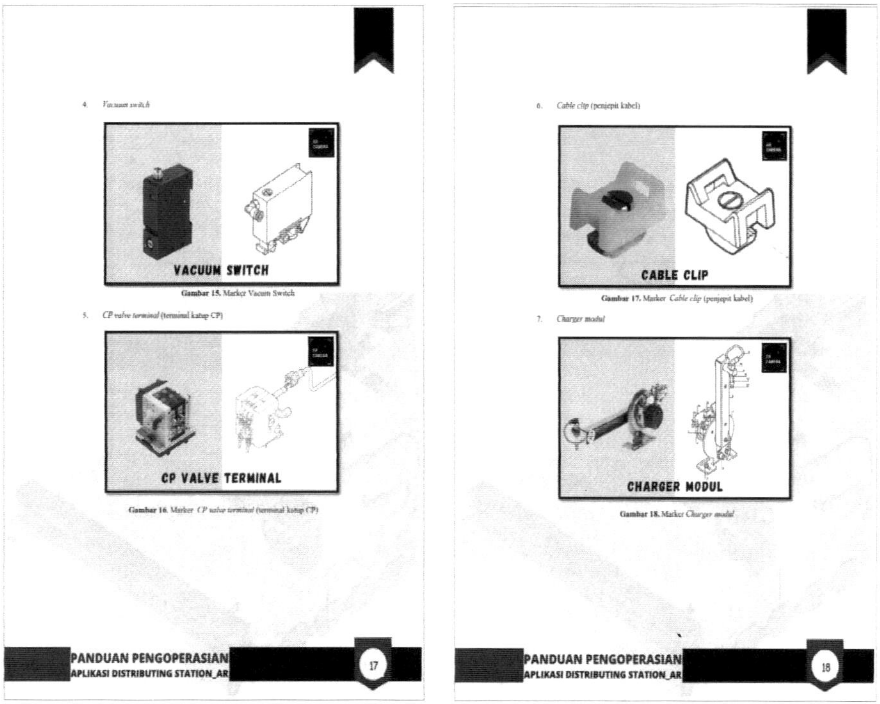

Figure 9. The workbook parts of the distributing station component identification

The distributing station workbook has been equipped with Marker-AR which can be scanned by smartphone. The purpose of adding Marker-AR is to clarify component images, so that component identification is more real because component images appear in a virtual environment in 3D. Figure 10 and Figure 11 show examples of Marker-AR in a workbook for identification of distribution station components.

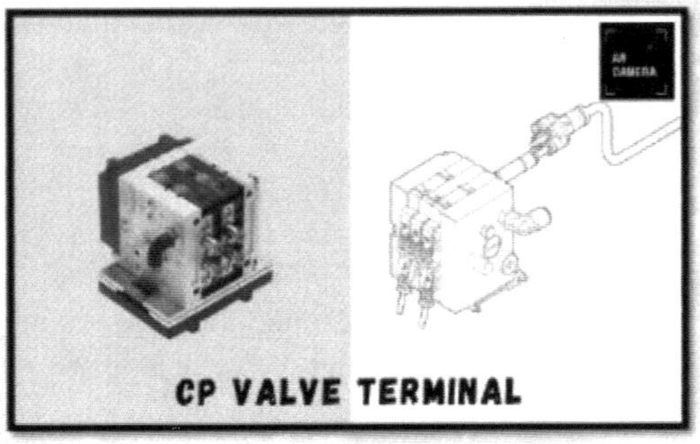

Figure 10. Marker-AR of valve terminal

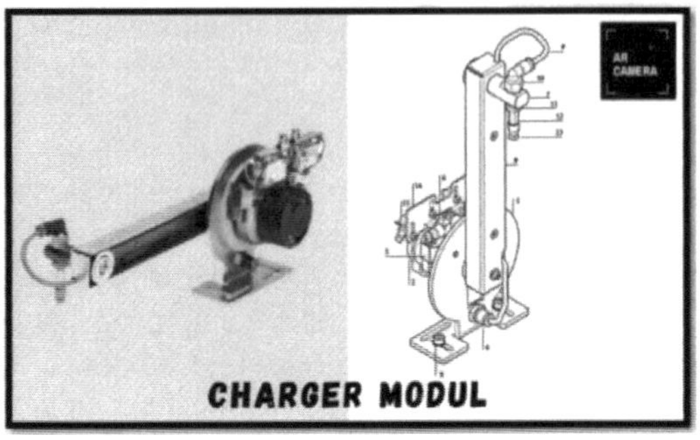

Figure 11. Marker-AR of changer module

4.2. The Distributing Station-AR

Distributing Station-AR is an android application with Augmented Reality technology that can project two-dimensional images into three-dimensional shapes. The Distributing Station-AR application is used to project component images on the distributing station so that it is easier to understand. Figure 12 is the icon for the Distributing Station-AR application that has been installed on a smartphone. Figure 13 is a user interface display on the welcome page. On the welcome page there are main menus, including the learning objectives menu, instructions menu, info menu, about menu, and AR-Camera menu.

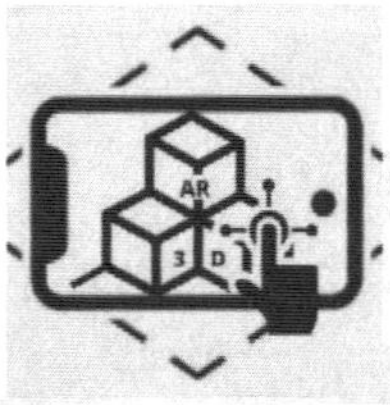

Figure 12. Icon application of distributing Station-AR

Figure 13. Distribution station component scan example

To use workbooks with AR technology, users must launch the Distributing Station-AR application, then allow the application to use the smartphone camera. In the AR scan menu, point the camera at the Marker-AR distributing identification workbook. On the display of the smartphone appears a three-dimensional model of the scanned component image. The three-dimensional model can be seen from various sides by moving the smartphone according to the side to be observed.

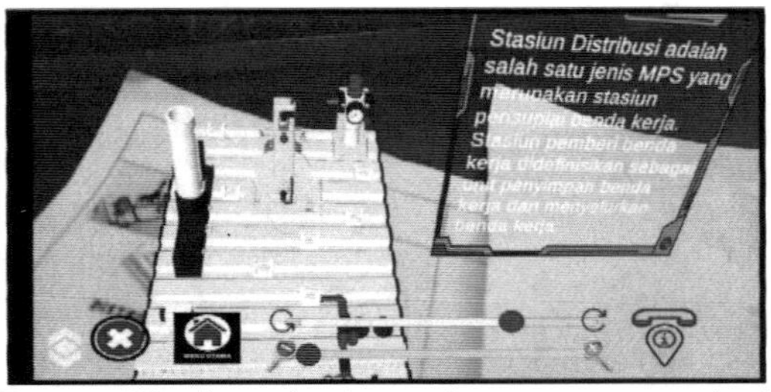

Figure 14. Camera-AR view when scanning a component in the workbook

4.3. Product Validation

The performance of the AR Distributing Station is known through the black box test. There are two aspects to the black box test, covering the application installation aspect and the application operation aspect. The test uses a questionnaire, consisting of 23 statements items with two answer choices, namely "0" and "1". Four statements on the application installation aspect and 19 items on the application operation aspect. A total of 3 respondents as test subjects. The results of the black box test are known to all respondents giving a maximum rating on all statement items. Thus, it can be concluded that the performance of the Distributing Station-AR application product is in the "Very Good" category with a percentage of 100%.

Product validation includes material validation about distribution station machines, software-based media validation, and first user response. Assessment by material experts consists of two aspects, including aspects of the suitability of the material and aspects of the usefulness of the material. The test uses a questionnaire, consisting of 24 statement items with a Likert scale of 1 - 4 as answer choices. there are two respondents as subject matter experts. The results of material validation are known to obtain an average assessment of 90.63% and are included in the "Very Good" category. Assessment by media experts consists of two aspects, including aspects of multimedia design and aspects of multimedia usefulness. The test uses a questionnaire, consisting of 18 statements with a Likert scale of 1 - 4 as the answer choices. There are two respondents as the subject of media experts. The results of media validation are known to obtain an average rating of 81.94% and are included in the "Very Good" category. The assessment by the first user consists of five aspects, including application design aspects, technical quality aspects, application usefulness aspects, material quality aspects, and AR technology aspects. The test uses a questionnaire, consisting of 28 statements with a Likert scale of 1 - 4 as the answer choices. There are 31 respondents as the subject of the first user. The results of the assessment are known to have an average rating of 90.01% and are included in the "Very Good" category.

5. Conclusion

The application of Distributing Station-AR has been developed by applying a marker-based tracking method. The application test has successfully displayed the components of Distributing Station using a marker. Component identification of Distributing Station can be displayed in a three-dimensional model using a smartphone camera. The three-dimensional model can be observed from various sides by rotating the smartphone according to the side to be observed. Product performance is known to be very good and product validation is also very good, including material validation, media validation, and first user responses. For further development, this application is expected to add animation of component working principle.

6. References

[1] T. H. T. Maryadi, T. Sukisno, A. N. Chandra, and A. W. Atmoko, "Augmented Reality-Based Instructional Media for Electrical Power Protection Learning," *J. Phys. Conf. Ser.*, vol. 1387, no. 1, 2019, doi: 10.1088/1742-6596/1387/1/012015.

[2] I. Mustaqim and N. Kurniawan, "Pengembangan media pembelajaran berbasis augmented reality," *J. Edukasi Elektro*, vol. 1, no. 1, 2017.

[3] G. R. Meilani, "Membangun Aplikasi Augmented Reality Dengan Unity," *Surabaya CV Garuda Mas Sejah.*, 2018.

[4] G. W. Ng, Y. B. Oon, H. Y. Lee, and E. H. Teoh, "An Augmented Reality System for Biology Science Education in Malaysia," *Int. J. Innov. Comput.*, vol. 6, no. 2, pp. 8–13, 2016.

[5] Y. S. Kim, J. O. Lee, and C. W. Park, "A hybrid learning system proposal for PLC wiring training using AR," *Proc. - 2011 5th IEEE Int. Conf. E-Learning Ind. Electron. ICELIE 2011*, pp. 90–94, 2011, doi: 10.1109/ICELIE.2011.6130032.

[6] G. Zhao, Q. Zhang, J. Chu, Y. Li, S. Liu, and L. Lin, "Augmented Reality Application for Plant Learning," *Proc. IEEE Int. Conf. Softw. Eng. Serv. Sci. ICSESS*, vol. 2018-Novem, pp. 1108–1111, 2019, doi: 10.1109/ICSESS.2018.8663953.

[7] S. Sendari, D. Anggreani, M. Jiono, A. Nurhandayani, and C. Suardi, "Augmented reality

performance in detecting hardware components using marker based tracking method," *4th Int. Conf. Vocat. Educ. Training, ICOVET 2020*, pp. 175–179, 2020, doi: 10.1109/ICOVET50258.2020.9229895.

[8] B. Bajči *et al.*, "Work-in-Progress: Development of Augmented Reality Application for Learning Pneumatic Control," *Lect. Notes Networks Syst.*, vol. 80, pp. 711–718, 2020, doi: 10.1007/978-3-030-23162-0_64.

[9] A. Suryanto, D. A. Kusumawati, and I. M. H. Sanhoury, "Development of Augmented Reality Technology Based Learning Media of Lathe Machines," *J. Pendidik. Teknol. dan Kejuru.*, vol. 24, no. 1, pp. 32–38, 2018, doi: 10.21831/jptk.v24i1.18245.

[10] R. M. Branch, *Instructional design: The ADDIE approach*, vol. 722. Springer Science & Business Media, 2009.

[11] A. Zaky, *Animasi karakter dengan Blender dan Unity*. Elex Media Komputindo, 2016.

[12] J. Enterprise, *Belajar VB, Visual C#, dan Phyton Menggunakan Visual Studio*. Jakarta: PT. Alex Media Komputindo, 2017.

[13] M. Fernando, "Membuat aplikasi android augmented reality menggunakan vuforia sdk dan unity," 2013.

[14] R. Meier, *Professional Android 4 application development*. John Wiley & Sons, 2012.

[15] B. Band, "Canva," *Sch. Libr.*, vol. 65, no. 3, p. 149, 2017.

Performance of Trainers kits for Industrial Automation Based on Programmable Logic Controllers

Sukir[1], Sa'adilah[2], Hartoyo[3], Sunaryo[3], and B. P. Saifullizam[5]

[1,2,3,&4]Lecturer in Electrical Engineering Education Department, Engineering Faculty of Yogyakarta State University.
[5]Lecturer of Fakulti Pendidikan Teknikal dan Vokasional, Universiti Tun Hussein Onn Malaysia

E-mail: sukir@uny.ac.id

Abstract. The objectives of this study include: (1) get results; and (2) know the performance of trainer kits for industrial automation based on Programmable Logic Controllers (PLC). The type of research is research and development that refers to the ADDIE model according to Robert Maribe Branch. Research steps include: Analyze, Design, Develop, Implement, and Evaluate. Data obtained through observation, interviews, and questionnaires, using instruments in the form of observation sheets, interview guidelines, and questionnaires. Data were analysed descriptively. The results showed that: (1) A PLC-based industrial automation trainer kit has been developed in the form of simulation of drilling objects, spraying paint on objects, applying stamps, and moving objects manually and automatically; and (2) PLC-based industrial automation trainer kit products have good performance, which is shown that 100% of the important components can work properly, and all job descriptions (100%) of the trainer kits can work as their function.

1. Introduction

The following lists the Industrial Revolution 4.0 is a phenomenon that applies the concept of automation. Production activities in the industry carry out automatically through industrial automation [1]. Industrial automation is characterized by the use of machines that work automatically with or without human assistance in the production or manufacturing process. To control the machines that work automatically, a controlling brain is needed, one of which is the Programmable Logic Controller (PLC).

In facing the automation in the industry as the demands of the industrial revolution 4.0, human resources are needed that are in charge of industrial automation and its control base. A college of Applied Electrical Engineering Study Program, as an educational institution to provide skilled human resources also faces these challenges. This educational institution must equip the students with competence in industrial automation and control base. So, they will be ready to face the demands of the industrial revolution 4.0 when they graduated.

Yogyakarta State University is an institution that administrates the Applied Bachelor (D4) of Electrical Engineering Study Program. The study of the Applied Bachelor Study Program at Yogyakarta State University was held in Wates, Kulon Progo, Special Region of Yogyakarta. The Study Program is new which began in the 2019/2020 academic year. The study program used the 2019 curriculum. The courses that expect to directly equip students based on PLC control and industrial automation development is the courses of Programmable Logic Controller Practice provided in semester 4 with 3 credits. Observations show that this practical courses has no trainer kits, measuring instruments, and materials to support practical activities. Also, the limitation found on observation is caused by the limited funds to buy the practical equipment which affects difficulties to provide an industrial automation trainer kit and an applicative PLC in the industry. As the alternative, they may later borrow a trainer kit and practical material from the Electrical Engineering Study Program at Yogyakarta State University, but the available trainer kit is only a simple PLC trainer.

The objectives of the study were to: (1)obtain product of PLC-based industrial automation trainer kit; and (2)find out the product performance of the PLC-based industrial automation trainer kit needed for learning PLC Practice in the Applied Bachelor of Electrical Engineering Study Program.

Learning media of real objects or artificial objects can be manifested into trainer kits [2]. The trainer kit as a learning media consists of teaching aids or tangible or artificial objects, and an explanation of the use of the media in a job sheet or learning module. Industrial automation is the use of various control systems to operate equipment, machinery, production processes, boilers, telephone network settings, steering and stabilization of ships, aircraft, robots, and the use others tools with minimal or diminished human intervention [3]. Industrial automation is an important part of the industrial revolution era 4.0 because the equipment, machinery, production processes, and other components in the industry work automatically. Programmable Logic Controllers (PLCs) are used in control functions in the industry [4]. Based on the previous opinions, it concluded that PLC-based industrial automation trainer kits are practical learning media in the form of teaching aids, real objects or imitations of machines, production processes, robots, and other equipment in industries that work automatically controlled by PLC.

The relevant research was conducted by Asuncion & Ronaldo that analyzed "Low Cost 2 Station Industrial Automation Trainer." The research produced a trainer kit with 2 stations, namely robot arms and sorting at a low cost. The trainer kit works for practical activities in the field of industrial automation and mechatronics in schools in the Philippine[5]. Gill, Kumar & Kumar researched the "Designing and fabrication of electro-pneumatic trainer kits," which produced electro-pneumatic trainer kits for learning in the Mechanical Engineering Department, Gulzar Group of Institutes, India [6]. Jamaluddin, Ghani, Rahman, & Deros analysed the "Development of training kit for learning Taguchi methods and design of experiments," showed that the Taguchi experiment was successfully carried out through the trainer kit and the tested factorial design are, in principle, based on information obtained from the Taguchi method [7]. Samanol, Hamid, & Ramli analysed "Development of pneumatic trainer kits for polytechnic students," which produces pneumatic trainer kits for learning the basics of pneumatics for students of the Department of Engineering, Seberang Perai Polytechnic, Malaysia. These studies differ from this particular research, in terms of trainer kit products and use [8].

The robotics also include in the field of industrial automation. The research was conducted by Sean that researched "BeRobot-the robotic scientific education development kits." The research produced a human-like robot trainer kit, called BeRobot, which used as a training tool for people who work or interest in the robot [9]. Other research conducted by Heilo Altin & Margus Pedate on "Leaming Approaches to Applying to Robotics in Science Education," shows that robotics may be seen as a tool in applying learning approaches [01]. Research conducted by Maria Maxirnova & YoungHwan Kirn on Research Trend Analysis on the Usage of Robotic in Education, shows that there are 3 main issues and challenges of robotics in education, namely the lack of quantitative research on the impact of robotics in education, no precise curriculum definition for students, and the use of robots in education [11]. Another study was carried out by Georg Rauter, et al. entitled 'When a robot teaches humans: Automated feedback selection accelerates motor learning." The result showed that the rowing robot simulator can accelerate feedback automatically on motor learning [12]. The study was also carried out by Hong, et.al. which examines the development of educational robotics training kits. The study shows that the use of educational robotic trainer kits in learning can save costs, facilitate understanding for students, and simple [13]. These studies differ from this particular research, in terms of trainer kit products and use.

Related to research on Programmable Logic Controllers (PLC), Akparibo, Appiah, & Antwi onducted a study entitled "Development of a programmable logic controller training platform for the industrial control of processes." The research obtained an interactive, cheaper and more portable PLC trainer kit for industrial process control simulations. The PLC Trainer kit makes it easy for students of the Department of Electrical and Electronic Engineering in Ghana to install and program all types of inputs and outputs according to their choice. The application of the PLC trainer kit in learning improves student achievement compared to before using the PLC trainer kit [14]. Another study was conducted by Bhise & Arnte on "Embedded PLC trainer kits with industry applications." The results obtained the Embedded PLC trainer kit for application simulations in the industry which has the characteristics of safe, low cost, having an interface with a personal computer, the PLC meets the standard, the programming language uses IEC 6-1131 and having an interface with an electrical component. The application of the trainer kit for learning at MTech National Institute of Electronics and Information Technology, Auranagabad, Maharashtra, India, shows the increasing learning outcomes achieved by students. These studies differ from this particular research, in terms of trainer kit products and use [15].

2. Research Method

The research type was research and development to produce a valid product of PLC-based industrial automation trainer kit. The development model refers to the ADDIE development model according to Robert Maribe Branch. According to Branch the ADDIE development model is the development of a product that has steps, namely: Analyze, Design, Develop, Implement, and Evaluate [16]. However, the Implementation step has not yet been carried out in the particular research due to time limitation.

Data collection techniques and instruments were: (1) in the needs analysis activities, data are collected using observations, interviews, and questionnaires, while the instruments are observation sheets, interview guidelines, and questionnaires; (2) in the evaluation step, data collection technique is observation, while the instrument is observation sheets; and (3) in the test of product performance, observation and questionnaires use to collect the data, while the instruments are observation sheets and questionnaires. The obtained data were analyzed descriptively to determine the needs and performance of the product.

3. Research Result

In this study, the development model used was the ADDIE development model according to Robert Maribe Branch, with steps according to the acronym, namely: Analyze, Design, Develop, Implement, and Evaluate. The results of the development based on these steps can be described as follows.

3.1. Survey Result of The Needs Analysis

Survey result of the needs analysis with students (58 persons) and industry practitioners (2 persons). The results showed that the needs chosen by most respondents in practical learning of PLC were: (1) PLC-based industrial automation trainer kits in the form of industrial process miniatures selected by 100% of industry practitioners and 86.21% of students; (2) the practical learning module was chosen by 100% of industry practitioners and 89.66% of students; and (3) the criteria for the trainer kit which were classified as indispensable, namely: (a) according to the competencies chosen by 100% of industry practitioners and 87.93% of students, (b) the size according to laboratory standards chosen by 100% of industry practitioners and 91.37% of students, (c) made of good components selected by 100% of industry practitioners and 93.10% of students, and (d) refers to work safety, which was chosen by 100% of industry practitioners and 94.83% of students.

3.2. Design of Trainer Kits for Industrial Automation Based on Programmable Logic Controllers.

Design of Trainer Kits for Industrial Automation Based on Programmable Logic Controllers is shown in Figure 1.

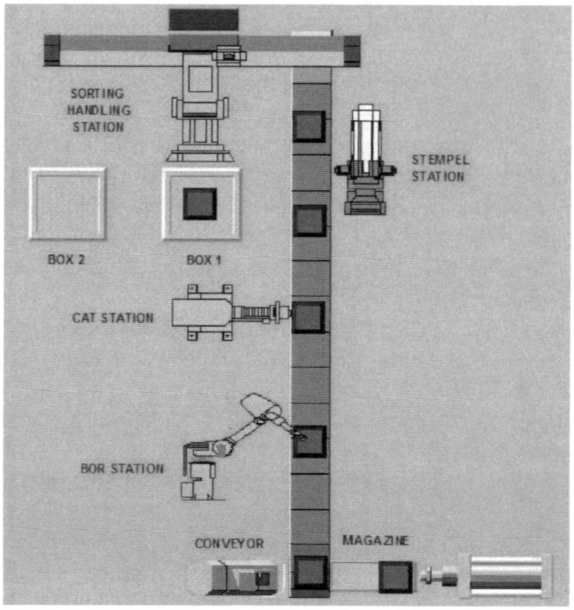

Figure 1. Design of Trainer Kits for Industrial Automation Based on PLC

3.3. Development Result of Trainer Kits Industrial Automation Based on PLC.
Development Result of Trainer Kits Industrial Automation Based on PLC is shown in Figure 2.

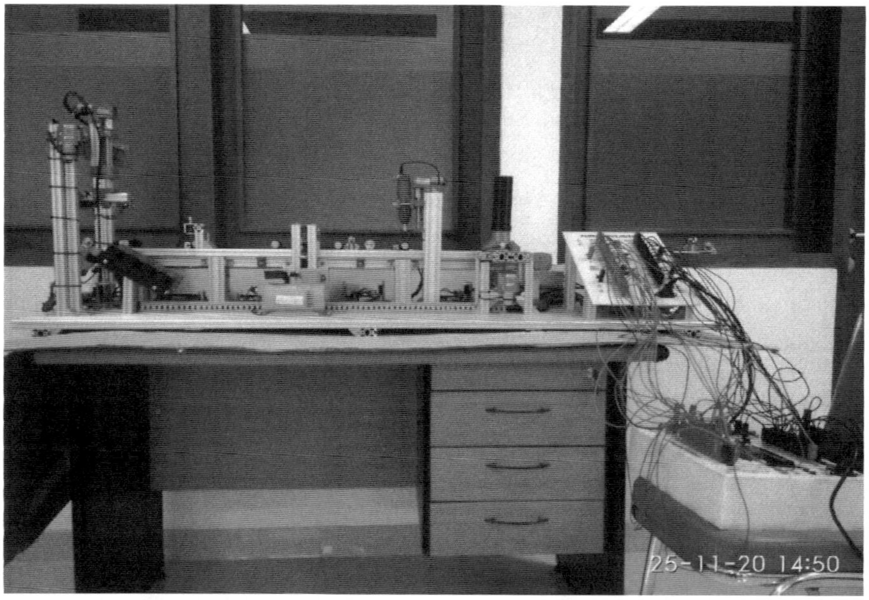

Figure 2. Development Result of Trainer Kits Industrial Automation Based on PLC

3.4. Performance Testing of Trainer Kits Industrial Automation Based on PLC.
Performance testing of Trainers kits Industrial Automation Based on PLC is shown in the following tables. Performance test of the power source and input components is shown in Table 1.

Table 1. Performance test of the power source and input components

Statement	Examiner & Results of Testing				
	E1	E2	E3	E4	E5
220 VAC and 24 VDC voltage source is connected properly.	√	√	√	√	√
The 12 VDC voltage source is well connected.	√	√	√	√	√
Limit Switch1B1 and 1B2 (Back and Front Magazine Back) is functioning properly.	√	√	√	√	√
Capasitive Proximity Sensor 1B3, 2B3, 3B1, 3B2, 4B1 (Object Sensor) is functioning properly.	√	√	√	√	√
Limit Switch2B1and 2B2 (Rear and Front Drill) is functioning properly.	√	√	√	√	√
Limit Switch5B1 (Rear sliding handling) works fine.	√	√	√	√	√
Limit Switch 5B2 (Front handling gerer) is functioning properly.	√	√	√	√	√
Inductive Proximity Sensor5B3 (Middle slide Handling Sensor) is functioning properly.	√	√	√	√	√
Optical Proximity Sensor5B4 (Color sensor) is working properly.	√	√	√	√	√
Capacitive Proximity Sensor 5B5, 5B6, 5B7 (Body sensor) is functioning properly.	√	√	√	√	√
Limit Switch6B1 (Rear Down Handling) is functioning properly.	√	√	√	√	√
Limit Switch 6B2 (Front drop handling) works fine.	√	√	√	√	√

Table 1 describes the Performance Test of Input Resources and Components, based on the data displayed, it can be concluded that the Industrial Automation Based on PLC Trainer Kits work well. Performance test of the output and load components is shown in Table 2.

Table 2. Performance test of the output and load components

Statement	Examiner & Results of Testing				
	E1	E2	E3	E4	E5
Motor Magazine Forward (1M1) works fine.	√	√	√	√	√
Motor Magazine Reverse (1M2) is functioning properly.	√	√	√	√	√
The Forward Drill Motor (2M1) is functioning properly.	√	√	√	√	√
Reversing Drill Motor (2M2) is functioning properly.	√	√	√	√	√
The Drill Motor (2M3) is functioning properly.	√	√	√	√	√
Solenoid 1 (2M4) works fine.	√	√	√	√	√
Cat motor (3M1) is functioning properly.	√	√	√	√	√
Solenoid Seals (4M1) worked fine.	√	√	√	√	√
Solenoid 2 (4M2) works fine.	√	√	√	√	√
The Forward Slide Handling Motor (5M1) is functioning properly.	√	√	√	√	√
The Reverse Handling Motor (5M2) is functioning properly.	√	√	√	√	√
Handling Motor Forward Down (6M1) is functioning properly.	√	√	√	√	√
Reverse Handling Motor (6M2) is functioning properly.	√	√	√	√	√
Solenoid Griper (6M3) works fine.	√	√	√	√	√

The motor conveyor (7M1) is functioning properly.	√	√	√	√	√

Table 2 describes the Performance test of the output and load components, based on the data displayed, it can be concluded that the Industrial Automation Based on PLC Trainer Kits work well. Assessment of work function test on initial conditions is shown in Table 3.

Table 3. Assessment of work function test on initial conditions

Action	Reaction	Examiner & Results of Testing				
		E1	E2	E3	E4	E5
MCB 1 = OFF MCB 2 = OFF	*Trainer* cannot be executed: - - HL1 OFF - HL2 OFF	√	√	√	√	√
MCB 1 = ON MCB 2 = ON	- *Trainer* indicates initial condition / not - HL1 ON - HL2 ON	√	√	√	√	√
Trainer not initials position. EM has not been suppressed	HL3 flashes 1 Hz	√	√	√	√	√
EM is suppressed. *Trainer* move to initial position	HL3 flashes 1 Hz	√	√	√	√	√
EM is released. *Trainer* initial position	HL3 OFF	√	√	√	√	√

Table 3 describes the assessment of work function test on initial conditions, based on the data displayed, it can be concluded that the Industrial Automation Based on PLC Trainer Kits work well. Assessment of work function test on run cycle is shown in Table 4.

Table 4. Assessment of work function test on run cycle

Action	Reaction	Examiner & Results of Testing				
		E1	E2	E3	E4	E5
Insert a random workpiece into the magazine	1B3 sensor ON	√	√	√	√	√
Press the Start button (SB1) at the initial position	The ON system is indicated by: HL3 ON 1M1 ON (forward magazine)	√	√	√	√	√
1B2 = ON	1M1 OFF (forward magazine) 1M2 ON (reverse magazine)	√	√	√	√	√
1B1 = ON	1M2 OFF (reverse magazine) 7M1 ON (conveyor)	√	√	√	√	√
2B3 = ON	7M1 OFF (conveyor) 2M3 ON (drill) 2M1 ON (drill down)	√	√	√	√	√
2B1 = ON	2M1 OFF (drill down) 2M2 ON (drill up)	√	√	√	√	√
2B1 = ON	2M3 OFF (drill) 2M2 OFF (drill up) 2M4 ON (sol 1) 7M1 ON (conveyor)	√	√	√	√	√
3B1 = ON	3M1 ON (compressor)	√	√	√	√	√

3B2 = ON	3M1 OFF (compressor)	√	√	√	√	√
4B1 = ON	7M1 OFF (conveyor) 4M1 ON (stemple)	√	√	√	√	√
Timer	4M1 OFF (stample) 7M1 ON (conveyor)	√	√	√	√	√
5B5 = ON	7M1 OFF (conveyor)	√	√	√	√	√

Table 4 describes the assessment of work function test on run cycle, based on the data displayed, it can be concluded that the Industrial Automation Based on PLC Trainer Kits work well. Assessment of work function test on black body condition is shown in Table 5.

Table 5. Assessment of work function test on black body condition

Action	Reaction	Examiner & Results of Testing				
		E1	E2	E3	E4	E5
5B4 = OFF 5B5 = ON	5M1 ON (sliding forward handling)	√	√	√	√	√
5B2 = ON	5M1 OFF (sliding forward handling) 6M1 ON (handling down)	√	√	√	√	√
6B2 = ON	6M1 OFF (handling down) 6M3 ON (gripper)	√	√	√	√	√
Timer	6M2 ON (handling up)	√	√	√	√	√
6B1 = ON	6M2 OFF (handling up) 5M2 ON (sliding reverse handling)	√	√	√	√	√
5B1 = ON	5M2 OFF (sliding reverse handling) 6M1 ON (handling down)	√	√	√	√	√
6B2 = ON 5B7 = ON	6M1 OFF (handling down) 6M3 OFF (gripper)	√	√	√	√	√
Timer	6M2 ON (handling up)	√	√	√	√	√
6B1 = ON	6M2 OFF (handling up) The cycle repeats when there is a workpiece on the magazine	√	√	√	√	√

Table 5 describes the assessment of work function test on black body condition, based on the data displayed, it can be concluded that the Industrial Automation Based on PLC Trainer Kits work well. Assessment of work function test on red object condition is shown in Table 6.

Table 6. Assessment of work function test on red object condition

Action	Reaction	Examiner & Results of Testing				
		E1	E2	E3	E4	E5
5B4 = OFF 5B5 = ON	5M1 ON (sliding forward handling)	√	√	√	√	√
5B2 = ON	5M1 OFF (sliding forward handling) 6M1 ON (handling down)	√	√	√	√	√
6B2 = ON	6M1 OFF (handling down) 6M3 ON (gripper)	√	√	√	√	√
Timer	6M2 ON (handling up)	√	√	√	√	√
6B1 = ON	6M2 OFF (handling up) 5M2 ON (sliding reverse handling)	√	√	√	√	√
5B3 = ON	5M2 OFF (sliding reverse handling)	√	√	√	√	√

	6M1 ON (handling down)					
6B2 = ON	6M1 OFF (handling down)	√	√	√	√	√
5B7 = ON	6M3 OFF (gripper)					
Timer	6M2 ON (handling up)	√	√	√	√	√
	6M2 OFF (handling up)					
6B1 = ON	The cycle repeats when there is a workpiece on the magazine	√	√	√	√	√

Table 6 describes the assessment of work function test on red object condition, based on the data displayed, it can be concluded that the Industrial Automation Based on PLC Trainer Kits work well.

4. Discussion

Based on the needs analysis, it shows that the Applied Undergraduate Study Program of Electrical Engineering, Yogyakarta State University, was really needed of a PLC-based industrial automation trainer kit in the form of a miniature industrial process for PLC practical learning. PLC-based industrial automation trainer kit in the form of a miniature industrial process can provide an overview of the application of PLC in industrial process control. This trainer kit can also increase students' attention, motivation, and enthusiasm for learning because in the use of this trainer kit there is movement control which is a simulation in the industry. The use of this trainer kit in practical PLC learning can increase student curiosity because of the movement of the production process in the industry. This of course can make learning more meaningful. In addition, the PLC-based industrial automation trainer kit that was really needed was a trainer kit that was in accordance with the proposed competencies, was sized according to laboratory standards, was made of quality components, and refers to occupational safety and health.

Based on the results of the needs analysis, a trainer kit design is carried out as shown in Figure 2. The design of a PLC-based industrial automation trainer kit consists of several main parts, namely: pushing objects from the initial box position, conveyors, drill stations, sterilization spraying stations, cap stations, shorting handling stations, placing goods in red boxes and black boxes. The PLC-based industrial automation trainer kit measures 160 cm in length, 60 cm in width, and is placed on a table measuring 160 cm long, 170 cm wide and 80 cm high.

The next step is the manufacture of a PLC-based industrial automation trainer kit which refers to the design as mentioned above. In general, the manufacture of PLC-based industrial automation trainer kits includes the manufacture of mechanical, electrical and programming constructions. The results of making a PLC-based industrial automation trainer kit are shown in Figure 3.

Furthermore, the performance of the industrial automation trainer kit based on the PLC was then tested for its performance. Testing the performance of the PLC-based industrial automation trainer kit as shown in Table 1, Table 2, Table 3, Table 4, Table 5, and Table 6. It turns out that all test items for the trainer kit work tested by 5 testers obtained good performance or can work according to its function.

The job description of the PLC-based industrial automation trainer kit can be outlined as follows. If there is a black object that falls or is placed in the initial box, then the On button is pressed, the Magazine moves forward to push the black object towards the conveyor. When the black object has reached the conveyor, the magazine's forward movement stops 1 second, then the magazine moves back and stops when the magazine is back in its starting position. Then the conveyor runs to deliver the black object to the drilling station, and the conveyor stops when the black object is right in the drilling position. Then the drilling machine moves down to drill the black object, after drilling for 10 seconds the drilling machine moves back up and stops at the initial position of the drilling machine. Then the conveyor works again to deliver the black object to the sterilizing agent spraying room. When the black object moving over the conveyor is in the sterilization room, the sterilizing agent spray compressor works to spray the sterilizing agent. If the black object moving above the conveyor comes out of the sterilizing agent spraying zone, the sterilizing agent spraying compressor stops working. Furthermore, the black object is still moving, carried by conveyors to the tasting station, When the black object arrives at the tasting station, the conveyor stops, then the cap lever moves down to stamp the black object, and then the stamp lever returns to the top. Then the conveyor moves again to deliver the black object to the place where the finished goods are lifted. When the black object arrives at the position where the goods are lifted, the conveyor stops and the Cran's moves forward to the place of lifting the goods. When the Cran's position has reached above the lifting area, the Cran's arm will move down and stop when the Cran's Gripper reaches the position of the black object. Next the Cran's Gripper grabs the black object and the Cran's arms move up again and stop when the Cran's arms arrive at the starting position. Next Cran moved backwards to bring the black goods to the black banda shelter box. When Cran is above the black object holding box, Cran's arm moves down and stops when the black object arrives at the position of the black object holding box, then the Cran's gripper releases the black object so that the black object is accommodated in the black object box. Next, Cran's arm moves up towards Cran's starting position. If there is a red object that falls or is placed in the initial box, then the On button is pressed, then the red object work process is the same as the black object work process as mentioned above, but when the red object arrives at the position where the goods are lifted, Cran will work to place it. the red object in the red object storage box.

After obtaining the performance test of the PLC-based industrial automation trainer kit which shows good performance, the next step is the validation of material experts, media experts and student assessments of PLC-based industrial automation trainer kit products. Referring to the validation results of material experts and media experts on the feasibility of a PLC-based industrial automation trainer kit, it shows that in all aspects, material experts give a score of 92.11%, and media experts give a score of 93.23%, both of which fall into the very feasible category. Likewise, in the implementation of the conveyor trainer kit in learning, it turned out that students gave an assessment of the PLC-based industrial automation trainer kit with a value of 83.55% which was also included in the very feasible category. The results of validation carried out by material experts, media experts, and assessments made by students of the feasibility of a PLC-based industrial automation trainer kit as mentioned above, it turns out that when viewed from all aspects, the achievement is 827 and the maximum score is 952, so the percentage of the achievement score to the maximum score of 86.87% which falls into the very feasible category. This means that the PLC-based industrial automation trainer kit meets the very appropriate category for use in PLC practical learning in the Applied Electrical Engineering Undergraduate Study Program, Faculty of Engineering, Yogyakarta State University.

Newly developed in this study is a PLC-based industrial automation trainer kit in the form of a production process simulation in industry with 4 stations, namely a drilling station, a sterilizing spray station, a tasting station, and a shorting handling station. This of course can contribute to the progress of learning in the Applied Electrical Engineering Undergraduate Study Program, Faculty of Engineering, UNY which simulates industrial equipment in the form of PLC-based industrial automation trainer kits into the PLC practical learning process. Thus it can be expected that students will get learning experiences that resemble technological developments in industry so that the meaning of learning can increase.

5. Conclusion

PLC-based industrial automation trainer kit has been developed in the form of simulation of drilling objects, spraying paint on objects, applying stamps, and moving objects manually and automatically. Beside that PLC-based industrial automation trainer kit products have good performance, which is shown that 100% of the important components can work properly, and all job descriptions (100%) of the trainer kits can work as their function.

6. References

[1] Adi Permana 2019 *Memahami Berbagai Perubahan di Era Revolusi Industri 4.0*. Taken on February 20, 2020 from: https://www.itb.ac.id/news/read/57037/home/memahami-berbagai-perubahan-di-era-revolusi-industri-40 p 1.

[2] Rauner, F., et. al. 2011 *Handbook of technical and vocational education and training research*. (Bremen: Springer Science Business Media) p 569.

[3] Mamodiya, U., & Priyanka 2014 *Review in Industrial Automation*. IOSR Journal of Electrical and Electronics Engineering (IOSR-JEEE)e-ISSN: 2278-1676,p-ISSN: 2320-3331, Volume 9, Issue 3 Ver. IV(May –Jun. 2014), PP 33-38www.iosrjournals.org.

[4] A. Michael, S.St.(2018). *What Is a PLC? An Introduction to Programmable Logic Controllers*. https://www.allaboutcircuits.com/technical-articles/what-is-a-plc -introduction-to-programmable-logic-controllers/

[5] Asuncion & Ronaldo. (2019). *Low Cost 2 Station Industrial Automation Trainer*. Available at SSRN: https://ssrn.com/abstract=3471693 or http://dx.doi.org/10.2139/ssrn.3471693

[6] Gill, K.J.S., Kumar, R., & Kumar, S. (2015). *Designing and fabrication of electro-pneumatic trainer kit. International Conference of Advance Research and Innovation (ICARI-2015)*, India, p.59-66.

[7] Jamaluddin, H., et.al. (2014). *Development of Training Kit for Learning Taguchi Method and Design of Experiments*. Jurnal Teknologi (Sciences & Engineering) 68:4 (2014), 19–25.Penerbit UTM Press, Malaysia.

[8] Samanol, S.B., Hamid, S.B.A, & Ramli, M.N.B. (2014). *Development of Pneumatic Trainer Kit For Polytechnic Student. Journal of Mechanical Manufacturing (J-Mfac)*. Vol. I. 2014. p. 178-183.

[9] Sean, C. (2015). *Be Robot-The Robotic Scientific Education Development Kits*. Lovotics, an open access journal, Volume 3, Issue 1, 2015. p. 1-4. ISSN: 2090-9888.

[10] Heilo Altin & Margus Pedate. (2013). *Learning Aproaches to Applying to Robotics in Science Education*. Journal of Baltic Science Education. Vol 12, No.3. 2013. ISSN 1648-3898.

[11] Maria Maximova & YoungHwan Kim. (2016*). Research Trend Analysis on the Usage of Robotic in Education*. Asia-Pasific Collaborative Education Journal. Vol 12. No. 1. 2016. pp. 45-60.

[12] Georg Rauter,et.al. (2019). *When a robot teaches humans: Automated feedback selection accelerates motor learning. Science Robotics*. Vol. 4, Issue 27, 20 Feb 2019.

[13] Hong, et.al. (2013). *Development of an Educational Robotic Training Kit*. Advanced Science Letters. Volume 19, Number 11, November 2013, pp. 3304-3306(3).

[14] Akparibo, A.R., Appiah, A., & Antwi, O.F. (2016). *Development of a Programmable Logic Controller Training Platform for the Industrial Control of Processes*. American Scientific Research Journal for Engineering, Technology, and Sciences (ASRJETS, (2016), Volume 15, No 1, pp 186-196.

[15] Bhise, K., & Amte, S. (2015). *Embedded PLC Trainer Kit with Industry Application. International Journal of Engineering Science and Innovative Technology (IJESIT)*. Volume 4, Issue 3, May 2015, ISSN. 2319-5967. P. 58-65.

[16] Branch, R.M. (2009). *Instruction design: the ADDIE approach*. New York: Springer Science + Business Media.

Journal of Physics: Conference Series **2111** (2021) 012041 doi:10.1088/1742-6596/2111/1/012041

Electric Machines Virtual Laboratory: Testing of DC Motor

R Badarudin[1], D Hariyanto[2], M Ali[3], and S Meunmany[4]

[1,2,3] Department of Electrical Engineering Education, Faculty of Engineering, Universitas Negeri Yogyakarta, Indonesia
[4] Department of Technical and Vocational Education, Ministry of Education and Sports, Lao PDR

E-mail: [1] rohjai.badarudin@uny.ac.id; [2] didik_hr@uny.ac.id; [3] muhal@uny.ac.id; [4] santiphap.m@moes.edu.la

Abstract. This article describes the realization of testing the characteristics of a DC motor on Simulink as a virtual laboratory. Testing of DC motors based on real conditions in a physical laboratory by combining two DC machines coupled to shafts including DC motors and DC generators. The types of DC motors tested include separately excited DC motors, shunt DC motors, and series DC motors. The implementation of the Simulink model adopts the wiring diagram in the real test. The test procedure is also adopted on the real test procedure. The data from the test results are made in DC motor characteristics graph. Virtual laboratory using Simulink can be used to realize DC motor testing, including test procedures, running no-load DC motors, increasing the load of DC motors with infinitely simulation features on Simulink. The results of the study show the characteristics of the experimental data for all types of DC motors tested according to the theory.

1. Introduction

Mechanical energy can be generated by converting electrical energy using an electric motor [1]. The application of DC motors in everyday life is very wide and is still used today. For the need for constant rotation, adjustable speed, a DC motor is a type of motor that is very easy to adjust the rotational speed of the rotor. DC motors produce mechanical energy by converting DC power source [2]. The types of DC motors include separately excited DC motors, shunt motors, series motors, and compound motors. Separately excited DC motors and shunt DC motors are widely used for purposes that require constant rotation, while series DC motors are widely used for purposes that require high torque. Selection of the type of DC motor is needed so that the characteristics of the DC motor are in accordance with the needs of the machine. The main advantage of the DC motor over other motors is the ease of controlling the armature speed. The DC motor armature speed regulation range can be set at a variable source voltage or varying the voltage across the field windings [3], [4].

Experimental simulation is a format for presenting concepts that try to match the dynamic processes that occur in the real world [5]. A virtual environment is a digitally replica version of something real. Replication of the real form into a digital model is made using computer software. The results of replication may not be similar to the real world, but it is enough to describe something abstract into reality [6], [7]. Simulation is used when the actual system cannot be involved for several reasons, including the system is inaccessible, the system is dangerous or unacceptable for human security, and the system is in the design stage. The concept of virtual simulation applied in this research is to test a DC motor. The characteristics of a DC motor will be tested, including separately excited DC motor, shunt DC motor, and series DC motor. The test was carried out using a simulation of the motor-generator model on Simulink [8]. The DC motor test procedure includes 1) running the DC motor from 0 rpm to nominal speed at nominal voltage, 2) loading the DC motor coupled to a DC generator, 3) the process of increasing the load gradually until data is obtained in the form of armature current, armature speed, and armature torque. The testing procedure in the simulation is adapted from the test procedure using real equipment, so it is hoped that the testing process will be as expected.

2. Literature Review

Mechanical energy can be generated by converting electrical energy using an electric motor. DC motors produce mechanical energy by converting DC power source [2]. The source of electrical energy uses a direct current power source. Mechanical force is generated based on the principle that a current-carrying conductor is in a magnetic field area, then the conductor experiences a mechanical force. The construction of a DC motor consists of a moving part or rotor and a stationary part or stator. There are five types of DC motors, namely separately excited DC motors, series DC motors, shunt DC motors, cumulative short compound DC motors, and cumulative long compound DC motors [1].

The characteristics of a DC motor are the relationship between armature torque, armature current, and armature speed. There are three characteristic relationships, namely 1) the characteristics between the armature torque and the armature current, 2) the characteristics of the armature speed and the armature current, and 3) the characteristics of the armature speed and the armature torque [1], [4]. The characteristics of a DC motor can be approximated by the following equation.

$$T_a \approx \emptyset \,.\, I_a$$
$$n \approx \frac{E_a}{\emptyset}$$

Ta : armature torque
Ia : armature current
n : armature speed
Ea : back of e.m.f.
Ø : magnetic flux

2.1. Separately Excited DC Motor

The separately excited DC motor is a DC electric motor with a magnetic winding connected to another voltage source that is different from the voltage source for the armature. The equivalent circuit of separately excited DC motor is shown in Figure 1. Based on Figure 1, the mathematical equation is also obtained.

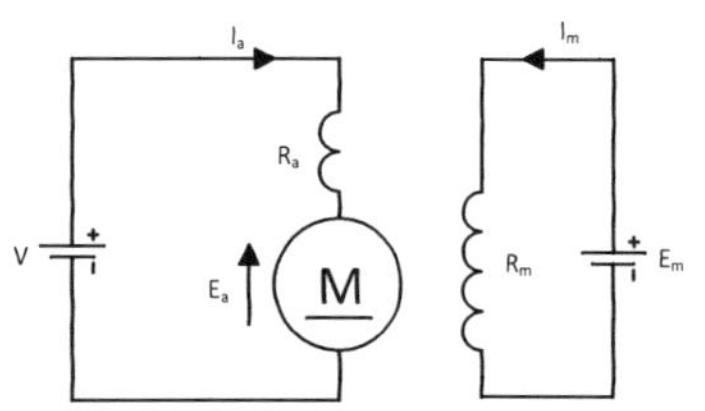

$I_a=I$;
$I_m=V_m/R_m$;
$V=E_a+I_a*R_a$

Figure 1. The equivalent circuit of separately excited DC motor [9].

The characteristics of a separately excited DC motor are identical to those of a shunt DC motor. Separately excited DC motors and shunt DC motors have a constant magnetic flux, so the characteristics of these two types of motors are identical. The discussion of the characteristics of a separately excited DC motor is presented in the discussion of the characteristics of a shunt DC motor.

2.2. Shunt DC Motor

The shunt DC motor is the same as a separately excited DC motor, except that the shunt winding is connected in parallel with the armature winding. The equivalent circuit of a DC shunt motor is shown in Figure 2. Based on Figure 2, the mathematical equation is also obtained.

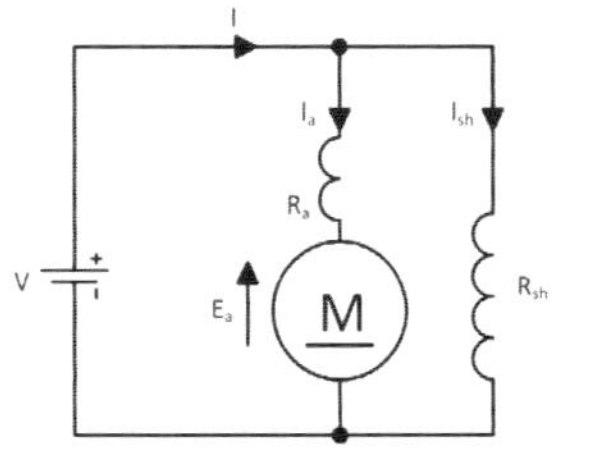

$I = I_a + I_{sh}$
$I_{sh} = V / R_{sh}$
$V = I_{sh} * R_{sh}$
$V = E_a + I_a * R_a$

Figure 2. The equivalent circuit of shunt DC motor [9].

Figure 1 is the equivalent circuit of a separately excited DC motor, while Figure 2 is the equivalent circuit of a shunt DC motor. Separately excited DC motor and shunt DC motor have a constant field current when connected to a DC source which is also constant, so the magnetic flux produced is also constant.

The characteristics of the armature torque to the armature current *[Ta=f(Ia)]* form a straight line (linear), ignoring the armature reaction based on the equation *Ta=Ø*Ia*, because is constant, when the armature torque increases, the armature current also increases *(Ta≈Ia)*. The characteristics of separate excitation DC motors and shunt DC motors for the relationship between armature torque and armature current are shown in Figure 3. The 0-A area is the no-load current to generate the starting torque.

The characteristics of the armature speed to the armature current *[n=f(Ia)]* are based on $n = \frac{E_a}{\emptyset}$, if the motor load increases, the armature speed decreases, because *Ea=V-Ia*Ra*. The decrease in armature speed is also influenced by the armature resistance *(Ra)*, so that Ra is made small so that the power loss is also small, so that the decrease in speed due to an increase in load is also small.

The characteristics of a separate excitation DC motor and a shunt DC motor for the relationship between armature speed and armature current are shown in Figure 4.

The characteristics of the armature speed to the armature torque *[n=f(Ta)]* form a linear curve. The magnetic flux is constant, so that the armature torque is proportional to the armature current *(Ta≈Ia)*. The characteristics of separate excitation DC motors and shunt DC motors for the relationship between armature speed and armature torque are shown in Figure 5.

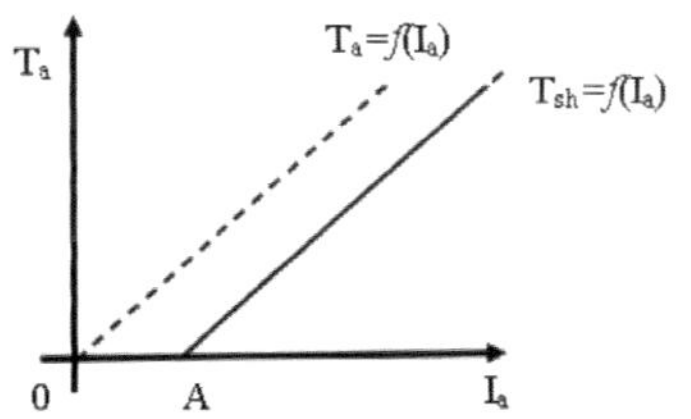

Figure 3. The characteristics between the armature torque and the armature current of the shunt DC motor [4].

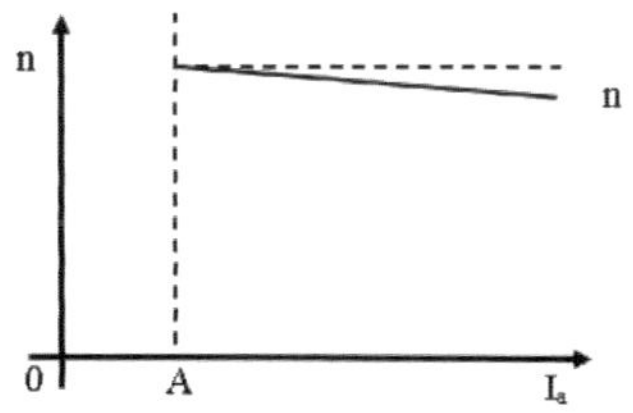

Figure 4. The characteristics of the armature speed and the armature current of the shunt DC motor [4].

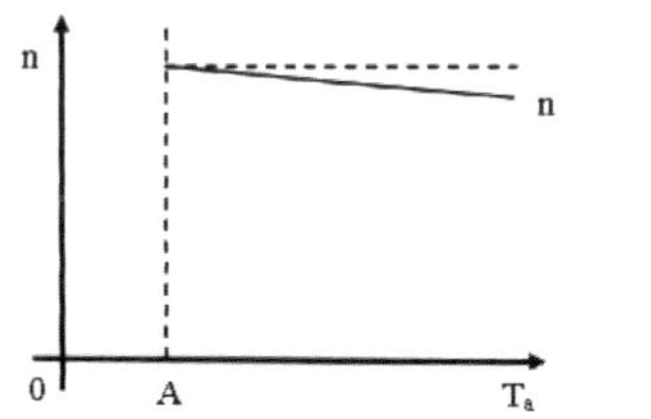

Figure 5. The characteristics of the armature speed and the armature torque of the shunt DC motor [4].

2.3. Series DC Motor

The series DC motor use series windings and connected in series with the armature windings. The equivalent circuit of a series DC motor is shown in Figure 6. Based on Figure 6, the mathematical equation is also obtained.

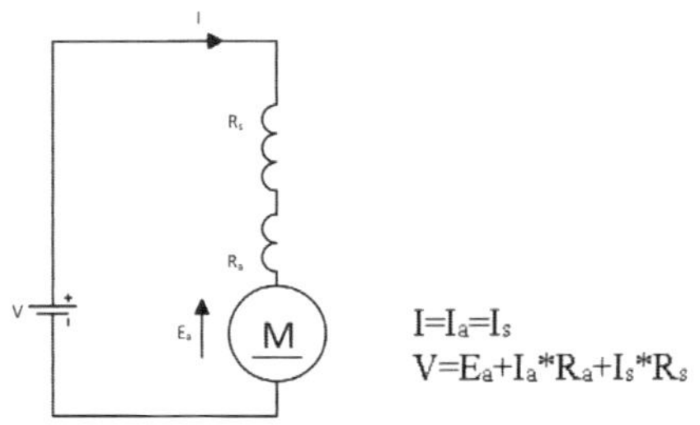

Figure 6. The equivalent circuit of series DC motor [9].

Figure 6 is the equivalent circuit of a series DC motor. The series field current is the same as the armature current, so when the armature current increases as the mechanical load increases, the series field current also increases *(I=Ia=Is)*.

The characteristics of the armature torque to the armature current *[Ta=f(Ia)]* form a curved line (quadratic) based on the equation *Ta=Ø*Ia*, because *Ø* is proportional to *Ia*, then *Ø=Ia*, so $Ta=Ia^2$. Figure 7 shows the characteristic curve of the armature torque relationship to the armature current for a series DC motor. The 0-A area is the no-load current to generate the starting torque

The characteristics of the armature speed to the armature current are based on *n=Ea/Ø,* if the motor load increases, the armature current also increases, and the series field also increases, so the armature speed decreases drastically, because $n = \frac{V - I_a . R_a}{I_a}$. Figure 8 shows the characteristic curve of the relationship between armature speed and the armature current for series DC motor.

Series DC motor produce high torque at low speeds, while at high speeds the armature torque is low. High armature torque requires high armature current. The high armature current means the series field current is also high, so the series magnetic flux is also high. Based on the equations *Ta=Ø.Ia* and *n=Ea/Ø,* the characteristic curve of armature speed to armature torque *[n=f(Ta)]* is identical to the curve *[n=f(Ia)]*. Figure 9 shows the characteristic curve of the relationship between armature speed and the armature torque for series DC motor.

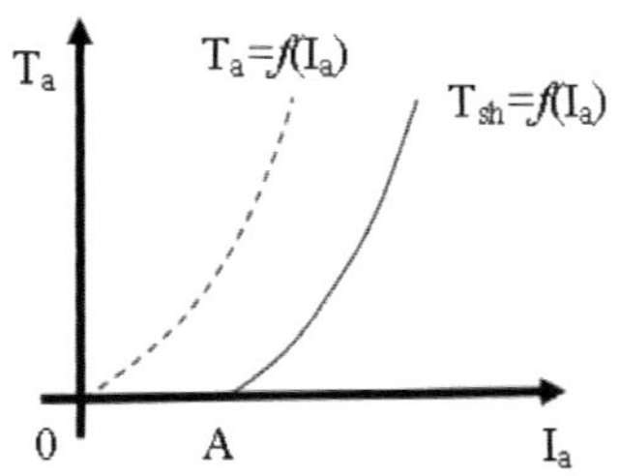

Figure 7. The characteristics between the armature torque and the armature current of the shunt DC motor [4].

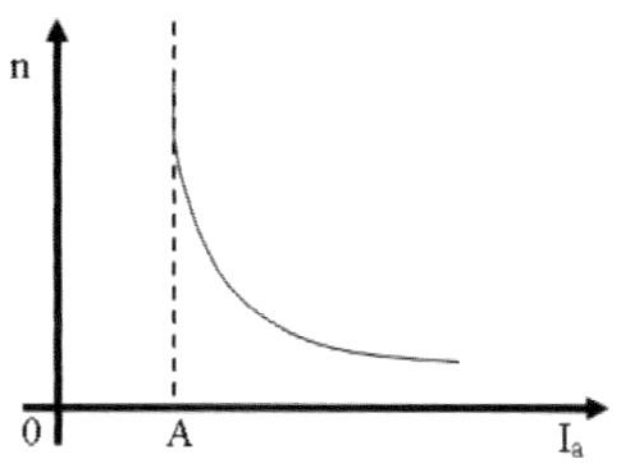

Figure 8. The characteristics of the armature speed and the armature current of the shunt DC motor [4].

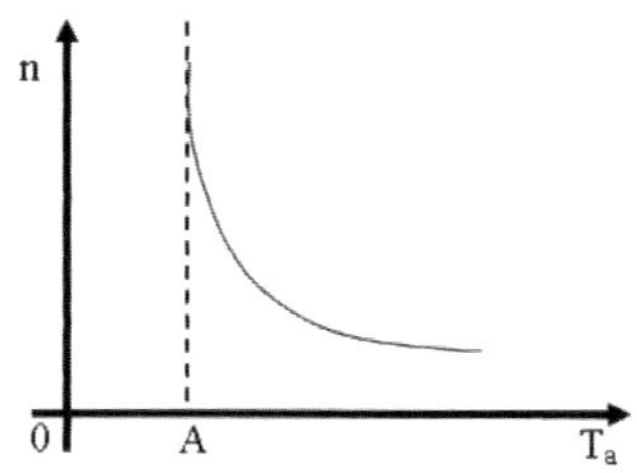

Figure 9. The characteristics of the armature speed and the armature torque of the shunt DC motor [4].

3. Experimental Setup and Simulink Models

Testing the characteristics of the DC motor is proven by running the DC motor on load. The DC generator is coupled to a DC motor as a load. By increasing the resistive load on the generator output voltage, the motor load also increases. For every increase in generator load, DC motor parameter data are recorded in the table including armature current, armature torque, and armature speed. Figure 10 shows the actual implementation of the characteristic test of a DC motor connected to a DC generator, while Figure 11 shows motor-generator shaft interconnection model using simulink. The generator speed input is connected to the motor speed output, while the motor torque input is connected to the generator torque output. [10].

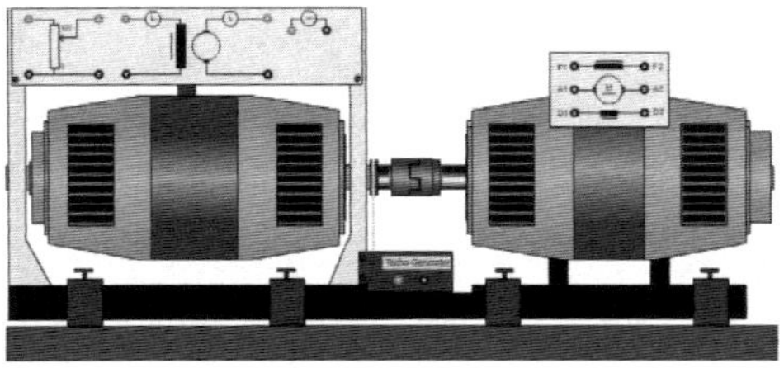

Figure 10. Motor-generator in real implementation.

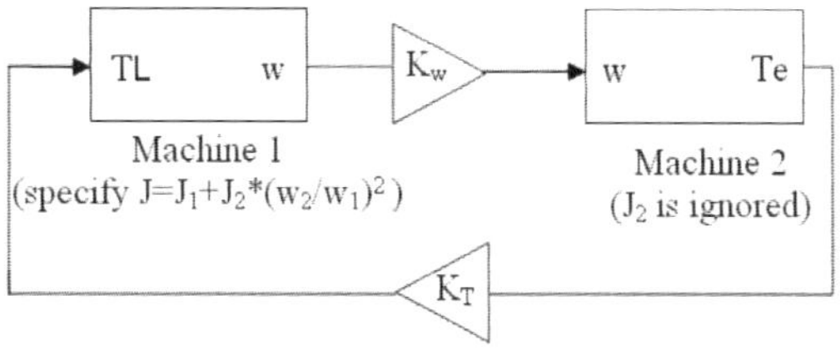

Figure 11. Model of motor-generator in virtual laboratory using Simulink [10].

3.1. Testing Separately Excited DC Motor

Implementation of a separately excited DC motor using Simulink is presented in Figure 12. Based on the actual implementation circuit in Figure 13, the characteristic test procedure can be implemented using Simulink. The separately excited DC motor requires two DC voltage sources, including fixed DC voltage source of 220V and variable DC voltage source of 0-220V. The fixed DC voltage source is to supply the field winding, while the variable voltage source is to supply the motor armature. Rheostat is needed to vary the field winding voltage to regulate the magnetic flux. The DC generator is coupled to a shaft with a DC motor. The DC generator is used as the load of the DC motor. When a DC motor rotates the DC generator, the DC generator generates a voltage which is connected to variable resistive load. By varying the resistance value of the resistive load variable, the generator output current increases, so that the generator torque also increases. The shaft coupling between the DC motor and the DC generator causes the torque generated to be the same between the two. With proper Simulink modeling, the output parameter of the DC motor model, the armature speed, is connected to the input speed of the DC generator model, and then the output parameter of the DC generator model, the armature torque, is fed back to the input of the DC motor model.

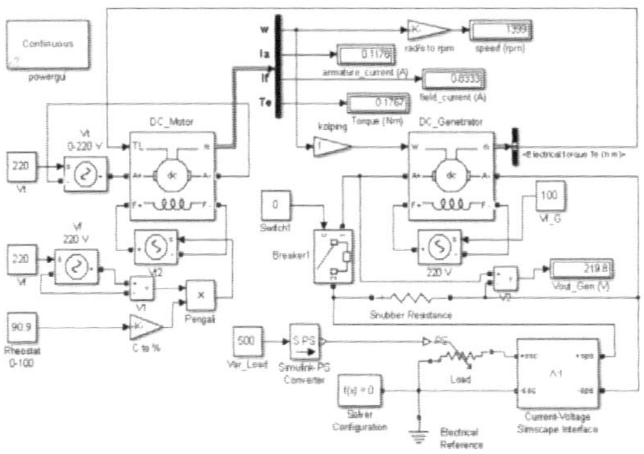

Figure 12. Simulink model for separately excited DC motor.

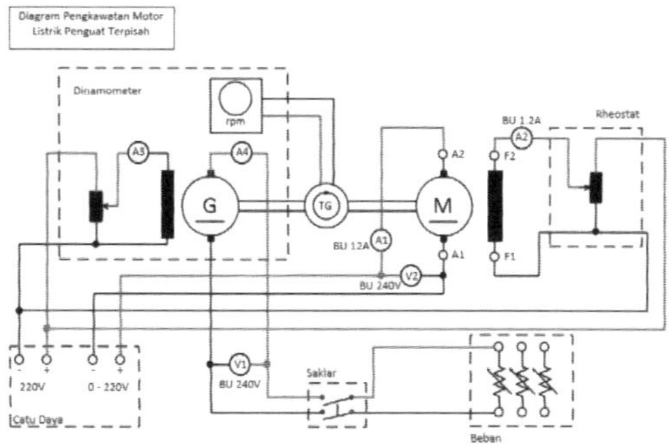

Figure 13. Circuit diagram for separately excited DC motor in actual implementation.

The process of running a DC motor can also be implemented on Simulink. With the feature of running an infinite simulation, starting the separately excited DC motor includes: 1) supplying the field winding so that a magnetic flux is generated in the DC motor, 2) supplying the DC motor armature gradually starting from 0V and slowly rising to a maximum limit of 220V, 3) adjusting the speed armature by varying the field winding voltage using a rheostat until the armature speed is 1400rpm no-load with an armature power supply of 220V.

The process of running a separately excited DC motor can also be implemented on Simulink. With the feature of running an infinite simulation, starting the DC motor includes: 1) supplying the field winding so that a magnetic flux is generated in the DC motor, 2) supplying the DC motor armature gradually starting from 0V and slowly rising to a maximum limit of 220V, 3) adjusting the speed armature by varying the field winding voltage using a rheostat until the armature speed is 1400rpm no-load with an armature power supply of 220V.

Starting from the no-load condition, the DC motor is loaded by adjusting the variable resistive load on the DC generator until the appropriate DC motor armature current is obtained. Every time the load increases on the DC generator, a step is made to increase the DC motor armature current of 0.5A then record each parameter including the armature speed and armature speed. Data collection is limited until the maximum armature current value is 5A.

3.2. Testing of Shunt DC Motor

Implementation of a shunt DC motor using Simulink is presented in Figure 14. Based on the actual implementation circuit in Figure 15, the characteristic test procedure can be implemented using Simulink. Unlike the separately excited DC motor, the shunt DC motor only needs one type of DC voltage source, namely a 0-220V variable DC voltage source. The power supply for the field winding is taken from the DC motor armature power supply. Rheostats are also needed in shunt DC motors to vary the field winding voltage to regulate the magnetic flux. There is an additional tool, namely a motor starter which is used to provide a 0-220V variable power supply pause to generate magnetic flux. Just as in the separately excited DC motor test, the DC generator is coupled to the shaft with a shunt DC motor.

The process of running a shunt DC motor can also be implemented on Simulink. In the same way, running the simulation feature indefinitely, starting the shunt DC motor includes: 1) initial conditions, the motor starter device is in an off or stop condition, the field winding rheostat is in 100% condition, 2) supplying the field winding using a variable power supply up to a maximum voltage of 220V so that the magnetic flux is generated in the DC motor, 3) supplying the motor armature by turning the motor starter from the stop condition to the run condition, so that the armature rotates, 3) adjusting the speed of the armature by varying the field winding voltage using a rheostat until the armature speed is 1400rpm no-load with 220V armature power supply.

The same steps are also carried out in testing the shunt DC motor, starting from the no-load condition, the DC motor is loaded by adjusting the variable resistive load on the DC generator until the appropriate DC motor armature current is obtained. Every time the load increases on the DC generator, a step is made to increase the DC motor armature current by 0.5A then record each parameter including the armature speed and armature speed. Data collection is limited until the maximum armature current value is 5A.

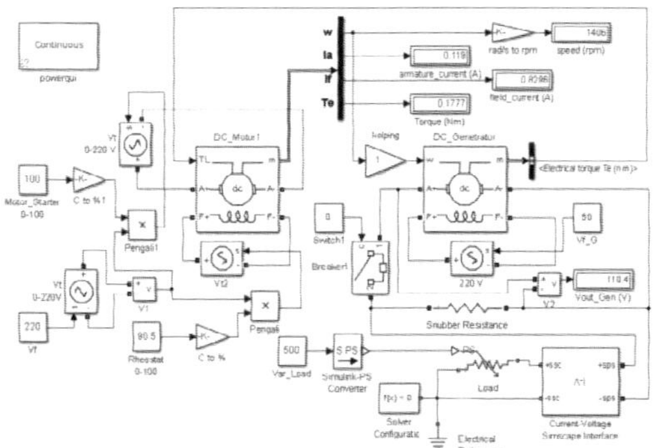

Figure 14. Simulink model for shunt DC motor.

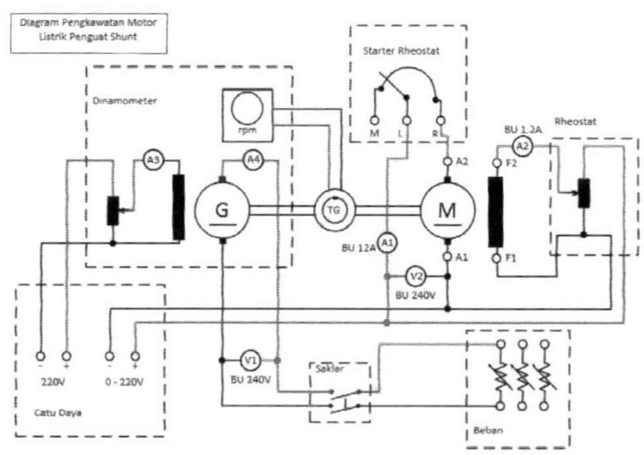

Figure 15. Circuit diagram for shunt DC motor in actual implementation.

3.3. Testing of Series DC Motor

Implementation of series DC motor using Simulink is presented in Figure 16. Based on the actual implementation circuit in Figure 17, the characteristic test procedure can be implemented using Simulink. Series DC motors must run with on-load because if run with no-load, the armature speed will be very high. This is because when the series DC motor is run at no-load condition, the armature current is very small, because the field winding is connected in series with the armature; the magnetic flux produced is also small so the armature speed will be very high. Adjusting the armature speed on a series DC motor is very difficult, so this type of series DC motor is avoided for purposes that require a relatively constant armature speed. The series DC motor produces high torque at low armature speeds, so that at high load increases, the series DC motor has a stronger torque.

The process of running a series DC motor can also be implemented on Simulink. In the same way, using the unlimited simulation feature, starting a series DC motor includes: 1) in the initial state, the DC generator is already in a loaded condition, meaning that the breaker leading to the variable resistive load is in the on condition and the field winding power supply on the DC generator is on. , 2) supplying the motor armature gradually starting from 0V and slowly rising to the maximum limit of 220V, 3) it is very difficult to get the desired armature speed, it can be obtained by adjusting the variable resistive load on the generator so that the motor armature speed is 1400rpm at the armature supply voltage 220V DC motor.

Starting at the on-load condition, the armature speed is 1400rpm, in the same way in the previous test, the load is increased until the appropriate DC motor armature current is obtained. Every time the load increases on the DC generator, a step is made to increase the DC motor armature current by 0.5A then record each parameter including the armature speed and armature speed. Data collection is limited until the maximum armature current value is 5A.

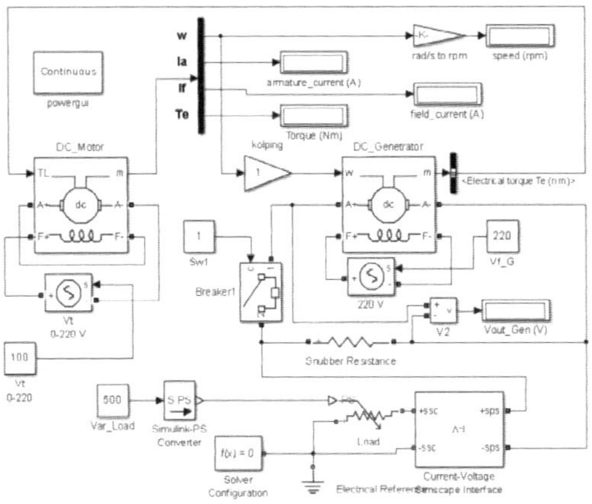

Figure 16. Simulink model for series DC motor.

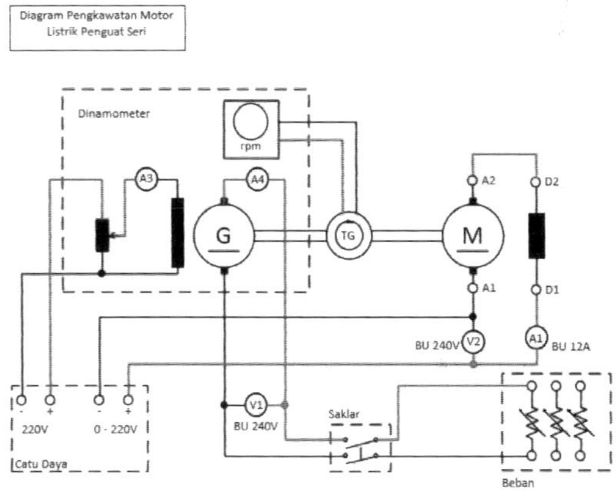

Figure 17. Circuit diagram for series DC motor in actual implementation.

4. Result and Discussion

The results of testing the characteristics of a DC motor based on the Simulink simulation model for separate, shunt, and series excitation DC motors are shown in Table x. Each test data includes armature current, armature speed, and armature torque as data to graph the characteristics of each type of DC motor.

Based on the data in Table 1, a graph of the characteristics of each DC motor can be made, including the characteristics of armature torque to armature current [Ta=f(Ia)], characteristics of armature speed to armature current [n=f(Ia)], and characteristics of armature speed to armature torque [n=f(Ta)]. The characteristics of a separate excitation DC motor are shown in Figure 18, the characteristics of a shunt DC motor are shown in Figure 19, and the characteristics of a series DC motor are shown in Figure 20.

Table 1. The results of the simulation of testing the characteristics of a DC motor.

Separately Excited			Shunt			Series		
Ia (A)	N (rpm)	Ta (Nm)	Ia (A)	N (rpm)	Ta (Nm)	Ia (A)	N (rpm)	Ta (Nm)
1,032	1397	1,535	1,021	1396	1,533	0,830	1400	1,242
1,536	1395	2,304	1,533	1394	2,301	1,605	721,2	4,637
2,056	1394	3,084	2,006	1392	3,011	2,070	557,7	7,714
2,546	1392	3,819	2,541	1390	3,815	2,506	460	11,303
3,031	1390	4,547	3,025	1388	4,542	3,005	382	16,253
3,497	1388	5,246	3,551	1386	5,329	3,503	327	22,01
3,995	1386	5,993	4,052	1384	6,083	4,003	285	28,845
4,525	1384	6,787	4,585	1382	6,883	4,506	253	36,554
5,015	1382	7,522	5,149	1380	7,731	5,003	227	45,06

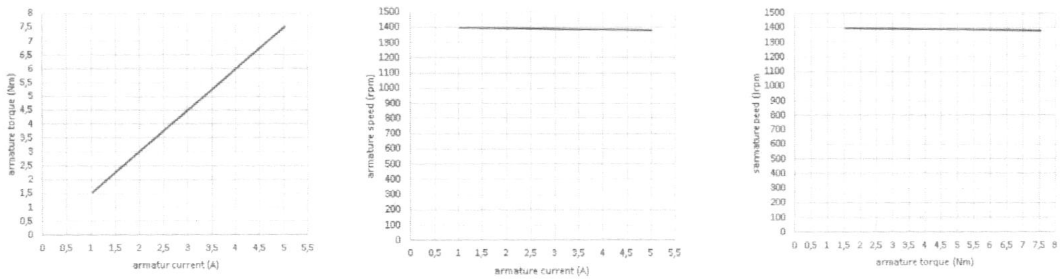

(a) (b) (c)

Figure 18. Separately excited DC motor characteristics graph, (a) Ta=f(Ia), (b) n=f(Ia), (c) n=f(Ta).

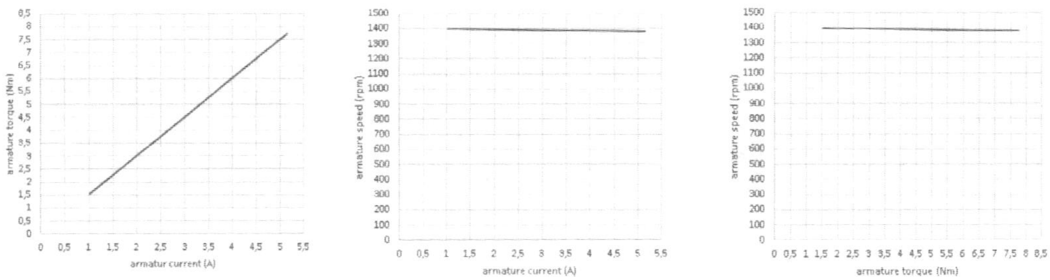

(a) (b) (c)

Figure 19. Shunt DC motor characteristics graph, (a) Ta=f(Ia), (b) n=f(Ia), (c) n=f(Ta).

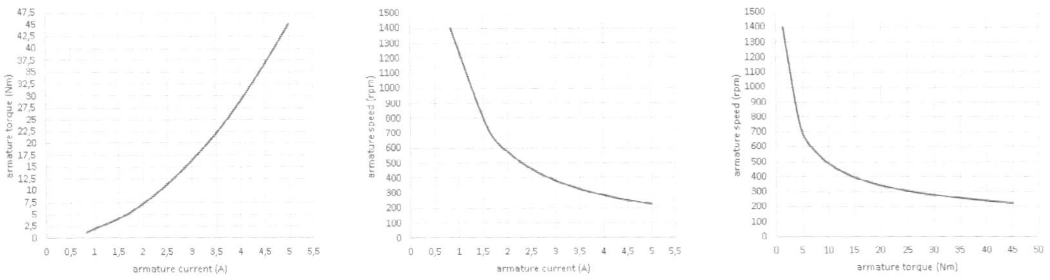

(a) (b) (c)

Figure 20. Series DC motor characteristics graph, (a) Ta=f(Ia), (b) n=f(Ia), (c) n=f(Ta).

Based on the test data, the DC motor is run from zero rpm to 1400rpm without load on a 220V voltage source. Except for the series DC motor, it runs from zero rpm to 220V under load. We know that a series DC motor should not be run under no-load condition, because the armature speed will be very high. Testing data retrieval starts from the no-load condition. Variations in the load on the generator by manipulating the resistive load variable cause an increase in the load on the DC motor so that the armature current also increases. The load variation is made in 0.5A increments of the motor armature current until an armature current of 5A is obtained. Every change in load increase or increase in armature current, data of armature speed and armature torque are recorded, so that three data are obtained at one time of increasing load, including armature current, armature speed, and armature torque.

The data that has been obtained on each type of DC motor, then graphed the characteristics of the relationship between armature torque and armature current, armature speed to armature current, and armature speed to armature torque. Three kinds of characteristics for each type of motor have been obtained and are in accordance with the theory discussed earlier. Thus, testing the characteristics of a DC motor coupled with a DC generator has been successful. Similar models can be used for testing other types of electrical machines such as DC generators, induction motors, synchronous motors, and synchronous generators.

5. Conclusion

In this study, testing the characteristics of a DC motor coupled with a DC generator as a load has been completed. The testing process is carried out by infinitely simulating so that the DC motor starting process starts from zero rpm to a no-load speed of 1400 rpm at a source voltage of 220V. From no-load conditions and variations in load increases, data on changes related to armature current, armature speed, and armature torque have been obtained. The graph of the characteristics of all types of DC motors results from the test is very in line with the theory.

6. References

[1] B. L. Theraja, *A textbook of electrical technology*. S. Chand Publishing, 2008.

[2] H. Djeghloud, M. Larakeb, and A. Bentounsi, "Virtual labs of conventional electric machines," *2012 Int. Conf. Interact. Mob. Comput. Aided Learn. IMCL 2012*, no. Imcl, pp. 52–57, 2012, doi: 10.1109/IMCL.2012.6396450.

[3] P. Verma, P. Gupta, and B. Kumar, "Graphical user interface (GUI) to study DC motor dynamic characteristics," *Proc. Int. Conf. Electron. Commun. Aerosp. Technol. ICECA 2017*, vol. 2017-Janua, pp. 649–654, 2017, doi: 10.1109/ICECA.2017.8212746.

[4] V. K. Mehta and R. Mehta, *Principles of electrical machines*. S. Chand Publishing, 2008.

[5] D. Daryanto, "Media pembelajaran peranannya sangat penting dalam mencapai tujuan pembelajaran," *Gava Media*, 2013.

[6] M. Roblyer and A. H. Doering, *Integrating educational technology into teaching*. USA: Pearson, 2007, 2013.

[7] A. K. Triatmaja, M. Muchlas, and Y. Wardana, "Virtual Laboratorium Teknik Digital Berbasis Mobile Virtual Reality," *J. Edukasi Elektro*, vol. 5, no. 1, pp. 1–13, 2021, doi: 10.21831/jee.v5i1.38705.
[8] S. Ayasun and C. O. Nwankpa, "Transformer tests using MATLAB/Simulink and their integration into undergraduate electric machinery courses," *Comput. Appl. Eng. Educ.*, vol. 14, no. 2, pp. 142–150, 2006, doi: 10.1002/cae.20077.
[9] Y. H. Ku, *Electric machinery*, vol. 319, no. 4. 1985.
[10] Mathworks, "DC Machine," 2021. https://www.mathworks.com/help/physmod/sps/powersys/ref/dcmachine.html (accessed Sep. 26, 2021).

Development of Electrical Circuit Learning Media Using Virtual Simulation

[1]E S Damarwan, [2]M L Hakim, [3]A S J Wardhana, [4]N Kholis

[1234] Department of Electrical Engineering Education, Faculty of Engineering, UNY

Email: ekoswie@uny.ac.id

Abstract. The Covid-19 pandemic has had a major impact on the world of education. One of the impacts is related to the learning process of electrical circuits. Online lectures make students have difficulty in studying electrical circuit courses. An appropriate learning media is needed as a solution to overcome these problems. This study aims to develop electrical circuit learning media using virtual simulations and determine the feasibility of applying the media. This type of research is research and development with the ADDIE model. The stages of development carried out are Analyze, Design, Development, Implementation, and Evaluation. The research was conducted at the Electrical Engineering Education Study Program, Faculty of Engineering, UNY. The virtual simulation used is a Falstad circuit simulation. This simulation can be accessed online or offline. In this research, the focus is on developing learning media using online Falstad simulation. The results showed that 1) the electric circuit learning media using virtual simulations can run well, 2) based on user trials, the electrical circuit learning media using virtual simulation is classified as very feasible.

Keywords: electrical circuits, virtual simulations, the covid 19 pandemic

1. Introduction

The Covid-19 pandemic affects all activities of human life, including in the field of education. Educational activities usually carried out face-to-face in class have now been stopped to avoid the spread of the Covid-19 pandemic. The cessation of face-to-face learning activities is shifted to online-based learning (distance) by utilizing technology that is connected to the internet. This distance learning system applies to all levels of education, including vocational education.

Vocational education is one of the formal educations with the aim of focusing more on increasing work-ready competencies in specific fields. One of the fields in vocational education is the electricity sector. A quality learning process is needed to prepare graduates with electrical competence who can compete in the world of work. The learning process carried out in vocational education is divided into three, namely theory, practice, and field. One of the electricity courses given is the Electrical Circuit Course. In the era of the COVID-19 pandemic, lectures became very difficult because they had to be carried out online. The electrical circuit course is the basic guideline for students in the electricity sector. This course studies the analysis of electrical circuits that require understanding in principle, theoretically, and practically. The learning process of electric circuits can be qualified if one of the course's learning objectives is achieved.

To achieve the learning objectives, it is essential media that can improve the ability of students [1]. Learning media can be used to stimulate the thoughts, feelings, and interests and willingness of students so that the learning process occurs to achieve learning objectives [2]. Based on the above opinion, it can be concluded that to achieve the learning objectives of electric circuits. Learning media is needed that can stimulate the thoughts, interests and feelings of students to improve their abilities.

Based on the results of observations of the implementation of the electrical circuit course, data were obtained if: (1) the impact of the covid-19 pandemic made it difficult for students to understand the basic concepts of electric circuits. It happens because learning is carried out online. (2) The need to increase student interest and motivation in the electric circuit learning process. Increased motivation can be through media that is easy to understand and interesting for students. One of the media that should be applied is virtual simulation-based media. Virtual simulations can prepare students to understand basic concepts and materials more efficiently [3]. The virtual simulation that will be used is the Falstad circuit. The development of electric circuit learning media using virtual simulation is the title of the research offered to increase student competence.

2. Main Theory

Learning media is a tool that can help the teaching and learning process so that learning objectives can be adequately achieved. Learning media are everything that can be used to channel messages to stimulate the thoughts, feelings, and interests and willingness of students to achieve learning objectives effectively [2]. Learning media can be understood as anything that can convey and distribute messages from sources in a planned manner to create a conducive learning environment [4]. Based on the description above, it can be concluded that learning media is anything that is used to clarify the meaning of the message conveyed so that it stimulates the thoughts, feelings, attention, and interests and willingness of students so that learning objectives are achieved effectively and efficiently.

2.1. Virtual Simulation

Virtual simulation is a medium that serves to convey ideas and concepts in the form of digital simulations. Virtual simulations can provide students with the opportunity to apply theoretical concepts to the simulated experience [5]. The application of virtual simulation media can improve students' mastery of concepts and higher-order thinking skills [6]. The use of virtual simulations in education is needed to explain a material or problem that is difficult for students to understand if they do not use teaching aids because the simulation can be an example of what should be done. Virtual simulation is very useful to be used to explain a complicated variable relationship about a teaching material so that it can increase the involvement of students actively in finding concepts and knowledge in learning. The virtual simulation that will be used is the Falstad Circuit which can be accessed via the page: https://www.falstad.com/circuit/circuitjs.html. Falstad simulation is one of the simulations that can be applied to electrical circuit courses. This simulation is easy to use and understand in solving electrical circuits. In general, virtual simulation media is beneficial in the practicum-based learning process because, with the virtual simulation, students can learn quickly in understanding the material that has been studied [7].

2.2. Electrical Circuit Course

Electrical circuit is a compulsory subject for electrical engineering students. This course studies the basics and analysis of electrical circuits. In this simulation-based learning media, three primary materials will be developed, namely: 1) node voltage analysis, 2) mesh analysis and 3) resonance in RLC series circuits. Node analysis is based on Kirchoff's Law I / KCL, where the amount of current entering and leaving the branching point will be equal to zero, where the voltage is an unknown parameter [8]. Mesh is a property of plane circuits and is not defined for non-plane circuits. Mesh analysis can be used only on circuits that lie in one plane. A planar circuit is a circuit on a plane surface so that no branches pass above or below other components [8]. To solve it by using the concept of mesh current and Kirchoff's 2nd Law (Kirchoff Voltage Law/KVL). Resonance occurs when the magnitude of XL = XC [9].

3. Methods

This study is included in research and development (R&D) research. The purpose of this development research is to develop an electric circuit learning media using virtual simulation. The development model used in this research is the ADDIE approach. The ADDIE approach includes: Analysis, design, develop, implementation, and evaluation [10]. The implementation chart of the ADDIE model is shown in Figure 1 below.

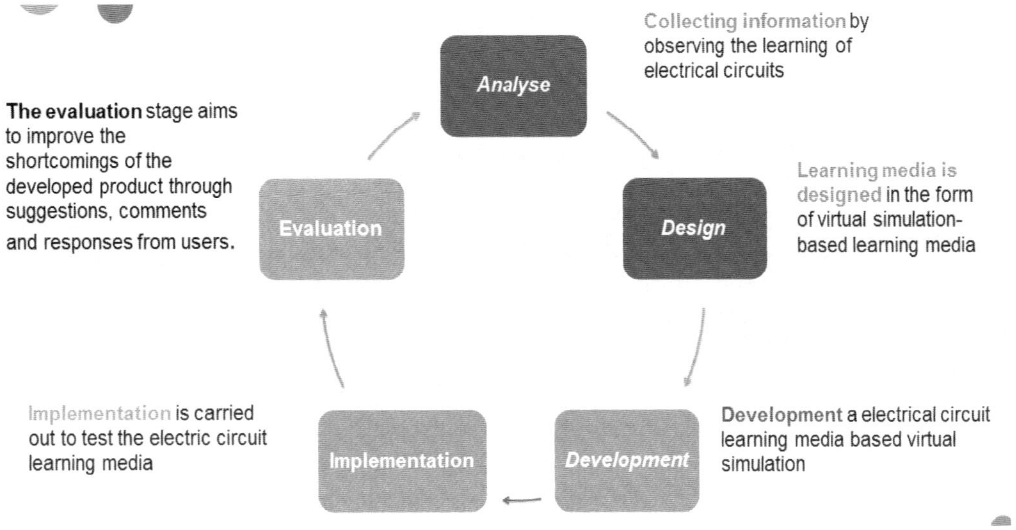

Figure 1. Implementation of the ADDIE Model

The analysis activity begins with identifying the problem of implementing learning in the electrical circuit course, looking for differences between learning objectives and the reality of the learning process that has been implemented. Analyzing the urgency of whether or not the development of learning media is necessary and adapted to various aspects of supporting the electric circuit learning media. The results of the analysis showed that: (1) the impact of the covid-19 pandemic made it difficult for students to understand the basic concepts of electrical circuits. This happens because learning is carried out online. (2) The need to increase student interest and motivation in the electric circuit learning process. The electrical circuit course is the basic guideline for students in the electricity sector. This course studies the analysis of electrical circuits that require understanding in principle, theoretically, and practically.

The next stage is design, which is an activity to design learning media products that will be developed starting from determining the material to be studied. Determine the design of the electrical circuit be developed, and adapt it to the Falstad virtual simulation. The design of the material to be studied is shown in Figure 2 below.

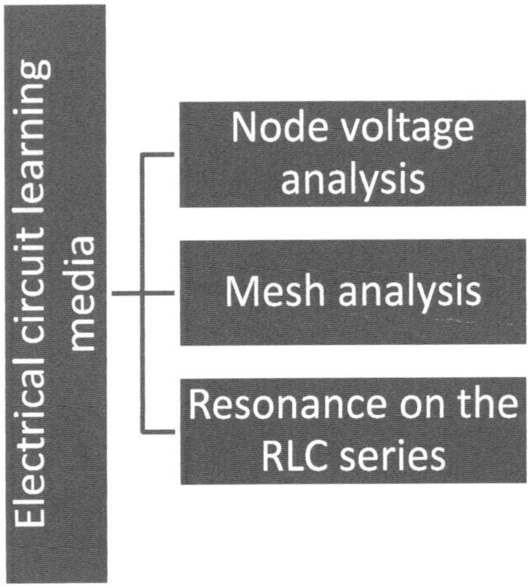

Figure 2. Material to be Designed

After the material is determined, the next step is to determine the electrical circuit that will be included in the electric circuit learning media. The circuit that will be included is an electrical circuit that will be analyzed using node analysis, mesh analysis, and the RLC series circuit shown in Figures 3, 4, and 5 below.

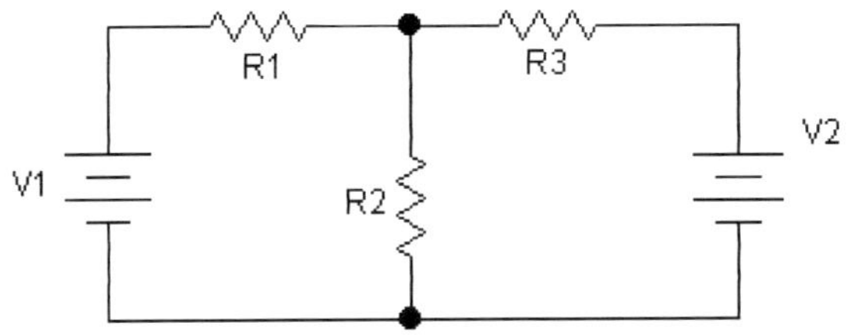

Figure 3. The Electrical Circuit for Node Analysis.

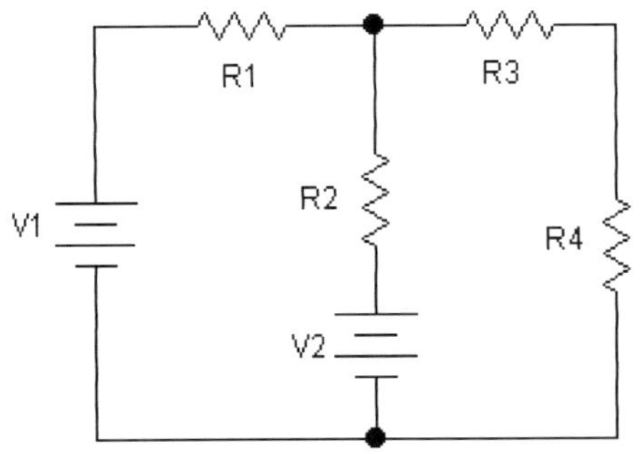

Figure 4. The Electrical Circuit for Mesh Analysis

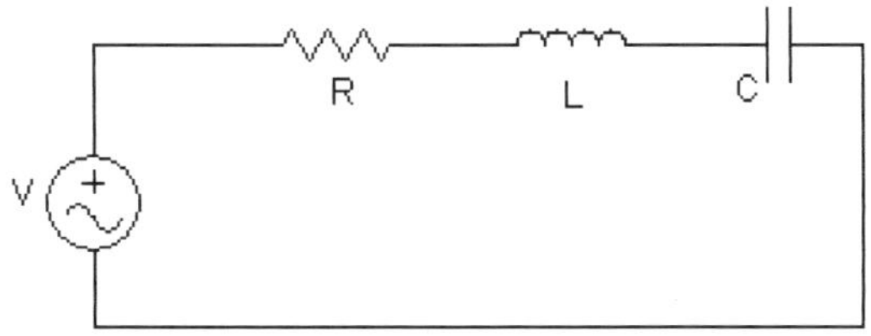

Figure 5. RLC Series Circuit

The next step is to make a theoretical circuit solution formula. To complete the theoretical node analysis, the steps taken are as follows. 1) Specifies the reference node. 2) Determine the Node voltage (V). 3) Create a circuit equation [11]. In Figure 3 the circuit equation according to the node analysis rules is as follows.

$$\frac{V-Vg}{R2} = \frac{V1-V}{R1} + \frac{V2-V}{R3} \tag{1}$$

To complete the mesh analysis theoretically is as follows. 1) Determine the type of circuit (planar circuit). 2) Make loop assumptions. 3) Make a circuit equation [11]. In Figure 4, the circuit equations according to the mesh analysis rule are as follows.
Loop 1:

$$-V1 + R1I1 + R3I3 + V2 = 0 \tag{2}$$

Loop 2:

$$R2I2 + R4I4 - V2 + R3I3 = 0 \tag{3}$$

To complete the resonance analysis on the RLC series circuit is as follows.

$$XL = XC \tag{4}$$

$$2\pi fL = \frac{1}{2\pi fC} \tag{5}$$

$$f^2 = \frac{1}{4\pi^2 LC} \tag{6}$$

The third stage is development, which is the stage used to realize the concept that has been designed into a product that is ready to be implemented. At this stage, the researchers developed an electric circuit learning media using virtual simulation. The virtual simulation used is the Falstad circuit.

The next stage is implementation, which is the stage to apply the product that has been developed in real conditions. Implementation is done on a limited basis. Respondents in the implementation were students of the UNY Electrical Engineering Education Study Program who took electrical circuit courses. Evaluations are the last stage. After a limited application, the electrical circuit learning media was evaluated to provide feedback on the following application.

4. Results and Discussion

At the development stage, the main step is to incorporate the circuit design into the virtual simulation. In this study, there are three electrical circuits that will be developed. The results of the node analysis in the virtual simulation are shown in Figure 6 below.

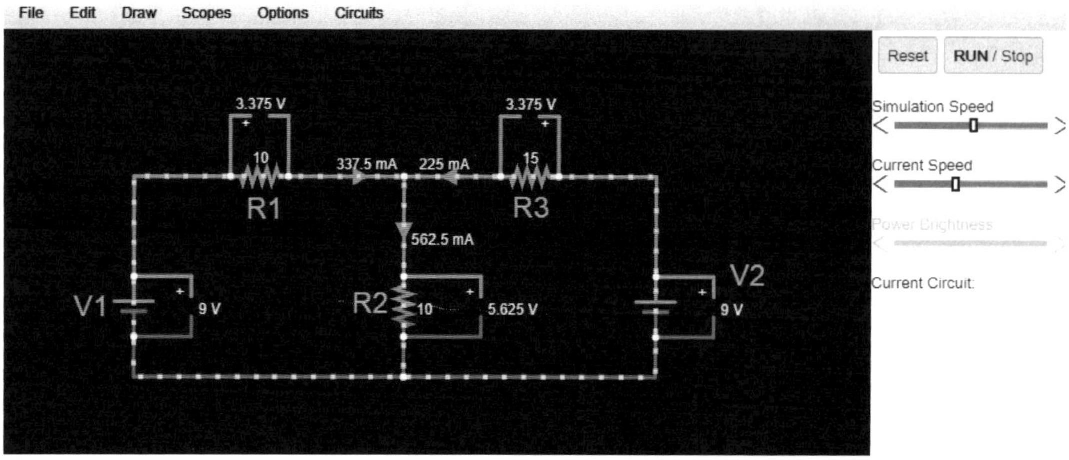

Figure 6. Virtual Simulation of Node Analysis

This circuit is composed of V1 =V2=9V, R1=R2=10Ω, R3 = 15Ω. Based on the results of the electric circuit learning media, it was found that there was no difference between theoretical calculations and virtual simulations. In detail, the simulation results are shown in Table 1 below.

Table 1. Result of Virtual Simulation for Analysis Node

VS		Calculation Voltage (V) & Current (A)						Simulation Voltage (V) & Current (A)					
V1	V2	VR1	VR2	VR3	IR1	IR2	IR3	VR1	VR2	VR3	IR1	IR2	IR3
9	9	3.375	5.625	3.375	0.3375	0.5625	0.225	3.375	5.625	3.375	0.3375	0.5625	0.225

The second material is mesh analysis in electrical circuits. The results of the mesh analysis in the virtual simulation are shown in Figure 7 below.

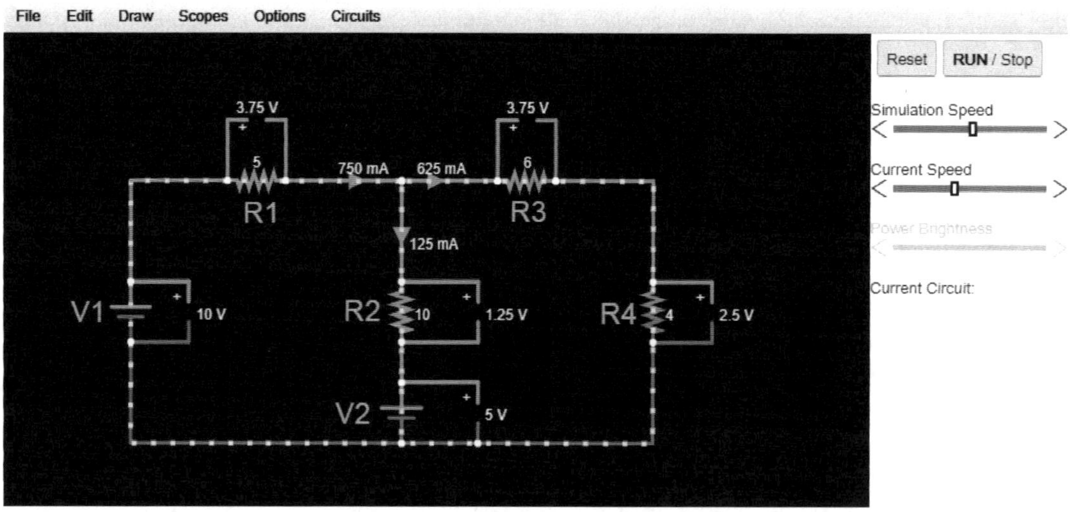

Figure 7. Mesh Analysis Simulation

This circuit is composed of V1=10V, V2=5V, R1=5Ω, R2=10Ω, R3=6Ω, R4=4Ω. Based on the results of the electric circuit learning media, it was found that there was no difference between theoretical calculations and virtual simulations. In detail, the results of the mesh analysis simulation are shown in Table 2 below.

Table 2. Result of Virtual Simulation for Analysis Mesh

VS		**Calculation Voltage (V) & Current (A)**							**Simulation Voltage (V) & Current (A)**						
V1	V2	VR1	VR2	VR3	VR4	IR1	IR2	IR3	VR1	VR2	VR3	VR4	IR1	IR2	IR3
10	5	3.75	1.25	3.75	2.5	0.75	0.125	0.625	3.75	1.25	3.75	2.5	0.75	0.125	0.625

The third material is resonance analysis in RLC series circuits. This circuit is composed of V source = 8VPP, R = 2K Ω, L = 300 mH, C = 0.01 μF. Based on theoretical calculations, the magnitude of the resonant frequency is 2907, 23 Hz. The results of the resonance analysis in the virtual simulation are shown in Figure 8 below.

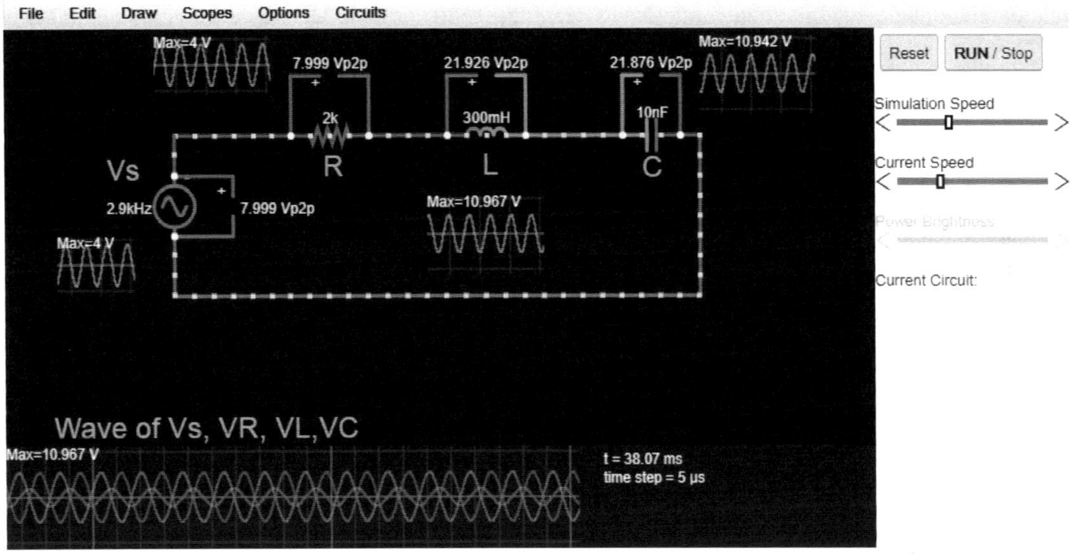

Figure 8. Virtual Simulation of Resonance in RLC Series Circuits

Based on the results of the electric circuit learning media, it was found that there was no difference between theoretical calculations and virtual simulations on resonance material. The experimental results of the resonance analysis are shown in Table 3 below.

Table 3. Resonance Analysis Results on RLC Series Load

Frequency (Hz)	Volt	Result			
	VR (Vp-p)	I Rms	XL	XC	Z
1000,00 Hz	1,129	199,60μA	1884Ω	15923,6 Ω	14181,3 Ω
2000,00 Hz	3,449	609,91 μA	3768 Ω	7961,8 Ω	4646,3 Ω
2907,23 Hz	**7,997**	**1,426mA**	**5477,2 Ω**	**5477,2 Ω**	**2000 Ω**
4000,00 Hz	3,901	690,299 μA	7536 Ω	3980, 9 Ω	4079,1 Ω
5000,00 Hz	2,426	429,916 μA	9420 Ω	3184,7 Ω	6548,2 Ω

IOP Publishing
2111 (2021) 012042 doi:10.1088/1742-6596/2111/1/012042

When there is a resonant frequency, the voltage across the resistor is equal to the source voltage. The effective current is at its peak and the impedance is equal to the resistance in the resistor. Figure 9 shows a graph between the XL, XC, and Z.

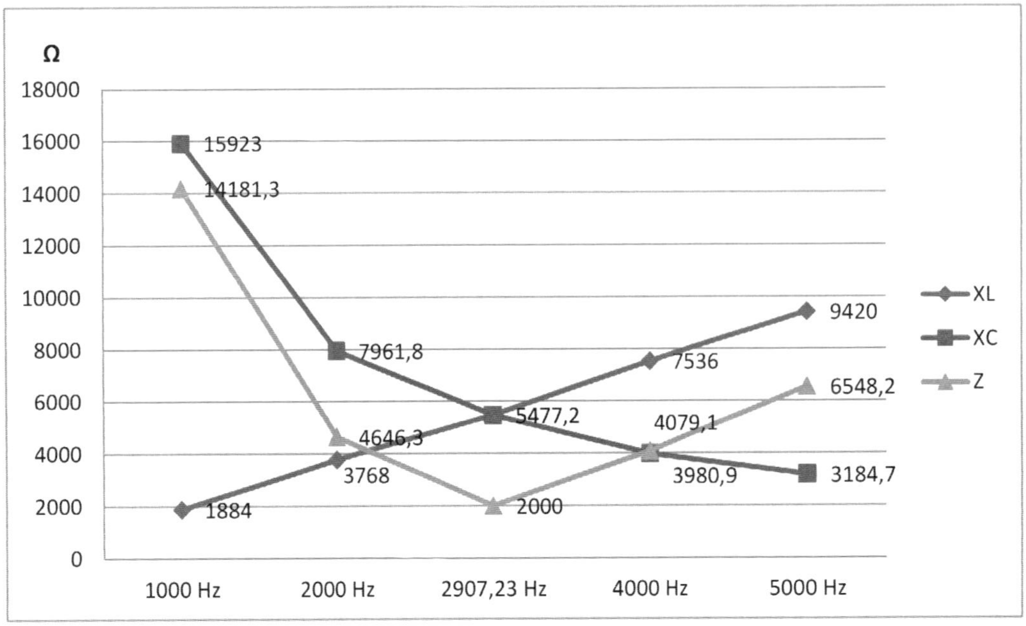

Figure 9. XL, XC, and Z Graphs on the RLC Series Circuit

After the development stage is complete, the next stage is implementation. Implementation is carried out on a limited basis by involving students. In detail, the feasibility of using electric circuit learning media is reviewed from four aspects, namely material, media, ease of use, and motivation. The detailed feasibility results are shown in Figure 10 below.

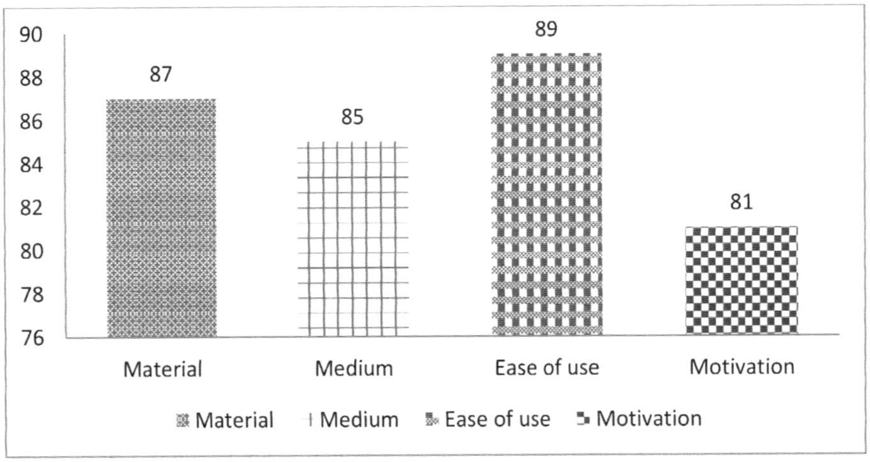

Figure 10. Feasibility Test Results

Based on Figure 10, it can be explained that the results of the implementation of electric circuit learning media using virtual simulations are very feasible to use. Virtual simulation media is beneficial in the learning process [7]. Students can learn quickly in understanding the material that has been studied.

5. Conclusion
Due to the COVID-19 pandemic, learning must be carried out online. Students find it difficult, especially in electrical circuit courses. Therefore, a breakthrough was made to develop electric circuit learning media. Development of electric circuit learning media using the ADDIE model. The stages of development carried out are Analyze, Design, Development, Implementation, and Evaluation. The electric circuit learning media is designed using Falstad circuit simulation. The simulation results show that node analysis, mesh analysis, and resonance analysis on RLC series loads are the same as theoretical calculations. In terms of the material aspect, media aspect, ease of use, and motivation, the electrical circuit learning media using virtual simulation is very feasible.

6. References
[1] Kartikasari. Pengaruh Media Pembelajaran Berbasis Multimedia Terhadap motivasi dan Hasil Belajar Materi System Pencernaan Manusia. *Jurnal Dinamika Penelitian (63)*. 2016. 63-68

[2] Sukiman. *Pengembangan Media Pembelajaran*. Yogyakarta: Pedagogia. 2012.

[3] Coyne, E., Calleja, P., Forster, E., & Lin, F. A review of virtual-simulation for assessing healthcare students' clinical competency. *Nurse Education Today*. Volume 96. 2020.10462. ISSN 0260-6917, https://doi.org/10.1016/j.nedt.2020.104623. (https://www.sciencedirect.com/science/article/pii/S0260691720314738)

[4] Munadi, Yudhi. *Media Pembelajaran: Sebuah Pendekatan Baru*. Jakarta: Referensi Melinda. 2013.

[5] Verkuyl, M., & Hughes, M. Virtual Gaming Simulation in Nursing Education: A mixed-methods study. *Clinical Simulation in Nursing*, Vol 29, 2019. Pp 9-14. ISSN 1876-1399, https://doi.org/10.1016/j.ecns.2019.02.001. (https://www.sciencedirect.com/science/article/pii/S1876139918302354)

[6] Miftahul Khirana Anisa, Niki Dian Permana P, & Theresia Lidya Nova. Penggunaan Simulasi Virtual pada Pembelajaran Fisika untuk Meningkatkan *Higher Order Thinking Skill* (Hots) Siswa: Meta Analisis. *Jurnal Kumparan Fisika*. 2020. *Vol 3, No 2 Agustus*. DOI: https://doi.org/10.33369/jkf.3.2.163-170

[7] Dewi SM, Harjono A, Gunawan G. Pengaruh Model Pembelajaran Berbasis Masalah Berbantuan Simulasi Virtual Terhadap Penguasaan Konsep dan Kreativitas Fisika Siswa SMAN 2 Mataram. *Jurnal Pendidikan Fisika dan Teknologi*. 2016. 2(3):123–8

[8] Alexander, C. K., & Sadiku, M. N. O. *Fundamentals of Electric Circuits*. (M. Lange, Ed.) (Fifth). New York: McGraw-Hill. 2012.

[9] Thomas L.Floyd & David L.Buchla. *Electronics Fundamentals Circuits, Devices and Application. Eighth Edition.* UK: Pearson Education. 2014.

[10] Lee, W. W., & Owens, D. L. *Multimedia-Based Instructional Design: Computer-Based Training, Web-Based Training, Distance Broadcast Training, Performance-Based Solutions*. John Wiley & Sons. 2004

[11] Mohamad Ramdhani. *Rangkaian Listrik*. Jakarta: Penerbit Erlangga. 2008.

The Design of Sensor and Transducer Learning Media based on Augmented Reality

D Hariyanto[1], Zamtinah[2], T Triasih[3], and M S Wati[4], and G K Kassymova[5]

[1,2,3,4] Department of Electrical Engineering Education, Faculty of Engineering, Universitas Negeri Yogyakarta, Indonesia
[5] Institute of Metallurgy and Ore Benefication, Satbayev University, Kazakhstan
[5] Abai Kazakh National Pedagogical University, Kazakhstan

E-mail: [1] didik_hr@uny.ac.id; [2] zamtinah@uny.ac.id; [3] tina.triasih@uny.ac.id; [4] miladiahsetio.2019@student.uny.ac.id; [5] g.kassymova@satbayev.university

Abstract. This study aimed at determining the design stage of the augmented reality-based learning media for sensors and transducers course. This study employed the Research and Development method with the ADDIE model with the stages of analysis, design, development, implementation, and evaluation. This article focuses on the design stage during the process of media development. The developed learning media can be used among vocational students by visualizing the sensor and transducer components like in the real laboratory setting. The design of the developed learning media contained application storyboards and animation storyboards for practicing the six sensors. This augmented reality learning media is designed to support online learning and students' autonomous learning in the practical course of sensors and transducers.

1. Introduction

COVID-19 has changed human activities including the whole educational system into online learning from elementary to tertiary level. This radical shift demands a reconstruction in the teaching and learning process [1]. The obstacles of online learning are mostly in the practical courses that emphasize skills development since it requires specific tools and equipment in the labs or workshop which is quite impossible to be done during the pandemic. The inadequate online learning media for practical activities has been another major issue that hinder the material mastery among students. The low quality of learning media will hamper the online learning process, especially the practical course since the existence of learning media can attract students' learning motivation [2].

This condition is widely open an opportunity to develop a learning media for the practical course. One of the gadget-based technologies that can be put forward is augmented reality (AR) that has been widely used in the game world where it can provide real experiences to gamers. Augmented Reality (AR) refers to an interactive technology that combines the physical environment with virtual elements where a virtual layer is placed between the physical environment and the user who can add images, textual information, videos, or other virtual components [3]. AR is suitable for the teaching and learning process because students can have the same virtual form that allows them to interact or work together on the same target [4]. AR technology can be used in the world of engineering education since it can make virtual objects like real objects and the results are displayed interactively in real-time. AR can also be a beneficial media for learning process, especially in practical activities [5], [6] and it can support autonomous learning since it can be used anytime and anywhere [7], [8].

Besides, the use of virtual technology allows students to develop critical thinking skills, innovation, and teamwork [9]. It indicates that AR is appropriate to be applied in the sensor and transducer course to provide a better learning experience among students during their online learning. Sensors and transducers are important components in electronic circuits. The sensor is a device that receives a stimulus or signal and can respond to the stimulus in the form of an electrical signal [10]. The sensor is a system element that functions to measure a variable and produce output in a certain form [11]. The output signal is in the form of an electrical signal, such as a voltage or current. The results of the sensor readings are still in the form of quantities based on the variables being measured so that these quantities need to be converted into electrical forms to be processed by the electronic system. This article aims to discuss the designing process of the Sensor and Transducer learning media based on the augmented reality technology.

2. Research Method

This study employed the Research and Development method based on the Analysis, Design, Development, Implementation, and Evaluation (ADDIE) development model according to Lee and Owens [12]. The ADDIE model began with an analysis phase consisting of a need assessment and front-end analysis. The data were obtained through observation and interviews. A needs analysis was done to determine the gap between the real situation and the expected state as well as determine the following action. Meanwhile, the front-end analysis was to obtain more detailed information including technical analysis, student analysis, situation analysis, task analysis, current issue analysis, objective analysis, media analysis, cost analysis, and data analysis.

Furthermore, the design stage was carried out by arranging the implementation schedule and the content plan to be included in the learning media. The development and implementation stages in the development of learning media were based on the plan and validated by the material and media experts. Meanwhile, the implementation stage was the field testing of the developed learning media. The last stage, the evaluation stage, was done to assess the effectiveness of the product. The flow of the ADDIE model development process can be seen in Figure 1 below.

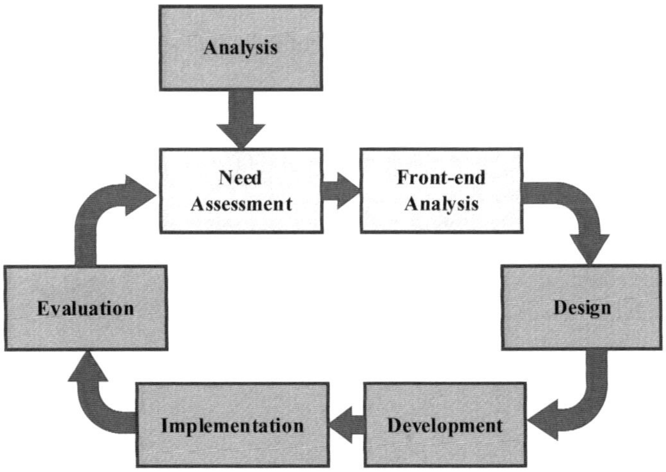

Figure 1. Development Stage of ADDIE Model

3. Research Results and Discussion

This research began with the analysis phase consisting of needs assessment and front-end analysis. The needs assessment was carried out in the form of observation to find out the gaps in the sensor and transducer courses. The current gap was that the students' competencies that were below the expected results due to the online learning system in the practical course. The gap indicated that, during their online learning, the students only carried out simulations that were different from the real practicum settings. Meanwhile, the front-end analysis involved technology analysis of which most students had already used smartphones. It means that the development of smartphone-based augmented reality applications can be applied in the learning process. The students' analysis included background identification, learning characteristics, and students' prerequisite skills. The analysis of the current situation indicated that the sensor and transducer practicum was carried out online which resulted in a lack of real practical experience or the condition that resembled the real situation. Task analysis contained the learning outcomes of sensors and transducers where students can assemble and operate sensors and transducers.

The analysis of the latest issues related to the current conditions of the Covid-19 outbreak urged educators to be more creative in preparing the learning media. The objective analysis was to identify the target to be achieved in the sensors and transducers learning where students must understand the various working principles of both components. Media analysis referred to the strategies for distributing media to students which can be implemented using the cloud and social media. The cost analysis was the development of augmented reality-based software which was cheaper than the procurement of practicum trainers. The data analysis employed the identification of learning materials, worksheets, syllabus, sensor, and transducer references. Based on the description above, it is clear that the development of augmented reality-based learning media can be realized to support online learning of sensors and transducers course.

The design stage was carried out through content planning including application design, animation design, and material content design. The contents of the material in the application included reed switch sensors, flame sensors, MQ-7 sensors, sound microphone sensors, PIR sensors, and soil moisture sensors. The animation needs to be added to make the display more attractive. The application design was done by making storyboards as presented in Table 1 below.

Table 1. Storyboard of Augmented Reality Applications for Sensors and Transducers

No	Interface	Explanation
1		**"Application Start"** A. Application icon containing an overview of the application. B. Application name "Sensors and Transducers".
2	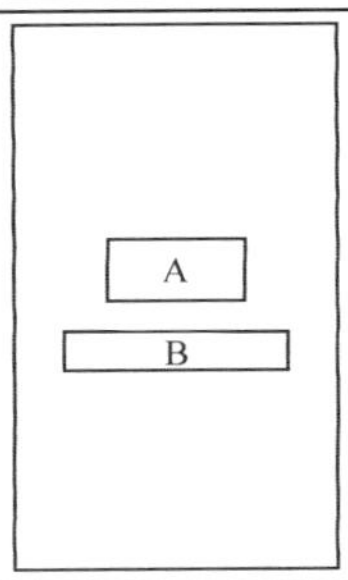	**"Loading"** A. Application logo. B. Loading process to provide information to users that the application can be run on the device.

No	Interface	Explanation
3	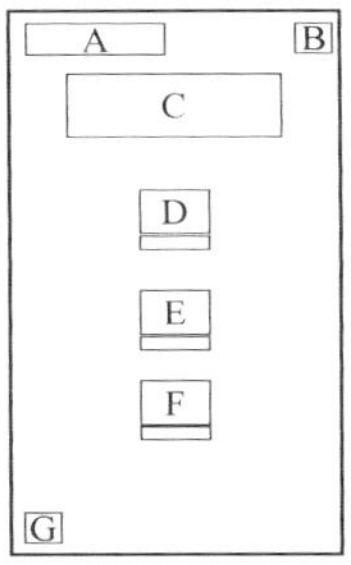	**“Menu Display”** A. Small size logo of Ristekdikti and UNY on the top left of the screen. B. Cross logo to exit from the application. C. Application Title “Sensors and Transducers”. D. Logo “Scan Augmented Reality” with a text explanation at the bottom. E. A logo that describes the developer or researcher with the text of “Over View” at the bottom. F. A logo that explains the instructions for the use of the Augmented Reality application with the text of "Instructions for Use" at the bottom. G. The music logo is crossed out and when the user presses the logo, the music will turn off and the logo changes to music without streaks.
4	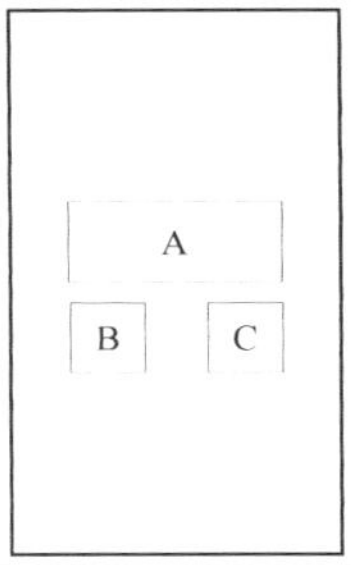	**“Cross logo to exit from the application”** A. The question “Are you sure you want to exit”. B. “Yes” statement. When the user presses the button, the application will be closed. C. “No” statement. When the user presses this button, it will return to the last open page.
5	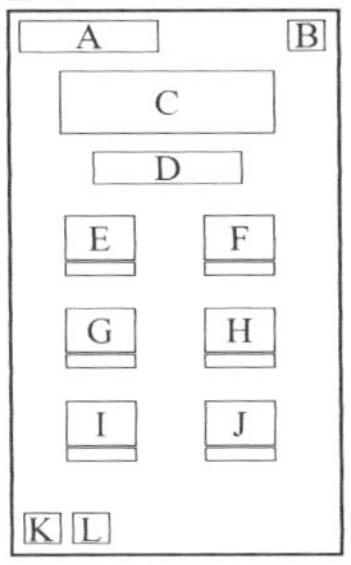	**“Scan Augmented Reality”** A. Small size logo of Ristekdikti and UNY on the top left of the screen. B. Cross logo to exit the application. C. Title of application “Sensors and Transducers”. D. The text of "Sensor Menu Options". E. “Reed Switch Sensor”. F. "Flame Sensor". G. “MQ-7 Sensor”. H. "Sensor Sound Activation". I. “PIR Sensor”. J. “Soil Moisture Sensor”. K. The music logo is crossed out and when the user presses the logo the music will turn off and the logo changes to music without streaks. L. Logo Home to return to the start page of the application.
6	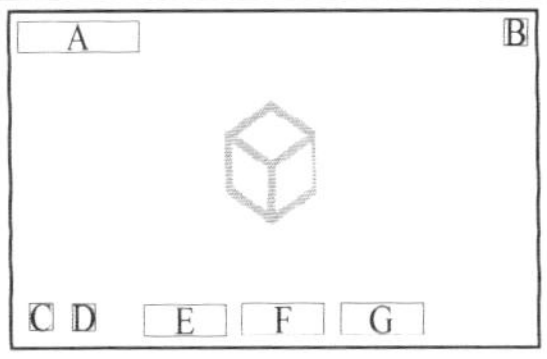	**“Reed Switch Sensor, Flame Sensor, MQ-7 Sensor, Sound Sensor Activation, PIR Sensor, and Soil Moisture Sensor”** A. Small size logo of Ristekdikti and UNY on the top left of the screen. B. Cross logo to exit the application. C. The music logo is crossed out and when the user presses the logo, the music will turn off and the logo changes to music without streaks. D. Logo Home to return to the Augmented Reality scan page. E. The “View Information” button displays information about the scanned AR. F. The "Rotation" button rotates the displayed 3D object rotates 360 degrees. G. The “Simulation” button simulates the displayed 3D object with the detecting 3D object.

No	Interface	Explanation
7	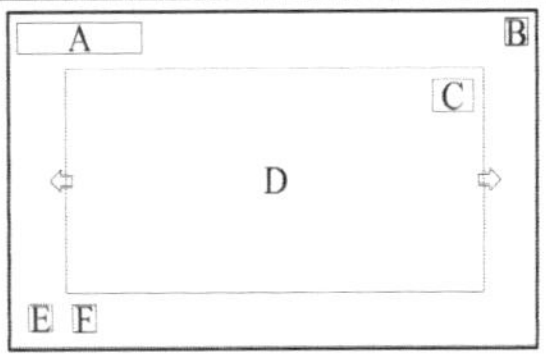	**"Information Check"** A. Small size logo of Ristekdikti and UNY on the top left of the screen. B. Cross logo to exit the application. C. Cross logo to close the information about the displayed 3D object and return to the marker scan position. D. Information about the displayed 3D Object. E. The music logo is crossed out and when the user presses the logo, the music will turn off and the logo changes to music without streaks. F. Home logo to return to the Augmented Reality scan page.
8	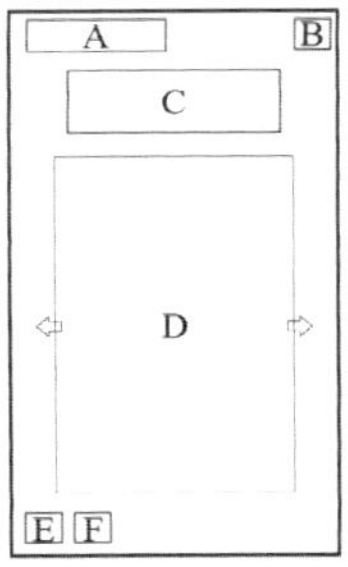	**"Over View"** A. Small size logo of Ristekdikti and UNY on the top left of the screen. B. Cross logo to exit the application. C. Title of application of "Sensors and Transducers". D. Information display related to the scope of competencies and the researchers. E. The music logo is crossed out and when the user presses the logo the music will turn off and the logo changes to music without streaks. F. Home logo to return to the start page of the application.

3.1. Reed Switch Sensor Animation

The animation of the reed switch sensor is shown in the form of a storyboard in Table 2. The animation of Reed Switch Sensor is for a safety indicator of motor rotation. When the motor stops, the indicator (light) will turn off. If the motor is on, there will be a magnetic field around the motor and the indicator will turn on to warn the workers around the working area.

Table2. Storyboard of Reed Switch Sensor Animation

No	Action	Duration	Explanation
1		00:00:03	There are 3 objects in the first state, i.e., the motor is off, lights, and sensors.
2		00:00:05	The motor starts (marked by the movement of the motor rotation) and it produces a magnetic field around the motor.
3	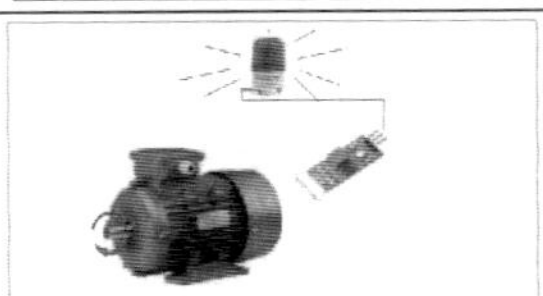	00:00:04	The sensor detects the magnetic field around the motor, then the indicator is on (lamp).
4	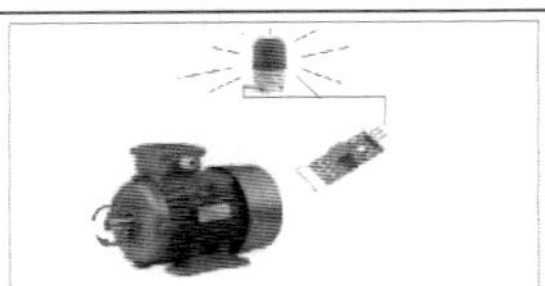	00:00:04	The motor is getting slow down.

No	Action	Duration	Explanation
5		00:00:02	The motor stops rotating, and the indicator light turns off showing the loss of the magnetic field generated by the rotation of the motor.

3.2. Flame Sensor Animation

The concept of the animation of the Flame Sensor is to show the indicator of the sparks in a room. If there is a spark in the room, the indicator will light up. Table 3 shows the storyboard of the sensor flame animation.

Table 3. Storyboard of Sensor Flame Animation

No	Action	Duration	Explanation
1		00:00:02	The sensor is installed on the roof of the room and the indicator light on the walls of the room, and the water sprayer on the ceiling.
2	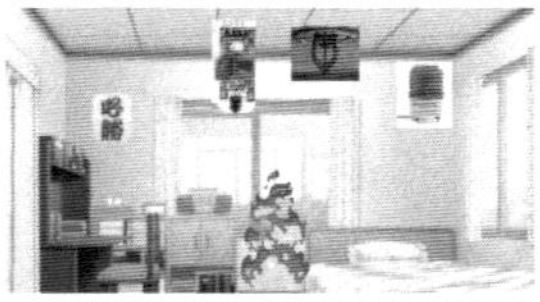	00:00:05	The sparks appear, and the sensor detects the incidence in the room.
3		00:00:04	Along with the sensor that detects a spark, it immediately turns on the indicator in the form of a lamp and water spray.
4		00:00:04	The fire began to be diminished and going out.
5		00:00:02	The indicator light and water spray turn off, and the sensor goes back in standby mode.

3.3. MQ-7 Sensor Animation

The animation of the MQ-7 sensor is presented in the form of a storyboard as shown in Table 4. The animation concept is a gas cylinder storage warehouse. When the sensor detects a leak from the gas, it will turn on the indicator and blows a warning sound.

Table 4. Storyboard of Sensor MQ-7 Animation

No	Action	Duration	Explanation
1	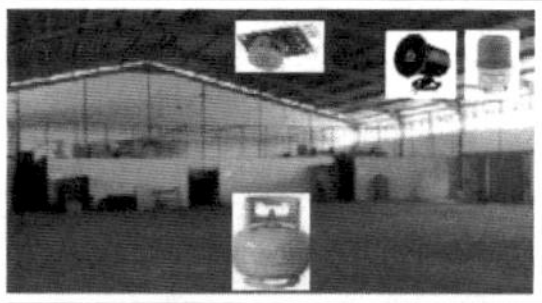	00.00.02	The sensors detect gas leaks in a storage warehouse.
2		00:00:04	The gas comes out due to a leak in the tube.
3		00:00:01	The gas leak has been detected and it signals the indicator and warning lights.
4	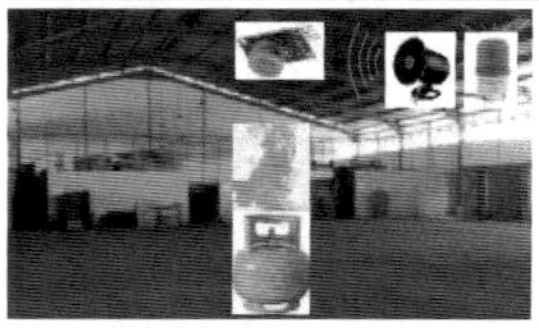	00:00:05	The indicator and warning lights are getting on.
5		00:00:04	The officers handle the leak in the gas cylinder.
6	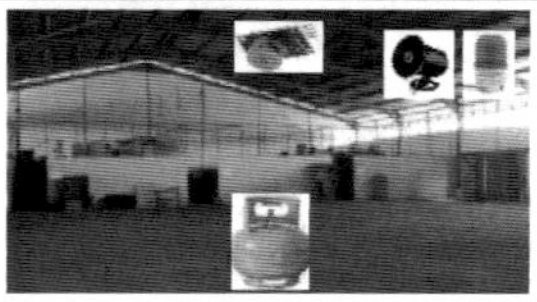	00:00:02	After all has been handled, the indicator light and siren turn off, and the sensor goes back in standby mode.

3.4. Animation of Sound Activation Sensor

The animation of the sound activation sensor is presented in the form of a storyboard as shown in Table 5. The concept is a room equipped with a sound activation sensor that detects hand clapping to turn on or turn off the room lights.

Table 5. Storyboard of Sensor Sound Activation Animation

No	Action	Duration	Explanation
1		00:00:02	The sensor is installed on the wall with the lamp as the output of the sensor.

No	Action	Duration	Explanation
2		00:00:02	Hand clapping.
3	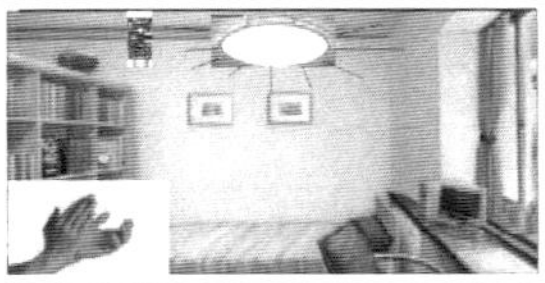	00:00:03	The sensor detects the sound and gives a signal to turn on the room lights.
4	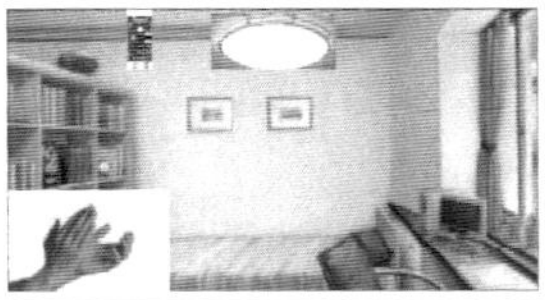	00:00:02	The following hand clapping is to turn off the lamp.

3.5. Animation of PIR Sensor

The animation of the PIR sensor in the form storyboard is presented in Table 6. The concept of this animation refers to the application of PIR sensors in hotel hallways that detects human movement. The lights will be on when a person is passing through the hallway. On contrary, if the PIR sensor does not detect any movements, the lights remain off.

Table 6. Storyboard of PIR Sensor Animation

No	Action	Duration	Explanation
1		00:00:02	The sensor is installed in the hotel hallway.
2		00:00:03	Someone appears and the sensor detects a movement.
3		00:00:04	The lamp is turning on.
4	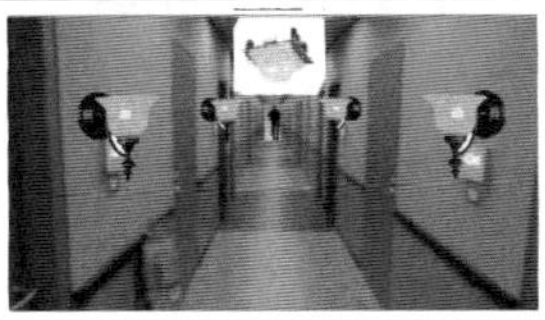	00:00:05	The person moves away from the hallway and no more movement are detected, so the sensor signals the light to turn off.

No	Action	Duration	Explanation
5	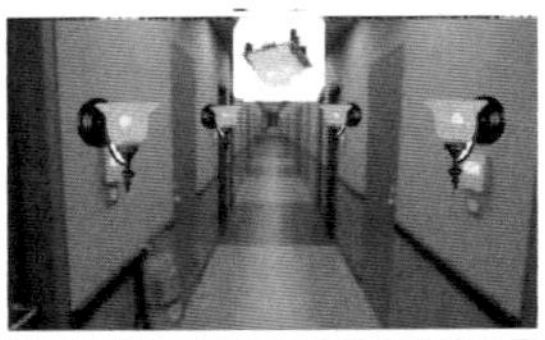	00:00:03	The sensor goes back in standby mode.

3.6. Animation of Soil Moisture Sensor

Animation of soil moisture sensor is presented in Table 7. The concept of this sensor is to monitor the humidity level in plants. When the plant lacks water, the soil color becomes light brown (dry). If it is watered and getting wet, the moisture level is good, and the soil color will be dark brown. The humidity level on this plant can be known from the given indicator in the upper right corner.

Table 7. Storyboard of Soil Moisture Animation

No	Action	Duration	Explanation
1		00:00:03	The sensor detects that the humidity decrease based on the indicator (from 80% to 20%), and the color of the soil becomes light brown (dry).
2		00:00:08	The water sprinkler comes out and pours water on the soil, and this indicator level changes (from 20% to 80%), so the soil color that was light brown (dry) becomes dark brown (wet).
3		00:00:02	The sprinkler moves away from the plant and back to its place.

4. Conclusion

The design of augmented reality-based learning media on the sensors and transducers course can be developed to support the practical learning. The design of this learning media is in the form of application display storyboards and animation storyboards. The designed storyboard will be used as a reference for the following stage. Further research is the development stage of the augmented reality-based learning media.

Acknowledgments

This research was supported by Innovation Research Group Grant from DIPA BLU UNY 2021 with the contract number of T/2.6/UN34.15/PT.01.02/2021.

References

[1] Upoalkpajor J L N and Upoalkpajor C B. 2020. *The Impact of Covid-19 on Education in Ghana.* Asian Journal of Education and Social Studies, Vol.9, No.1, p.23-33. https://doi.org/10.9734/ajess/2020/v9i130238

[2] Akrim. 2018. *Media Learning in Digital Era. Advances in Social Science, Education, and Humanities Research.* Proceedings of the 5th International Conference on Community Development (AMCA 2018), Vol.231, p.458-460. https://doi.org/10.2991/amca-18.2018.127

[3] Javornik A. 2016. *Augmented reality: Research agenda for studying the impact of its media characteristics on consumer behavior*. Journal of Retailing and Consumer Services, Vol.30, p. 252-261. https://doi.org/10.1016/j.jretconser.2016.02.004.

[4] Santos M E C, et al. 2016. *Augmented reality as multimedia: the case for situated vocabulary learning*. Research and Practice in Technology Enhanced Learning, Vol.11, No.4, p.1-23. https://doi.org/10.1186/s41039-016-0028-2

[5] Mustaqim, I. 2017. *Pengembangan media pembelajaran berbasis augmented reality*. Jurnal Edukasi Elektro, 1(1). https://doi.org/10.21831/jee.v1i1.13267

[6] Triatmaja, A. K., Muchlas, M., & Wardana, Y. 2021. *Virtual Laboratorium Teknik Digital berbasis Mobile Virtual Reality*. Jurnal Edukasi Elektro, 5(1). https://doi.org/10.21831/jee.v5i1.38705

[7] Jones, P R. 2016. *The Structure and Pedagogical Style of the Virtual Developmental Education Classroom: Benefit or Barrier to the Developmental Learning Process?*. International Journal of Language and Literature, Vol. 4, No. 1, p. 43-48. https://doi.org/10.15640/ijll.v4n1a5

[8] Badarudin, R., & Hariyanto, D. (2021, March). *Visual validation of PLC program using virtual simulator*. In Journal of Physics: Conference Series (Vol. 1833, No. 1, p. 012063). IOP Publishing. https://doi.org/10.1088/1742-6596/1833/1/012063

[9] Lynch T and Ghergulescu L. 2017. *Review of Virtual Labs as The Emerging Technologies For Teaching Stem Subjects*. INTED2017 Proceedings: International Technology, Education and Development Conference, p.6082-6091. https://doi.org/10.21125/inted.2017.1422

[10] Patel B C, et al. 2020. *Introduction to sensors*. Advances in Modern Sensors: Physics, design, simulation and applications, p. 1-1 to 1-21. https://doi.org/10.1088/978-0-7503-2707-7ch1.

[11] Syam R. 2013. *Dasar-dasar Teknik Sensor* (Universitas Hasanuddin Makassar).

[12] Lee, W. W., & Owens, D. L. 2004. *Multimedia-based instructional design: computer-based training, web-based training, distance broadcast training, performance-based solutions*. John Wiley & Sons.

The Effects of Rain Intensity and Water Elevation with Energy Productivity of Kracak Hydro Power Plant, Bogor Regency – West Java

I A Rahardjo[1], Suyitno[1], M Subekti[1], W Gunawan[1], R P Wijayanto[2], I Mahir[3], and Sakirah[4]

[1]Electrical Engineering Education Program, Universitas Negeri Jakarta, Rawamangun, Indonesia
[2]BPPT, Indonesia
[3]Mechanical Engineering Department, Universitas Negeri Jakarta, Rawamangun, Indonesia
[4]Electrical Engineering Program, Taruna Bangsa of Vocational High School, Bekasi, Indonesia

Email: [1]imam_ar@unj.ac.id

Abstract. This paper was aimed to know the effects of rain intensity and water elevation with energy productivity of Kracak Hydro Power Plant, Bogor Regency – West Java. The method used multiple regression analysis with a quantitative approach to describe the effects of rain intensity and water elevation with energy productivity of Kracak Hydro Power Plant. Based on the data, the highest rain intensity was in February of 13,35 mm with water elevation of 346,09 cm and produced electrical energy of 198.296 kWh. The lowest rain intensity was in July of 0,86 mm with water elevation of 194,02 cm and produced electrical energy of 49.772 kWh. The average rain intensity was 8,9 mm with water elevation of 324,12 cm and produced electrical energy of 156.010 kWh. The results of testing the effect of rain intensity with energy productivity at the Kracak hydropower plant resulted in a significance value of 0.002 ($p < 0.05$) and a t value of 4.306. This value indicates that the significance value is below 0.05 and has a positive coefficient direction. It means that rain intensity has a significant positive effect with energy productivity at the Kracak hydropower plant. And the results of testing the effect of water elevation with energy productivity at the Kracak hydropower plant resulted in a significance value of 0.003 ($p < 0.05$) and a t value of 3.864. This value indicates that the significance value is below 0.05 and has a positive coefficient direction. It means that water elevation has a significant positive effect with energy productivity at the Kracak hydropower plant. The conclusion on this research is the rain intensity and water elevation have effects with energy productivity of Kracak Hydro Power Plant.

1. Introduction

The Kracak Hydro Power Plant is a reservoir type hydropower plant from the upstream of the Cianten River with an area of the Cikuluwung dam and river [1]. This plant is capable of producing an average of 150,000 kWh of electrical energy in one year, equivalent to 0.01% of Indonesia's hydroelectric production [2],or around 0.0001% of national energy production [3].

The electrical energy generated by hydroelectric power is closely related to climate change, rain [4-10] and water elevation [11-13]. Geographically, the Kracak Hydro Power Plant is in a fairly high rain area, located in Kracak village, Leuwiliang sub-district, Bogor regency.

The rain intensity can affect the water elevation in the reservoir which has an impact on the production of electrical energy. Therefore it is important to know the effects of rain fall and water elevation towards Kracak Hydro Power Plant, Bogor Regency – West Java.

2. Method

The method used in this study is a descriptive method through a quantitative approach. The data of rain fall, water elevation, and Power Plant Productivity were collected from BMKG and Kracak hydroelectric, then analyzed with application of SPSS IBM 24. The analysis results obtained from the SPSS application are concluded the effects of rain intensity and water elevation with energy productivity of Kracak Hydro Power Plant. The flow chart of methods was shown in Figure 1.

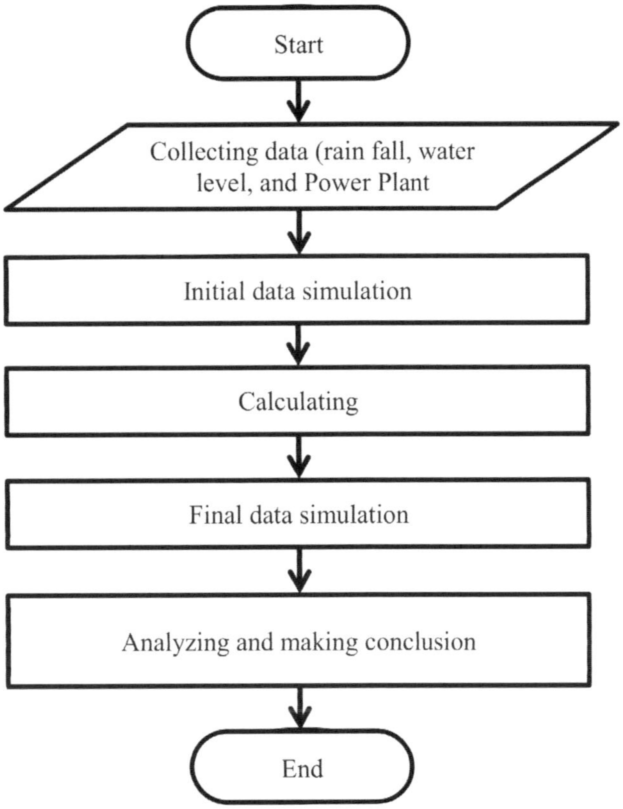

Figure 1. Flow chart of method

3. Results And Discussion

3.1. Result

The rain intensity, water elevation, and energy productivity of Kracak Hydro Power Plant during was listed in the Figure 2.

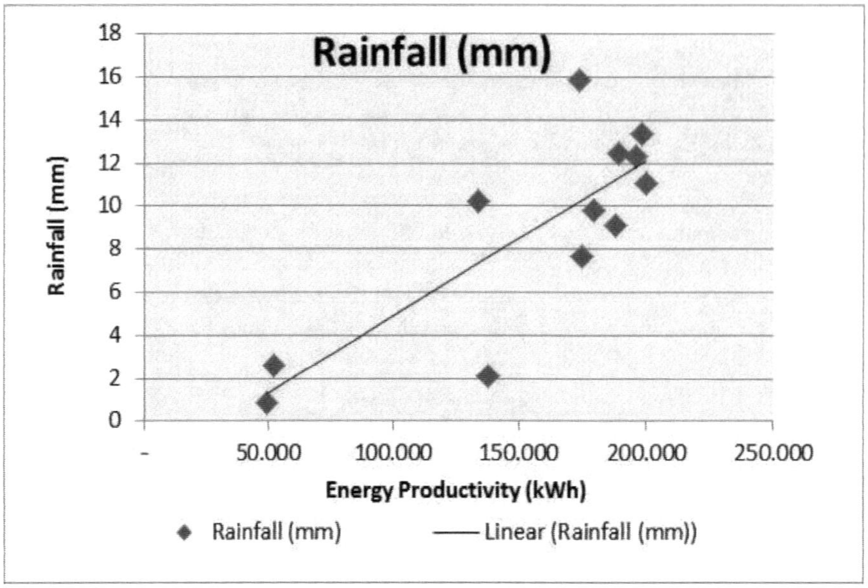

(a)

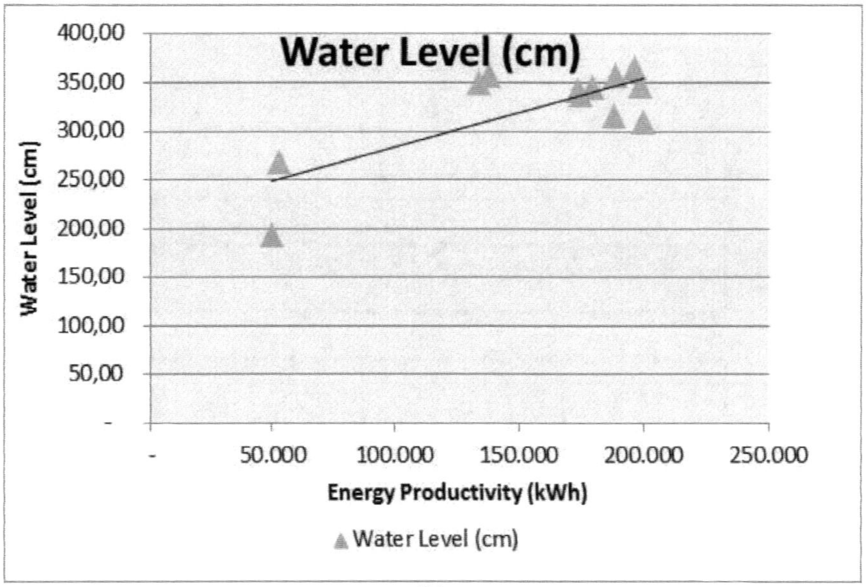

(b)

Figure 2. (a). Graph effect of rain intensity with energy productivity; (b). Graph effect of water elevation towards energy productivity

Based figure 2, the highest rain intensity was in February of 13,35 mm with water elevation of 346,09 cm and produced electrical energy of 198.296 kWh. The lowest rain intensity was in July of 0,86 mm with water elevation of 194,02 cm and produced electrical energy of 49.772 kWh. The average rain intensity was 8,9 mm with water elevation of 324,12 cm and produced electrical energy of 156.010 kWh.

The rain intensity and water elevation have a positive effect on increasing energy productivity. This means that the higher the rain intensity and water elevation, the higher energy productivity of Kracak hydroele power plant

3.2. Discussion

The effect of rain intensity with energy productivity was shown in Figure 3.

Model Summary

			Adjusted R	Std. Error of the	Change Statistics				
Model	R	R Square	Square	Estimate	R Square Change	F Change	df1	df2	Sig. F Change
1	1,000[a]	1,000	1,000	,00000	1,000	15518012,260	2	8	,000

a. Predictors: (Constant), Temperature, Distance

Coefficients[a]

Model		Unstandardized Coefficients		Standardized Coefficients			95,0% Confidence Interval for B		Correlations			Collinearity Statistics	
		B	Std. Error	Beta	t	Sig.	Lower Bound	Upper Bound	Zero-order	Partial	Part	Tolerance	V
1	(Constant)	75,429	21,022		3,588	,005	28,589	122,270					Doubl ad
	Rainffal	8,994	2,089	,806	4,306	,002	4,340	13,649	,806	,806	,806	1,000	1,000

a. Dependent Variable: Produksi_Energi

Figure 3. The Effect of rain intensity with energy productivity

The results of testing the effect of rain intensity with energy productivity at the Kracak hydropower plant resulted in a significance value of 0.002 ($p < 0.05$) and a t value of 4.306. This value indicates that the significance value is below 0.05 and has a positive coefficient direction. It means that rain intensity has a significant positive effect with energy productivity at the Kracak hydropower plant.

The effect of rain intensity with energy productivity was shown in Figure 4.

Coefficients[a]

Model		Unstandardized Coefficients		Standardized Coefficients			95,0% Confidence Interval for B		Correlations			Collinearity Statistics	
		B	Std. Error	Beta	t	Sig.	Lower Bound	Upper Bound	Zero-order	Partial	Part	Tolerance	VIF
1	(Constant)	265,645	25,077		10,593	,000	209,771	321,520					
	Rainffal	6,610	2,492	,643	2,653	,024	1,058	12,163	,643	,643	,643	1,000	1,000

a. Dependent Variable: Water_Height

Figure 4. The effect of rain intensity with water elevation

The results of testing the effect of rain intensity with water elevation at the Kracak hydropower plant resulted in a significance value of 0.0024 ($p < 0.05$) and a t value of 2.653. This value indicates that the significance value is below 0.05 and has a positive coefficient direction. It means that rain intensity has a significant positive effect with water elevation at the Kracak hydropower plant

The effect of water elevation with energy productivity was shown in Figure 5.

Coefficients[a]

Model		Unstandardized Coefficients		Standardized Coefficients			95,0% Confidence Interval for B		Correlations			Collinearity Statistics	
		B	Std. Error	Beta	t	Sig.	Lower Bound	Upper Bound	Zero-order	Partial	Part	Tolerance	VIF
1	(Constant)	-116,764	71,331		-1,637	,133	-275,699	42,172					
	Water_Height	,840	,217	,774	3,864	,003	,355	1,324	,774	,774	,774	1,000	1,000

a. Dependent Variable: Produksi_Energi

Figure 5. The Effect of Water elevation towards Energy Productivity

The results of testing the effect of water elevation with energy productivity at the Kracak hydropower plant resulted in a significance value of 0.003 ($p < 0.05$) and a t value of 3.864. This value indicates that the significance value is below 0.05 and has a positive coefficient direction. It means that water elevation has a significant positive effect with energy productivity at the Kracak hydropower plant.

4. Conclusion

This article studied, the highest rain intensity was in February of 13,35 mm with water elevation of 346,09 cm and produced electrical energy of 198.296 kWh. The lowest rain intensity was in July of 0,86 mm with water elevation of 194,02 cm and produced electrical energy of 49.772 kWh. The average rain intensity was 8,9 mm with water elevation of 324,12 cm and produced electrical energy of 156.010 kWh.

The results of testing the effect of rain intensity with energy productivity at the Kracak hydropower plant resulted in a significance value of 0.002 ($p < 0.05$) and a t value of 4.306. This value indicates that the significance value is below 0.05 and has a positive coefficient direction. It means that rain intensity has a significant positive effect with energy productivity at the Kracak hydropower plant.

The results of testing the effect of rain intensity with water elevation at the Kracak hydropower plant resulted in a significance value of 0.0024 ($p < 0.05$) and a t value of 2.653. This value indicates that the significance value is below 0.05 and has a positive coefficient direction. It means that rain intensity has a significant positive effect with water elevation at the Kracak hydropower plant.

The results of testing the effect of water elevation with energy productivity at the Kracak hydropower plant resulted in a significance value of 0.003 ($p < 0.05$) and a t value of 3.864. This value indicates that the significance value is below 0.05 and has a positive coefficient direction. It means that water elevation has a significant positive effect with energy productivity at the Kracak hydropower plant.

References

[1] Sub Unit Hydroelectric Power Plant Kracak. (2018): Bogor

[2] ESDM, K., & Ketenagalistrikan, D. (2019). Statistik Ketenaga Listrikan Tahun 2018. *Journal of Chemical Information and Modeling*, *53*(9), 1689–1699.

[3] PT. Perusahaan Listrik Negara. (2019). Rencana Usaha Penyediaan Tenaga Listrik 2019-2028. *PT. Perusahaan Listrik Negara*, 2019–2028.

[4] Falchetta, G., Gernaat, D. E. H. J., Hunt, J., & Sterl, S. (2019). Hydropower dependency and climate change in sub-Saharan Africa: A nexus framework and evidence-based review. *Journal of Cleaner Production*, *231*, 1399–1417. https://doi.org/10.1016/j.jclepro.2019.05.263

[5] Mishra, S. K., Veselka, T. D., Prusevich, A. A., Grogan, D. S., Lammers, R. B., Rounce, D. R., … Christian, M. H. (2020). Differential Impact of Climate Change on the Hydropower Economics of Two River Basins in High Mountain Asia. *Frontiers in Environmental Science*, *8*(March). https://doi.org/10.3389/fenvs.2020.00026

[6] Plants, H. (2017). Forecast Rain for Power Production Management of Namkhan 2 and 3 Hydropower Plants. *International Journal of Technology and Engineering Studies*, *3*(4). https://doi.org/10.20469/ijtes.3.40003-4

[7] M. M. Hasan and G. Wyseure, "Impact of climate change on hydropower generation in Rio Jubones Basin, Ecuador," *Water Science and Engineering*, vol. 11, no. 2, pp. 157–166, 2018.].

[8] A. Perera and U. Rathnayake, "Impact of climate variability on hydropower generation in an un-gauged catchment: Erathna run-of-the-river hydropower plant, Sri Lanka," *Applied Water Science*, vol. 9, no. 3, p. 57, 2019.

[9] B. Khaniya, H. G. Priyantha, N. Baduge, H. Azamathulla, and U. Rathnayake, "Impact of climate variability on hydropower Mathematical Problems in Engineering 7 generation : a case study from Sri Lanka," *ISH Journal of Hydraulic Engineering*, vol. 26, no. 3, pp. 301–309, 2020.

[10] Wagner, T., Themeßl, M., Schüppel, A., Gobiet, A., Stigler, H., & Birk, S. (2017). Impacts of climate change on stream flow and hydro power generation in the Alpine region. *Environmental Earth Sciences*, *76*(1). https://doi.org/10.1007/s12665-016-6318-6

[11] Otuagoma, S., Ogujor, E., & Kuale, P. (2015). Determination of Head for Small Hydropower Development: a Case Study of River Ethiope At Umutu. *Nigerian Journal of Technology*, *35*(1), 190. https://doi.org/10.4314/njt.v35i1.26

[12] Sritram, P., & Suntivarakorn, R. (2017). Comparative Study of Small Hydropower Turbine Efficiency at Low Head Water. *Energy Procedia*, *138*, 646–650. https://doi.org/10.1016/j.egypro.2017.10.181

[13] Perera, A., & Rathnayake, U. (2020). Relationships between Hydropower Generation and Rain-Gauged and Ungauged Catchments from Sri Lanka. *Mathematical Problems in Engineering*, *2020* (Dm). https://doi.org/10.1155/2020/9650251

Web-Based Recruitment Selection Application Development with Analytical Hierarchy Process (AHP) Method

G J T Maruanaya[1], R F Maruanaya[2], and D Hariyanto[3]

[1] Department of Informatics Engineering, Faculty of Information Technology, Budi Luhur University, Indonesia
[2] Institute of Vocational Education and Vocational Didactics, Faculty of Education, TU Dresden, Germany
[2] Department of German Language Education, Pattimura University, Indonesia
[3] Department of Electrical Engineering Education, Faculty of Engineering, Yogyakarta State University, Indonesia

E-mail: [1]1611501576@student.budiluhur.ac.id;
[2]rita_fransina.maruanaya@tu-dresden.de; [3]didik_hr@uny.ac.id

Abstract. This research aimed to design and develop a support system to determine the selection of recruitment, especially assistant for the Integrated ICT Laboratory at Budi Luhur University according to predetermined criteria. The method used in developing the application was the Analytical Hierarchy Process (AHP) method, which was to rank decision alternatives and accept the excellence one for several criterion cases that mixes qualitative and quantitative components in the whole assessment of available alternatives. The results showed that there were 6 (six) criteria in the selection of the Integrated ICT Laboratory Assistant, including Attendance Value, Competency Value, Project Value, Hardware Value, Interview Value, and Psychological Test Value, which were in accordance with the criteria in the Integrated ICT Laboratory at Budi Luhur University. The system can speed up the process of accepting assistants, because the calculation process, as well as printing the final score are in one system, according to the terms and criteria specified, making it more effective and efficient.

1. Introduction

Currently, Information and Communication Technology has experienced progress, which is very rapid and has entered all fields. Technology cannot be separated from almost all aspects of human activity. With the use of such information technology, humans, being able to get information, are able to make an agency, both public and private, able to process data more effectively and efficiently compared to the manual method. For an agency the use of computers is an advantage. The information system was created with the aim of making it easier for an agency to process and store data so that it can provide accurate and precise information. That way, an accurate and precise information system can make an agency's performance more efficient because it can reduce unwanted errors.

The Integrated Information and Communication Technology Laboratory (Lab ICT) is a computer laboratory used to support practicum courses at Budi Luhur University, such as teaching and learning activities, Mid-Semester Examinations, and Final Semester Examinations for practicum courses. Each activity is supervised by at least one assistant. Assistants are students who help the ICT Lab operations. Before becoming an assistant, students who register to become assistants will undergo a training period of approximately six months and are referred to as prospective assistants. During the training period, prospective assistants will be assessed on the basis of craftsmanship, neatness, given projects, and so on. The project is given based on the existing business processes in the Integrated ICT Lab, such as creating systems and servers.

So far, the process of measuring the value of the assistant candidates is made manually by the Integrated ICT Lab supervisor and that is why this process can take long time. In order to make the process of calculating the value of prospective assistants easier and faster, on this basis it is needed to design a Decision Support System to assess prospective assistants who are eligible to be appointed as assistants using the Analytical Hierarchy Process (AHP) method. The AHP method brings benefit of being able to consider both qualitative and quantitative assessments.

2. Theoretical review

2.1. Information System

A system consists of various elements that complement each other in achieving goals and objectives. The elements that are enclosed in the system are termed subsystems, which should be linked each other and be involved through appropriate communications in order to enable the effectivity and efficiency of the system [1]. The Information means the data which is operated into a form and can be utilized by everyone who desires it. Data as a plural form of datum is classified as a source of information that has a function as a presenter of events that actually happened [2].

Furthermore, Iswandy states that the information system is a structured series that is used to support various activities that are useful in daily life such as to support official operational activities in institutions and provide information that is certainly very useful for users. In addition, this information system is defined as a collection of several parts of a complex element that are intertwined with one another and cannot be separated. These elements can greatly assist an institution in terms of processing the data needed to support operational activities. After the data processing process is complete, this system will distribute the data and store the information so that it allows users to still be able to access the data at any time if they need it. In short it can be said that the design of an information system is carried out to assist users in facilitating their work where this system can provide a detailed description and complete design so that it can make it easier for programmers or computer experts to process the data they need. [1].

2.2. Decision Support System

Before the concept of decision support or what is usually abbreviated as DSS is famous as it is today, this system was first developed by Michael S. Scott Morton in the 1970s [3]. At that time Morton did not use the term DSS but rather the term management decision system which aims to assist users in organizing and making decisions within the scope of management in various institutions. The background of the emergence of this concept was at that time the management system had difficulty in solving unstructured problems. Furthermore, Moore and Chang characterize this system as a system that has the ability to process data accurately, can make it easier for users to make decisions, is sustainable and very reliable to use when users experience doubts in making decisions [4].

As the name implies, this decision support system certainly functions in terms of providing support to the party in charge of an institution or organization in making a decision, especially if the related parties above are trapped in an unfavorable situation, namely when the party does not have a standard and structured reference. However, this decision support system can only be used as an alternative decision provider, because the final decision will be returned to the decision maker such as company leaders, supervisors and others. In other words, it can be said that the results achieved by this decision support system need to be reviewed again by the responsible parties so that the decisions taken can be more reliable. As it is known that everyone wants to produce the best decision and with the support of calculation results based on DSS, the decision-making process will be more systematic and consistent and especially not subjective [5].

According to Wira et al [6], the stages carried out in the Decision Support System process are as follows:

1. Problem Definition.
2. Collection of relevant data or information elements.
3. Processing data into information in the form of reports graphics and writing.
4. Find alternative solutions.

The components of the Decision Support System as shown in Figure 1 consist of [7]:

1. Data management is one part of the DSS that was developed with the aim of carrying out data organizing tasks, starting from providing all the data needed in an organizational system to the data storage stage. The place to store data in this system is called the Database Management System (DBMS). The purpose of storing data in this DBMS is to make it easier for users to access data again and extract the data in a very short time.
2. The Management Model is a component that helps users in managing the financial system of a particular organization or institution. In addition, this model can also be used in terms of statistical management or various models in the form of numbers.
3. Communication or user dialogue subsystem is a system that allows users to interact. In this case the user has access to control the DSS where they can design the interface.
4. Knowledge management is an optional part of the system developed in order to provide support to other parts of the system.

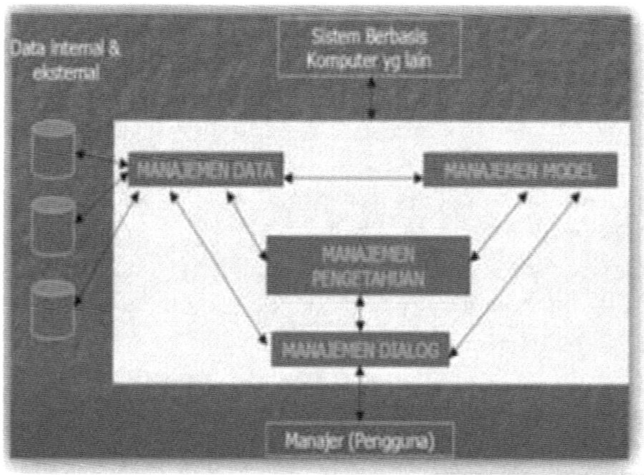

Figure 1. The component of the Decision Support System [7]

2.3. Analytical Hierarchy Process (AHP)

The Analytical Hierarchy Process (AHP) method is a system developed by a mathematician named by Dr. Thomas L. Satty [8]. This system is designed to assist users in providing a scale for various indicators. Usually, an organization has difficulty in determining the priority scale in making decisions, especially when they are faced with a number of unstructured components. One example is when you have to give weight to each component that is a criterion in decision making. With the help of this AHP, it is certain that the process of determining the weight of each component will be more structured, systematic and reliable [5].

The analysis system with AHP is very possible to be used in solving problems that can be measured by numbers, namely problems related to assessment. Usually these problems are related to statements of attitudes or opinions in the form of words, but the level of high or low or good and bad can be measured by giving a weighting scale [9].

3. Application Design

3.1. Screen Design

In making an application, layer design is a very important thing. The layer design must be easy to understand, so that admins feel comfortable using the application of the Analytical Hierarchy Process (AHP) Algorithm Implementation to determine the selection of labor recruitment.

3.1.1. Login Form Screen Design. Figure 2 is a screen design of the Login Form for admins to login. Users who want to enter the AHP Algorithm Implementation application must enter the correct username and password in the textfield provided on the Login Form in this application. If the username and password are correct, the user will enter the main menu which can access the menu in the application.

Figure 2. Login Form Screen Design

3.1.2. Main Menu Screen Design. Figure 3 is the layer design of the admin main menu. The main admin menu can access the master data menu, transaction data analysis and reports. For the main admin sub menu, the admin can access the master data sub menu which includes: Value Data, Criteria Data, Alternative Data, User Data for the Data Analysis Transaction sub menu which consists of Criteria Analysis, Alternative Analysis, Revenue Results and reports and Logout for the profile menu is at the top right of the user's photo. In order to access the menus and sub menus in the main admin menu, the admin must first log in.

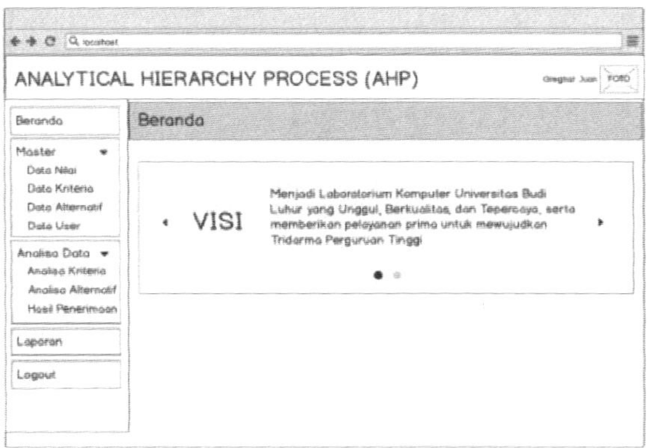

Figure 3. Main Menu Screen Design

3.1.3. Value Data Master Form Screen Design. Figure 4 displays the Value Master Data Form for paired value comparisons. The Value Data Master Form consists of nine scales, the modal button is added to add data, the modal button is to change to change the data, and the modal button is to delete to delete the data.

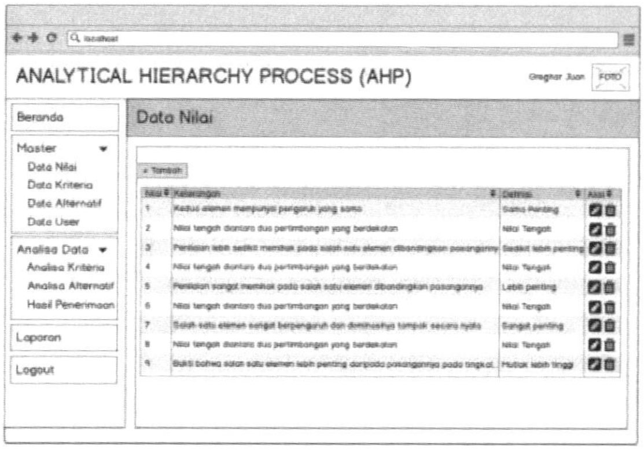

Figure 4. Value Data Master Form Screen Design

3.1.4. Criteria Master Data Form Screen Design. Figure 5 displays the Criteria Master Data Form to determine the weight value in the criteria master data form. In the Criteria Master Data Form there is a modal add data button to add criteria data, a mod button modal to change the data in the Criteria Master Data, and a delete button to delete data in the Criteria Master Data.

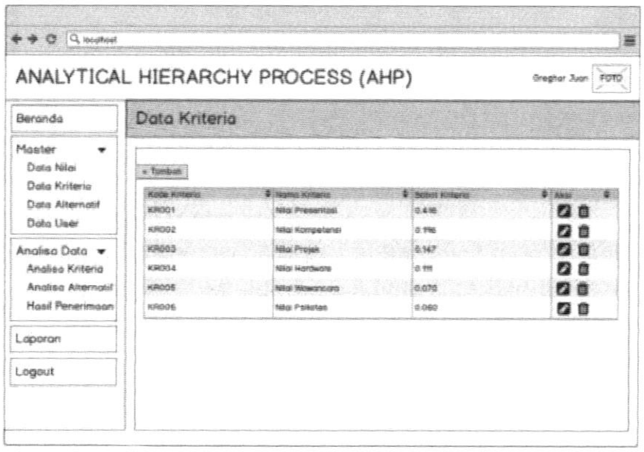

Figure 5. Criteria Master Data Form Screen Design

3.1.5. Report Form Screen Design. The design of the Report Form layer can be seen in Figure 6 below.

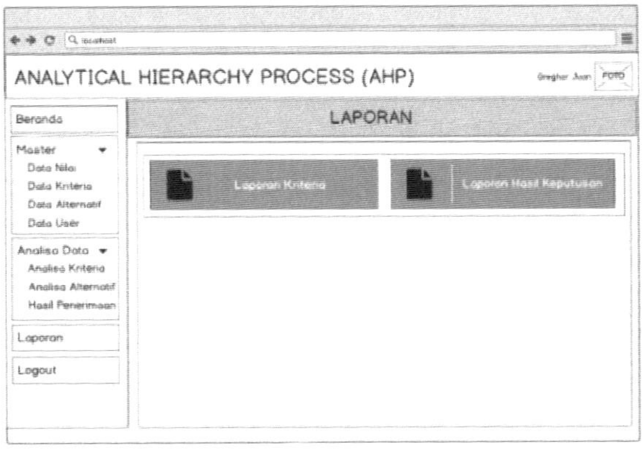

Figure 6. Design of the Report Form Layer

Criteria report screen design will print criteria data, namely criteria code, criteria name and criteria weights. Figure 7 shows the Draft Criteria Report:

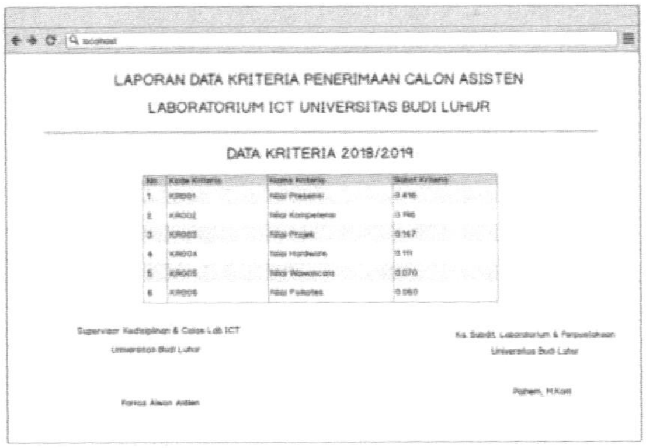

Figure 7. Criteria report screen design

Figure 8 shows the decision result report that will print Presence Value, Competency Value, Project Value, Hardware Value, Interview Value, Psychological test score on each alternative, total score and decision description.

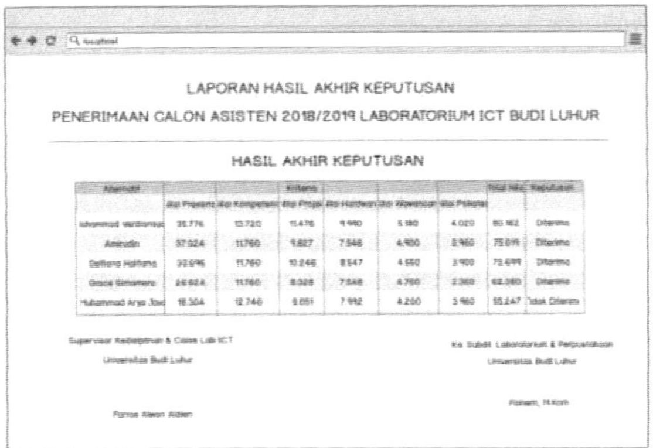

Figure 8. The Decision Result Report Screen Design

4. Implementation

4.1. Program Implementation
So that the application implementation of the Analytical Hierarchy Process algorithm (AHP). Can run well, the specifications used for implementation of this application must also support. The following specifications can support this system, including:

4.1.1. Hardware. Hardware used for implementation of this application is as follows: a. Processor AMD E1-6010 1.4Ghz b. RAM / Memory 10 GB c. Laptops d. 500GB Hard Drive.

4.1.2. Software. Software used for implementation of this application is as follows: a. Microsoft Windows 7 Operating System b. MySQL c. PHP 7.2.12 d. Xampp e. Google Chrome f. Notepad++.

4.2. Screen Display
In this section, we describe the application screen display for the implementation of the Analytical Hierarchy Process (AHP) algorithm from the first run to completion. The following will provide an explanation of each display in the application implementation of the Analytical Hierarchy Process (AHP) algorithm.

4.2.1. Login Form Screen Display. Before being able to access the Main Menu, the User is required to fill in their username and password. Display the login form as shown in Figure 9. If an error occurs when filling out the login form, the application will display an error message as shown in Figure 10 and if the username is no longer active, the application will display a message as shown in Figure 11 and if it is successfully logged in it will display a message as shown in Figure 12 then go to the Main Menu.

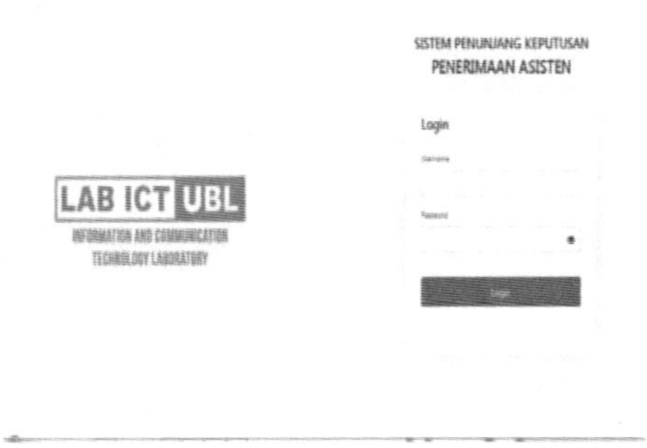

Figure 9. Login Form Screen Display

Figure 10. Error Message Screen Display

Figure 11. No Active Username Screen Display

Figure 12. Succeed Login Screen Display

4.2.2. Value Data Master Screen Display. Figure 13 displays Master Data Value Data for paired value comparison. The Value Data Master Form has nine scales, there is a modal add button to add data, a delete data button to delete the value data you want to delete, there is modal button change to change data.

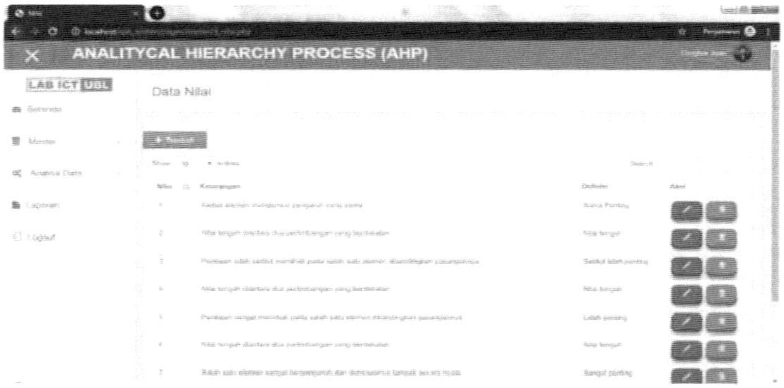

Figure 13. Value Data Master Screen Display

4.2.3. Report Screen Display. Figure 14 displays two report options, including: Criteria Report and Decision Result Report.

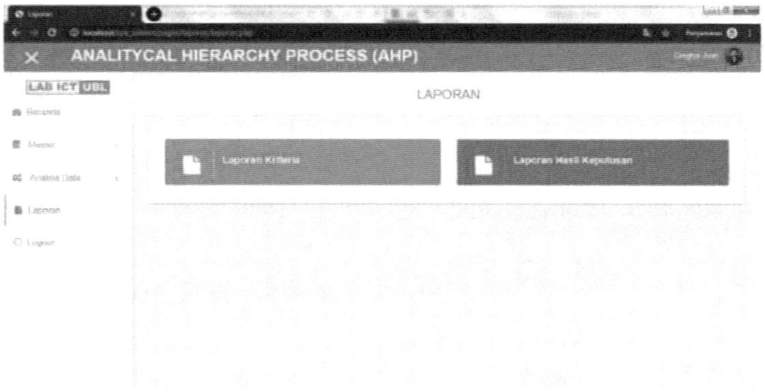

Figure 14. Report Screen Display

The results of the Criteria Report will print criteria data containing the Criteria Code, Criterion Name and Criteria Weight as shown in Figure 15 below.

Figure 15. Criteria Report Screen Display

The results of the Decision Result Report will print the final score data on each alternative and a decision statement for the acceptance of alternatives or prospective assistants. Figure 16 is the result of the display of the Decision Result Report.

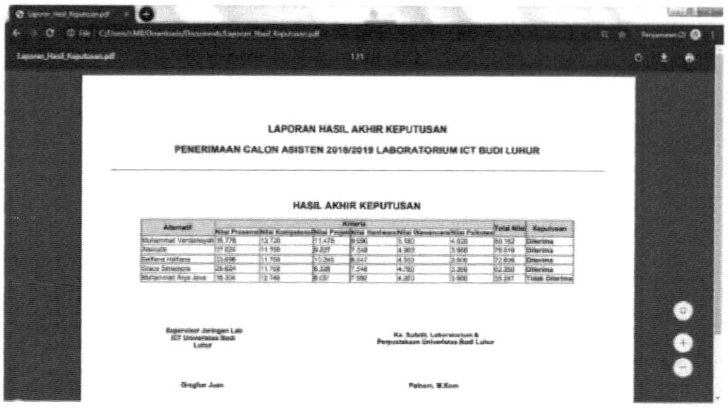

Figure 16. Decision Result Report Display

4.2.4. Profile Form Screen Display. Figure 17 displays the Profile Form in which there is a profile that is useful for knowing the User Profile who is in control of the system. And in it there is a Username, User Name, Position, and Password. Then there is a change profile button which is used to change the user profile.

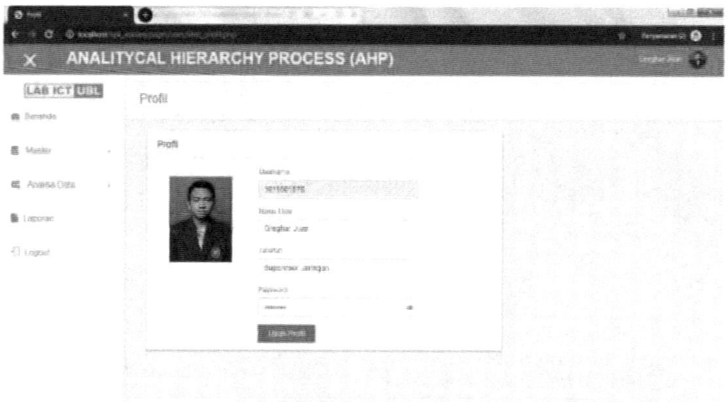

Figure 17. Profile Form Screen Display

5. Conclusion and Suggestion

5.1. Conclusion

There are several conclusions which can be taken in accordance with the research objectives, including:

- There are six criteria in the selection of the Integrated ICT Laboratory Assistant at Budi Luhur University, they were Attendance Value, Competency Value, Project Value, Hardware Value, Interview Value, and Psychological Test Value, which are in accordance with the criteria in the Integrated ICT Laboratory of Budi Luhur University. The system can speed up the process of accepting assistants, because the calculation processes, as well as printing the final score are in one system, according to the terms and criteria specified, making it more effective and efficient.

- This system produces a report on the value of the Analytical Hierarchy Process (AHP) which has the final score for the assistant candidate along with a statement of acceptance, so that it can assist the selection team for the Integrated ICT Laboratory of Budi Luhur University in the decision-making process. The final result of the Implementation of the Analytical Hierarchy Process (AHP) Application is only to assist decision makers in accepting alternatives from many existing alternatives, so that the assessment becomes faster with accurate data. However, the final decision is still determined by the decision maker.

5.2. Suggestion

For the sake of the conclusions above, the authors provide suggestions that are considered useful in the future, including:

- This application can be developed in line with the development of system user needs so as to improve system performance.
- This application can be developed by comparison using the Simple Additive Weighting (SAW) or Weighted Product (WP) method so that the accuracy of the data can be seen
- Further development of the system is to build a more user-friendly system by taking into account the aspects of interaction and computers.

6. References

[1] Iswandy E 2015 *Sistem Penunjang Keputusan untuk Menentukan Penerimaan Santunan Sosial Anak Nagari dan Penyaluranya bagi Mahasiswa dan Pelajar Kurang Mampu di Kenagarian Barung-Barung Balantai Timur* Jurnal Teknoif vol 3 (2) p 70–79.

[2] Junianto E, Primaesha Y 2015 *Perancangan Sistem Tracking Invoice Laboratorium pada PT Sucufindo (Persero) Bandung* Jurnal Informatika vol 2 (2) p 442-452.

[3] McCosh, A.M. and Morton, M.S.S., 1978 *Management decision support systems* Springer.

[4] Moore, J.H. and Chang, M.G., 1980 *Design of decision support systems* ACM SIGOA Newsletter, 1(4-5), pp.8-14.

[5] Sasongko A, Astuti I F, Maharani S 2017 *Pemilihan Karyawan Baru dengan Metode AHP (Analytic Hierarchy Process* Jurnal Informatika Mulawarman vol 12 (2) p 88-93.

[6] Wira D, Putra T, Epriyono M 2017 2015 *Sistem Pendukung Keputusan Pemilihan Sepeda Motor Jenis Sport 150cc Berbasis Web Menggunakan Metode Analytical Hierarchy Process (AHP)* Jurnal Teknoif vol 5 (2) p 16-24.

[7] Darmanto E, Latifah N, Susanti N 2014 *Penerapan Metode AHP (Analytic Hierarchy Process) untuk Kualitas Gula Tumbu* Jurnal Simetris vol 5 (1) p 75–82.

[8] Saaty, T.L., 1990 *The analytic hierarchy process in conflict management.* International Journal of Conflict Management.

[9] Ahmad R 2018 *Penggunaan Sistem Pendukung Keputusan dengan Menggunakan Metode Analytical Hirarchy Process (AHP) dalam Menyeleksi Kelayakan Penerima Beasiswa* Jurnal Metik vol 2 (1) p 28–33.

Face Tracking for Flying Robot Quadcopter based on Haar Cascade Classifier and PID Controller

A S Priambodo[1], F Arifin[2], A Nasuha[3] and A Winursito[4]

[1,2,3,4] Electronics and Informatics Engineering of Education Department, Universitas Negeri Yogyakarta, Indonesia

E-mail: ardyseto@uny.ac.id

Abstract. The fundamental aim of this research is to develop a face detection system for a quadcopter in order to follow the face object. This research has two main stages, namely the face detection stage and the position control system. The face detection algorithm used in this research is the haar cascade method which is run using the python and OpenCV programming languages. The algorithm worked well, getting around 16fps on a low spec computer without a GPU unit. The results of the face detection algorithm are proven to be able to recognize faces from the camera installed on the DJI Tello mini drone. The mini drone was chosen because it is small and light, so it is harmless, and testing can be carried out indoors. Besides, the DJI Tello can be programmed easily using the python programming language. The drone's position is then compared with the set point in the middle of the image to obtain errors so that control signals can be calculated for up/down, forward/backward, and right/left movements. From the testing results, the response speed that occurs in the right/left and up/down movements is less than 2 seconds, while for the forward/backward movement, it is less than 3 seconds.

1. Introduction

The development of science and technology that is increasingly advanced in recent years can benefit all aspects of human life. Especially the rapid developments in computers, electronics, mechanics, and robotics over the last few years convince us that more robot-human interactions are not a dream but a possibility that could happen in the future. A robot is a set of mechanical devices that can perform physical tasks with human supervision and control or a predetermined program.

Robots help humans a lot, including doing dangerous work to reduce the risk that can injure humans [1]. Robotics technology has been widely applied to production machines to produce products in large quantities. Robots have several advantages, including being fast, thorough, working full time and automatically, and producing uniform products regardless of fatigue and time [2]. However, robots sometimes require periodic maintenance at certain times so that the results are carried out optimally. Many types of robots have been interesting lately; one type of robot that is widely researched and studied is a flying robot. This type of flying robot is widely researched because it brings many application opportunities.

An unmanned Aerial Vehicle (UAV) is a vehicle that can fly without a human crew when carrying out flight missions. UAVs are grouped into two types, namely the fixed-wing aircraft type and the rotary-wing multirotor type. Multicopter is a Vertical Take-Off and Landing (VTOL) vehicle because the UAV can take off and land vertically without the need for a large runway [3]. One type of VTOL UAV is a

Quadcopter with four motors that have the characteristics of good maneuverability and high stability. Quadcopter has many advantages such as low energy consumption, efficient system, and simple operation. The quadcopter can be controlled remotely, so it can be used to replace the human role. Multicopter UAVs in mapping, aerial monitoring, and handling natural disasters have better stability than aircraft-type UAVs [4].

The computing capabilities of computers have increased, and the price-to-performance ratio of computer systems has decreased, thereby contributing to the development of broader computer applications. Humans have the sense of sight, one of the natural sensors imitated in developing computer capabilities. The camera is a sensor that can replicate how the eye works, which can be implemented in computer applications. The data captured by the camera is raw and meaningless information. Then the next step is to develop a computer application to understand the information in the image captured by the camera. The utilization of images like this is included in the field of computer vision.

One of the research topics in computer vision that is often studied in the last few decades is face detection. Face detection is used to determine the presence or absence of a face in an image, and this section is the first step in the identification process, so its existence is vital [5]. Detecting faces is very easy for humans; however, this task is very complicated for the computer because there is some complexity associated with location, viewing angle, light, and occlusion. Of the many face detection methods, the most frequently used method is the Viola-Jones method [6]. The Viola-Jones method was created by Paul Viola and Michael Jones in 2001. The face detection process is performed by classifying an image based on simple feature values through a classifier formed from training data. This method combines the concept of Haar, Integral Image, and AdaBoost features, which are then processed into a Cascade Classifier [7]. The main part of the Viola-Jones algorithm is computation and feature selection. Using Integral Image to extract Haar features has accelerated computation time compared to a per-pixel calculation. The main advantage of the Viola-Jones algorithm is that it is robust, which has a high detection for face tracking in images with a low error rate.

Several studies applying image processing to quadcopters are the development of quadcopters as traffic controllers, air pollution monitoring, and area mapping [8]. In addition, other studies have also used quadcopters to detect, localize, and recognize trees. In this study, we propose a quadcopter system that can follow faces that have proven to be effective in real-time experiments on a computer with limited specifications. Quadcopter motion control uses a proven PID control algorithm that can work well on a comprehensive system with good results.

2. Material and Methods

2.1. Materials

This research uses a commercial mini quadcopter made by DJI in collaboration with Ryze Robotics, the DJI Tello, shown in figure 1. This mini quadcopter was chosen because it offers the best solution when flying around people without giving danger to the people around them. The small size of the DJI Tello with dimensions of 98×92.5×41 mm and a weight of 80 grams makes this quadcopter also practical and easy to fly indoors. The DJI Tello is equipped with a propeller guard to protect the propeller when it hits an object and protect object. The DJI Tello is equipped with a 3-axis gyroscope and accelerometer sensor that functions to obtain the angular position and speed of the vehicle, a barometer sensor that is used for altitude from the vehicle, and a downward-facing infrared sensor to maintain a more precise flight altitude compared to a barometer.

Figure 1. Mini Quadcopter DJI Tello

The main advantage of this small drone is the extra features such as the Vision Positioning System (VPS) and an onboard camera. Combining these features with an advanced control system makes this mini quadcopter hover in a quiet position. The onboard camera contained in this mini drone has specifications of 5-megapixel photos and 720p live video streams. The maximum flight time is 13 minutes, with a maximum distance of about 100m. This drone has been equipped with a failsafe protection feature, which means that when the connection is lost, the drone will land safely. In addition, another safety feature is that when the drone hits an object hard enough, the drone motor will stop rotating. This drones have several fascinating features which make them the perfect candidate for the drone, such as affordability, relatively smaller size, programmable with python and swift, embedded camera, Intel processor for stable flight and turbulence reduction.

This drone control uses a wifi network with an application installed on an Android or iPhone. DJI released a Software Development Kit (SDK) with various read and write commands over UDP communication. Thanks to Damià Fuentes Escoté for porting the SDK into the python programming language [9]. By utilizing the djitellopy library, we can access sensor data on the DJI Tello and give orders to move according to the control signal we provide. There are two basic network streams: a dual-ended full-duplex connection for sending and receiving commands between the laptop and drone and a one-way half-duplex connection for streaming video from the drone to a laptop connection. After establishing a fairly stable connection with the live streaming function, python and OpenCV were used for facial recognition on video streaming from the drone.

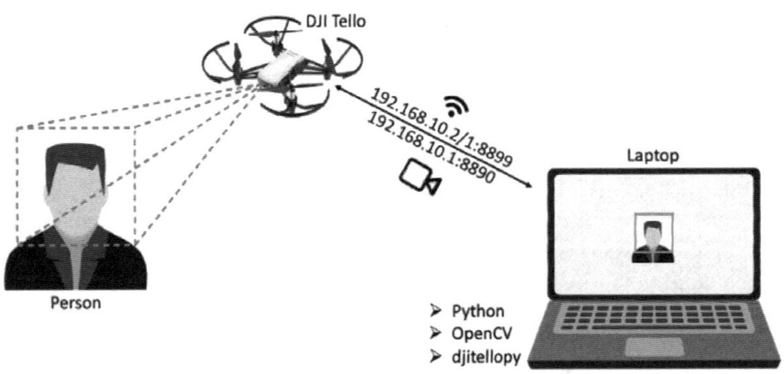

Figure 2. System Proposed Overview

The complete proposed system is shown in Figure 2. A DJI Tello drone that can be communicated via a Wi-Fi network is connected to a laptop. The camera from DJI Tello captures an image transmitted to a laptop for further processing using a computer vision algorithm whose details are discussed in the method section. The computer used in this project has detailed specifications, as shown in Table 1.

Tabel 1. Computer Specification

CPU	Intel i5-7200U (2C 4T, Max 3.10 Ghz)
RAM	DDR4 12GB (8GB+4GB)
Storage	SSD SATA 500GB
Connectivity	WiFi 2.4Ghz
Webcam	720p
Operating System	PopOS! 20.04 LTS

2.2. Methods
Face detection is a way of localizing and extracting facial areas for facial recognition purposes. Facial recognition is a technology in image processing (computer vision) that can identify a person's identity or information from the face. This technology is still very broad in its use in other fields, including security, robotics, or the health sector.

Face detection in this research uses the Haar Cascade classifier algorithm. The Haar Cascade algorithm applies a cascade function to train images through 4 main stages: (1) determining Haar features, (2) creating integral images, (3) Adaboost training, and (4) classifying using a cascading classifier. This process is shown in figure 3. The important thing that must be considered in using this algorithm is that it requires many datasets with positive facial images and non-face negative images for the classification training process.

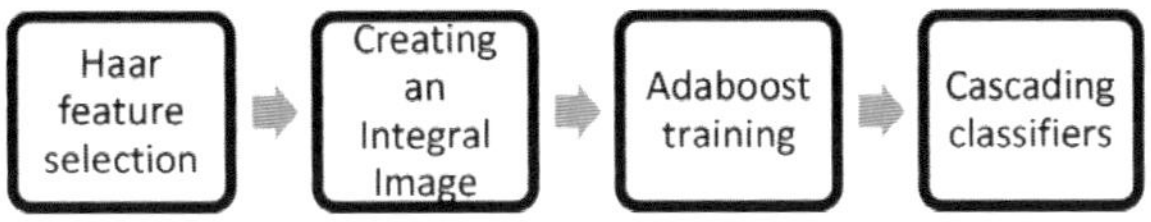

Figure 3. Stages of Haar Cascade Classifier

The first step in the haar cascade algorithm is to collect the haar features. Haar feature is a calculation performed on the adjacent square area at a certain location in a detection window. This calculation involves adding up the pixel intensities in each section and calculating the difference between the summed results. Some types of haar features are Edge features, line features, four-rectangle features. This feature will experience problems when executed for large images, so an integral images process is needed because the number of operations is reduced.

Integral image aims to speed up the process of calculating haar features. In general, an integral image is an image whose value for each pixel is an accumulation of the values of the top and left pixels. It should be noted that almost all haar features used have low performance when detecting objects. So we need haar features that match the object to be detected. The use of Adaboost here is the answer to selecting the optimal haar feature that can be used. The AdaBoost algorithm is an algorithm that builds strong classifiers by combining several simple (weak) classifiers. Weak classification results are created by moving the window over the input image and calculating the Haar feature for each subsection. This difference is compared with the studied threshold that separates non-objects from objects. Since this is a "weak classifier", many Haar features are required for accuracy to form a strong classifier.

The combination of Cascade of Classifier is the last stage in the haar cascade method. By combining classifications in a cascade structure or Cascade of Classifier, the speed of the detection process can be increased, namely by focusing on areas in the image that are likely to be. This classifier is done to determine where the object is looking for is in an image. The classification in this algorithm consists of three levels where each level produces a sub-image that is believed to be not an object. This process is done because it is easier to judge that the sub-image is not the object wanted to detect than whether the sub-image is the object wanted to detect.

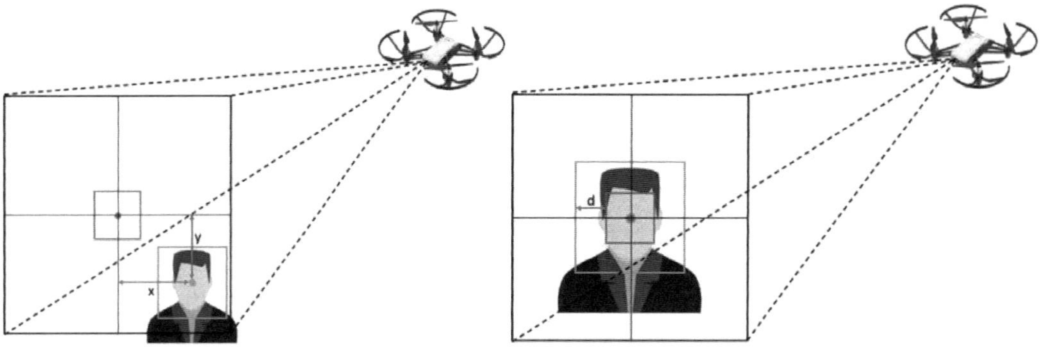

Figure 4. Error Calculation Methods (a) Error Calculation in X and Y Axis; (b) Error Calculation in Distance Between Drone and Face

In order to be able to do tracking, a parameter is needed that becomes a reference for the quadcopter in making movements. The method proposed in this study is to create a set point in the form of a bounding box with a certain size located in the center of the image, as shown in figure 4. The object detection bounding box results are compared with the middle bounding box so that the error rate of the quadcopter's position against the object's face is known. There are three parameters, namely x, y, and d, representing the distance on the x and y axes, and d or distance, the distance between the object and the quadcopter. This illustration is shown in Figure 4. This error is then used as a reference in calculating the PID control system to generate a control signal.

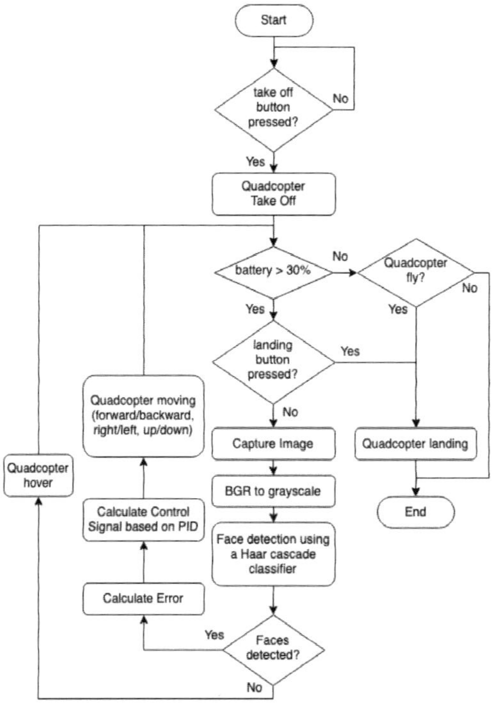

Figure 5. Flowchart of the face tracking algorithm for quadcopter

In a PID control system, a variable e represents the error obtained from the difference between the actual value of the position and the desired set point r(t). The output of the control system is denoted by y(t). each component of each Proportional, Integral, and Derivative control is added to produce a control signal. The general form of the equation of the PID controller [10] is written as follows:

$$u(t) = K_p e(t) + K_i \int_0^t e(\tau) d\tau + K_d \frac{de(t)}{dt}$$

The algorithm used in this study is shown in Figure 5. In general, it is an iterative process of face recognition, and calculation of control signals used to control forward/backward, right/left, and up/down movements. The API has covered the process of requesting images from the camera and sending motion commands on djitellopy via Wi-Fi communication.

3. Result and Discussion

The face detection system is built using the python programming language with Open-CV and djitellopy libraries. The test is carried out in several poses and conditions. The results of this test are shown in Figure 7. It can be seen in the test results of 4 poses that the Haar Cascade algorithm can detect faces well.

Figure 6. Face Detection Testing for Some Poses

The next test is an error test that applies the method described in the method section. There are three error variables in the X, Y axes and the distance between the drone and the face. The error results are displayed on the screen to make it easier to observe, denoted by eX, eY, and eD. It can be seen in Figure 7 that there is an error on the X-axis of -24, which means to the right 24 pixels; the error on the Y-axis is 8, which means upwards of 8 pixels. Error at a distance is calculated using To calculate the distance error using a comparison of the area of the reference box with the results of the face reading. The test results show that the eD value is -86, which means the drone is too close to the face.

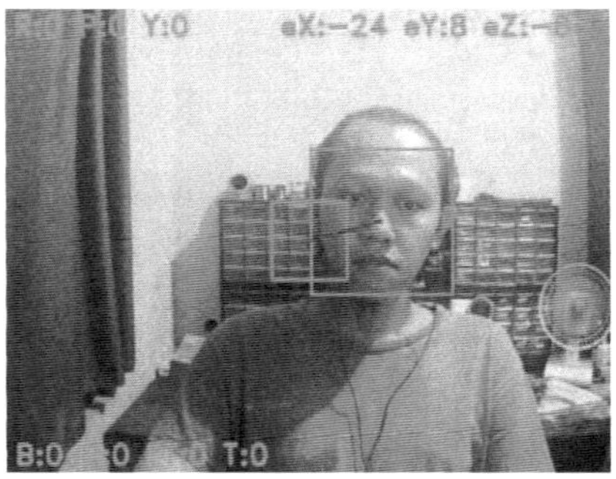

Figure 7. Face Detection and Error Calculation (eX, eY, and eD) Test Using Webcam and Image

After getting the error value from the actual condition of the drone, the control signal can be calculated so that the drone can follow the face. PID control system is used to deal with this problem. The test results show the optimal parameter values of Kp, Ki, and Kd in table 3. It should be noted that each movement on the quadcopter has its control system so that there are nine parameters. The movement results in each action can be seen in Figure 8, where when a change occurs, the error enlarges, and as soon as possible, the drone will correct the error so that it goes to zero. From Figure 8, it can be seen that the corrections made by the designed control system can work well.

Tabel 2. PID Parameter

Kp – forward / backward	1.05
Ki – forward / backward	0.02
Kd – forward / backward	0.1
Kp – right / left	0.65
Ki – right / left	0.004
Kd – right / left	0.08
Kp – up / down	1
Ki – up / down	0
Kd – up / down	0.2

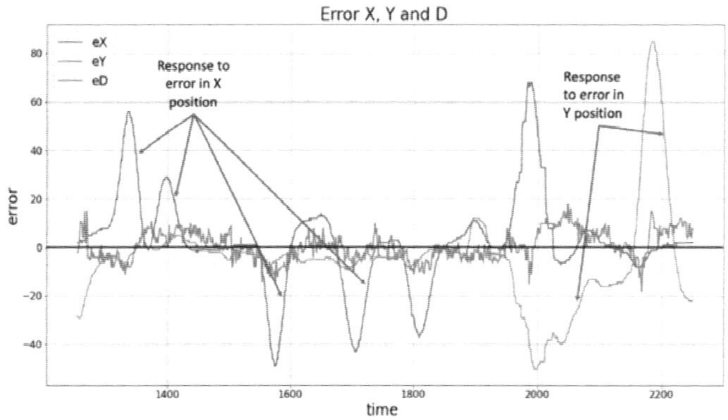

Figure 8. Error in X, Y and D Position

The last test is to test the system with various displacements from the face and then observe the drone's movement whether it can follow the face or not. For observations to be made, additional features are needed, namely the storage of video results. The process of storing video results needs a little adjustment related to the defined fps parameter. When the fps parameter is too large compared to the actual fps, the video recording will experience acceleration. The defined fps parameter must match the actual fps.

The results of several tests are shown in Figures 9 - 11, wherein Figure 9 observations are made on right-left movements, and the results are that the drone can follow the face. In Figure 10, observations are made on the up-down movement by changing the position of the face down, and it is proven that the drone can follow these changes. The last test is related to the distance between the face and the drone, as shown in Figure 11. The drone can move back and forth by adjusting the set point that has been determined.

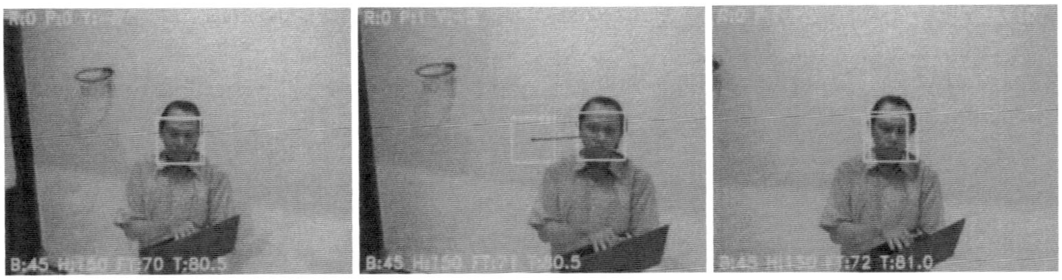

Figure 9. Face Tracking Movement of Quadcopter in Right/Left (a) Initial Position; (b) Face Change Position to The Right; (c) Quadcopter Follow the Face

Figure 10. Face Tracking Movement of Quadcopter in Up/Down (a) Initial Position; (b) Face Change Position to The Bottom; (c) Quadcopter Follow the Face

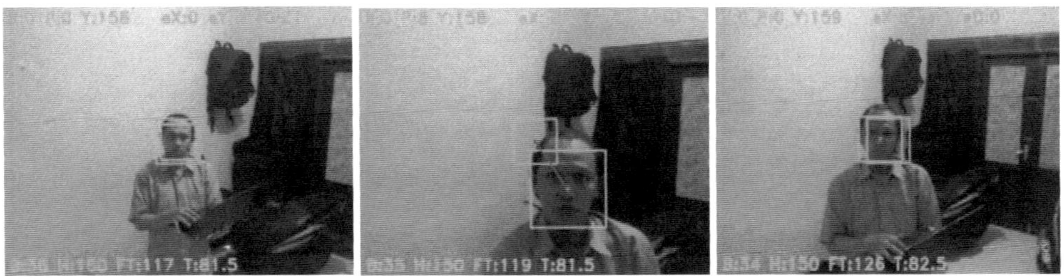

Figure 11. Face Tracking Movement of Quadcopter in Forward/Backward (A) Initial Position; (B) Face Change Position Closed to the Drone; (C) Quadcopter Follow the Face

4. Conclusion

Haar cascade classifier method can be used for face detection in images from webcams with low computer specifications because this method is simple and lightweight classification method. The haar cascade algorithm is designed using a dataset of facial images so that only facial images can be detected and other objects cannot be detected. Sometimes, the face is not detected due to poor lighting, or the algorithm does not get the feature is looking for at that position. The drone position is calculated based on the actual distance to the set point located in the center of the image and then becomes the basis for calculating control signals for up/down, forward/backward, and right/left movements. The PID control system that is designed to work well in real-time is proven from drone testing results that it can follow objects when faces change positions. Many object detection algorithms have better performance but require large computations, so the selection of the haar cascade method is due to the limitations of the computing system owned. Improvements that can be made are to add a recognition algorithm so that the drone can follow a predetermined face object.

5. References

[1] Galin, R.R. and Meshcheryakov, R.V., 2020. Human-robot interaction efficiency and human-

robot collaboration. *In Robotics: Industry 4.0 Issues & New Intelligent Control Paradigms* (pp. 55-63). Springer, Cham.

[2] Stano, G. and Percoco, G., 2020. Additive manufacturing aimed to soft robots fabrication: A review. *Extreme Mechanics Letters*, p.101079.

[3] Rashad, R., Goerres, J., Aarts, R., Engelen, J.B. and Stramigioli, S., 2020. Fully actuated multirotor UAVs: A literature review. *IEEE Robotics & Automation Magazine*, 27(3), pp.97-107.

[4] Dündar, Ö., Bilici, M. and Ünler, T., 2020. Design and performance analyses of a fixed wing battery VTOL UAV. *Engineering Science and Technology, an International Journal*, 23(5), pp.1182-1193.

[5] Liu, Z., Qi, X. and Torr, P.H., 2020. Global texture enhancement for fake face detection in the wild. In *Proceedings of the IEEE/CVF Conference on Computer Vision and Pattern Recognition* (pp. 8060-8069).

[6] Viola, P. and Jones, M.J., 2004. Robust real-time face detection. *International journal of computer vision,* 57(2), pp.137-154.

[7] Soo, S., 2014. *Object detection using Haar-cascade Classifier.* Institute of Computer Science, University of Tartu, 2(3), pp.1-12.

[8] Ostojić, G., Stankovski, S., Tejić, B., Đukić, N. and Tegeltija, S., 2015. Design control and application of quadcopter. *International Journal of Industrial Engineering and Management*, 6(1), pp.43-48.

[9] D. F. Escoté, djitellopy, (2021), GitHub repository, https://github.com/damiafuentes/DJITelloPy

[10] Priambodo, A.S., Cahyadi, A.I. and Herdjunanto, S., 2017. Perancangan Sistem Kendali PD untuk Kestabilan Terbang Melayang UAV Quadcopter. Departemen *Jurnal Teknik Elektro dan Teknologi Informasi*. Fakultas Teknik, Universitas Gadjah Mada.

Smart Air Filter Based on Activated Charcoal and Aloe Vera Using Fuzzy Logic Algorithm

F Arifin[1], W R Gusti[2], A Nasuha[3], A S Priambodo[4], and A Winursito[5]

[1,2,3,4,5]Electronic Engineering, Faculty of Engineering, Yogyakarta State University

E-mail: [1]fatchul@uny.ac.id

Abstract. Smoking is one of the main risk factors of several chronic diseases such as lung cancer, upper respiratory tract cancer, heart disease, stroke, bronchitis, emphysema and others, even smoking can cause death. Cigarette smoke inhaled by passive smoking is the main cause of lung cancer in people who are not smokers. The risk of lung cancer increases by 20-30% in non-smokers but always surrounded by cigarette smoke, compared to non-smokers who are not exposed to smoke. This study aims to develop a smart device to filter cigarette smoke so that the air released from this device is free from the dangers of cigarette smoke. The materials used in this device as a cigarette smoke filter are activated charcoal and aloe vera, the pores of the charcoal open and can become adsorbents. This device uses fuzzy logic as a control method that can provide decisions that resemble human decisions. Arduino is used to adjust the fan rotation in absorb cigarette smoke and the LCD to display the level of cigarette smoke.

Keywords: cigarette smoke, smart air filter, arduino, fuzzy logic

1. Introduction

In Indonesia, smoking is a serious health problem. The impact caused by smoking is a major risk factor for several chronic diseases such as lung cancer, upper respiratory tract cancer, heart disease, stroke, bronchitis, emphysema and so on, even cases of death can be caused by smoking. People who smoke are divided into two categories, namely active smokers, and passive smokers [1]. Active smokers are people who intentionally smoke cigarettes themselves. Passive smokers are people around active smokers and also inhale cigarette smoke scattered in the air.

Active and passive smokers inhale cigarette smoke which contains gaseous and particle components. Cigarette smoke gas components consist of nitrogen and hydrocarbon compounds. Particle components consist of nicotine, tar, phenol, benzopyrene and cadmium. There are three main toxic components of cigarette smoke, including carbon monoxide, nicotine, and tar. Fifty percent of smokers cause death, almost killing six million people each year who are ex-smokers and 600,000 of them are passive smokers [2]. If not followed up, this mortality and morbidity rate will continue to increase every year.

The cigarette smoke in the air will result in clean air pollution so that the air quality becomes less good. Cigarette smoke inhaled into the lungs can causes respiratory problems [3]. In addition of air pollution caused by cigarette smoke in the room, the decline in air quality also occurs due to poor air circulation. This causes a lot of dust and germs to spread in the room [4].

Air pollution that has been exposed mainly due to cigarette smoke can be overcome by a developed device named smart air filter. The air output from this device is free from the dangers of compounds contained in cigarette smoke. This device uses Arduino UNO with an ATmega328 microcontroller which is equipped with the ability to interact with other devices through its input/output (I/O). Arduino

UNO has 14 I/O pins (0–13) which can be used to control sensors, motors and various other types of accumulators [5]. This device uses a gas sensor which it triggers a fan to absorb cigarette smoke and also uses fuzzy logic so that the fan will spin automatically according to the gas volume in the air and power efficiency will occur. This device is equipped with filters from activated charcoal and aloe vera.

Fuzzy logic was updated in 1985 by Takagi-Sugeno Kang. Sugeno corrected the weakness of the pure fuzzy system. It is in simple mathematical calculations–the THEN section was added. The fuzzy system, especially IF-THEN which has been modified, contains a weighted average value. Mathematical calculations that are not appropriate so that they cannot provide a natural framework when representing human thinking. The problem of pure fuzzy logic then has limitations in the application of different principles, so that the uncertainty of the fuzzy system cannot be processed properly [6].

Activated charcoal and aloe vera which filter of the device have an important role in filtering cigarette smoke. Activated charcoal can be made from coconut shell with chemical activation of $ZnCl2$ and $Na2CO3$ [7]. Coconut shells go through the combustion process so that they become active. Dey [8] in a study on the benefits of activated charcoal proved that cigarette filters using activated charcoal is more effective in filtering substances from cigarettes. The type of active filter developed in the laboratory shows that the filter can protect smokers from cytotoxicity due to cigarette smoke, protein modification, apoptosis, and emphysema.

Furthermore, the aloe vera filter contained in the developed device can prevent lung disease. Aloe Vera has the potential to modulate smoke-induced changes in lung tissue that may influence the management of smoking-related diseases [9]. So, the use of aloe vera as a filter on the device is very possible to filter the air from the dangers of cigarette smoke pollution.

2. Method

This study uses Research and Development (R&D) methods, the steps in R&D as follows (1) potential and problems, potential is anything that when utilized will have added value, problems are deviations between what is expected and what is happening; (2) gather information, after the potentials and problems can be factually demonstrated and updated, then it is necessary to collect information that can be used as material for planning certain products that are expected to overcome these problems; (3) product design, products are designed to solve existing problems; (4) design validation, design validation is a product design assessment activity by experts who are competent in their fields; (5) design improvements. after the product design, assessed through discussions with experts and other experts, the weaknesses will be identified, these weaknesses are reduced by improving the design; (6) product trial, the product was tested on a limited group; (7) product revision, the product weaknesses found in the device, then corrected to obtain a more perfect product; (8) Trial usage. The product is tested on a wider group [10].

This research was conducted in the Electronic Engineering Education Class of Yogyakarta State University. This research is a continuation of the final project of the intelligent control system course.

3. Result and Discussion

The developed device aims to filter indoor air, especially from the dangers of cigarette smoke. This device uses Arduino UNO as a controller board using an ATmega processor with the ability to interact with other devices via its input/output (I/O) pins.

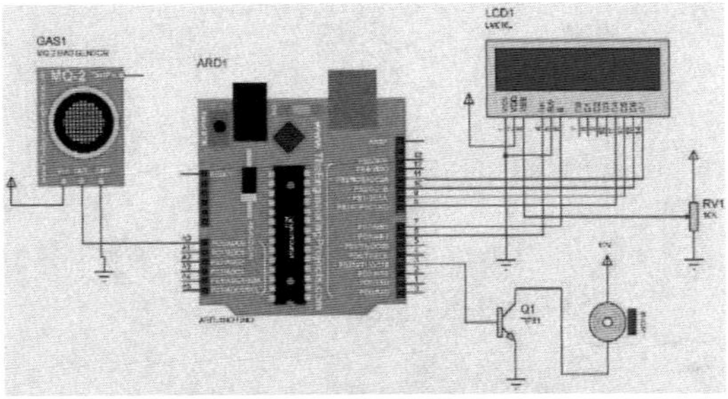

Figure 1. Electronic Circuit

Arduino board with fuzzy logic has an important role to process of determining the fan rotation speed. The fuzzification process is based on the levels of CO gas and smoke in the air. The sensor used to detect the gas content is the MQ2 gas sensor. In addition, to display the levels of gas contained in cigarette smoke uses the LCD. The electronic circuit can be seen in Figure 1.

Figure 2 is a flowchart and figure 3 is a block diagram of the device's working system include software and hardware based on figure 1. Starting from the gas sensor will detect the content of smoke. If detected, the fan will rotate to absorb the smoke according to the intensity of the amount of gas content of smoke. After the smoke is absorbed, the filtering process begins and in the end the air coming out from the device is clean from harmful compounds of cigarette smoke. The smoke level will be displayed on the LCD.

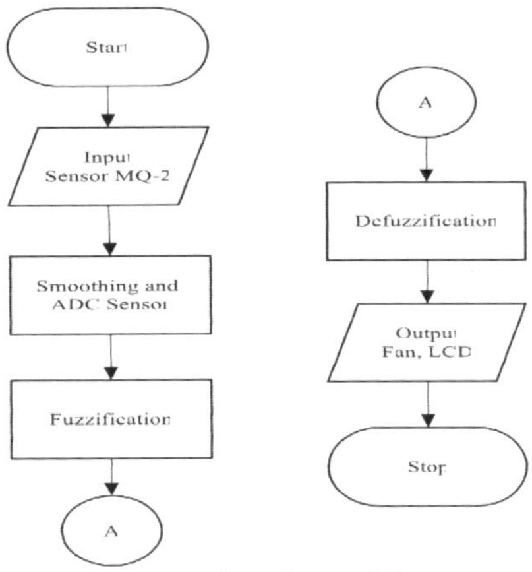

Figure 2. Flowchart of System

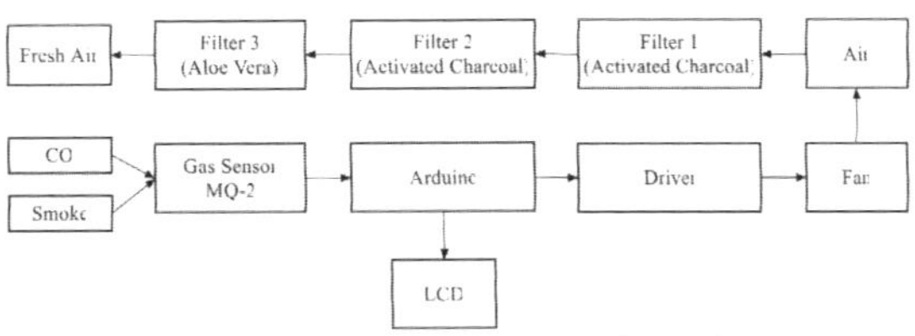

Figure 3. Block Diagram of Hardware

Based on the block diagram in Figure 3, the device model can be seen in Figure 4 and Figure 5.

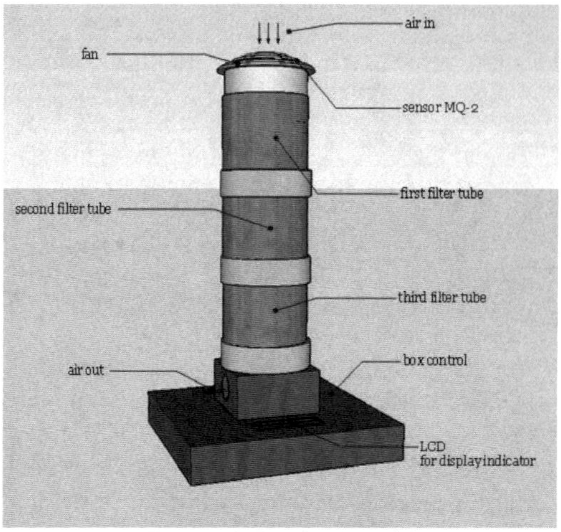

Figure 4. Device Model

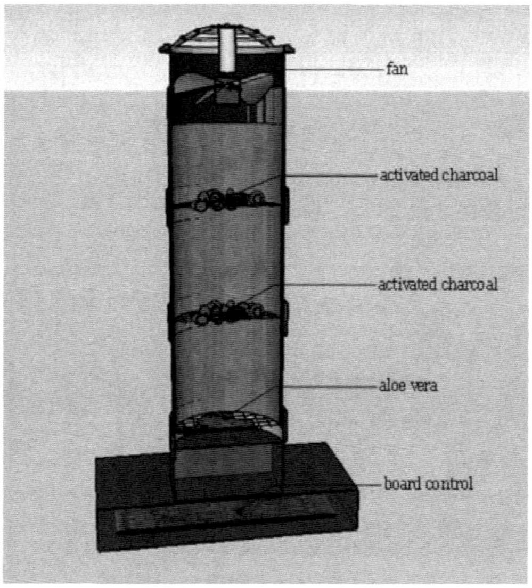

Figure 5. Cut View

This device is used to protect passive smokers from the dangers posed by active smokers. The device is placed close to an active smoker in a room. So, the cigarette smoke emitted by smokers is filtered by this device. The fuzzy logic used in processing output from the MQ2 sensor are CO and Smoke gas variables with low, medium, and high scales. Fan rotation speed settings from low, medium, fast, and very fast. There are four steps that must be done as follows:

3.1. Fuzzifications

Fuzzification is a process to convert a firm value into a fuzzy set or change a numeric variable into a linguistic variable. At this stage there are two numerical variables that will be converted into linguistic variables, namely:

3.1.1. CO Variable. CO inputs have 3 membership functions with 3 triangular membership functions.

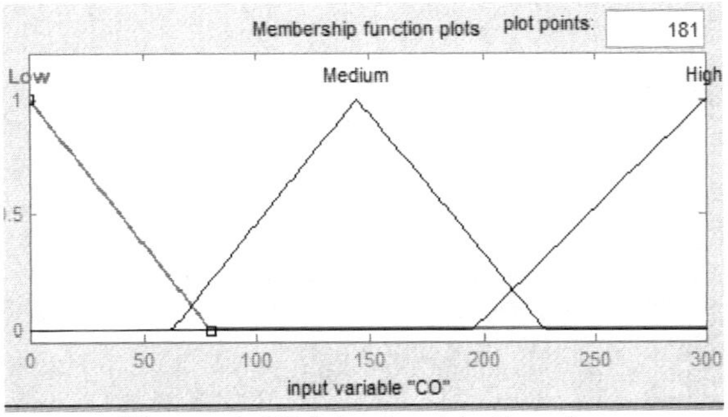

Figure 6. CO Membership Function

3.1.2. Variabel Smoke. Smoke inputs have 3 membership functions with 3 triangular membership functions.

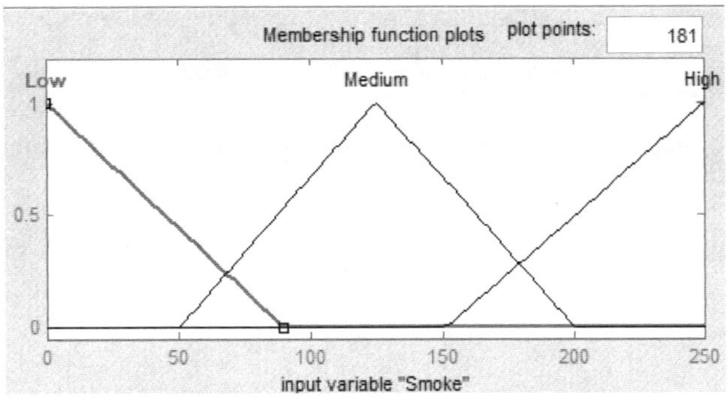

Figure 7. Smoke Membership Function

3.2. Establishment of Ground Rules

After determining membership function of the variable, then the fuzzy rules are formed. The rule is made to express the relation between fuzzy input and fuzzy output. The operator that will be used to connect the fuzzy inputs is AND and IF-THEN to describe the input-output Based on the existing fuzzy inputs, a fuzzy rule is made.

Table 1. Rule

CO↓/ SMOKE→	Low	Medium	High
Low	Low	Medium	Medium
Medium	Medium	Medium	Fast
High	Medium	Fast	Very Fast

From the two detected CO and Smoke, a rule is made as shown in table 1. Membership function of the gas content is categorized into 3, namely low, medium, and high. The input variable is connected to the fan output which has 4 membership functions, the output rotates speeds namely low, medium, fast, and very fast.

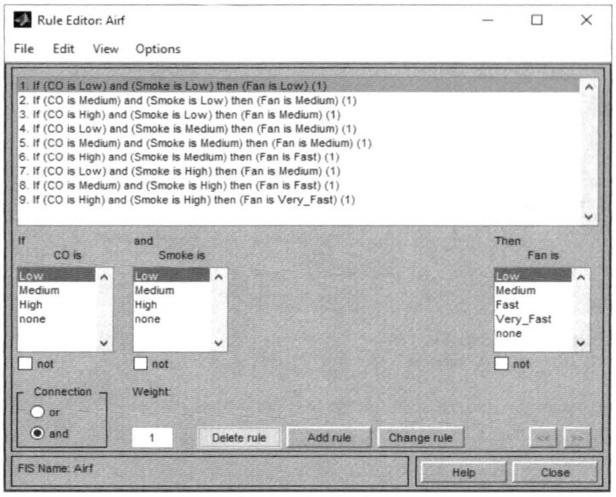

Figure 8. Rules

3.3. Interface Engine

After creating the fuzzy basic rules, the composition of the rules will be carried out to get the weights of the basic rules using the MIN implication function.

$$Wi = min\ (I1, I2, I3, \ldots\ Ii)$$

Explanation:

Wi : The weight of the implication to – i; i = 1, 2, 3...i

Ii : Fuzzy set input value from basic rule to –i.

3.4. Defuzzification

After doing the inference, defuzzification will be carried out to get the output in the form of a firm value. This firm value will be converted into the fan rotation speed. Defuzzification is complete using the weight average method.

$$Output = i = 1\ N\ Wi\ Zi = 1\ N\ W\ i$$

Explanation:

N = Number of rules.

Wi = The weight of the implication to -i

Zi = Behavioral output value of rule-i

The test is carried out by burning a cigarette, then the MQ2 sensor detects the levels of Smoke and CO gas in the air with units of Part Per Million (PPM). The output is detected on the Arduino IDE named clock scale that generates Pulse Width Modulation (PWM) with a maximum clock of 8 bits, range 0 – 255. The input has a maximum limit that has been calibrated according to the distance between the device and the smoker, in other words, as the range of the device. The maximum input for CO gas is 300 PPM and Smoke maximum input is 250 PPM which will be processed in a fuzzy algorithm. The delay for each reading is 3 seconds. PWM monitoring using a serial monitor on the Arduino IDE Software gives the results in Table 2.

Table 2. Test Result

NO	INPUT		OUTPUT (0 – 255)
	Gas CO (PPM)	Smoke (PPM)	
1	0	0	34,6
2	56	25	39,2
3	36	109	107
4	134	98	107

5	257	105	189
6	280	243	254
7	251	72	154
8	27	15	36,1
9	300	250	254
10	250	250	254

Arduino UNO has a 10-bit analog input and an 8-bit analog output. PWM manipulates the digital output in such a way that it produces an analog signal. The microcontroller shifts the digital output to HIGH and LOW alternately with a certain time portion for each output value. The time duration for the HIGH value is called the pulse width or pulse length. Variations in analog output values are obtained from changes in pulse length given over a period of time and repeated. Analog voltage output can be calculated by the formula:

$$\text{Vout} = \frac{clock\ \text{8bit scale}}{255} \text{ x } 5 \text{ V}$$

Table 3. Convert To Analog Output

NO	Skala (0 – 255)	Analog Output (Volt)
1	34,6	0,68
2	39,2	0,77
3	107	2,10
4	107	2,10
5	189	3,71
6	254	4,98
7	154	3,02
8	36,1	0,71
9	254	4,98
10	254	4,98

The reading on the MQ2 sensor shows a fluctuate value. This refers to the levels of CO gas and smoke levels in the air that are read. Cigarette smoke that enters the device then passes through 2 times the activated charcoal filter and an aloe vera filter. After going through 3 filter processes, the air produced is free from cigarette smoke pollution. The results of the readings on the device's output are often zero.

4. Conclusion

Air pollution especially due to cigarette smoke, can be overcome with a developed filter. After making and testing smart air filter based on activated charcoal and aloe vera using fuzzy logic algorithm, the following conclusions are (a) the device developed using fuzzy logic as signal processing obtained from the gas sensor requires three stages, including fuzzification, manufacture rules, and defuzzification. (b) In making the device software, it begins with making a flowchart of the workings of the tool that will be made with a fuzzy logic algorithm which is then written using the Arduino language. (c) In making hardware, MQ2 sensor is used as input, Arduino as processor, fan as output and LCD as a display of CO2 and smoke detected. The materials used in the cigarette smoke filter tube are aloe and coconut shell charcoal. (d) The way of this device works is when the MQ2 sensor detects cigarette smoke, arduino processes and activates a fan that functions absorb and insert cigarette smoke around the device into the filter tube which has been given aloe and coconut shell charcoal. (e) Smart air filter based on fuzzy logic using activated charcoal and aloe vera works quite well with filter output as expected.

This developed device still has several shortcomings, suggestions for improvement for the perfection of this tool are: (a) The filter tube size is enlarged so that the smoke filter results are maximized (b) The tube design uses stronger materials. (c) The fan used has a stronger suction

Journal of Physics: Conference Series **2111** (2021) 012047 doi:10.1088/1742-6596/2111/1/012047

capability so that the absorb range of the device is wider. (d) The bulkhead used to put aloe and coconut shell charcoal is strengthened and made neater.

5. References

[1] Jaya, Muhammad, "Pembunuh Berbahaya Itu Bernama Rokok," Yogyakarta: Riz'ma, 2009.

[2] Prignot, J, "Quantification and chemical markers of tobacco exposure," Eur. J. Respir Dis, vol. 70, no. 1, pp.1– 7 , January 1987.

[3] Baker, R. R, "Smoke generation inside a burning cigarette: Modifying combustion to develop cigarettes that may be less hazardous to health," Progress in Energy and Combustion Science, vol. 32, no. 4, pp. 373–385, December 2006.

[4] Daher, Nancy, Radwan Saleh, Ezzat Jaroudi, & Hiba Shaheitli, "Comparison of Carcinogen, Carbon Monoxide, and Ultrafine Particle Emissions From Narghile Water Pipe and Cigarette Smoking: Sidestream Smoke Measurements and Assessment of Second-Hand Smoke Emission Factors. J. Atmospheric Environment, vol. 44, no.1, pp. 8–14, January 2010.

[5] Arduino, "Datasheet Arduino UNO," [online: arduino.cc].

[6] Kurnia, Dona & Febby Kesumananingtyas, "Penerapan Fuzzy Logic dalam Pencarian Jalur Terbaik Menuju Lokasi Wisata di Kota Bukittinggi," J. TEKNOIF, vol. 5, no. 2, pp. 1–7, October 2017.

[7] Pambayun, Gilar S., Remigius Y. E. Yulianto, M. Rachimoellah, & Endah M. M. Putri, "Pembuatan Karbon Aktif dari Arang Tempurung Kelapa dengan Aktivator ZnCl2 Dan Na2CO3 Sebagai Adsorben Untuk Mengurangi Kadar Fenol Dalam Air Limbah," J. Teknik POMITS, vol. 2, no. 1, pp. 116–120, 2013.

[8] Dey, Neekkan, Archita Das, Arunava Ghosh & Indu B. Chatterjee, "Activated Charcoal Filter Effectively Reduces P-Benzosemiquinone from The Mainstream Cigarette Smoke and Prevents Emphysema," J. Bioscience, vol. 35, no. 2, pp. 217–230, may 2010.

[9] Koul, Ashwani, Sashi Bala, Yasmeen Goyal & Naveen Arora, "Aloe vera Affects Changes Induced in Pulmonary Tissue of Mice Caused by Cigarette Smoke Inhalation", Environmental Toxicology, vol 30, no. 989-1013, February 2014.

[10] Sugiyono, "Metode Penelitian Pendidikan," Bandung: CV. Alfabeta, 2015.

Design of Robust Heart Abnormality Detection System based on Wavelet Denoising Algorithm

A Winursito[1], F Arifin[2], A Nasuha[3], A S Priambodo[4], Muslikhin[5]

[1,2,3,4,5] Electronics and Informatics Engineering Education Department, Universitas Negeri Yogyakarta, Indonesia

E-mail: anggunwinursito@uny.ac.id

Abstract. The technology that continues to be developed by many researchers today is an automatic heart attack detection system based on an Electrocardiogram (ECG) signal. Several other studies have been carried out to build an Internet of Things (IoT) based heart abnormality detection system. Based on the analysis of related studies that have been carried out previously, several researchers have developed an ECG signal-based heart abnormality detection system using clean ECG signal data. While the reality of the concept of an IoT-based detection system, the process of recording ECG signal data on the sensor, the process of sending data to the server, and the process of retrieving data from the server are vulnerable to noise exposure. ECG signal containing noise will greatly affect the accuracy of system detection. This paper proposes the development of a noise-resistant heart condition detection system using a wavelet denoising algorithm. The process of denoising ECG signals using the Wavelet algorithm is generally able to improve the accuracy of detecting noisy ECG signals. The most significant increase in accuracy is seen in the low SNR value. The Daubechies 4 (db4) denoising algorithm is the best-performing algorithm. The ECG signal classification method uses the Artificial Neural Network (ANN) algorithm. Detection system hardware is also designed in this research using the concept based on the Internet of Things.

1. Introduction

Heart attack disease is one of the deadliest diseases in the world. Not a few cases of heart disease sufferers become increasingly severe because of the delay in detecting the disease. More than three-quarters of these deaths occur in low- and middle-income countries [1]. A large number of patients with heart disease has become a priority for many researchers in various fields to continue to develop technology that can help detect symptoms of heart disease. One technology that continues to be developed by many researchers is an automatic heart abnormality detection system based on an Electrocardiogram (ECG) signal. The ECG signal represents the electrical activity of the four chambers of the heart, which can be analyzed from a series of P waves, QRS complexes, and T waves [2]. Several other studies have been conducted to build an Internet of Things-based heart abnormality detection system. Rangappa, et al [3] conducted a study by designing an ECG signal classification algorithm for cardiac abnormalities. Researchers used 3 steps in the ECG signal detection process. The system is used to recognize normal and abnormal conditions of the human heart based on the characteristics of the ECG signal obtained through the MIT/BIH database. The researcher also compared the classification methods of KNN, Naïve Bayes, Random Forest, Neural Network, and Support Vector Machine (SVM) with the same test data. The results show that the KNN method produces the best accuracy compared to the others. Khan, et al [4] conducted a study on the recognition of heart disease based on ECG signals using the Deep Neural Network. Researchers focused on the system algorithm used to detect abnormalities in the human heart based on ECG signal images generated from electrodes attached to 12 points of the human body. The detection results using the proposed algorithm are quite capable of distinguishing 4 heart conditions that have been previously labeled when testing the detection system. Liu, et al [5] conducted a study on the classification of ECG signals to detect human heart abnormalities using the Wavelet Scattering Transform algorithm. Researchers also focused on detection system algorithms to classify 4 human heart conditions, namely non-ectopic (N), supraventricular ectopic (S), ventricular ectopic (V), and fusion (F). The results showed that the highest accuracy was achieved by the KNN algorithm with a constant value of k=4 with a 10-fold Cross-Validation validation test.

Based on the exposure to related research that has been done previously, several researchers have developed an ECG signal-based heart abnormality detection system. Several studies have been carried out using the concept of the Internet of Things in the detection system for heart abnormalities, but the data used is only limited to the clean ECG signal. While, the process of recording ECG signal data on the sensor, the process of sending data to the server, and the process of retrieving data from the server is vulnerable to noise exposure. ECG signal that contains noise will greatly affect the accuracy of system detection. This paper proposes the development of a robust heart abnormality detection system using the ANN algorithm and wavelet-based denoising. Several other studies have used wavelet algorithms in the process of denoising speech signals [6], [7]. Besides being used in the sound signal denoising process, the wavelet algorithm has also been used in the image denoising process [8], [9]. In this paper, a wavelet algorithm is proposed to be implemented in the ECG signal denoising process. The wavelet algorithm will be combined with the ANN detection algorithm. Several parameters of the wavelet algorithm and ANN will be analyzed to produce the best performance. Furthermore, the development algorithm will be designed in a detection system based on the Internet of Things.

2. Methodology

The development of a heart abnormality detection system is designed in several stages. The first process is the initial processing which aims to standardize the ECG signal data. The ECG signal data used is obtained from the PTB Diagnostic ECG Database dataset which has two conditions, namely normal and abnormal (myocardial infarction). The ECG signal data is further divided into 11641 training data and 2911 test data. Noise is mixed with the test data in some SNR values. The addition of noise to the test data aims to test the development system using the wavelet transform algorithm. The next step is denoising the ECG signal using a wavelet algorithm. Three types of wavelet algorithms used in this study are daubechies 4 (db4), daubechies 5 (db5), and symlet 8 (sym8). After the denoising process, the next process is the detection of the ECG signal using the Artificial Neural Network (ANN) algorithm. The method of developing a human heart abnormality detection system is shown in Figure 1. The wavelet transform algorithm is quite widely used in the image and video denoising process. In this study, several types of wavelet algorithms will be used for the signal denoising process. The denoising process using the wavelet algorithm is shown in Figure 1, starting from calculating the DWT value, thresholding in the wavelet domain, then calculating the inverse wavelet transform.

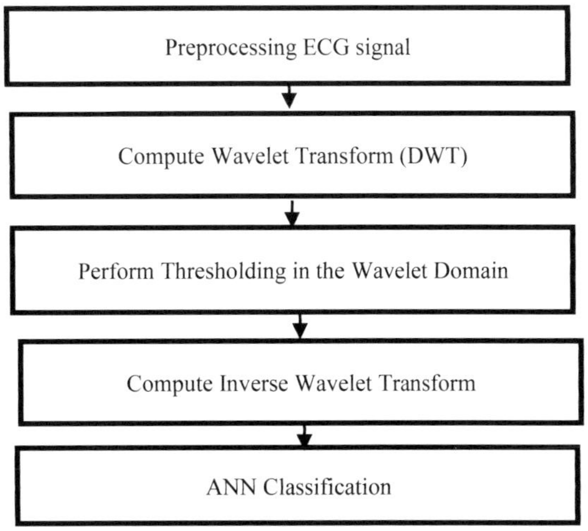

Figure 1. Proposed robust heart disease detection process

Several parameters need to be set in the denoising process. The level of decomposition, the type of wavelet algorithm, and the thresholding method are important parameters that need to be analyzed. Two types of thresholding processes are soft and hard thresholding operations [5]. The soft or hard thresholding is applied to the coefficient yielded from the wavelet decomposition process [6]. The respective mathematical formula of hard and soft thresholding can be seen in Equations (1) and (2),

$$y_{hard}(t) = \begin{cases} x(t), & |x(t)| > \delta \\ 0, & |x(t)| < \delta \end{cases} \tag{1}$$

$$y_{\text{soft}}(t) = \begin{cases} sgn\big(x(t)\big).(|x(t) - \delta|), & |x(t)| > \delta \\ 0, & |x(t)| < \delta \end{cases} \tag{2}$$

with x represents the noise coefficient, δ is the threshold, and y is the estimated output.

After the denoising process, the next process is the detection of the ECG signal using the Artificial Neural Network (ANN) algorithm. An overview of the ANN algorithm used as a signal pattern detection method is shown in Figure 2. In the ANN algorithm, neurons are arranged in several layers. The number of layers used can be adjusted to the needs of the system. The test data results from the denoising process using the wavelet algorithm and then the detection process is carried out using the ANN algorithm. Several important parameters have been determined in the ANN algorithm for the training process and prediction of ECG signal data.

Table 1. ANN parameter setting

Parameter	Detail
Number of hidden neurons	20
Activation function	Leaky ReLu
Learning rate	0.001
Number of Epoch	50

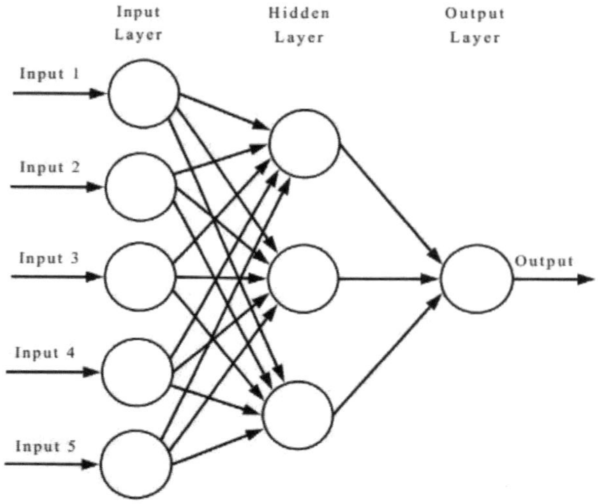

Figure 2. Artificial Neural Network Diagram

The heart abnormality detection system is also designed in the form of internet of things (IoT)-based hardware. The hardware design of an IoT-based heart abnormality detection system is shown in Figure 3. The sensor used to read the ECG signal is AD8232, then as a controller and sender of ECG data to the server using an ESP32 microcontroller. Monitoring and detection system using a laptop with python programming. Web application-based monitoring system using the Flask web framework. The output of the heart abnormality detection process is in the form of a PDF report file containing the patient's identity, the identity, and notes of the medical staff, the display of the patient's ECG signal, the results of monitoring the patient's heart rate, as well as the results of system detection in the form of normal or abnormal heart conditions (myocardial infarction).

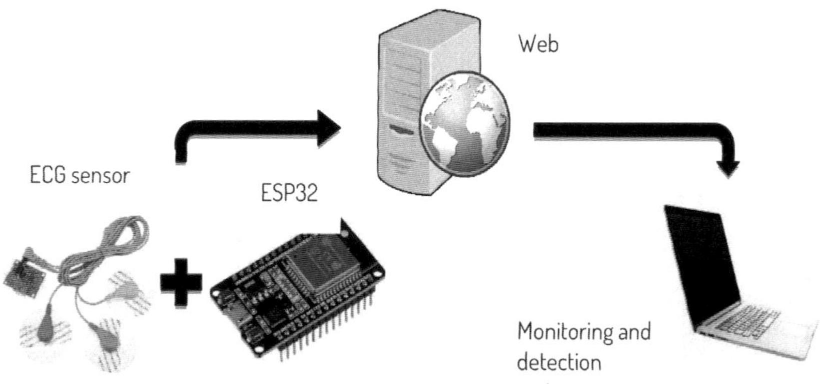

Figure 3. Hardware detection system-based Internet of Things

3. Results & Discussions

The heart abnormality detection system based on the ECG signal that is designed has several stages, starting from recording the ECG signal, preprocessing, and the detection process using the ANN algorithm. In addition, the denoising process using the Wavelet Transform is added to this system. The use of the Wavelet algorithm is carried out to eliminate noise that can interfere with the detection system. In this research, noise is added manually to test the algorithm performance of the development system. The ECG signal-based heart abnormality recognition system using the ANN algorithm has produced high accuracy. System accuracy will decrease if the ECG signal gets noise interference. Some noise with varying SNR intensities is added to the ECG signal. The effect of noise on the ECG signal on the conventional recognition system using the ANN algorithm is shown in Table 2.

Tabel 2. Comparison of Recognition Accuracy on Clean and Noisy ECG signals

ECG signal	SNR	Recognition Accuracy (%)
Clean signal	-	97,05
Noisy signal	15	80,50
	10	78,98
	8	77,84
	5	76,40
	3	73,65
	2	70,94

Table 2 above shows that the heart abnormality recognition system using the ANN algorithm has produced a high accuracy of 97.05%. The detection system does have high accuracy on clean ECG signals, but on noisy ECG signals it turns out that the detection accuracy has decreased significantly. A noisy ECG signal with a value of SNR = 15 resulted in a decrease in accuracy of 16.55% from the detection system that uses a clean ECG signal. Meanwhile, the lower the SNR value, the accuracy of the detection system also decreases.

The ECG signal denoising system is designed using a Wavelet transform algorithm. Three types of wavelet algorithms tested in this study are Daubechies 4 (db4), Daubechies 5 (db5), and Symlet 8 (sym8). Each wavelet algorithm is used for the ECG signal denoising process, then a comparison is made to determine the best type of wavelet algorithm in this study. The results of the comparison of the three wavelet algorithms in the ECG signal denoising process are shown in Table 3.

Table 3. Comparison of the performance of the three Wavelet algorithms in the ECG signal denoising process

No	SNR	Recognition Accuracy (%)		
		db4	db5	sym8
1	15	82,27	83,89	81,76
2	10	81,72	79,49	81,38
3	8	79,66	75,30	78,53
4	5	76,92	76,78	74,30
5	3	76,61	73,24	74,58
6	2	76,23	70,18	72,55

Table 3 shows the comparison of the three wavelet algorithms in the ECG signal denoising process. The results showed that the greater the SNR value of the ECG signal, the greater the accuracy of the detection system in the three wavelet algorithms. The db4 wavelet algorithm has the best performance in almost all SNR values (2, 3, 5, 8, & 10). At the SNR value of 15, the db5 algorithm produces the best performance, with an accuracy of 83.89%. Figure 4 shows the comparison of the accuracy of the conventional detection system with the development of a robust detection system using the three wavelet denoising algorithms (db4, db5, & sym7).

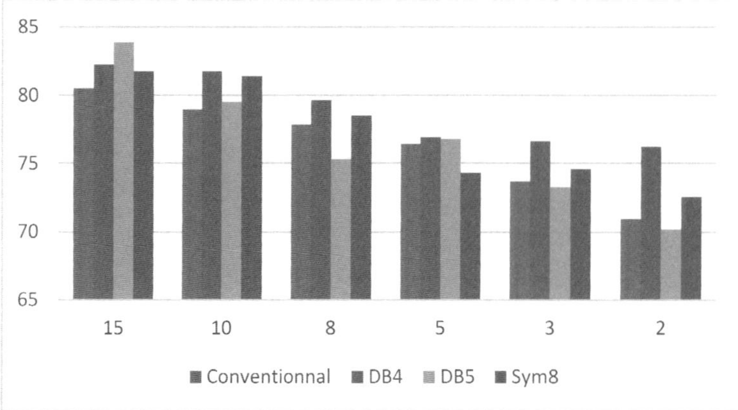

Figure 4. Comparison of recognition accuracy of all methods

Based on the comparison of the accuracy of several methods shown in Figure 4, the denoising process of ECG signals using the Wavelet algorithm is generally able to increase the detection accuracy of noisy ECG signals compared to conventional detection methods. The most significant increase in accuracy is seen at low SNR values. The denoising Daubechies 4 (db4) algorithm is the algorithm that produces the best increase in accuracy at SNR 2 to SNR 10. While at the SNR 15 value, Daubechies 5 (db5) produces the highest accuracy compared to other algorithms. Based on the experimental results, the Daubechies 4 algorithm was chosen as the best algorithm in the ECG signal denoising process. The heart abnormality detection system is then designed using the ANN development algorithm and the db4 wavelet. The detection system development algorithm is then implemented on the internet of things-based system hardware. The AD8232 sensor is used to pick up the patient's ECG signal. The ECG signal data will then be sent by the ESP32 microcontroller to the webserver. The detection system is embedded in a computer at the location of the medical officer. When the ECG signal data has been sent to the webserver, the medical officer's computer will retrieve the ECG signal data and then carry out the detection process on the detection system algorithm that has been developed previously.

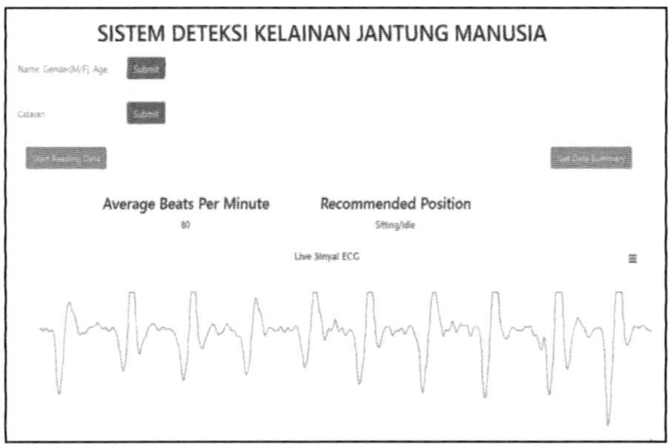

Figure 5. Display of data input and monitoring system.

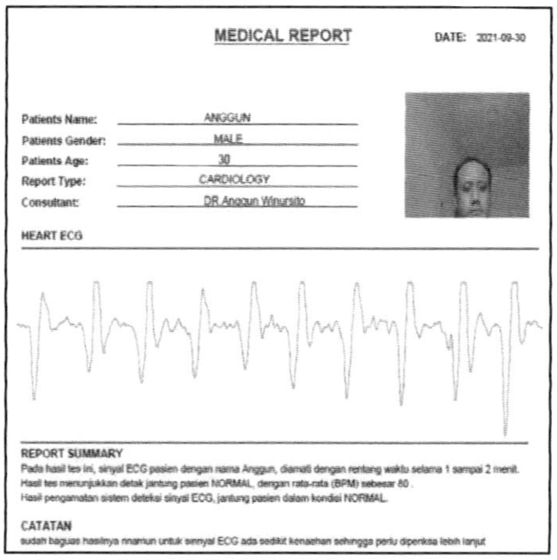

Figure 6. Display of file report.

4. Conclusion

The development of a human heart abnormality detection system is carried out by designing a system that is resistant to noise. The denoising algorithm using wavelet has been successfully implemented in the detection system. The process of denoising ECG signals using the Wavelet algorithm is generally able to improve the detection accuracy of noisy ECG signals. The most significant increase in accuracy is seen at low SNR values. The denoising Daubechies 4 (db4) algorithm is the best performing algorithm that can overcome noise at SNR 2, 3, 5, 8 & 10.

The system development algorithm has been implemented on the internet of things-based system hardware. The AD8232 sensor is used to pick up the patient's ECG signal. The ECG signal data is then sent by the ESP32 microcontroller and stored on the webserver. The detection system is embedded in computers at the location of medical personnel. The results of the detection process for the patient's ECG signal are in the form of a pdf report file with information about the patient's identity, the identity and notes of the medical staff, the display of the patient's ECG signal, the results of monitoring the patient's heart rate, as well as the results of system detection in the form of normal or abnormal heart conditions (myocardial infarction).

References

[1] S. A. A. Yusuf, "Analisis Metode Ekstraksi Fitur Pada Isyarat EKG Menggunakan Discrete Wavelet Transformation dan Mel- Frequency Cepstral Coefficient," Universitas Gadjah Mada, Yogyakarta, Indonesia, 2020.

[2] N. Ahmed and Y. Zhu, "Early Detection of Atrial Fibrillation Based on ECG Signals," *Bioengineering*, vol. 7, no. 1, p. 16, Feb. 2020.

[3] V. Rangappa, S. Prasad, and A. Agarwal, "Classification of Cardiac Arrhythmia stages using Hybrid Features Extraction with K-Nearest Neighbour classifier of ECG Signals," *Int. J. Intell. Eng. Syst.*, vol. 11, no. 6, pp. 21–32, Dec. 2018.

[4] A. H. Khan, M. Hussain, and M. K. Malik, "Cardiac Disorder Classification by Electrocardiogram Sensing Using Deep Neural Network," *Complexity*, vol. 2021, pp. 1–8, Mar. 2021.

[5] Z. Liu, G. Yao, Q. Zhang, J. Zhang, and X. Zeng, "Wavelet Scattering Transform for ECG Beat Classification," *Comput. Math. Methods Med.*, vol. 2020, pp. 1–11, Oct. 2020.

[6] R. Hidayat and A. Winursito, "A Modified MFCC for Improved Wavelet-Based Denoising on Robust Speech Recognition," *Int. J. Intell. Eng. Syst.*, vol. 14, no. 1, pp. 12–21, Feb. 2021.

[7] M. A. Oktar, M. Nibouche, and Y. Baltaci, "Speech denoising using discrete wavelet packet decomposition technique," in *2016 24th Signal Processing and Communication Application Conference (SIU)*, Zonguldak, Turkey, May 2016, pp. 817–820.

[8] L. Li, J. Wu, and X. Jin, "CNN Denoising for Medical Image Based on Wavelet Domain," in *2018 9th International Conference on Information Technology in Medicine and Education (ITME)*, 2018, pp. 105–109.

[9] Y. Qian, "Image Denoising Algorithm Based on Improved Wavelet Threshold Function and Median Filter," in *2018 IEEE 18th International Conference on Communication Technology (ICCT)*, 2018, pp. 1197–1202.

[10] P. V. Lavanya, C. V. Narasimhulu, and K. S. Prasad, "Transformations analysis for image denoising using complex wavelet transform," in *Innovations in Information, Embedded and Communication Systems (ICIIECS), 2017 International Conference on*, 2017, pp. 1–7.

[11] X. Liu, "A New Wavelet Threshold Denoising Algorithm in Speech Recognition," in *2009 Asia-Pacific Conference on Information Processing*, 2009, pp. 310-313.

Prototype Development Of Distance Detection System Based On The Internet Of Things Using Esp 8266 Wifi Nodemcu Module

[1]Deny Budi hertanto', [2]Rustam Asnawi, [3]Faranita Surwi, [4]Nurman Setiawan

[1,2,3,4] Department of Electrical Engineering Education, Faculty of Engineering, UNY

Email: [1]denybudi@uny.ac.id, [2]rustam@uny.ac.id, [3]faranita@uny.ac.id, [4]nurman.setiawan@uny.ac.id

Abstract. In the previous article about detecting train arrivals [4], we discussed the data transmission process which is the weakness of the system. This article discusses how to overcome these shortcomings. Distance detection systems need to be added to devices that are more powerful and faster in transmitting data. The prototype is built based on the development of the previous prototype. The initial product is a distance detector that has a motion sensor and a data transmission module in the form of Lora 400 MHz and a GSM module. Product development includes the addition of the NodeMCU WIFI module to existing devices. System development aims to: (1) Develop a distance detection system equipped with a WIFI module; and (2) Increase the data transmission speed of the distance detection system equipped with the WIFI NodeMCU module. The implementation method uses research and development techniques [3]. System development takes 6 months, with the stages of completion being system requirements analysis, hardware, and software design, system prototyping, and tool testing. Checklist data and delivery time are used as test data. Furthermore, the data is analyzed quantitatively to determine the achievement of results according to predetermined indicators. The device is equipped with a GPS module, a Lora-R02 receiver and transmitter module, and a Nodemcu ESP8266 module as a replacement for the 900A SIM module which transmits data faster. When the device is being tested, all the sensors of the tool work well at range of 125 meters (previously less than 100 meters). While the Lora module can react when the object has reached a distance of 300 meters. Data transmission previously using GSM modules took 10-13 seconds. After using the WIFI module, data transmission only takes 1-3 seconds.

Keywords: prototype, distance detection, wifi module

1. Background

Previously, the author has built a tool that detects the arrival distance of an object. The device is equipped with SIM900A and Lora 400mHz modules. These devices have difficulty communicating due to network constraints and obstruction of radio waves by buildings or objects that absorb radio waves. So that the data transmission signal cannot be sent quickly. If the data is not sent quickly, then the tool cannot function as it originally intended, which is to detect the arrival of an object quickly.

Various efforts need to be made, such as strengthening the signal and replacing the SIM module with the latest module that supports 4G technology. Another way to amplify the sending signal is to convert the GSM signal into a WIFI signal. Therefore, an additional module is needed in the form of a WIFI NODEMCU module which is expected to speed up signal transmission to the receiver.

In the next section, it will be discussed whether the performance of the NodeMCU module can speed up signal transmission so that the purpose of making the tool can be achieved.

2. Literature Review

The 900A SIM module is a GSM module that runs on a 2/3G network. This module is no longer suitable for use in Indonesia, which is already using the 4G network and is even starting to explore 5G networks. However, the presence of cellular signals in the Indonesian Region is still not evenly distributed.

Internet for Everything (English: Internet of Things, or IoT) is a concept that aims to expand the benefits of continuously connected internet connectivity. As for capabilities such as data sharing, remote control, and so on, including objects in the real world. [1]

Basically, the Internet of Things refers to objects that can be uniquely identified as virtual representations in an Internet-based structure. How the Internet of Things works is by utilizing a programming argument where each argument command produces an interaction between machines that are connected automatically without human intervention and at any distance.

Arduino is an open-source single-board microcontroller, derived from the Wiring platform, designed to facilitate the use of electronics in various fields. It is also an open hardware list aimed at anyone who wants to create interactive electronic prototypes based on flexible and easy-to-use hardware and software. The hardware has an Atmel AVR processor, and the software has its own programming language [2].

For flexibility, programs are loaded via the bootloader although there is an option to bypass the bootloader and use the downloader to program the microcontroller directly through the ISP port. One of the Arduino modules is Arduino GSM. Arduino GSM module allows devices to be controlled via cellular signal.

2.1 Modul NodeMCU ESP8266

This module is a development derivative module of the ESP8266 family of IoT (Internet of Things) platform modules, type ESP-12[1]. The ESP8266 module can be learned from the previous article. Functionally, this module is almost similar to the Arduino module platform, but what makes it different is that it is devoted to "Connected to the Internet".
For now, there are 3 versions of the NodeMCU [1] module, including:

2.1.1 NodeMCU 0.9

This version (v0.9) is the first version to have 4 MB flash memory as its (System on Chip) SoC and the ESP8266 used is the ESP-12. The weakness of this version is in terms of the size of the wide board module, so if you want to make a prototype using this version of the module on the breadboard, the pins are used up only for this module.

2.1.2 NodeMCU 1.0

This version is a development of version 0.9. And in version 1.0, the ESP8266 used is the ESP-12E type which is considered more stable than the ESP-12. In addition, the size of the module board is reduced so that it is compatible with making project prototypes on the breadboard. And there are dedicated pins for SPI (Serial Peripheral Interface) and PWM (Pulse Width Modulation) communication which are not available in version 0.9.

2.1.3 NodeMCU 1.0 (unofficial board).

It is said to be an unofficial board because this module product was produced unofficially with the approval of the NodeMCU Official Developer.

In general, the NODEMCU based on the ESP8266 microprocessor has a very low current consumption of between 15 A and 400 mA which can be further reduced with deep sleep mode enabled to 0.5 A. The current consumption is then a factor of 70000 higher in deep sleep mode for Arduino Uno with 35 mA.

3. Development Method

The device is developed using the research and development method [3]. The tool development stage consists of: evaluating the previous device, integrating the tool with the WIFI NodeMCU module, software adjustment, testing data transmission, and completing the device. The test results will be compared with tools that previously used the 900A SIM Module.

Based on the evaluation of the previous device, there are delays in data transmission. Next, the suggestion is to use a supplement module that can accelerate data transmission. Using WIFI NodeMCU, the system is expected to improve the ability of the early warning system by increasing the ability to transmit the data. This system aims to be able to detect the arrival of objects at a certain distance from the receiver, then turn on the siren and the lights. Detection data is broadcast from the moving object continuously while it is running.

Prototyping this system requires several components: (i) Arduino devices in both devices, both receivers at the intersection point and in moving objects, (ii) Nodemcu wifi ESp8266 for internet communication between two devices via cellular signals, (iii) prototype system power supply devices, (iv) LCD and alarm, (v) Siren, and (vi) Lora Radio Module.

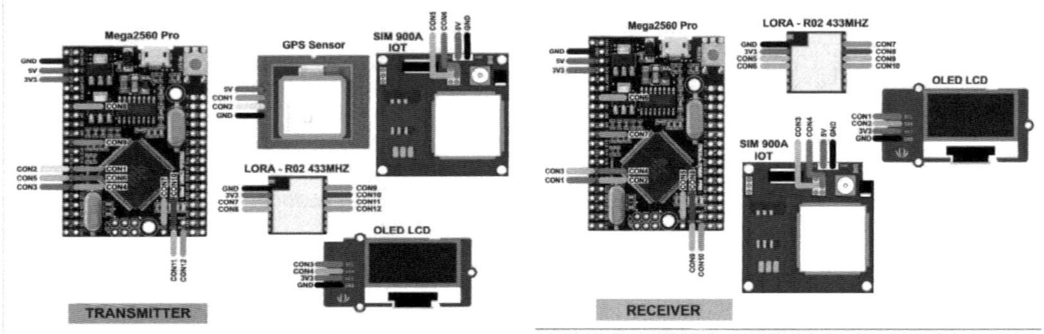

Figure 1. Transmitter and Receiver prototype design

As shown in Figure 1, all the modules will be assembled into a single unit. The finished tool will be connected to the computer. The computer will set the initial position of the coordinates to be determined by the GPS module. The computer program will arrange that when the object moves, the coordinates of the object are immediately sent from the transmitter to the receiver via the internet. Internet connection is established through a 4G-signal WIFI transmitter which will be captured by the NodeMCU. The process of sending data and sending time will be recorded by the computer. The delivery data will be used as evaluation material whether the tool is functionally an early detection of the arrival of objects or not.

The test includes black-box testing, which is testing the function of the arrival detection system of an object. Observations were made to perform functional tests on the prototype. The instrument used is a system functional test checklist. The instrument used in the validation is also a checklist or validation format. The data obtained were analysed descriptively.

4. Result and Discussion

The workings of the tool are: (i) The tool is first connected to WIFI, (ii) On one side of the tool, the motion sensor detects movement, (iii) Once an object is detected, the transmitter equipment will send data to the receiver, (iv) The Receiver will accept the signal, (vi) the device will turn on warning siren.

The test begins by determining the starting point of departure where the latitude is -7.744106, and the longitude is 110.373986. The object then moves. When crossing the reference point, the receiver device will ring a buzzer for a few seconds, then the indicator light will change from red to green when the simulated transmitter enters the specified distance. The reference distance is set at a radius of 125 meters from the receiver.

The distance (S) between transmitter and receiver is calculated using the following formula:

$$S = ACOS(COS(RAD(90 - L_1))COS(RAD(90 - L_2)) + SIN(RAD(90 - L_1))SIN(RAD(90 - L_1) \quad (1)$$

The test results are shown in Table 1. Observations show that in the functional test, the WIFI module, GPS module, 433MHz Lora module, LED, and Buzzer function as desired. At a distance of 307.4501867 meters, the WIFI module will transmit telemetry data to the receiver. In range 98.21 meters, the LED turning from red to green, indicating that position of the transmitter has reach 100 meters range. Then the Buzzer will beep for a few seconds.

Table 1. Some of the results of the tool test

Time	GPS Status	Coordinat of Transmitter	Coordinat of Receiver	Distance in meter	Status of IOT	Status of Lora 433 MHz	Status of Led	Status of Buzzer
05:33:14	Lock	-7,744106, 110,373986	-7,746076, 110,372028	307,4501867	Success	Failed	off	off
05:33:16	Lock	-7,744111, 110,373963	-7,746076, 110,372028	305,3587329	Success	Failed	off	off
05:33:16	Lock	-7,744111, 110,373963	-7,746076, 110,372028	305,3587329	Success	Success	off	off
05:35:09	Lock	-7,745209, 110,372344	-7,746076, 110,372028	102,5004502	Success	Success	off	off
05:35:10	Lock	-7,745245, 110,372233	-7,746076, 110,372028	98,21151785	Success	Success	on	on

In establishing a connection to the internet, the WIFI module maintains a stable and reliable speed when initiating IP addresses and data transmission protocols. This causes the process of sending data to be faster, both from the side of the receiving device and from the side of the sending device. The GSM module has difficulty initiating a bond to the internet due to poor signal problems so that the communication protocol and sending address are somewhat disrupted. The comparison of data transmission speed can be seen in Table 2.

Table 2. Comparison of Sensor Data Delivery Between GSM Module (GSM) and WIFI NodeMCU Module (MCU)

Module	Time	Initiation Time	Transmitting Time	Receiving Time	Total Time
GSM	05:33:14	3 seconds	5 seconds	2 seconds	10 seconds
MCU		1 seconds	1 seconds	0,5 seconds	2,5 seconds
GSM	05:35:09	4 seconds	5 seconds	2,5 seconds	11,5 seconds
MCU		1 seconds	1 seconds	0,5 seconds	2,5 seconds
GSM	05:39:24	3 seconds	3 seconds	2 seconds	8 seconds
MCU		1 seconds	1 seconds	0,5 seconds	2,5 seconds
GSM	05:51:10	4 seconds	6 seconds	2 seconds	12 seconds
MCU		1 seconds	1 seconds	0,5 seconds	2,5 seconds

5. Summary

According to the results and discussion, it can be concluded that: (1) All devices work completely well, (2) Data is sent by WIFI module 3-4 times faster than sending using GSM Module, (3) Receiver receives a signal when Transmitter reaches distance less than 300 meters. The receiver then will activate the led, buzzer, and siren when the object reaches a range of 100 meters, and (4) the device of the distance detection system is successfully developed with some shortcomings.

References

[1] Janssen, Cory. *Internet of Things: IoT*. Diakses dari situs techopedia pada 9 November 2013.
[2] *Getting Started with Arduino;* Massimo Banzi, Michael Shiloh*; 262 pages; 2014; ISBN 1-4493-6333-4.*
[3] Pressman, Roger S. (2012). *Rekayasa Perangkat Lunak. Yogyakarta: ANDI OFFSET*
[4] D B Hertanto, R Asnawi, F Surwi. *"Early Warning System Prototype of Train Arrival Based on Internet of Things",* Journal of Physics: Conference Series, 2021

Utilization of Used Electric Vehicle Batteries for The Design of Mini-Generating Systems

[1]Khairunnisa', [2]Muhammad Afif Amalul Arifidin, [3]Hartoyo

[1,2,3] Department of Electrical Engineering Education, Faculty of Engineering, UNY

Email: [1]khairunnisa@uny.ac.id, [2]afifamalul@uny.ac.id, [3]hartoyo@uny.ac.id

Abstract. One of the main objectives of this research is to reduce battery waste from electric vehicles, namely by processing the battery waste to be used as electrical energy storage which is then used in mini generators. Therefore, it is necessary to have methods and stages to do it all. The method used with BMS (Battery Management System) with the Battery Cell Balancing model. It aims to determine the classification of the total battery capacity and battery life time, and determine which batteries can still be reused. The results of the balancing of the battery cells will be used as recommendations and are more economical in power generation storage systems and also provide practical solutions in the management of used electric vehicle battery waste to prevent pollution of active ingredients in battery cells, and also become an alternative for energy storage choices. for reuse. This research can be considered in the future for the electric vehicle industry to be able to reuse used batteries from electric vehicles, as well as the design design of the use of used batteries for mini generation systems.

Keywords: Battery Management System, Cell Ballancing Model, Energy Storage System, Electric Vehicle, Solar Power Plant

1. Introduction

Advances in the use of technology are increasing every year, especially the increase in transportation by using a variety of the latest technologies, ranging from machinery to transportation energy storage. Moreover, in Indonesia, the use of electric vehicles has begun to be found in many areas, such as one made by the nation's children, the GESITS electric motor. nevertheless, with the increasing number of users, exactly, the battery usage will also increase, and give rise to new problems, namely, there is damage to the battery or a decrease in the performance of the battery itself, thus making users switch to replacing the old battery with a new battery. On the other hand, the battery from the use of electric vehicles can be transferred to a new smart grid [1]. In reference [1-3] mentions that the use of used electric vehicle batteries will be reduced by 70-80% of their original capacity, and that is considered the end of their first life, with a typical battery capacity of around 5- 24 kWh [mild hybrid electric vehicle (MHEV) 5 kWh–electric vehicle battery (BEV) 18–24 kWh battery]. Where the estimated battery life for ordinary users is around 10 years, with 40 minutes of battery use per day with a DoD (Depth of Discharge) 35% with 200 annual charges.

Lithium-Ion batteries are the most widely used type of battery, because they are rechargeable and are widely used in various applications ranging from electronic products to electric vehicles and also as energy storage for power generation scales, where the advantages are high energy / high power, low self-discharge, and also has a long-life cycle [4]. Battery life can be extended by balancing the cells, thus requiring a battery management system [Battery Management System (BMS)] [5-7].

The aim of this research is to reduce the waste of used electric vehicle batteries by utilizing the used batteries into electrical energy storage in PV power plant system using the battery management system method with a battery balancing model, and also to find out to what extent these used batteries can be reused and find out which batteries are still worth reusing.

2. Research Method

In this study the battery from the electric vehicle will be tested with a battery tester, to see the capacity of the battery and the voltage on the battery, because the main function of the BMS is to monitor and balance the battery cells in the battery pack. There are 2 things that cause and affect the imbalance of the cell battery, namely intrinsic and extrinsic. Where these intrinsic factors are related to the manufacturing process, causing the number of variations of the active ingredients, and internal barriers. While the extrinsic factors include how to connect (series/parallel), charging/discharging currents, and hot condensation.

A battery pack are balanced if at multiple SOCs, all cells are at the same SOC. The BMS will balance the battery cells to ensure that charging is maximized so that the battery can deliver the required amount of charge each time, otherwise it will cause battery damage problems such as heating effect, charge reduction, reduced energy efficiency and accelerated degradation [6].

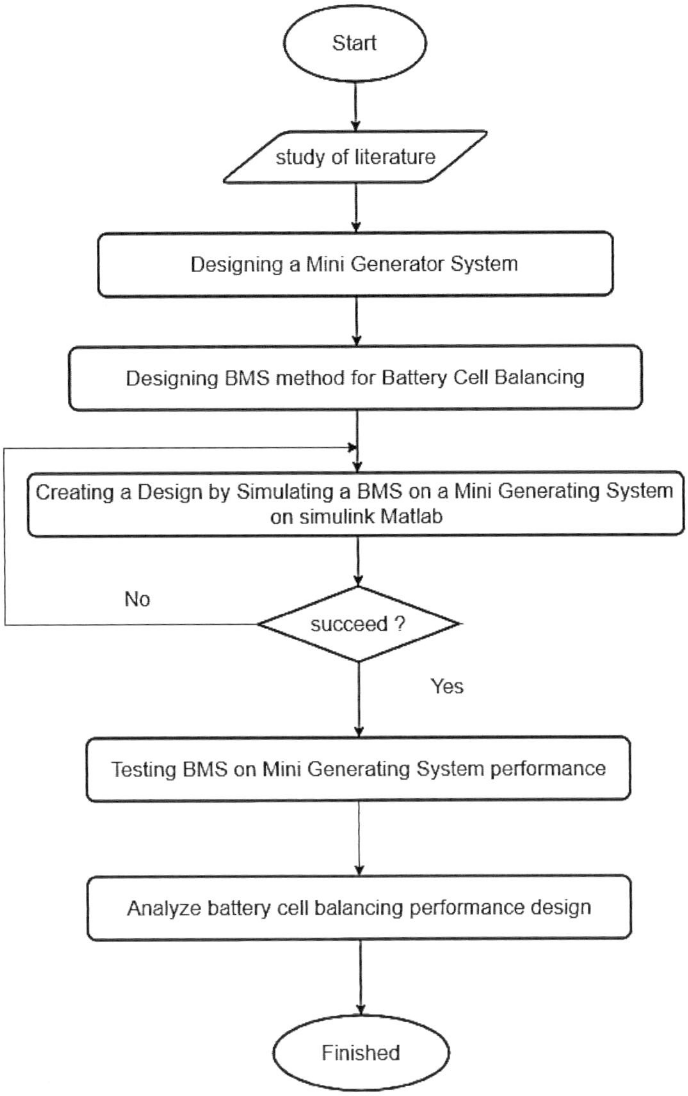

Figure 1. Flow Chart Method

Figure 1. It is a research flow chart diagram, after conducting a literature study related to the second use of electric vehicle batteries, then a mini generator system design will be made first, after that apply the BMS method with battery cell balancing, the BMS method that will be applied will be simulated in Matlab Simulink, by dividing 3 modes, namely Standby Mode, Charge Mode and Discharge Mode, which are described in Figure 2 [3]. The following steps will be carried out based on Figure 1:

1. The first stage is conducting a literature study by looking for references from previous research. And direct data retrieval to Ultima Automotive Design and Braja Electric Motor Workshop, a GESITS motorcycle team, in Surabaya
2. The second stage is to design a Mini Generating System, which includes a new renewable energy generation unit, namely a Solar Power Plant. This stage uses Simulink Matlab to design and test the simulation. Then design the BMS method for battery cell balancing.
3. If it doesn't work, it will repeat the next step, after it is successful, the BMS method will be implemented on the power plant system.

4. The last stage is to analyze the performance of the BMS performance on the power generation system

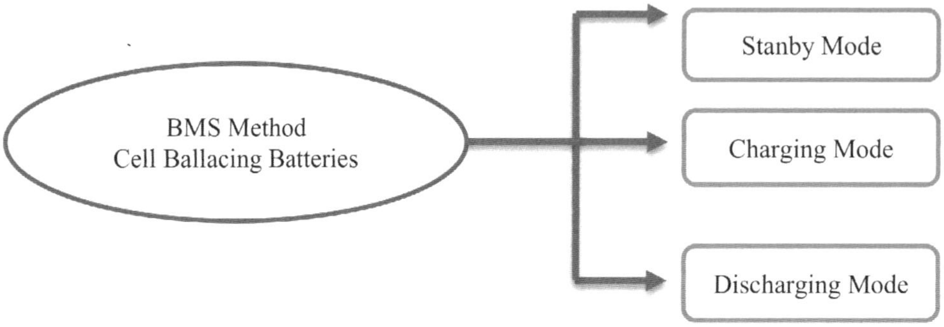

Figure 2. Battery Cell Balancing Scenario

The BMS method in Figure 2 has 3 modes, the standby mode is set if the battery capacity is less than 25%, while the charging mode will be active if the battery capacity is 20%, and the last discharge mode will be active when the battery is in a condition more than 85%, for that constraint on this BMS is SoC and Voltage. There are 3 conditions for this BMS method:

a) Batteries with unbalanced total capacity, where one cell is highly charged
b) One of the highly charged cells
c) Full charge condition with safe SoC

3. Research Results and Discussion

The scenario from Figure 2 will be designed using Matlab Simulink, by starting to create categories from the 3 existing modes, the initial assumption is to create a battery group according to the results of the battery capacity test. The type of battery used is a Lithium-Ion with capacity 2.6 Ah, 3.6 volt, specifications as shown in Figure 3. Figure 4. There are several batteries to be tested, using a battery tester in Figure 5.

Figure 3. Battery Spesifications

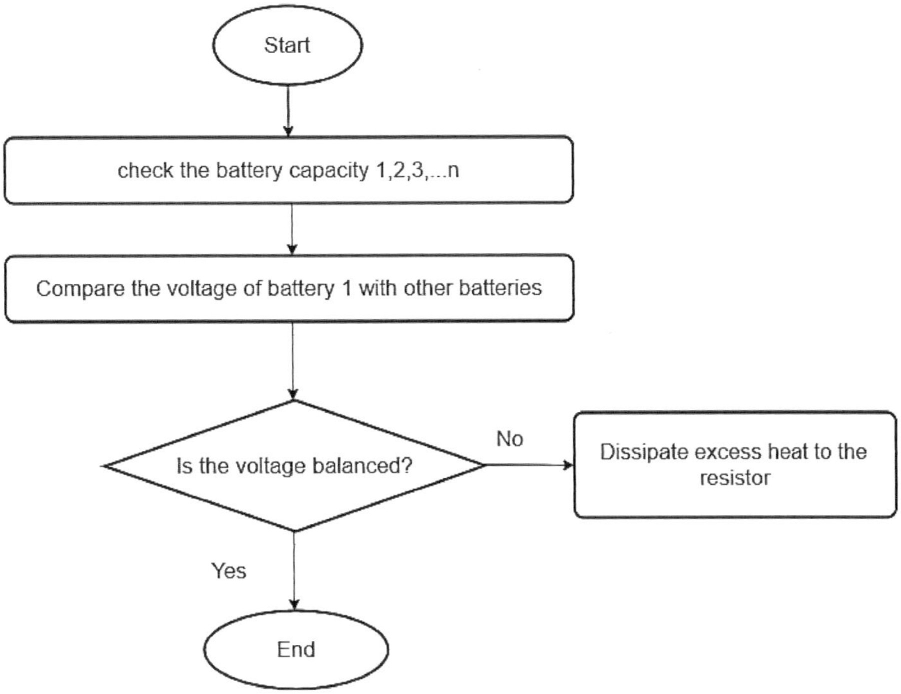

Figure 4. Battery Cell Balancing Step

After balancing, the battery will be retested using a battery tester, the results of checking the battery tester are parameters for the design of Simulink Matlab, as shown in Figure 5. The tester battery is given a load of 5 watts of the resistor. Before knowing the final capacity results, the battery is first emptied, and then recharged until the timer on the display stops or is in a constant position, then the value that appears is the current battery capacity. Figure 6 is a battery that has been checked and separated by capacity. The range value of the battery capacity is 2.4 Ah – 2.6 Ah.

Figure 5. Batteries after checking the capacity

Figure 6. Battery Tester

The measurement results using a battery tester will be used as input parameters in Simulink Matlab. Before designing, it is necessary to know that the BMS block diagram will be made as follows as shown in Figure 7. The BMS is in charge of regulating when the battery will be used and also managing the control circuit. In summary, Figure 7 represents the overall design of the method design, but for the initial stage, the results of the battery tester measurements are used as input values in Simulink Matlab as shown in Figure 8.

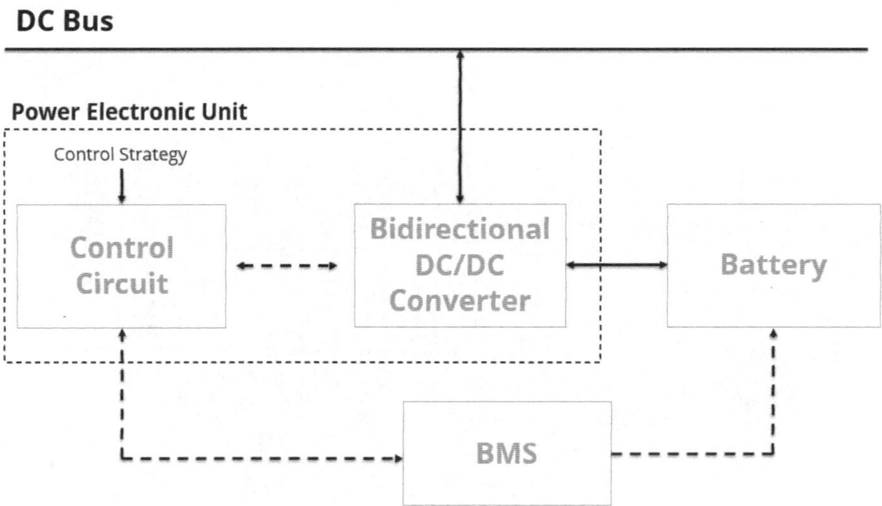

Figure 7. Block Diagram BMS

Journal of Physics: Conference Series **2111** (2021) 012050 doi:10.1088/1742-6596/2111/1/012050

From Figure 2 it has been explained previously that there are 3 conditions where these conditions are obtained:

$$V_{bat_1} + V_{bat_2} + \cdots + V_n = 1$$

a) Full charge condition with the same Soc

$$\text{if}\,(V_{bat_{total}} = 1)$$

b) Batteries with unbalanced total capacity where one of them is highly charged

$$\text{if}\,(V_{bat} > 1)$$

c) One battery one is low charged

$$\text{if}\,(V_{bat} < 1)$$

The three conditions above are implemented into Simulink Matlab as shown in Figure 8, the condition of the battery cell balancing. The parameters used are SoC, voltage, and current on the battery. The condition in Figure 8 is a condition full charge.

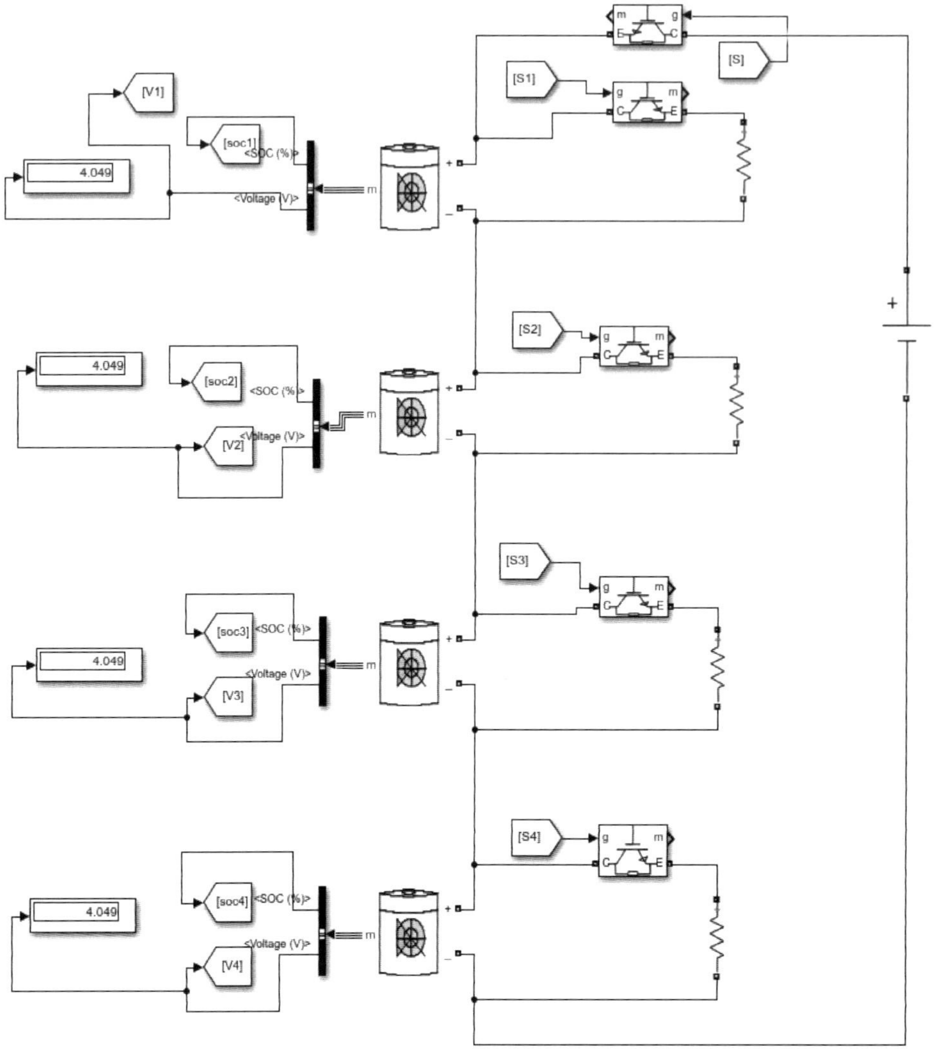

Figure 8. Cell Ballancing Condition

For conditions, B and C are represented in Figure 9. With the same parameters, and conditions are made below 20%, with one of the conditions of the battery being low. after the design of the BMS method is made, the BMS method is included or combined with the design of the mini solar power plant system design. In the design of the mini generator system, the battery is made in a pack. So that each previously balanced battery cell will be made into one image (In the mini generator system circuit there is 1 battery, where the battery represents each cell that has been previously balanced) so that the form of the BMS implementation design in the mini generator system is represented by a figure 10.

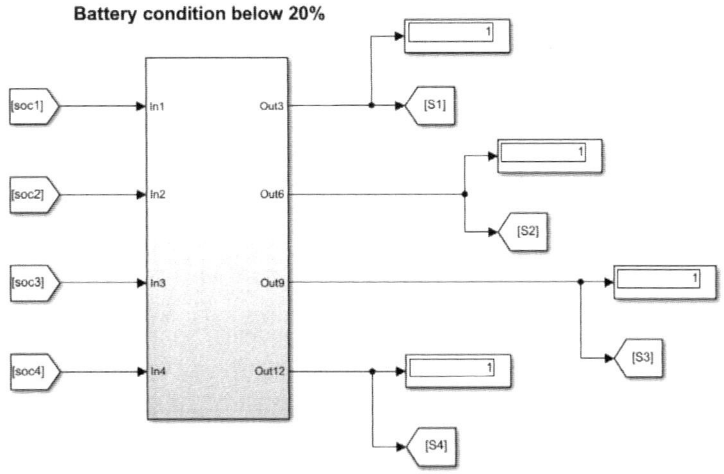

Figure 9. Charging Condition Mode

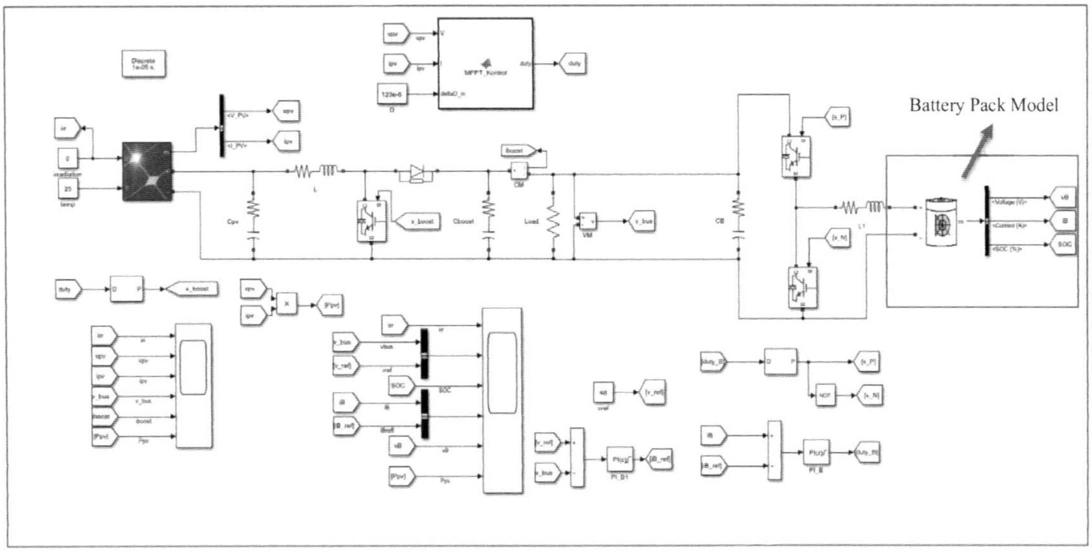

Figure 10. Implementation of the BMS method on the mini generator system

4. Conclusion

The second battery life of electric vehicles is generally not reused, this study intends to be able to reuse used batteries from electric vehicles as a store of electrical energy in the generating system. Therefore, it is necessary to check whether the battery is feasible or not to be used as a source of energy reserve storage, one of which is the cell balancing battery model. The Future research is the testing stage for the household scale.

Acknowledgments

This research was supported by Innovation Research Group Grant from DIPA BLU UNY 2021 with the contract number of T/2.6/UN34.15/PT.01.02/2021

Refrences

[1] V. V. Viswanathan and M. Kintner-Meyer, "Second Use of Transportation Batteries: Maximizing the Value of Batteries for Transportation and Grid Services," in IEEE Transactions on Vehicular Technology, vol. 60, no. 7, pp. 2963-2970, Sept. 2011, doi: 10.1109/TVT.2011.2160378.

[2] A. Keeli and R. K. Sharma, "Optimal use of second life battery for peak load management and improving the life of the battery," 2012 IEEE International Electric Vehicle Conference, Greenville, SC, 2012, pp. 1-6, doi: 10.1109/IEVC.2012.6183276.

[3] D. Strickland, L. Chittock, D. A. Stone, M. P. Foster and B. Price, "Estimation of Transportation Battery Second Life for Use in Electricity Grid Systems," in IEEE Transactions on Sustainable Energy, vol. 5, no. 3, pp. 795-803, July 2014, doi: 10.1109/TSTE.2014.2303572.

[4] F. Hossein, F.-F. Mahmud, and M.-A. Moein, "A practical scheme to involve degradation cost of lithium-ion batteries in vehicle-to-grid applications," IEEE Transactions on Sustainable Energy, vol. 7, no. 4, pp. 1730–1738, Oct 2016.

[5] Catherine Heymans, Sean B. Walker, Steven B. Young, Michael Fowler, "Economic analysis of second use electric vehicle batteries for residential energy storage and load-levelling," Energy Policy, Volume 71, 2014, Pages 22-30, ISSN 0301-4215, https://doi.org/10.1016/j.enpol.2014.04.2016

[6] Z. B. Omariba, L. Zhang, and D. Sun, ``Review on health management system for lithium-ion batteries of electric vehicles," *Electronics*, vol. 7, no. 5, p. 72, 2018.

[7] M. Dubarry, A. Devie, and B. Y. Liaw, ``Cell-balancing currents in parallel strings of a battery system," *J. Power Sources*, vol. 321, pp. 36_46, Jul. 2016.

Development of Unmanned Aerial Vehicle (UAV) Fixed-Wing for Monitoring, Mapping and Dropping applications on agricultural land

ML Hakim[1], H Pratiwi[2], AC Nugraha[1], S Yatmono[1], ASJ Wardhana[1], ES Damarwan[1], T Agustianto[2], S Noperi[2]

[1]Electrical Engineering Education Dept., Universitas Negeri Yogyakarta, Indonesia
[2]Mechanical Engineering Education Dept., Universitas Negeri Yogyakarta, Indonesia

E-mail: luthfihakim93@uny.ac.id

Abstract. UAV technology is used in various fields, namely the military, to strengthen defence, surveillance, dropping logistics. In addition, UAV applications are also used in agriculture. In this study, a fixed-wing UAV will be developed as a monitoring, mapping and dropping process. Several things were carried out in the development process, namely design, manufacturing, installing electronic components, and test flights. The results showed that the vehicle could carry out the mission well. The monitoring process is carried out to review land security. Besides that, it can monitor if there is a fire in agricultural land. In the mapping process, photo data processing taken in the context of the mapping is carried out into an orthophoto map. Meanwhile, in the dropping stage, the aircraft managed to drop a payload weighing 0.5 kg with a distance of 20m from a predetermined centre point. The dropped payload is also equipped with a parachute for increased safety and reduced speed when the payload is dropped.

1. Introduction

An Unmanned Aerial Vehicle (UAV) is a generic aircraft designed to operate without a human pilot. In recent years, unmanned aerial vehicle technology has developed very rapidly. Applications of UAVs are used in the fields of industry, agriculture, military, and others. UAV can also be used as a monitoring and mapping technology. Based on research conducted by Nugroho et al. on the use of UAVs to monitor and map disaster areas, it resulted in 3 conclusions: settlements at risk of tsunamis, no evacuation routes, and no green zones around the coastline [1]. In some cases, UAV technology is also used for monitoring access to high-risk locations [2].

Unmanned Aircraft Vehicle (UAV), which has remote sensing instrumentation, can be utilized in disaster situations. When the UAV obtains photogrammetric data with appropriate image metadata, the UAV's capabilities for disaster management can be further implemented. Another application is for volcano monitoring and coastal monitoring using unmanned aerial vehicles [3]. Xi Wang also researched the use of UAVs to monitor hazardous environments, where experimental results show that the probability of UAVs completing missions in dangerous environments is higher than in operator mode within a specific time [4].

UAV technology is applied in various fields, namely the military, to strengthen the defence, as surveillance, dropping logistics. In addition, in the civil sector, UAVs are used in the agricultural sector, namely for monitoring, mapping and shipping goods [5]. Research on the application of drones was also carried out by Hakim et al., namely using drones to deliver goods for Covid-19 patients [6]. In this study, we will discuss the role of UAVs to support agricultural processes, namely for monitoring agricultural land, mapping the area, and shipping goods.

2. Method

This research on the application of UAV as monitoring, mapping, and dropping was conducted in Pracimantoro, Central Java, Indonesia. Furthermore, several stages were carried out to make the UAV, namely the design process, manufacturing, installation of electronic components, and flight tests, as shown in Figure 1.

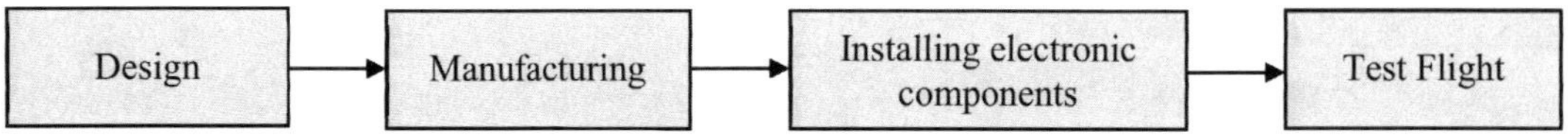

Figure 1. UAV manufacturing flow

The next step is to determine the dropping area, where the location is 5 km from the initial location or the location of the Ground Control Station (GCS). The UAV will fly towards the dropping zone with an autopilot system; during the flight process towards the dropping, the camera will take photos and videos for monitoring and mapping purposes. After arriving at the waypoint, the UAV will drop the payload at a predetermined location, or what we know as dropping. The weight of the payload used is 0.5 kg. Figure 2 shows an illustration of the flight mission of a drone.

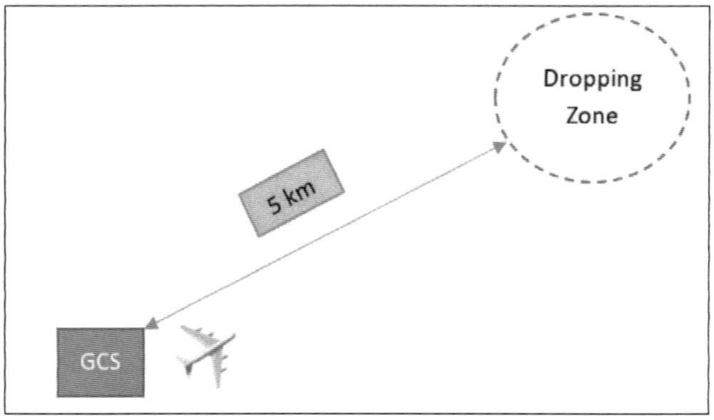

Figure 2. Illustration of a drone flight mission

3. Manufacturing

Making a UAV aircraft consists of several stages: design, manufacture of aircraft parts, installation of electronic circuits, and assembling into an aircraft vehicle that can fly.

3.1. Design

At the design stage, simulations were also carried out to know the system analysis and modelling, operation data collection and observation data analysis. According to Pei-Hsiang Chung, the design process is used to determine performance requirements, including stall speed, cruising altitude, maximum speed, absolute ceiling, and turning radius and speed [7]. The dimensions of the UAV aircraft are 2m long and 1m wide. The type of aircraft used is a twin-boom. Twin-boom aircraft have the advantage of having high efficiency; a double boom can distribute the load along the wingspan and strengthen the overall structure. Twin booms also have flexibility in setting the centre of gravity (CG) of the load so that the aircraft's performance will be equal to or without payload. The twin-boom design also has high stability due to the flat bottom airfoil design. Figure 3 shows the design of the aircraft used. The design of the aircraft is made aerodynamic so that it makes the aircraft more energy efficient. The twin-boom also has high stability due to its flat bottom airfoil design.

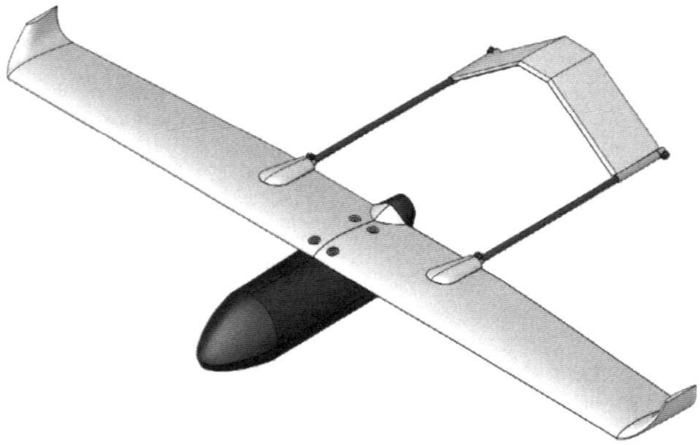

Figure 3. Airplane Design

3.2. Manufacturing of aircraft parts

Making parts of the aircraft vehicle is done using a CNC machine to drive the vehicle precisely or close to the design drawing. The material used for the manufacture of the aircraft is rigid foam and coated with fibre and carbon fibre so that the aircraft is stronger and not easily destroyed when a crash occurs. The production methods we use are moulding and layup. You will get a sturdy and lightweight frame with the moulding method because a little resin is used, but it is enough to make the body sturdy. Figures 4 and 5 are the wings and fuselage that have been made using high-density rigid foam.

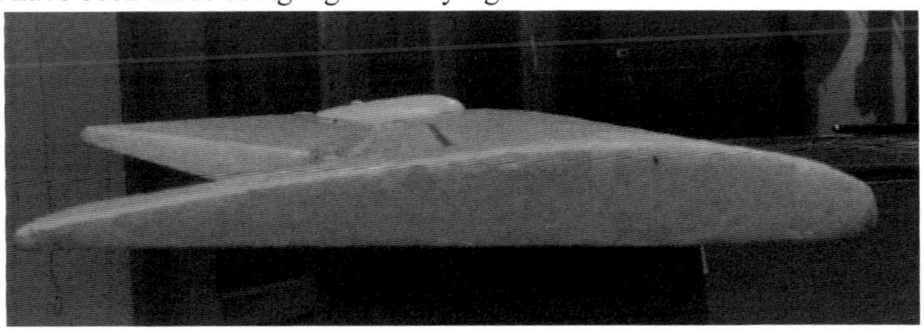

Figure 4. Airplane Wing

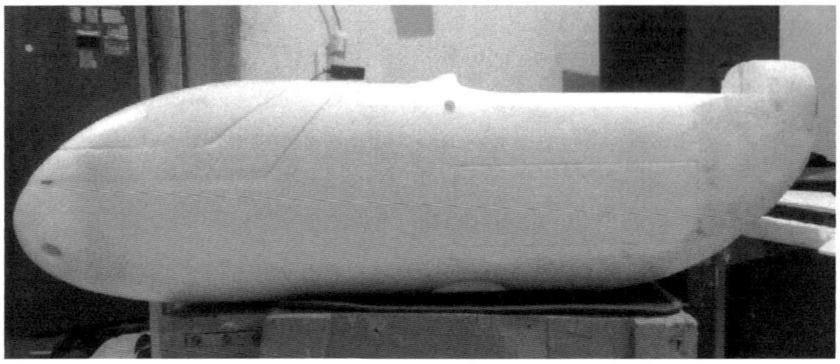

Figure 5. Aircraft Body

3.3. Electronic circuit installation

The vehicle is equipped with an autonomous system that can carry out missions outside of take-off and landing. The rides use batteries as a power source. The specifications for the electronic components of the UAV include a brushless electric motor and radio control system (transmitter and receiver) with a frequency of 433 Mhz, telemetry with a frequency of 433 MHz with a maximum power of 200 mW and a video transmitter with a frequency of 5.8 MHz with a maximum capacity of 1W. Figure 6 shows the process of assembling an electronic circuit with aircraft components.

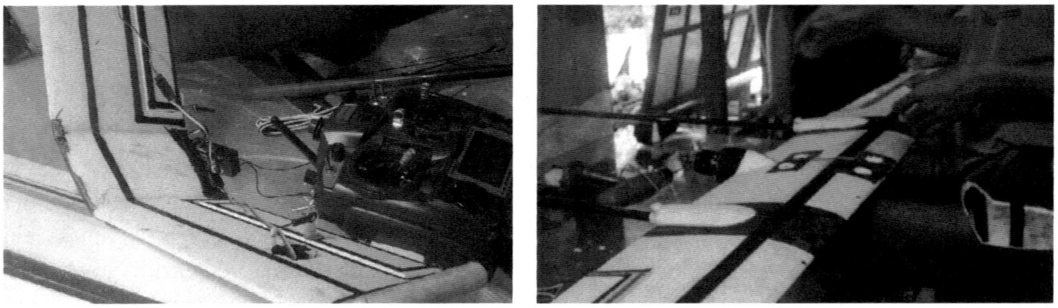

Figure 6. Installation of electronic circuits

3.4. Airplanes

Figure 7 is an aircraft vehicle that is ready for a flight test. The aircraft can carry out flying missions well. Furthermore, the aircraft can carry out the process of monitoring, mapping and dropping on agricultural land.

Journal of Physics: Conference Series **2111** (2021) 012051 doi:10.1088/1742-6596/2111/1/012051

Figure 7. Aircraft rides

4. Results and Discussion

4.1. Monitoring

UAV applications for monitoring systems have been widely used globally. Several examples of the use of monitoring technology have been carried out by Shults, namely for monitoring warehouse development [8]. This technology is used to monitor the warehouse excavation work process. The monitoring process is carried out to review land security; besides, it can watch a fire in agricultural land. Figure 8 shows the results of monitoring from the aircraft vehicle.

Figure 8. Results of monitoring aircraft rides

4.2. Mapping

The use of mapping techniques using unmanned aircraft has several advantages: mapping that is more accurate than using satellite imagery [9]. Mapping techniques are also very effective in terms of costs and time required. In general, the mapping technology concept consists of planning, installation of GCP (ground control point), data acquisition, data processing, reporting. Figure 9 shows the mapping results from the aircraft, and the mapping process is taken as far as 5 km.

Figure 9. The results of the mapping of aircraft rides

4.3. Dropping

In the dropping stage, the aircraft will drop a payload weighing 0.5 kg. In order to increase the safety of the dropped payload, the payload is parachuted to reduce the speed of the payload as it falls. The development of this research can also be used to send packages in quarantine areas for the Covid-19 infectious disease epidemic. This is done to ensure that monitoring activities and delivery of aid packages can be carried out quickly and effectively, which is expected to reduce the risk of disease transmission to other areas and aid workers. In addition, package delivery using UAV technology can be used to deliver fertilizer to agricultural land.

Figure 10. The results of dropping aircraft rides

5. Conclusion

The research results on the development of unmanned aerial vehicle (UAV) fixed-wing for monitoring, mapping and dropping applications on agricultural land show that the vehicle has been able to carry out all missions well. The monitoring process is carried out to review land security; besides, it can monitor a fire in agricultural land. In the mapping process, photo data processing taken in the context of the mapping is carried out into an orthophoto map. Meanwhile, in the dropping stage, the aircraft managed to drop a payload weighing 0.5 kg at a distance of 20m from a predetermined centre point. The dropped payload is also equipped with a parachute for increased safety and reduced speed when the payload is dropped.

References

[1] G. Nugroho, Z. Taha, T. S. Nugraha, and H. Hadsanggeni, "Development of a Fixed Wing Unmanned Aerial Vehicle (UAV) for Disaster Area Monitoring and Mapping," *J. Mechatronics, Electr. Power, Veh. Technol.*, vol. 6, no. 2, pp. 83–88, 2015, doi: 10.14203/j.mev.2015.v6.83-88.

[2] V. Baiocchi, D. Dominici, and M. Mormile, "Uav Application in Post – Seismic Environment," *Int. Arch. Photogramm. Remote Sens. Spat. Inf. Sci.*, vol. XL-1/W2, no. September, pp. 21–25, 2013, doi: 10.5194/isprsarchives-xl-1-w2-21-2013.

[3] S. M. Adams and C. J. Friedland, "A survey of unmanned aerial vehicle (UAV) usage for imagery collection in disaster research and management BT - 9th international workshop on remote sensing for disaster response," vol. 8, no. February, 2011.

[4] X. Wang, C. OuYang, XianglingShao, and HuiXu, "A method for UAV monitoring road conditions in dangerous environment," *J. Phys. Conf. Ser.*, vol. 1792, no. 1, 2021, doi: 10.1088/1742-6596/1792/1/012050.

[5] M. R. Sirojuddin, S. B. Wibowo, and G. Nugroho, "Perancangan dan Pengujian Terbang Pesawat Tanpa Awak Lokeswara," *Semin. Nas. Inov. dan Apl. Teknol. di Ind.*, pp. 334–338, 2019.

[6] M. L. Hakim, S. Yatmono, A. C. Nugraha, and M. Khairudin, "The Effects of Vertical and Horizontal Distance on the Performance of QR Code Detection System," *J. Phys. Conf. Ser.*, vol. 1737, no. 1, 2021, doi: 10.1088/1742-6596/1737/1/012029.

[7] P. H. Chung, D. M. Ma, and J. K. Shiau, "Design, manufacturing, and flight testing of an experimental flying wing UAV," *Appl. Sci.*, vol. 9, no. 15, 2019, doi: 10.3390/app9153043.

[8] R. Shults, K.-K. Kassymkanova, S. Burlibayeva, D. Skopinova, R. Demianenko, and Y. Medvedskyi, "UAV Monitoring of Excavation Works," *11th Int. Conf. Environ. Eng. 11th ICEE Sel. Pap.*, no. July, pp. 0–6, 2020, doi: 10.3846/enviro.2020.696.

[9] S. Sugeng, R. A. Putra, R. F. Muslim, and Y. Septianto, "Unmanned Aerial Vehicle (UAV) for Mapping Plantation Area," *Telekontran J. Ilm. Telekomun. Kendali dan Elektron. Terap.*, vol. 7, no. 1, pp. 79–89, 2019, doi: 10.34010/telekontran.v7i1.1642.

Analysis of Energy Consumption and Energy Efficiency in Government Office Buildings of Bengkulu, Indonesia

I A Rahardjo[1], M Djaohar[1], M Subekti[1], Parjiman[1], I Zakir[1], I Mahir[2], Sakirah[3], A Lubis[4]

[1]Electrical Engineering Education Program, Universitas Negeri Jakarta, Rawamangun, Indonesia
[2]Mechanical Engineering of Department, Universitas Negeri Jakarta, Rawamangun, Indonesia
[3]Electrical Engineering Program, Taruna Bangsa of Vocational High School, Bekasi, Indonesia
[4]Mechanical Engineering Department, University of Indonesia, Rawamangun, Indonesia

Email: [1]imam_ar@unj.ac.id

Abstract: This article aimed to analyze the energy consumption and energy efficiency in government office buildings of Bengkulu, Indonesia. This research was conducted using a quantitative descriptive method. The research step begins by observing the initial conditions of Energy Consumption Intensity (ECI) to determine the category of buildings that are efficient or not in terms of energy use, then collecting data on the profile of the use of the building and its rooms, recapitulation of the use of energy sources (both electrical energy and other chemical energy), a list of equipment that consumes electrical energy (lighting, air conditioning, and others). Furthermore, measuring the performance of systems and equipment that consumes energy and conducting efficiency analysis to then make recommendations with operating settings, to replace inefficient equipment with more efficient equipment. The final step is to draw conclusions about the final Energy Consumption Intensity (ECI) obtained based on the results of the analysis and recommendations obtained on the previous Energy Consumption Intensity (ECI) value. It can be concluded that the final Energy Consumption Intensity (ECI) of government office buildings of Bengkulu based on the results of the analysis and recommendations such as performing preventive maintenance on air conditioners, the temperature of the air conditioner is set to always be at 24°C, replacing the type of lamp in the artificial lighting system that still uses TLD lamps with LED lamps, and efforts in organizing energy management systems will decrease around 13 percent from the previous Energy Consumption Intensity (ECI) or down from 40.9 kWh/m^2/year to be 35.6 kWh/m^2/year.

Keywords: *Energy, Consumption, Efficiency, Government Office Building*

1. Introduction

Along with economic growth and population increase energy consumption. The high level of energy consumption which is inversely proportional to energy availability requires every energy user to make electrical energy efficient to reduce electrical energy consumption [1-3]. This is because the available energy sources are very limited. This limitation will certainly be a problem if it is not addressed with appropriate actions

Energy efficiency measures begin with conducting an energy audit [4] and evaluating the use of electrical energy [5]. Energy audit of actual energy use performance is used to assess the intensity of energy consumption. The value of the actual energy consumption intensity is compared with the standard to assess the actual energy use performance accompanied by recommendations for energy saving follow-up to achieve the planned energy use. Several energy efficiency follow-ups were carried out by managing building operations [6-7], implementing the installation of occupational sensors, replacing magnetic ballasts with electronics, and installing dimmers [8], turning off loads outside of working hours, setting the level lighting, setting the thermostat temperature regulator [9], to energy management [10-11] and implementation of government policies [12]. Energy efficiency follow-up can certainly be done by adjusting the conditions. However, in terms of energy efficiency, it should not interfere with the comfort of use [13-14]. This is because convenience is related to productivity [15].

The importance of analyzing energy consumption and energy efficiency in government office buildings of Bengkulu is felt to be important for research.

2. Method

To analyze energy consumption and energy efficiency in government office buildings of Bengkulu, it can be explained using a quantitative descriptive method according to the steps described in the research method flow chart in Figure 1 below.

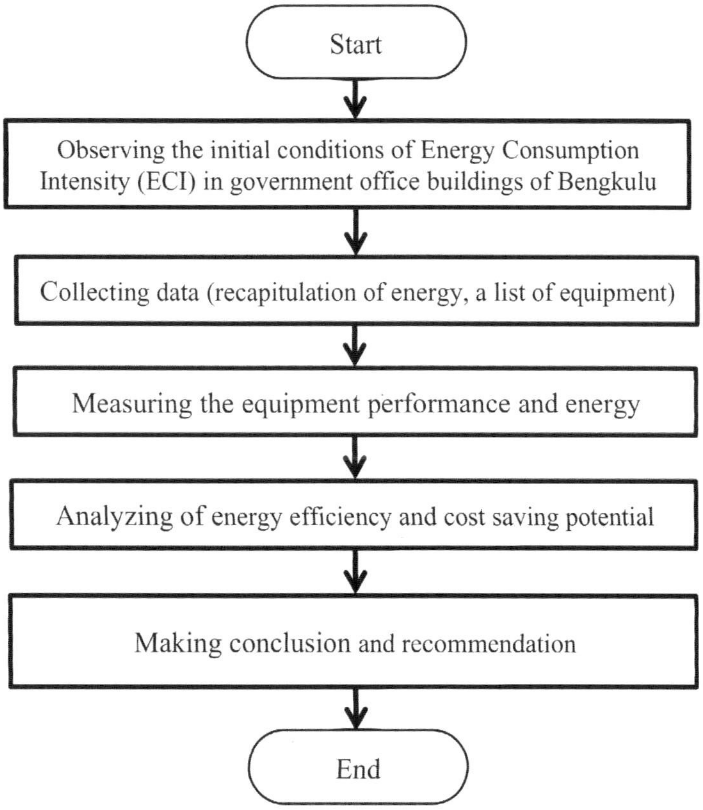

Figure 1. Research Method Flowchart

In Figure 1, observing the initial conditions of Energy Consumption Intensity (ECI) is the first step of research in determining the category of energy efficient or not in buildings, then collecting data on the profile of the use of the building and its rooms, recapitulation of the use of energy sources (both electrical energy and other chemical energy), a list of equipment that consumes electrical energy (lighting, air conditioning, and others). Furthermore, measuring the performance of systems and equipment that consumes energy and conducting efficiency analysis to then make recommendations with operating settings, to replace inefficient equipment with more efficient equipment. The final step is to draw conclusions about the final Energy Consumption Intensity (ECI) obtained based on the results of the analysis and recommendations obtained on the previous Energy Consumption Intensity (ECI) value.

3. Results and Discussion

3.1. Results

The Energy Consumption Intensity (ECI) in government office buildings of Bengkulu is 40.9 kWh/m^2/year or 3.7 kWh/m^2/month. This value is obtained based on the total value of energy consumption per year (37,348 kWh) divided by the building area (37.348 m^2) or the average value of energy consumption per month (3,395 kWh) divided by the building area (37.348 m^2). The value of ECI in government office buildings of Bengkulu is pretty good. The results of total energy consumption in government office buildings of Bengkulu are shown in table 1.

Table 1. Energy Consumption in Government Office Buildings of Bengkulu

Month	Energy (kWh)
November	3,840
December	3,198
January	3,600
February	3,343
March	3,585
April	3,496
May	3,310
June	3,608
July	3,263
August	3,198
September	2,907
Total	37,348
Average	3,395
Minimum	2,907
Maximum	3,840

Referring to Table 1, the average energy consumption in government office buildings of Bengkulu was 3.395 kWh, the lowest energy consumption in government office buildings of Bengkulu was 2.907 kWh in September, and the highest energy consumption in government office buildings of Bengkulu was in November.

Potentials recommendations of energy efficiency in the government office buildings of Bengkulu are among others.

1. Performing preventive maintenance on air conditioners
2. The temperature of the air conditioner is set to always be at 24°C.
3. Replacing the type of lamp in the artificial lighting system that still uses TLD lamps with LED lamps.
4. Efforts in organizing energy management systems

3.2. Discussion

In the discussion will discuss the calculation of energy efficiency for electric equipments. Table 2 shows the results of energy efficiency calculations by replacing TLD lamps with LED lamps.

Tabel 2. Energy Efficiency Calculations by Replacing TLD Lamps of 18 Watts with LED Lamps of 10 Watts

No.	Item	LED 10 W	TLD 18 W
I.	Qty lampu	32,00	32,00
II.	General Data		
	Lighting running hour/day (hour)	12,00	12,00
	Actual Power consumption (watt)	10,20	27,67
	Unit Price of lighting (Rp)	200.000,00	20.000,00
	Tarif/kwh (Rp)	800,00	800,00
III.	Ligthing Replacement		
	Lighting running hour/year (hour)	4.380,00	4.380,00
	Lifetime (Year)	9,13	0,46
IV.	Ligthing Cost/lamp		
	Energy used / lamp-year (kwh)	44,69	121,29
	Energy cost / lamp-year (Rp)	35.755,10	97.033,85
V.	Pay back period/lamp (year)		
	Cost of lamp replacement/year (Rp)	21.900,00	43.800,00
	Cost of energy/year	35.755,92	97.033,85
	Total running cost/lamp-year	57.655,12	140.833,85
	Saving/lamp-year	83.178,74	
	Pay back period/lamp (year)	2,40	
VI.	Ligthing Cost Total		
	Energy used total lamp-year (kwh)	1.430,20	3.881,35
	Energy cost total lamp-year(Rp)	1.144.163,27	3.105.083,08
VII.	Pay back period Total lamp (year)		
	Cost of lamp replacement total year (Rp)	700.800,00	1.401.600,00
	Cost of energy total year	1.144.163,27	3.105.083,08
	Total running cost total lamp-year	1.844.963,27	4.505.683,08
	Saving total lamp-year	2.661.719.81	
	Pay back period total lamp (year)	2,40	

Referring to Table 2, the replacement 32 unit lamps of 18 W TLD lamp with 10 W LED lamp will affect the efficiency of 2.451,15 kWh/year or 6,6% of total energy consumption energy consumption in government office buildings of Bengkulu.

The results of potentials recommendations of energy efficiency in government office buildings of Bengkulu are shown in table 3.

Table 3. Potentials Recommendations of Energy Efficiency in Government Office Buildings of Bengkulu

No	Potentials recommendations of energy efficiency	Energy Efficiency (kWh/ year)	Percentage of Energy Efficiency (%)	Cost Saving (Rp/ year)
1	The temperature of the air conditioner is set to always be at 24°C	327	0,88	309.430
2	Performing preventive maintenance on air conditioners	327	0,88	309.430
3	Replacing TLD lamps of 18 watts with LED lamps of 10 watts	2.451	6,56	1.960.920
4	Efforts in organizing energy management systems	1.867	5,00	1.618.456
	Total	4.972	13	4.198.236

Referring to Table 3, the total percentage of energy efficiency was found to be 13% of total energy consumption in government office buildings of Bengkulu or energy efficiency of 4,972 kWh/year with the annual cost savings of Rp 4,198,236,-.

The final Energy Consumption Intensity (ECI) in government buildings of Bengkulu by applying potential energy efficiency recommendations will decrease around 13 percent from the previous Energy Consumption Intensity (ECI) or down from 40.9 kWh/m^2/year to be 35.6 kWh/m^2/year.

4. Conclusion

The results of potentials recommendations of energy efficiency in government office buildings of Bengkulu are performing preventive maintenance on air conditioners, the temperature of the air conditioner is set to always be at 24°C, replacing the type of lamp in the artificial lighting system that still uses TLD lamps with LED lamps, and efforts in organizing energy management systems with total energy efficiency of 4,972 kWh/year.

The final Energy Consumption Intensity (ECI) in government buildings of Bengkulu by applying potential energy efficiency recommendations will decrease around 13 percent from the previous Energy Consumption Intensity (ECI) or down from 40.9 kWh/m^2/year to be 35.6 kWh/m^2/year.

Acknowledgements

The authors would like to thank the International Conference Committe of ice-elinvo 2021 which has accepted this article to be published.

References

[1] Dini Mulyani, Djoni Hartono, "Pengaruh Efisiensi Energi Listrik pada Sektor Industri dan Komersial terhadap Permintaan Listrik di Indonesia," Jurnal Ekonomi Kuantitatif Terapan Vol. 11 No.1 Februari 2018

[2] E. Sterling, C. Collett, S. Turner, C. Downing, Commissioning to avoid indoor air quality problems, in: Proceedings of the ASHRAE Transactions: Symposia, 1994, p. 867.

[3] N. Apergis, G. C. Aye, C. P. Barros, R. Gupta, and P. Wanke, "Energy efficiency of selected OECD countries: A slacks based model with undesirable outputs," *Energy Econ.*, vol. 51, pp. 45–53, 2015, doi: 10.1016/j.eneco.2015.05.022.

[4] Anna Kubule, Kristaps Ločmelis, Dagnija Blumberga, Analysis of the results of national energy audit program in Latvia,Energy, Volume 202, 2020, 117679, ISSN 0360-5442, https://doi.org/10.1016/j.energy.2020.117679.

[5] Fahrizal Tunjung Kresnadi, Didik Aribowo, Desmira, "Evaluasi Penggunaan Listrik dengan Metode Konservasi Energi untuk Efisiensi Energi di Gedung FKIP UNTIRTA," Energi dan Kelistrikan: Jurnal Ilmiah Vol. 12, No. 1, Januari - Juni 2020, P-ISSN 1979-0783, E-ISSN 2655-5042 https://doi.org/10.33322/energi.v12i1.949
[6] C. Marinosci, G. L. Morini, G. Semprini, and M. Garai, "Preliminary energy audit of the historical building of the School of Engineering and Architecture of Bologna," *Energy Procedia*, vol. 81, pp. 64–73, 2015, doi: 10.1016/j.egypro.2015.12.060.
[7] E. Mills, et al., The Cost-effectiveness of Commercial Buildings Commissioning: A Meta-analysis of Energy and Non-energy Impacts in Existing Buildings and New Construction in the United States, Lawrence Berkeley National Laboratory, LBNL-56637, 2004, http://tinyurl.com/2yh7d3 or http://eetd.lbl.gov/Emills/PUBS/PDF/Cx-Costs-Benefits.pdf.
[8] K. Hassouneh, A. Al-Salaymeh, and J. Qoussous, "Energy audit, an approach to apply the concept of green building for a building in Jordan," *Sustain. Cities Soc.*, vol. 14, no. 1, pp. 456–462, 2015, doi: 10.1016/j.scs.2014.08.010.
[9] H. H. Sait, "Auditing and analysis of energy consumption of an educational building in hot and humid area," *Energy Convers. Manag.*, vol. 66, pp. 143–152, 2013, doi: 10.1016/j.enconman.2012.10.005.
[10] Al-Saadi, S. N. J., Ramaswamy, M., Al-Rashdi, H., Al-Mamari, M., & Al-Abri, M. (2017). Energy Management Strategies for a Governmental Building in Oman. *Energy Procedia*, *141*, 206-210. https://doi.org/10.1016/j.egypro.2017.11.039
[11] J.S. Li, A study of energy performance and efficiency improvement procedures of Government Offices in Hong Kong Special Administrative Region, Energy and Buildings 40 (2008) 1872–1875.
[12] Hakimul Batih, Dhami Johar Damiri, "Estimasi Dampak Implementasi Kebijakan Terhadap Potensi Konservasi Energi Di Sektor Industri," Energi dan Kelistrikan: Jurnal Ilmiah Vol. 11, No. 1, Januari - Juni 2019, P-ISSN 1979-0783, E-ISSN 2655-5042
[13] M.M. Rahman, Building energy conservation and indoor air quality assessment in a subtropical climate. Master of engineering thesis. Central Queensland University, Rockhampton, Queensland, Australia, 2009
[14] E. Sterling, C. Collett, S. Turner, C. Downing, Commissioning to avoid indoor air quality problems, in: Proceedings of the ASHRAE Transactions: Symposia, 1994, p. 867.
[15] A. Alajmi, "Energy audit of an educational building in a hot summer climate," *Energy Build.*, vol. 47, pp. 122–130, 2012, doi: 10.1016/j.enbuild.2011.11.033

The Effect of Inclination Angle of The Solar Panel on The Resulting Output Voltage

[1]A S J Wardhana, [2]E S Damarwan, [3]ML Hakim

[1,2,3]Electrical Engineering Education Dept., Universitas Negeri Yogyakarta, Indonesia

E-mail: [1]alexwardhana@uny.ac.id, [2]ekoswie@uny.ac.id, [3]luthfihakim93@uny.ac.id

Abstract. Renewable energy is alternative energy for the future. One renewable energy that is easy to develop is solar power generation. In this study, we will discuss the effect of the tilt angle on the solar panel and the value of the intensity of sunlight on the resulting output voltage. The data collection process was carried out at the Mechanical and Electrical Power System Laboratory, Faculty of Engineering, Universitas Negeri Yogyakarta. The data collection stage was carried out at an angle of 0^0, 30^0, and 60^0. The results showed that the smaller the value of the intensity of sunlight, the smaller the output voltage produced. The tilt angle of the solar panel at 30^0 has a maximum output voltage more significant than the angle of 60^0 and 0^0.

1. Introduction

Most of the global energy comes from fossil fuels, of which these sources are finite and will run out in 50 to 75 years, as most studies show [1]. Renewable energy is alternative energy for the future [2]. The shift towards new and renewable, and environmentally friendly energy is an important issue today. The high dependence on fossil fuels converted into electrical energy or others causes environmental damage and human health [3]. New and renewable energy (EBT) has enormous potential to be applied in Indonesia in the next few years. Examples of renewable energy used in Indonesia are biomass energy, water energy, geothermal energy, wind energy, solar energy and biofuel energy. The energy produced from fossils cannot last long, and the world will run out of fossil energy reserves shortly. So there is a need for renewable energy whose availability is guaranteed for a long time [4]. One example of energy that is easily converted into electrical energy is solar energy.

The sun is a source of free and inexhaustible energy for planet Earth. Solar power plants are renewable energy-based plants that can contribute the second-largest power after wind power plants [5]. Earth receives energy released by the sun about 1366 kW/day. This is a source of energy that will never run out. The benefit of solar energy compared to other energies is that sunlight can be directly converted into electrical energy using photovoltaic (PV) technology.[6]. Solar energy also has another advantage, namely affordable installation costs compared to other power plants.

Based on the application and configuration, solar power plants are classified into three, namely systems where solar power plants are not connected to the network (off-grid solar power plants), solar power generation systems that are connected to the network (on-grid solar power plants), and Hybrid solar power generation system is a solar power system which in its installation is combined with other types of power plants to supply consumer loads. Based on data from [7] [8], it is explained that the PLTS system, when viewed from the capacity of the electrical energy produced, is classified into 3, namely Small Scale 10 kW, Medium Scale 10–500 kW and Large Scale 500 kW.

As a tropical country, Indonesia should be able to optimize solar power plants as an alternative to power generation. Based on this background, it will be investigated about the effect of the tilt angle of the solar panel and the value of the intensity of sunlight on the resulting output voltage.

2. Method

Research on the effect of the inclination angle of the solar cell on the output voltage was carried out at the Mechanical and Electrical Power System Laboratory, Faculty of Engineering, Universitas Negeri Yogyakarta. The tools and components used include solar cell panels, battery control regulator (BCR), inverters, connecting cables, batteries and lux meters. Figure 1 shows the frame design of the solar cell unit. For any location on Earth, the PV cell has an angle of incidence that depends on the incidence of optical solar radiation [1]. This angle can be measured theoretically and experimentally and used to improve the array result. The angle of incidence of the sun varies according to time and season [9]. Experiments on the effect of the slope angle on the PV output voltage were carried out quickly to avoid the effect of time variations.

The solar panels used in this study were three solar panels with a power of 50 watts each. The maximum power produced by a series of solar cells in this study is 150 watts. The characteristic parameters of solar PV are shown in Table 1.

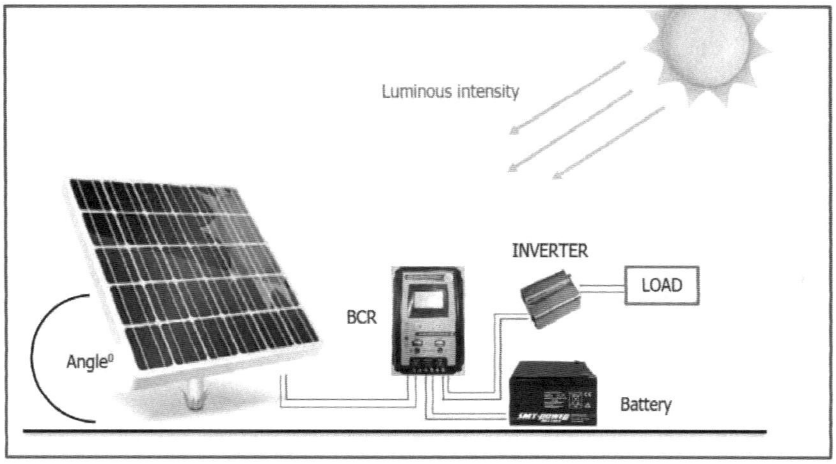

Figure 1. Solar cell circuit design

Table 1. Characteristics of Solar PV

Product Name	Voltage Level	Rated Current
Silicon Solar PV Module 50W	Rated = 12 volt V max power = 18 volt V open circuit = 22.19 volt	I max power = 2.78 A I short circuit = 3A

The solar cell used has a voltage rating of 12 V with a maximum power voltage of 18 V and current maximum power of 2.78 A. In this study, the solar cell module was connected in series to determine the maximum voltage that could be generated.

The inclination angle of the solar panels starts from the range of 0^0, 30^0 and 60^0. The data collection time for each slope angle is 45 minutes with a data collection interval of 5 minutes. The data generated from this research is the output voltage and the intensity of sunlight.

3. Results and Discussion

The data collection process can be seen in Figure 2. The measuring instrument used is an arc to adjust the slope angle, a voltmeter to measure voltage, and a lux meter to measure the amount of light intensity. The data collection process was carried out starting at 09.00 West Indonesian Time.

Figure 2. Data collection process

From the results of data retrieval, the size of the resulting voltage depends on the intensity of sunlight; if the lux is high, then the voltage is high and vice versa. In Figure 3, it can be seen that every time there is a decrease in light intensity, there will be a decrease in voltage at the 25th minute.

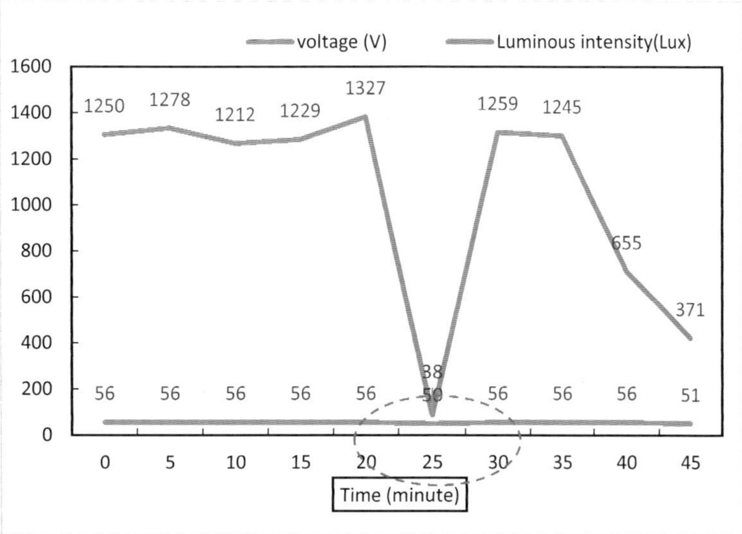

Figure 3. The results of data collection at an inclination angle of 0^0

The same thing also happened when the data was taken at an angle of 30^0 (Figure 4), where there was a decrease in the voltage from the previous 60 V to 54 V at 20 minutes. The decrease influenced the decrease in voltage in the value of the intensity of sunlight. At an angle of 30^0 the maximum voltage that can be obtained is 60 V.

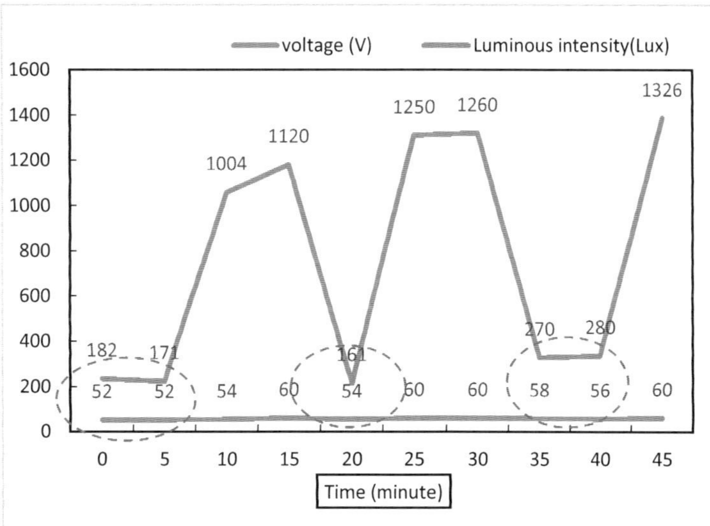

Figure 4. Results of data collection at an inclination angle of 30^0

Furthermore, in the experiment, the angle of inclination is 60^0 (Figure 5). When there is a decrease in the intensity of sunlight, it is always offset by a decrease in the output voltage. There was a decrease in voltage from the previous 56 V to 48 V at 25 minutes. The decrease influenced the decrease in voltage in the intensity of sunlight.

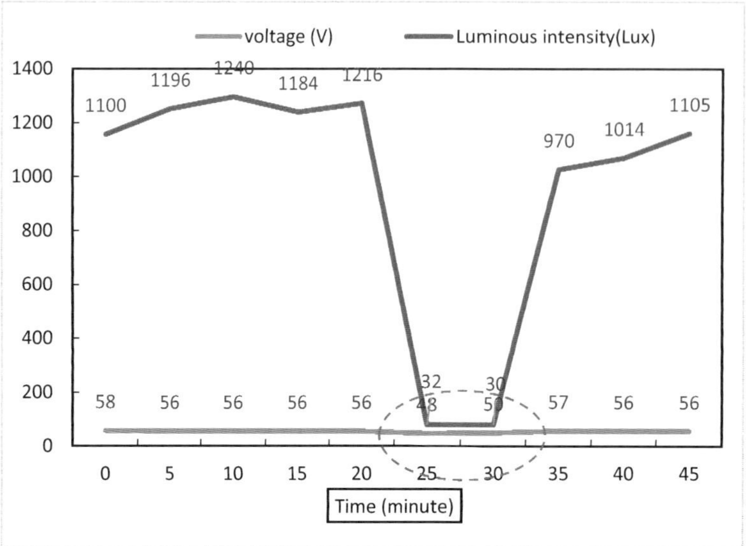

Figure 5. The results of data collection at an inclination angle of 60^0

Based on data collection, it can be seen that the intensity of sunlight will affect the magnitude of the output voltage. The higher the light intensity, the greater the voltage generated by the solar power plant. At 09.00 WIB, a slope angle of 30^0 can produce a higher maximum output voltage when compared to an angle of 0^0 or 60^0. This is in line with [10], which explains that the higher the light intensity, the greater the output voltage produced by PLTS. The most appropriate angle setting for the solar panel module is facing the sun.

4. Conclusion

The results of this study indicate that the smaller the value of the intensity of sunlight, the smaller the output voltage produced. Changing the tilt angle of the PV panels is better than fixing the cells at a specified angle, and the slope of the cells has to be changed over time. One of the factors that decrease the value of the intensity of sunlight is influenced by the blocking of sunlight by clouds or due to the presence of dust. In this study, the more dominant influence is due to the influence of clouds. From this study, it can also be seen that the voltage drop due to a decrease in the value of light intensity is in the range of 2-6 V. At an angle of 30^0 it can produce the highest maximum voltage when compared to an angle of 0^0 and 60^0.

References

[1] M. J. Basher, "The Impact of Solar Photovoltaic Cell Tilt Angle on Its Performance," *World Wide J. Multidiscip. Res. Dev.*, vol. 4, no. 8, pp. 44–50, 2018.

[2] M. L. Hakim, N. Yuniarti, S. Sukir, and E. S. Damarwan, "Pengaruh Debit Air Terhadap Tegangan Output Pada Pembangkit Listrik Tenaga Picohydro," *J. Edukasi Elektro*, vol. 4, no. 1, pp. 75–81, 2020, doi: 10.21831/jee.v4i1.32607.
[3] M. T. Chaichan and K. A. H. Al-Asadi, "Environmental Impact Assessment of Some Vehicles Pollutants in Sohar, Oman," *Int. J. Sci. Eng. Res*, vol. 6, no. 7, pp. 493–496, 2015.
[4] B. A. Supian, C. Ekaputri, and W. Priharti, "Increasing the output power of solar panel by using cooling system," *IOP Conf. Ser. Mater. Sci. Eng.*, vol. 830, no. 3, 2020, doi: 10.1088/1757-899X/830/3/032039.
[5] E. Kabir, P. Kumar, S. Kumar, A. A. Adelodun, and K. Kim, "Solar energy : Potential and future prospects Solar energy : Potential and future prospects," no. September, 2017, doi: 10.1016/j.rser.2017.09.094.
[6] M. Rizwan, S. Shaikh, S. Labade, and A. Tekale, "A Review Paper on Electricity Generation from," no. September, 2017, doi: 10.22214/ijraset.2017.9272.
[7] Institute of Electrical and Electronics Engineer (IEEE) standard 929-2000 .
[8] W. Omran, "Performance Analysis of Grid-Connected Photovoltaic Systems, University of Waterloo," p. 196, 2010.
[9] A. I. A. AL-Musawi, A. Taheri, A. Farzanehnia, M. Sardarabadi, and M. Passandideh-Fard, "Numerical study of the effects of nanofluids and phase-change materials in photovoltaic thermal (PVT) systems," *J. Therm. Anal. Calorim.*, vol. 137, no. 2, pp. 623–636, 2019, doi: 10.1007/s10973-018-7972-6.
[10] Hariningrum, R. Analisa Pengaruh Sudut Kemiringan Panel Surya100 WP Terhadap Daya Listrik. *Marine Science and Technology Journal*, *1*(2), 67-76. 2021

Performance efficiency of virtual laboratory based on Unity 3D and Blender during the Covid-19 pandemic

M.A. Hamid[1]*, S.A. Rahman[1], I.A. Darmawan[1], M. Fatkhurrokhman[1], M. Nurtanto[2,3]

[1]Department of Electrical Engineering Vocational Education, Universitas Sultan Ageng Tirtayasa, Indonesia
[2]Department of Mechanical Engineering Vocational Education, Universitas Sultan Ageng Tirtayasa, Indonesia
[3]Graduate School of Technical and Vocational Education, Universitas Negeri Yogyakarta, Indonesia

Corresponding author: abi.mustofa@untirta.ac.id

Abstract. Testing the performance efficiency aspect was carried out to test the performance efficiency of the Unity 3D and Blender-based virtual laboratory media during the COVID-19 pandemic at the Electrical Engineering Vocational Laboratory. This test is carried out to test the performance of the media that has been created. The aspects tested are access speed, process speed, and simulation speed when run. Tests were conducted to measure processor and memory consumption through real time monitoring using MSI Afterburner. Divided into 2 stages of testing, namely time behavior and resource utilization. Time-behavior is focused on how long it takes the media or software to provide a response time to perform an action from a certain function. Resource-utilization is the degree to which software uses some resources when doing something under certain conditions.

1. Introduction

Online learning is learning with accessibility, connectivity, flexibility, and visibility in various forms of interactive learning using the internet [1]. The Government of Indonesia is transforming the learning process in the pandemic period through online learning and establishing regulations that regulate the online learning process to run optimally. This policy is intended to control the rate of transmission of the Covid-19 virus in the educational environment. It requires educators to create new models in online learning that are innovative, effective, and efficient. These are all challenges that must be faced in the implementation of education in the current era. [2]. This online learning must be carefully planned through lesson plan [3], [4]. Various innovations or distance learning platforms were successfully created by educational researchers and practitioners. Among them, using MOOCs platform [5], various learning media [6], modules [7] and innovative e-module learning [8], [9], trainer kits capable of supporting learning [10], and ICT-based learning assessment instruments [11].

Therefore, virtual media has been created to overcome the problems that exist during this covid-19 pandemic. In virtual laboratories, students work with computer representations of equipment to observe phenomena and make measurements [12]. In another study, virtual experiments allowed users to focus on conceptual explanations because virtual programs were able to track details and allow users to focus on the 'big picture [13] [14]. The virtual learning environment, of course, must stimulate the learning development of both students and teachers if we want to succeed in realizing the goals of the educational process with virtual methods [15]. Therefore, the features presented must also be able to stimulate learning, namely by presenting it in 3D (3 Dimensional) form. Practical practicum allows students to do it virtually with the support of the completeness of tools and materials presented in 3D so that it will not reduce the essence of the lecture material provided. 3D Virtual is a platform that users can develop by applying their imagination [16]. The ability of the 3D virtual world has a big hand in the development of technology. Virtual 3D was originally only for entertainment and gaming purposes, now it is also used for educational purposes [17]. Virtual 3D allows users to design interactive environments with specified or desired content [18]. Virtual 3D also makes it possible to view a given problem from a different perspective and can include virtual activities that are difficult to safely practice in real life.

To make it happen is to use software to create a virtual laboratory. In this case, the researcher uses Blender 3D and Unity 3D software. The combination of the two software is considered to make the laboratory more optimal in the process. Blender is used to design 3D animation and visual effects of an object or laboratory room. And to integrate it, Game Developing Software is used, namely Unity 3D. related to this, a performance test is carried out on the media that has been made. Performance testing is carried out to determine the quality of the software performance engine. This test assesses memory usage, battery consumption, CPU and load on the server under various conditions. In the ISO 25010 standard [19], a software meets the performance efficiency aspect if the device in the testing process is carried out at least 5 times by calculating the average response time of the application to run each existing feature and is said to meet the criteria if the results of the calculation of the average response time are less of 9 seconds. In ISO 25010 [19] there are 3 indicators to assess performance efficiency, namely time-behavior, resource-utilization, and capacity. Time behavior is the degree to which the software responds to the time it takes to perform an action of a particular function. Resource utilization is the level of software using some resources when doing something under certain conditions. Capacity is the degree to which the software has the maximum limitation of a product meeting certain requirements.

Aspects of performance efficiency measured using MSI Afterburner is a tool provided by MSI to monitor hardware and software. MSI Afterburner is also a graphics card overclocking utility that gives the user complete control over the graphics card or VGA being used. This tool provides a very detailed description of the graphics card hardware used and comes with several additional features such as customizable fan profiles, benchmarking, and video recording. There are two applications in the installation package, MSI Afterburner and Riva Statistics Server. The optimization options available are core voltage, power limit, temperature limit, core clock, memory clock and fan speed.

Based on the description above, the performance efficiency aspect to test the virtual media laboratory based on Unity 3D and Blender is based on ISO 25010, namely time-behavior, resource-utilization. The resource-utilization aspect focuses on memory and CPU requirements which are tested using MSI Afterburner.

2. Methodology

The purpose of this research is to develop a virtual laboratory media using Unity 3D and Blender in the basic electronics practicum course. Based on the objectives, this research uses the R&D (Research and Development) method to develop media that can meet the software quality standards in accordance with the objectives. According to [20] R&D is a research on the manufacture of certain products and testing the effectiveness of these products. And based on a review of the background and objectives, this media is made using a software development model, namely by using the development of the Linear sequential model or commonly called the Waterfall Model.

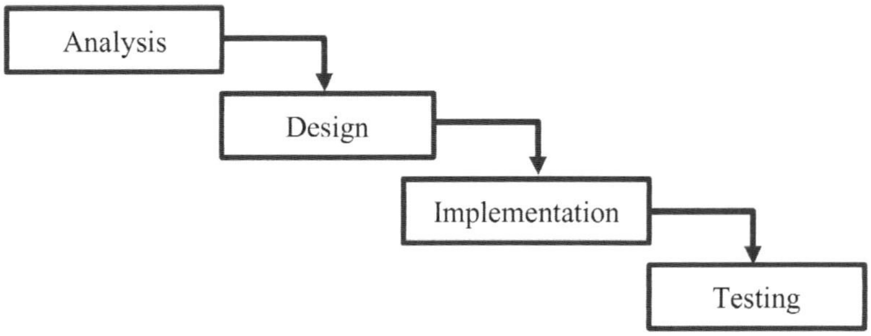

Figure 1. Waterfall Development Model

According to [21] this model provides a conceptual approach to software development systematically and sequentially starting at the level and progress of the system in analysis, design, code, testing and maintenance [21]. Therefore, this media development research aims to create learning media products that are ready to be used for lecture activities, especially practicum activities. This research is also focused on product design and product quality testing based on Unity 3D and Blender in the basic electronics practicum course.

3. Results and Discussion

The performance efficiency aspect is tested using MSI Afterburner which is tested in real time. In this test, time-behavior is tested, resource-utilization in the form of CPU and memory usage. The following are the results of testing aspects of performance efficiency:

3.1. *Time behaviour*

Time-behavior is focused on how long it takes the media or software to provide a response time to perform an action from a certain function. This test includes the use of time in the application launching process and execution tests. The use of the time required on the virtual lab media to carry out these processes is described in the specifications of different computer/laptop devices that are tested 5 times because a software meets the performance efficiency aspect if the device is in the testing process carried out at least 5 times by calculating the average response time of the application to run each existing feature and is said to meet the criteria if the results of the calculation of the average response time are less than 9 seconds. The devices were tested from Windows 7 to Windows 10 OS with a RAM capacity of 2 GB to 4 GB.

3.1.1. *Acer Travelmate P243*

This laptop provides performance performance through an Intel Core i3 2370M 2.4 GHz. It is supported by 2 GB of RAM and the OS used is Windows 7. In addition, it is equipped with a large enough HDD storage space of 500 GB. The results of the time behavior performance test that were tested 5 times were as follows.

Table 1. Media time usage on Acer Travelmate P243

No.	Process/ Media activity	The time required for the session					Avg
		Session 1	Session 2	Session 3	Session 4	Session 5	
1	Launcing application	7,20 s	7,47 s	7,81 s	7,27 s	7,54 s	7,45 s
2	Test execution	1 min 23 s	1 min 29 s	1 min 25 s	1 min 34 s	1 min 22 s	1 min 26.6 s

Based on the table above, it shows that in launching the application, the results of the calculation of the average response time are quite fast, which is 7.45 seconds. In addition, test execution is also carried out. Test execution is the time required to test each function contained in the virtual lab media. In the form of starting the simulation, zooming in and out, opening the options page, adding all the tools and materials to the simulation workbench.

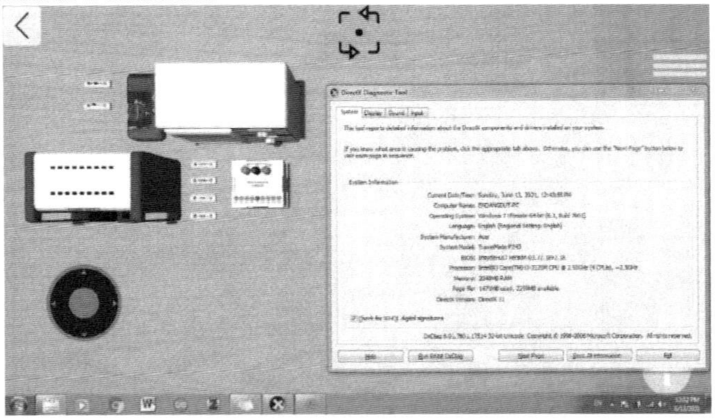

Figure 2. Test Execution on the Acer Travelmate P243

Based on the table, it shows that the response time is quite fast, which is an average of 1 minute 26.6 seconds. And also in the execution process there is also no system lag or slowness. So it can be concluded that the virtual lab media that is run on the ACER Travelmate P243 Laptop with an Intel Core i3 processor, 2 GB RAM and Windows 7 OS can run well without any problems in every process shown during testing.

3.1.2. Lenovo Ideapad G40/45

The performance of this laptop comes from an AMD A8-6410 Quad Core 2.0-2.4GH processor with 4GB RAM specifications and offers 500 GB of storage capacity. The OS used in this laptop is Windows 7, the results of the time behavior performance test were tested 5 times, namely as follows.

Table 2. Media time usage on Lenovo Ideapad G40/45

No.	Process/ Media activity	The time required for the session					Avg
		Session 1	Session 2	Session 3	Session 4	Session 5	
1	Launcing	6,20 s	6,40 s	6,15 s	6,49 s	6,35 s	6,31 s
2	Test execution	1 min 27 s	1 min 23 s	1 min 24 s	1 min 23 s	1 min 26 s	1 min 24 s

Based on the table above, it shows that in launching the application, the results of the calculation of the average response time are quite fast, namely 6.31 seconds. In addition, test execution is also carried out. Test execution is the time required to test each function contained in the virtual lab media. In the form of starting the simulation, zooming in and out, opening the options page, adding all the tools and materials to the simulation workbench.

Based on the table, it shows that the response time is quite fast, which is an average of 1 minute 24 seconds. And also in the execution process there is also no system lag or slowness. So it can be concluded that the virtual lab media running on the Lenovo Ideapad g40/45 Laptop with special specifications for the AMD A8 processor, 4GB RAM and Windows 7 OS can run well without any problems in each process shown during testing.

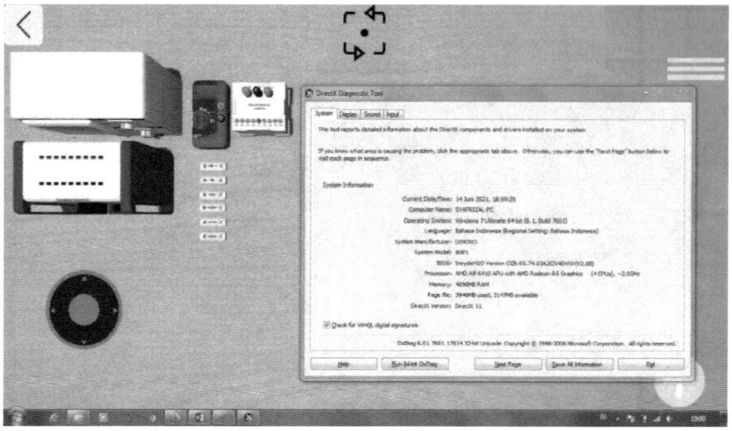

Figure 3. Test Execution on Lenovo Ideapad G40/45.

3.1.3. Acer Aspire 35 411

This laptop from Acer is equipped with an Intel Core Celeron N2830 Quad Core 2.16 GHz with 2 GB RAM and 500 GB HDD storage. The OS used in this laptop is Windows 8, the results of the time behavior performance test were tested 5 times, namely as follows.

Table 3. Media time usage on Acer Aspire e5 411

No.	Process/ Media activity	The time required for the session					Avg
		Session 1	Session 2	Session 3	Session 4	Session 5	
1	Launcing application	6,63 s	7,83 s	7,74 s	7,79 s	7,56 s	7,51 s
2	Test execution	1 min 27 s	1 min 29 s	1 min 20 s	1 min 25 s	1 min 26 s	1 min 25,4 s

Based on the table above, it shows that in launching the application, the results of the calculation of the average response time are quite fast, which is 7.51 seconds. In addition, test execution is also carried out. Test execution is the time required to test each function contained in the virtual lab media. In the form of starting the simulation, zooming in and out, opening the options page, adding all the tools and materials to the simulation workbench.

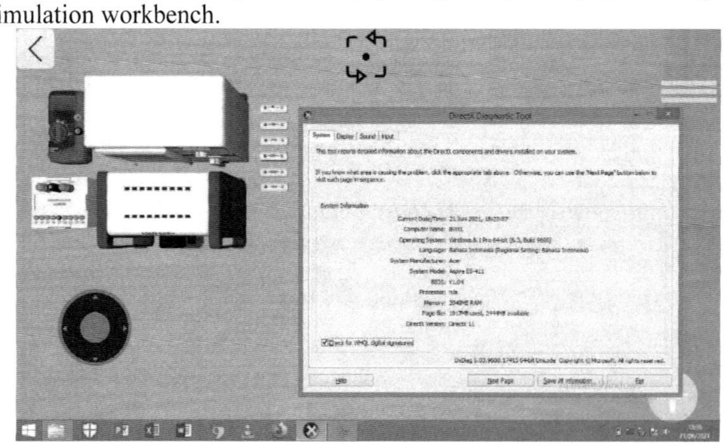

Figure 4. Test Execution on Acer Aspire e5 411

Based on the table, it shows that the response time is quite fast, which is an average of 1 minute 25.4 seconds. And in the execution process there is also a system lag or slowness when exploring the virtual lab room. So it can be concluded that the virtual lab media running on the ACER Travelmate P243 Laptop with an Intel Core i3 processor, 2 GB RAM and Windows 7 OS can run well even though there is system lag when exploring the virtual lab room.

3.1.4. Asus a46Cb

This laptop from ASUS is equipped with an Intel Core i5 processor with Quad Core 1.70 GHz with 4 GB RAM and 500 GB HDD storage. The OS used in this laptop is Windows 8, the results of the time behavior performance test were tested 5 times, namely as follows.

Table 4. Media time usage on Asus a46Cb

No.	Process/ Media activity	The time required for the session					Avg
		Session 1	Session 2	Session 3	Session 4	Session 5	
1	Launcing application	5,11 s	5,43 s	5,40 s	5,31 s	5,27 s	5,3 s
2	Test execution	1 min 16 s	1 min 15 s	1 min 22 s	1 min 26 s	1 min 11 s	1 min 18 s

Based on the table above, it shows that in launching the application, the results of the calculation of the average response time are quite fast when compared to the others, namely 5.3 seconds. In addition, test execution is also carried out. Test execution is the time required to test each function contained in the virtual lab media. In the form of starting the simulation, zooming in and out, opening the options page, adding all the tools and materials to the simulation workbench.

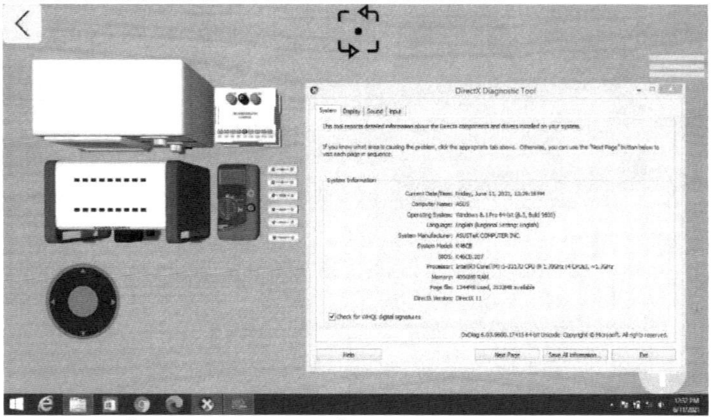

Figure 5. Test Execution on Asus a46Cb

Based on the table, it shows that the response time is quite fast, which is an average of 1 minute 18 seconds. And also in the execution process there is also no system lag or slowness. So it can be concluded that the virtual lab media that is run by the Asus a46Cb Laptop with special specifications for the Intel Core i5 processor, 4 GB RAM and Windows 8 OS can run well without any problems in every process shown during testing.

3.1.5. Asus E202SA

This laptop from ASUS is equipped with an Intel Celeron N3060 processor with Quad Core 1.60 GHz with 2 GB RAM and 500 GB HDD storage. The OS used in this laptop is Windows 10, the results of the time behavior performance test were tested 5 times, namely as follows.

Table 5. Media time usage on ASUS E202SA notebook

No.	Process/ Media activity	The time required for the session					Avg
		Session 1	Session 2	Session 3	Session 4	Session 5	
1	Launcing application	7,41 s	7,78 s	7,51 s	7,83 s	7,59 s	7,62 s
2	Test execution	1 min 16 s	1 min 31 s	1 min 20 s	1 min 22 s	1 min 17 s	1 min 21.2 s

Based on the table above, it shows that in launching the application, the results of the calculation of the average response time are 7.62 seconds. In addition, test execution is also carried out. Test execution is the time required to test each function contained in the virtual lab media. In the form of starting the simulation, zooming in and out, opening the options page, adding all the tools and materials to the simulation workbench.

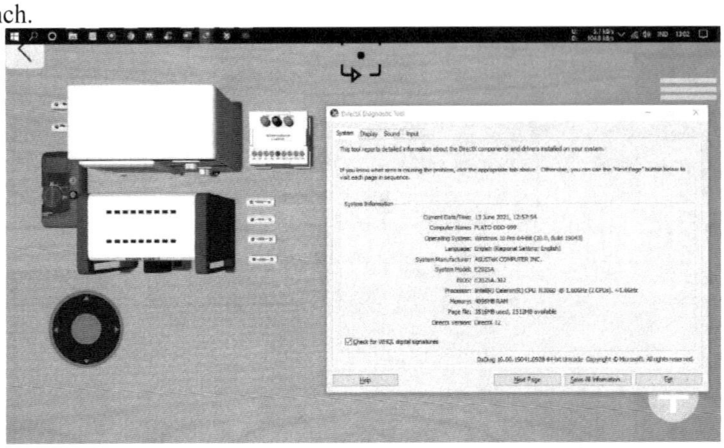

Figure 6. Test Execution on notebook ASUS E202SA

Based on the table, it shows that the response time is quite fast, which is an average of 1 minute 21.2 seconds. And also in the execution process there is also no system lag or slowness. So it can be concluded that the virtual lab media running on the ASUS E202SA Notebook with special specifications for the Intel Celeron N3060 processor, 2 GB RAM and Windows 10 OS can run well without any problems in each process shown during testing.

3.1.6. Asus E202SA

This all-in-one computer or desktop is powered by an Intel Celeron J3355 series processor equipped with 4GB RAM and 500 GB HDD storage. The OS used on this computer is Windows 10, the results of the time behavior performance test were tested 5 times, namely as follows.

Table 6. Media time usage on computer/desktop All In One Lenovo 310-20IAP (F0CL000KID)

No.	Process/ Media activity	The time required for the session					Avg
		Session 1	Session 2	Session 3	Session 4	Session 5	
1	Launcing application	8,23 s	8,43 s	8,28 s	8,30 s	8,47 s	8,34 s
2	Test execution	1 min 48 s	1 min 53 s	1 min 42 s	1 min 53 s	1 min 50 s	1 min 49.2 s

Based on the table above, it shows that in launching the application, the results of the calculation of the average response time are 7.45 seconds. In addition, test execution is also carried out. Test execution is the time required to test each function contained in the virtual lab media. In the form of starting the simulation, zooming in and out, opening the options page, adding all the tools and materials to the simulation workbench.

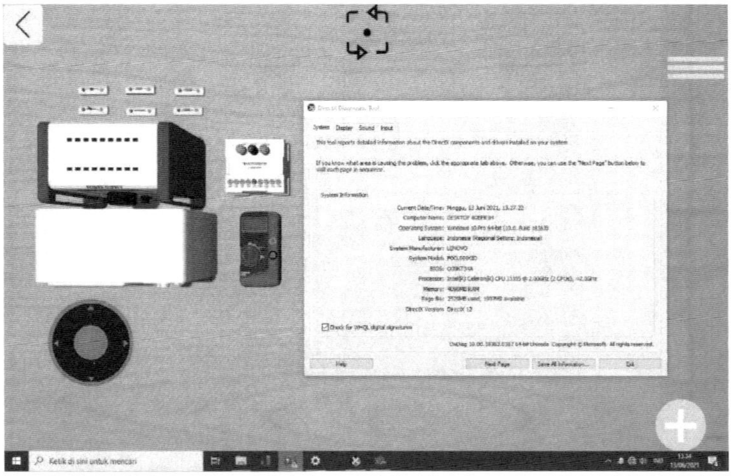

Figure 7. Test Execution pada komputer/ dekstop All In One Lenovo 310-20IAP (F0CL000KID)

Based on the table, it shows that the response time is slightly slower compared to previous tests with other laptops, which is an average of 1 minute 49.2 seconds because there is a lag when opening media or launching. In addition, in the execution process there is also no system lag or slowness. So it can be concluded that the virtual lab media that runs on the Lenovo All In One Desktop 310-20IAP (F0CL000KID) with special specifications for the Intel Celeron J3355 series processor, 4 GB RAM and Windows 10 OS can run well, there is only lag when opening the media. In addition, there are no more obstacles in each process shown during testing.

3.2. Resource utilization

Resource-utilization is the degree to which software uses some resources when doing something under certain conditions. This test shows how much memory usage is used by the system to run this virtual lab media application, as well as CPU usage shows how much CPU performance is in running media. The results of the calculation of memory and CPU usage in calculations on MSI Afterburner are shown in the following figure.

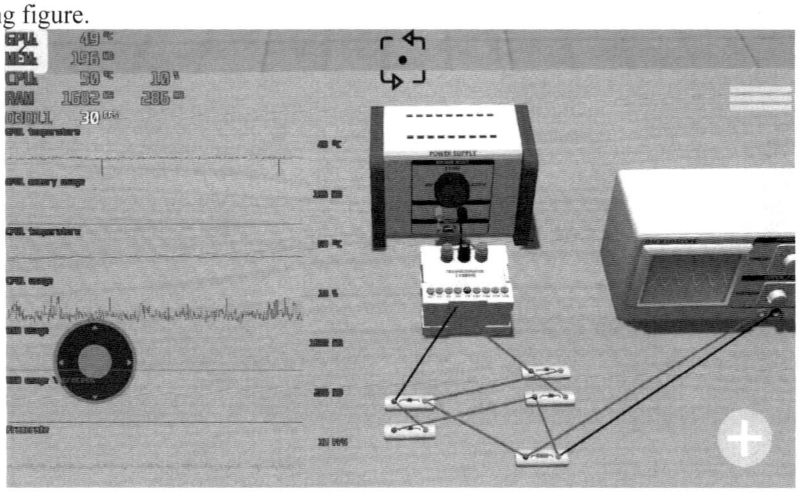

Figure 8. Grafik penggunaan memory dan CPU

Based on the observation data above, it shows that this virtual lab media runs without experiencing a memory shortage which causes the media to stop due to a memory leak which can cause a force close or launch fail. Meanwhile, the average CPU usage is below 20%. This figure is still below the safe limit set by Little Eye (mobile app analysis tools) which is 25%. In addition, the use of GPU Temperature (Graphics Processing Unit), GPU memory, and FPS (Frames per second) framerate is also displayed.

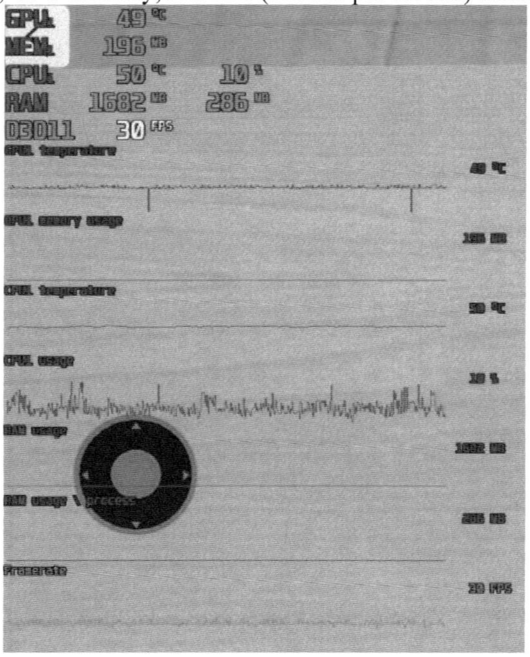

Figure 9. Grafik penggunaan memory dan CPU Zoom in

Based on these data, GPU usage is still in normal condition, namely 49°C, GPU memory usage is 196 mb, and with FPS 30, you can say it's good for use in media because it is able to provide a graphical display that is comfortable for the eyes. The following is the conclusion of the performance efficiency aspect test which is explained from the above description.

Table 7. The results of testing aspects of performance efficiency

No.	Desired results Test Results	Test result
1.	The media is able to run the application launching process, and test execution within a certain time without experiencing problems	corresponding
2.	This media can run without running out of memory	corresponding
3.	Maximum CPU usage in application does not exceed 25% percentage	corresponding

Based on the results of testing the performance efficiency aspect in the table above, the results are in accordance with the standards, so that it can be said that the virtual lab media based on Unity 3D and Blender is at a good level of performance efficiency.

References

[1] J. L. Moore, C. Dickson-Deane, and K. Galyen, "e-Learning, online learning, and distance learning environments: Are they the same?," *The Internet and Higher Education*, vol. 14, no. 2, pp. 129–135, Mar. 2011, doi: 10.1016/J.IHEDUC.2010.10.001.

[2] J. Simarmata *et al.*, *Pendidikan Di Era Revolusi 4.0: Tuntutan, Kompetensi & Tantangan*. Medan: Yayasan Kita Menulis, 2020.

[3] M. Hamid, M. Nurtanto, and M. Fawaid, "The Analysis of Learning Implementation Plan in Vocational Subjects Based on 2013 Curriculum," *Innovation of Vocational Technology Education*, vol. 14, no. 1, p. 17, Apr. 2018, doi: 10.17509/invotec.v14i1.11042.

[4] R. Ramadhani *et al.*, *Belajar dan Pembelajaran*. Medan: Yayasan Kita Menulis, 2020.

[5] F. Fajrillah *et al.*, *MOOC: Platform Pembelajaran Daring di Abad 21*. Medan: Yayasan Kita Menulis, 2020.

[6] M. A. Hamid *et al.*, *Media Pembelajaran*. Medan: Yayasan Kita Menulis, 2020.

[7] M. A. Hamid, D. Aribowo, and D. Desmira, "Development of learning modules of basic electronics-based problem solving in Vocational Secondary School," *Jurnal Pendidikan Vokasi*, vol. 7, no. 2, p. 149, Jul. 2017, doi: 10.21831/jpv.v7i2.12986.

[8] M. A. Hamid, L. Yuliawati, and D. Aribowo, "Feasibility of electromechanical basic work e-module as a new learning media for vocational students," *Journal of Education and Learning (EduLearn)*, vol. 14, no. 2, p. 199, May 2020, doi: 10.11591/edulearn.v14i2.15923.

[9] L. Yuliawati, D. Aribowo, and M. A. Hamid, "Analisis Kebutuhan Pengembangan Media Pembelajaran E-Modul Berbasis Adobe Flash pada Mata Pelajaran Pekerjaan Dasar Elektromekanik," *JUPITER (JURNAL PENDIDIKAN TEKNIK ELEKTRO)*, vol. 5, no. 1, pp. 35–42, Mar. 2020, doi: 10.25273/JUPITER.V5I1.6197.

[10] M. A. Hamid, E. Permata, D. Aribowo, I. A. Darmawan, M. Nurtanto, and S. Laraswati, "Development of cooperative learning based electric circuit kit trainer for basic electrical and electronics practice," *Journal of Physics: Conference Series*, vol. 1456, p. 12047, 2020, doi: 10.1088/1742-6596/1456/1/012047.

[11] M. A. Hamid, "Pengembangan Instrumen Penilaian Hasil Belajar Siswa Berbasis TIK pada Pembelajaran Dasar Listrik Elektronika," *VOLT : Jurnal Ilmiah Pendidikan Teknik Elektro*, vol. 1, no. 1, pp. 37–46, 2016.

[12] S. B. Nolen and M. D. Koretsky, "Affordances of Virtual and Physical Laboratory Projects for Instructional Design: Impacts on Student Engagement," *IEEE Transactions on Education*, vol. 61, no. 3, pp. 226–233, Aug. 2018, doi: 10.1109/TE.2018.2791445.

[13] R. Oser and B. J. Fraser, "Effectiveness of virtual laboratories in terms of learning environment, attitudes and achievement among high- school genetics students," *Curriculum and Teaching*, vol. 30, no. 2, pp. 65–80, 2015, doi: 10.7459/CT/30.2.05.

[14] S. T. Cohn and B. J. Fraser, "Effectiveness of student response systems in terms of learning environment, attitudes and achievement," *Learning Environments Research 2015 19:2*, vol. 19, no. 2, pp. 153–167, Oct. 2015, doi: 10.1007/S10984-015-9195-0.

[15] N. Popović and M. B. Naumović, "Virtual laboratory and learning management system in optimal control theory education:," *http://dx.doi.org/10.1177/0020720916639321*, vol. 53, no. 4, pp. 357–370, Apr. 2016, doi: 10.1177/0020720916639321.

[16] I. Reisoğlu, B. Topu, R. Yılmaz, T. Karakuş Yılmaz, and Y. Göktaş, "3D virtual learning environments in education: a meta-review," *Asia Pacific Education Review 2017 18:1*, vol. 18, no. 1, pp. 81–100, Feb. 2017, doi: 10.1007/S12564-016-9467-0.

[17] I. Duncan, A. Miller, and S. Jiang, "A taxonomy of virtual worlds usage in education," *British Journal of Educational Technology*, vol. 43, no. 6, pp. 949–964, Nov. 2012, doi: 10.1111/J.1467-8535.2011.01263.X.

[18] N. Omale, W.-C. Hung, L. Luetkehans, and J. Cooke-Plagwitz, "Learning in 3-D multiuser virtual environments: Exploring the use of unique 3-D attributes for online problem-based learning," *British Journal of Educational Technology*, vol. 40, no. 3, pp. 480–495, May 2009, doi: 10.1111/J.1467-8535.2009.00941.X.

[19] M. Q. Abbasi, J. Weng, Y. Wang, I. Rafique, X. Wang, and P. Lew, "Modeling and evaluating user interface aesthetics: Employing ISO 25010 quality standard," *Proceedings - 2012 8th International Conference on the Quality of Information and Communications Technology, QUATIC 2012*, pp. 303–306, 2012, doi: 10.1109/QUATIC.2012.39.

[20] Sugiyono, *Metode penelitian pendidikan: pendekatan kuantitatif, kualitatif dan R&D*. Bandung: Alfabeta, 2015.

[21] R. S. . PRESSMAN, *Rekayasa Perangkat Lunak*. Yogyakarta: ANDI, 2012.

Development of Ball Detection System with YOLOv3 in a Humanoid Soccer Robot

A C Nugraha[1], M L Hakim[2], S Yatmono[3] and M Khairudin

[1234]Electrical Engineering Education Dept., Universitas Negeri Yogyakarta, Indonesia

E-mail: [1]ariadie@uny.ac.id

Abstract. One of the practical researches of humanoid robots is research on the use of humanoid robots to play soccer. Research in this field is also encouraged by the existence of various humanoid robot soccer competitions. In humanoid robots for soccer, one of the important aspects is the robot's ability to detect the ball, goal, field boundaries and other players, both friend players and opposing players. This study focuses on the ball detection system which is a basic ability that humanoid robots need to have. The ball detection system developed in this study uses the YOLOv3 method. The test results show that the system built and trained with 3000 image samples can detect balls at a distance of 50 to 900 cm. The time it takes to detect the ball is about 0.033 seconds.

1. Introduction

Humanoid robot is a robot that has a body structure like a human so that it can walk like a human [1]. Humanoid robots also have senses like humans, so they are increasingly popular as research media. Many researchers around the world conduct research on various aspects of humanoid robots such as the ability to move on two legs, manipulate with hands, audio-visual sensing, interaction between robots and humans, and adaptive robot control systems [2].

One of the applications of humanoid robots is the use of humanoid robots to play soccer. In Indonesia, research in this field is also encouraged by the Indonesian Humanoid Soccer Robot Contest which is held annually by the National Achievement Center under the Ministry of Education, Culture, Research and Technology. The ability that a soccer humanoid robot needs to have is the ability to detect the ball, goal, field boundaries and other players, both friend players and opposing players. The robot's detection capability is supported by the development of a technology called computer vision. Computer vision is a field of knowledge that focuses on artificial intelligence systems and image acquisition and processing [3].

Computer vision that is applied to robots is usually called robot vision. Robot vision is the ability of a robot to extract, characterize and interpret information from a three-dimensional object [4]. In principle, the robot eye (camera) can distinguish the color of an image (object) it sees. Data from the image captured by the camera sensor will provide information about the color of the object, so that the robot can know the object it sees. After the object is seen by the robot, the robot will chase the object. In soccer robots, the detection of these objects is carried out, among others, by color segmentation, morphology, and "blob detection" [5]. With the development of machine learning, object detection methods are developed using artificial neural networks, one of them is a method developed by Joseph Redmon et al called You Only See Once (YOLO). The YOLO system has three stages, the first stage is to change the image size to 448 x 448, the second stage is to give the image to a single convolutional network, and the last stage will process the results of the network based on a certain threshold to determine what objects are in the image. the image [6]. The YOLO method was later developed into several versions, including YOLOv3 developed by Joseph Redmon and Ali Farhadi [7] and YOLOv4 by Alexey Bochkovskiy et al [8].

The focus of this study is to build and test the ability of a ball detection system, considering that ball detection ability is the most important ability that a soccer humanoid robot must possess. The method that will be used in this study is YOLOv3. It is hoped that this method will allow the humanoid soccer robot camera device to better recognize objects. This study aims to test that the YOLOv3 method can be applied to humanoid soccer robots, so that humanoid soccer robots can detect ball objects quickly and accurately.

2. Method

The design of the humanoid soccer robot was developed from the mechanical design of ROBOTIS-OP3. The humanoid robot uses NVIDIA Jetson Nano as the main controller and OpenCR as a sub-controller that controls the servo motors. A soccer humanoid robot made using a Logitech C525 camera device to obtain object images. Data from the camera is then forwarded to the Jetson Nano, to be processed using the YOLO method. Furthermore, the main controller will control the servo motors through the sub-controller so that the robot can perform the desired action based on the results of ball detection, for example walking towards the ball or kicking the ball. The algorithm for the decision-making process of the robot can be seen in Figure 1.

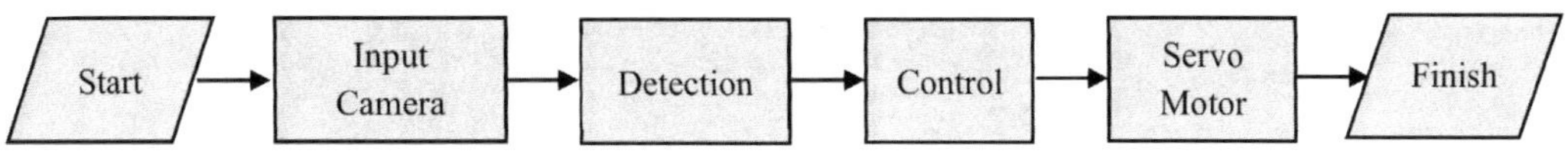

Figure 1. Robot Control System Flowchart

The scheme of the humanoid robot detection system in this study consisted of the stages of sampling, labeling, training, optimization, and running (Figure 2). Sampling is taking an image of the object to be detected using a digital camera [9]. Objects are positioned under several possible conditions, In this study, 3000 image samples were used. The training process requires a training dataset. The dataset is in the form of images and image labels. The label in the image refers to a bounding box that has the properties of position (x,y), height and width, and the appropriate label. These labels are determined manually by us at the labeling stage with the help of labeling software. Furthermore, the training process is carried out so that YOLO can recognize objects that have been previously determined in the labeling process. This process generates a detection model.

The training model generated from the training stage is too complex to run smoothly on Jetson Nano. So that the model can be run on Jetson Nano more smoothly, the output model of the training results needs to be optimized using Tensor RT. After being optimized, the training result model can be used in the humanoid robot control program (inference stage). This inference process requires image input from the camera and produces an output in the form of a bounding box. This property of the bounding box will later be used to determine decisions in the robot control system, in this case detecting whether there is a ball or not. The program code for YOLOv3 used in this study is based on the code published by Alexey Bochkovskiy at https://github.com/AlexeyAB/darknet.

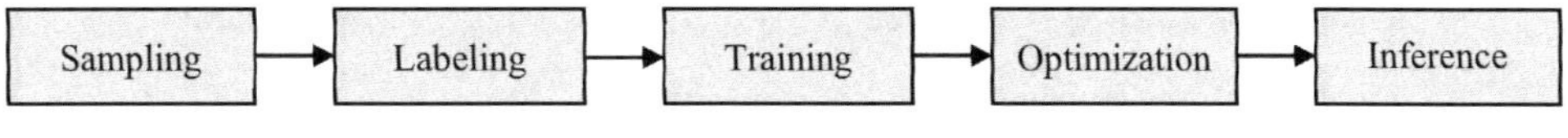

Figure 2. Stages of Object Detection System

The data to be taken in this study is the minimum and maximum distance of the robot in detecting the presence of the ball. A similar study has been conducted by Hakim et al, namely detecting the presence of a QR Code using a quadcopter [10]. Other data that is also observed is data on whether the robot can detect the presence of a ball that is still moving.

The data collection process was carried out three times and then the values were averaged. The purpose of data collection is to determine the maximum distance that can be detected by the humanoid robot detection system. In addition, this study also aims to determine the time required for the robot to detect the presence of the ball.

3. Results and Discussion

The procedure for data collection of the ball detection process against the ball can be seen in Figure 3. To simplify the data collection process, the distance taken in this study is the distance measured from the center position of the humanoid robot's foot to the ball.

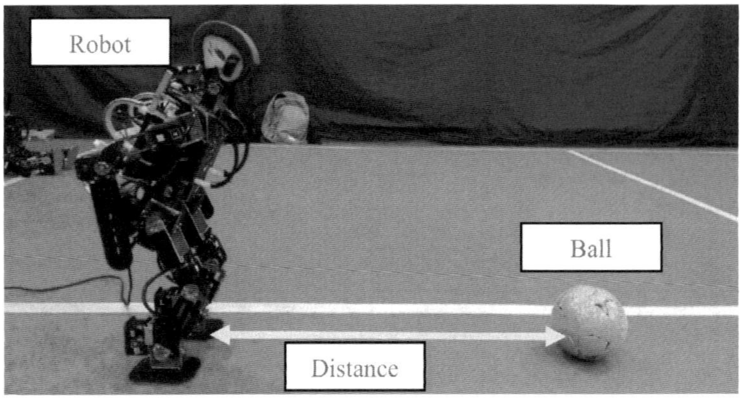

Figure 3. Procedure to test ball detection ability

3.1. Ball Detection Ability

The ball detection method uses YOLO which is specifically made to detect the presence of the ball. In this experiment, the humanoid robot will focus on detecting objects (balls) by labeling samples of ball images. With this training, the robot is expected to be able to recognize objects in various situations that may occur. Figure 4 and Figure 5 show the robot in detecting the presence of the ball when the ball is fully visible and when the ball is partially visible. The two images show that the robot can detect the presence of a stationary ball in various circumstances, so it can be concluded that the ball detection system has worked optimally.

Figure 4. Robot detects a fully visible ball

Figure 5. Robot detects partially visible ball

However, Figure 6 and Figure 7 shows the system has not been able to detect the ball accurately when the ball is moving. The ability to detect a moving ball is needed in actual match conditions, the robot needs to move closer to the ball early so it doesn't lag behind the opposing players. Further research related to the detection of moving balls still needs to be done.

Figure 6. The robot does not clearly detect a slow moving ball.

Figure 7. The robot does not clearly detect a ball that is moving slightly fast.

3.2. Ball Detection Distance and Time

The ball detection distance test is carried out starting from a distance of 50 cm, then 100 cm, and then increasing by 100 cm. Tests show the ball can be detected up to a maximum distance of 900 cm, and at a distance of 950 cm the ball cannot be detected. Figure 8 and Figure 9 show the two conditions at a distance of 50 cm and 900 cm.

Figure 8. The robot detect ball with 50 cm distance

Figure 9. The robot detect ball with 900 cm distance

Figure 10 shows the relationship between the robot's detection time and the distance of the ball from the robot. In general, the detection time required for the robot to detect the presence of the ball is relatively constant. From the picture, it can be seen that the robot can detect the presence of the ball at a distance of 50-900 cm. The average time required for the robot to detect the presence of the ball is 0.033 s.

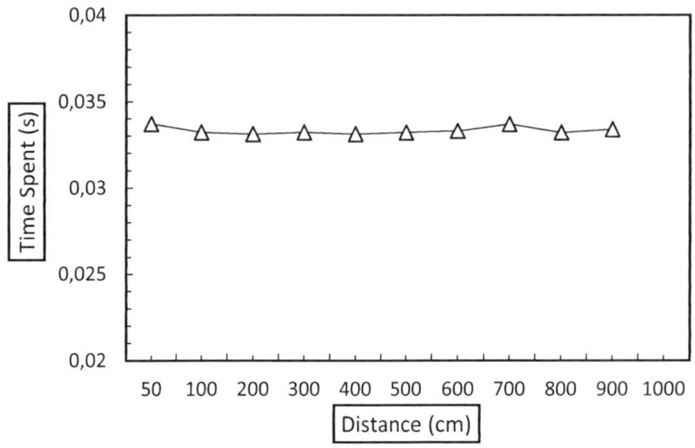

Figure 10. Relationship between ball distance and detection time

4. Conclusion

The ball detection system on the humanoid robot has been successfully built using NVIDIA Jetson Nano as the main computational unit. The test results show that the system built and trained with 3000 image samples can detect balls at a distance of 50 to 900 cm. The average time needed to detect the ball is 0.033 seconds. However, further research related to the detection of moving balls still needs to be done. Furthermore, this system needs to be developed to detect goals, field boundaries and other players.

References

[1] S. Soim, B. Joni, J. Junaidi, and F. Damsi, "Perancangan Robot Humanoid Berbasis Mikrokontroler ATmega 32," *J. Ampere*, vol. 1, no. 2, p. 22, Feb. 2017.

[2] S. Behnke, "Humanoid Robots – From Fiction to Reality ?," *KI-Zeitschrift*, vol. 4, no. December, pp. 5–9, 2008.

[3] M. Harahap *et al.*, "Sistem Cerdas Pemantauan Arus Lalu Lintas Dengan YOLO (You Only Look Once v3)," *Semin. Nas. APTIKOM*, p. 2019, 2019.

[4] B. Rahmat and D. Prayogi, "Perancangan Robot Pensortir Benda dengan Pengenalan Pola Warna Menggunakan Kamera," *J. Appl. Microcontrollers Auton. Syst.*, vol. 4, no. 1, 2018.

[5] A. K. Rizgi *et al.*, "Improving Field and Ball Detector for Humanoid Robot Soccer EROS Platform," in *2018 International Electronics Symposium on Engineering Technology and Applications (IES-ETA)*, 2018, pp. 266–269.

[6] J. Redmon, S. Divvala, R. Girshick, and A. Farhadi, "You Only Look Once: Unified, Real-Time Object Detection," *Proc. 12th Int. Conf. Distrib. Smart Cameras*, pp. 1–6, Jun. 2015.

[7] J. Redmon and A. Farhadi, "YOLOv3: An Incremental Improvement," Apr. 2018.

[8] A. Bochkovskiy, C.-Y. Wang, and H.-Y. M. Liao, "YOLOv4: Optimal Speed and Accuracy of Object Detection," Apr. 2020.

[9] M. Khairudin, S. Yatmono, A. Nugraha, M. Ikhsani, A. Shah, and M. Hakim, "Object Detection Robot Using Fuzzy Logic Controller Through Image Processing," *J. Phys. Conf. Ser.*, vol. 1737, no. 1, p. 012045, Jan. 2021.

[10] M. L. Hakim, S. Yatmono, A. C. Nugraha, and M. Khairudin, "The Effects of Vertical and Horizontal Distance on The Performance of QR Code Detection System," *J. Phys. Conf. Ser.*, vol. 1737, no. 1, p. 012029, Jan. 2021.

Smart Infusion and Web Based Monitoring Infusion Fluids in Isolation Room Based on Fuzzy Logic

M Yamin[1], S A Habir*[2], W O S Nur Alam*[3], and L Surimi[4]

[1,2] Department of Informatics Engineering, Faculty of Engineering, Halu Oleo University, Kendari, Indonesia
[3] Department of Electrical Engineering, Faculty of Engineering, Halu Oleo University, Kendari, Indonesia
[4] Department of Computer Science, Faculty of Mathematics, and Natural Sciences, Halu Oleo University, Kendari, Indonesia

E-mail: [1]muh_yamin@uho.ac.id, [2]salmanalhabir@gmail.com, [3]wdsitinuralam@uho.ac.id, and [4]lasurimi@uho.ac.id

Abstract. High-tech improvements with reference for health advance to develop numerous medical tools have been build to support doctors and nurses performances. This analysis research applied web-based system of infusion monitoring and microprosessor or microcontroller to grease performances of nursers and pharmaceutical helping in clinics or public health centers of the district. In this research, smart infusion web-based is builded for providing for the infusions by applying Node MCU and ESP 8266 Wi-Fi module. Smart Infusion aimed at helping nursers and medical helpers handling infusions of contagious complaint cases or Covid-19 cases to provide for infusion situation in isolation room or emergency room at hospital also public health centers at the district. Detecting the infusion weight, volume and percentage by applying a digital scale and the Load Cell sensors. This monitoring tools or devices is created to count the display from LCD and infusion weight the volumes and percentage of infusion with 3 colors alert display on the Website such as red color, yellow color, and green color depend on a web developed by php language, C programming with Arduino by C language embedded in microprocessor or microcontroller and based on fuzzy logic. Testing results have been achieved, monitoring device can work efficiently and completely when calculating weight, percentage, and volume. Scanning data of success accuracy of Load Cell sensors is 98,00%.
Keywords – *Microprocessor or Micro-controller, Fuzzy Logic, Infusion.*

1. Introduction

Previous studies, [2] and [6], showed that the design of a microcontroller-based infusion fluid monitoring system to assist inpatients in hospitals, which utilizes Wi-Fi module as an integration into the system can monitor the amount of patient's infusion fluids. The analyzing procedure of the infusion fluid is still hand-operated for nurses making decision to visit the patient room for controlling of infusion at any moment. If there are several infected patients in a clinic or environmental health care with toxic disease or Covid-19 patients, authority nurses of the infusion must be in isolation room immediately in purpose for monitoring infusion condition of isolation room of hospital.

Another problem is that nurses lean to be negligent to control infusion fluids for each patient and as result there are critics for the provided services by reason of delays of infusion fluids replacement. Medical staff can cause mistakes for providing intravenous fluids and this is probability that may appear. The error of installation of the infusion can cause obstacle or running out of intravenous fluid with unconscious medical staff. Several cases that are not considered can quickly be fatal to patients. The patient blood pressure in infusion tube will not be balance and it can be blocked by reason of blood clots [6].

Mistakes cannot be underestimated because the latency in infusion fluids replacement can lead to negative risk toward the patient if that is not changed immediately. An monitoring system of intravenous fluid has been built with providing purpose of medical staff with information on the status of the IV fluid in tube. Micro-controller-based monitoring system of infusion fluids to serve patients in hospitals has also been designed. This research is designed for a residual fluid monitoring system based on the Wemos D1 R2 micro-controller. It is able to recommend way for overwhelming delays of replacing infusions fluids [8].

Failures ofinstallation of the infusion fluids cause obstacle or running out of fluid volumes without notification by medical staff. If this is not detected immediately, it will be fatal to patient. Patient blood pressure in infusion tube will not be balanced and can be blocked by reason of blood clots [12].

The devices design uses Node MCU ESP 826612E micro-controller as main control and provides info over Wi-fi network and IR Obstacle Avoidance sensor module. Drip infusion detector can be designed by power supply with +12V and +5V DC output and then making voltage lower using DCDC LM 2596 Step Down Module as a voltage-lowering circuit from DC12VtoDC3.3V [8].

The study refer to presentations of valid investigation of experts and users in valid category. Infusion fluid data can be sent by the sensor to web page with a error measurement of 2,46% from the manual measurement results. MySQL database store data and send data on each 24 seconds and auto-refresh web pages on each 5 seconds [11].

In this research, monitoring device of infusion fluid with module of ESP 8266 Wi-Fi will be created to check remotely infusions without urgency for nurses to visit to patient room and patients will mention nurses without need to get up from bed.

Monitoring system apply Load Cell sensors where Load Cell sensors is a identifying tool for weight. Weight information can be showed at web. Based on web information, staff will take first movement in form of infusion fluid replacement. Based on mention background of problem, author intend to conduct a final project entitled "Smart Infusion and Web-Based Monitoring Infusion fluids in Isolation Room Based on Fuzzy Logic".

2. Method

2.1. Infusion

The requirement of various functional and painless to use medical equipment are very needed in the same way the sophisticated development of technology nowadays, with the development of medical technological instrumentation to help demands of medical devices for serving patients. Infusion fluids for patients is one of them. Infusion is a medical equipment operated under certain conditions providing periodically fluids into patient body. Infusion contain fluids, drugs or nutrients infused intravenously by direct infusion over vein network.

Infusion is chemical liquid infused through bloodstream via intravenous route in certain period. Intravenous fluid is commonly delivered to patient who has fluids reduction of body that requires liquid medicines as treatment. Infusion is applicable antibiotic due to effortlessly consumed by the human body cells by reason of liquid form. The infusion occasionally is used as a backup for patient feeding [10].

Infusion therapy mostly perform method for patients who undergo in hospitalization as a treatment of intravenous IV therapy, blood sampling or blood product, drugs, and fluids.

2.2. Load Cell Sensors

The Load Cell is a Transducer and it is a force that acts found on the foundation of a material transformation by reason of mechanical voltage, reorganize mechanical force in electrical signal. Automatic voltage is based on the inventions of "Robert Hooke" and then he invents link between mechanical voltage and results of deformation named strain. Strain occurs in top layer of material so that it is possible assessed by strain sensor or Strain Gauge [9].

Load Cell is fundamental unit of weighing digital system. It is sensor which develops electrical signal where the magnitude of the signal is evaluated proportionally. Furthermore, developing an output proportionally for the applied loads and forces that can supply detailed measurements of loads and forces, it is converted the strain to the metals to resistance [9] [11].

Table 1. Load Cells Sensors specifications.

Maximum load	Range of Output voltage	Impedance the Input	Impedance the Output	Input Voltage maximum	Range of Operating temperature	Size of Material Aluminum Alloy
5000 grams	a. 0,1mV ~1,00mV/V. b. 1:1000 scale to input voltage, errors margin≤ 1,5%.	1066 Ω ± 20,0 %.	1000 Ω ± 10,0 %.	10 Volt DC	-20 °C ~+ 65,00 °C	60x12,8x12,8 mm, and weight 23gram

2.3. HX711 module

HX711 module is integrated devices unit of a 24bit Avia Semiconductor for converting analog to digital. ADC is developed for weigh digital sensors and industrial control applications connecting to Wheatstone bridge sensors. HX711 module is a weighing module with working principle such as [3]:

- To convert measured changes in resistance changes.
- Then change them to voltages capacity over existing circuit.

The Specifications of HX711 module, such as:

a. There are two ADC channels can be operated for two loading cells with synchronized TTL output serial, SCK and DI.
b. 5Volt DC of voltage operating.
c. Various voltage input ±40mV at full scales 24bit of data accuracy or 24bit ADC.
d. Scanning of frequency and refresh rate 80Hz for consumption less than 10mA.
e. Capacity 38x21mm with a weight of 20grams.

2.4. Node MCU ESP 8266

Node MCU ESP 8266 module is electronic devices board on the ESP 8266 chip with capability of activate micro-controller operations and the connection of Internet by Wi-Fi. There are a few input-output pins in case they can be build up for IOT projects of monitoring and controlling applications. Node MCU ESP 8266 can be set up with Arduino compiler by Arduino IDE. The environmental form of Node MCU ESP 8266. There is mini-USB port in caset making to code programming easily. Node MCU ESP 8266 is a development derivative module of Internet of things platform of ESP 8266 type ESP 12 [7].

ESP 8266 Functions is additional device of Wi-Fi module for microprocessor or micro-controllers such as Arduino so that it can join directly to Wi-Fi to build up TCP/IP connections. This modules require about 3,3volt of power and have three Wi-Fi modes and access point. This modules equipped processor, memory and GPIO where pin numbers depend on ESP 8266 type. The module can stand-alone without any microprocessor or micro-controller because it is already having equipment like the microprocessor or micro-controller, but it can also operate an another microprocessor or micro-controller in this case the Arduino controller [7].

Journal of Physics: Conference Series **2111** (2021) 012056 doi:10.1088/1742-6596/2111/1/012056

Table 2. The following features of the ESP 8266 Wi-Fi module.

a. Software-AP. b. Wi-fi Direct or point to point. c. 802.11 b/g/n. d. 32 bit CPU, 1 MB Flash Memory.	a. LNA, Integrated TR switch, power amplifier, balloon. b. Regulator, DCXO, and Integrated PLL, power management unit. c. Integrated TCP / IP stack protocol	a. SPI, UART, SDIO 1,1 / 2,0. b. 1 × 1 MIMO, 2 × 1 MIMO, STBC	A-MSDU aggregation and 0,4 ms of guard interval and A-MPDU.	a. Standby power consumption < 1,0 mW or DTIM3. b. 19,5dBm output power in 802.11b mode.

2.5. C Language

By using are Microsoft and Linux/Unix OS or perhaps like Cygwin on Windows OS that has preinstalled compilers which are applicable for the C Programming language. C language on various platforms has a lot of compilers for this, serving as [1]:

GCC	CC on Linux/Unix (Turbo C and Miracle C)	Microsoft Visual C++ on Linux/Unix

There are many compilers that can be operated like GCC and it uses a text editor, such as:

Notepad	Emacs	Pico	Vim	Kwort	Nano	Vi	Gedit

The C language structure can be viewed as compilation of various functions, such us: The first operation that must exist in C language is main(). The functions of C language established with a pair curly brace ({) and (}). C language statements can be written between the curly braces [1].

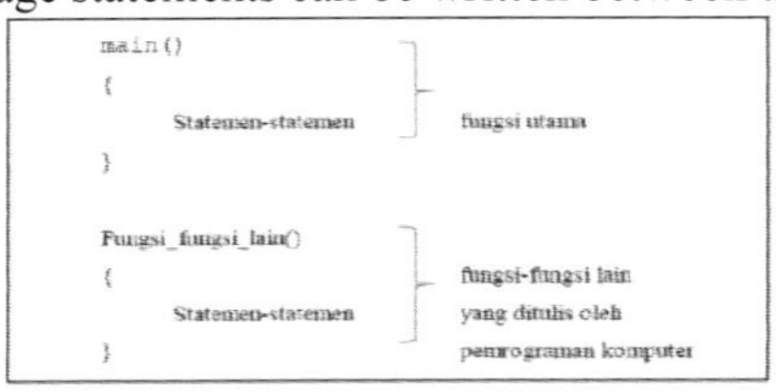

Figure 1. Structure of C Language

2.6. Arduino IDE

Arduino IDE devices is developed by java programming language and is needed with languages of C/C++ library commonly called wiring having function to input-output operations also specifically for programming with Arduino Arduino is developed from processing software remodeling into Arduino IDE. There are three parts of The IDE software, consists of [5]:

1. The function of program editor tool is editing and writing program code in processing languages system. Sketches is the program listing of code on Arduino.
2. The function of compiler tool is a module having duties to replacement the program code in processing languages system into binary and bit code because it is the only system of programming language understood by Arduino IDE at the microprocessor or microcontroller.
3. Upload tool is a module having operations to add bit and binary codes to the memory of the microprocessor or microcontroller.

There are two parts of command structure on Arduino basically, consists of:

- Void setup is contains commands that will be executed only once since Arduino is turned on.
- and Void loop is containing commands that will be executed repeatedly if Arduino is powered on.

2.7. General system for infusion monitoring flowchart

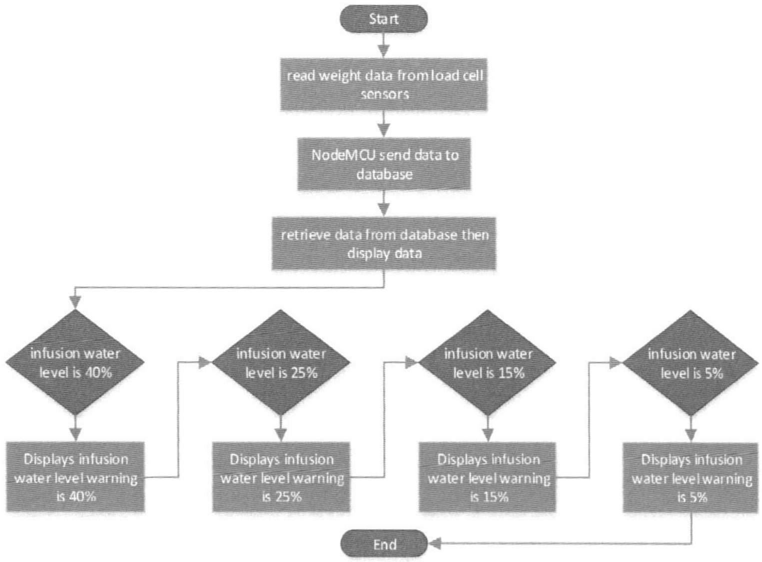

Figure 2. The Infusion monitoring flowchart

3. Results And Discussion

3.1. Fuzzy Logic

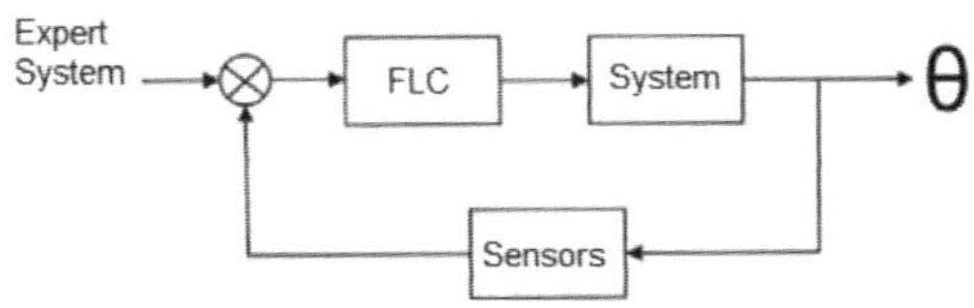

Figure 3. Fuzzy Logic diagram

The fuzzy logic-based infusion monitoring system has input from the expert system, the expert system as fuzzy Logic Control materials after that make the system as PWM input. Before becoming output, the sensors adjust the opening degrees in advance as feedback from the system. This feedback required as a correction factor for the degree of aperture [4].

3.2. Fuzzy Logic Control Design and Mamdani Method

Fuzzified parameters are inputs in the form of drops and weight. Drop values and body weight are divided into four linguistic forms, namely Slow, Medium1, Medium2, Fast and L, M1, M2, H. While the defuzzification result parameter, in the form of output, is the degree (deg) to rotate the hose clamp infusion. The output parameter has the same linguistic form as the input value. Inference process using Mamdani method [4].

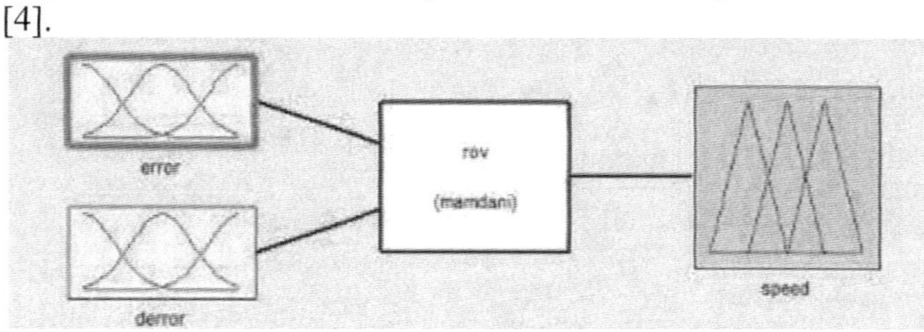

Figure 4. Fuzzy Logic Schematic for infusion using Mamdani method

ICE-ELINVO 2021 IOP Publishing
Journal of Physics: Conference Series **2111** (2021) 012056 doi:10.1088/1742-6596/2111/1/012056

d 2 . d .
d 2 r r. d
2 . d " 11. "
11 D D
D
r r

r d " D" r rd
" RD" r "1 2.1 . .2"
r r r r
d
r r d D
. 1

-
. 2 r . r
r d r
D " r " r d
. " 1 2.1 . . r D . " . dd d r "
" " r r d d"
d . D d r
r d . r r d
r . r d r r r d r . r d
r r r d r . r " r "
r d " " . d
d R d

r D d r 1
r . r " "

r D d r 1
r . r "M "
d 1
d r 1
r . r "M " d 1

3.3. *Smart Infusion Implementation and Application*

In this research, smart infusion implementation the function of the tools or devices main by part will be explained in the Figure 5, as follows.

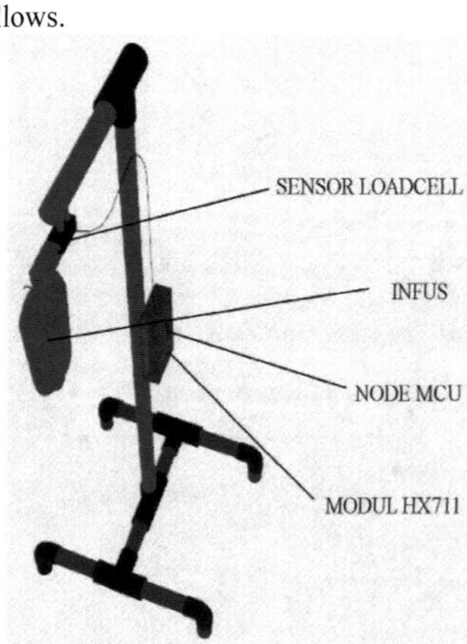

Figure 5. The Device Display and Position of Node MCU

The Load Cell sensors is placed at the top because testing start scanning data from Load Cell sensors like as a weight tool of total infusion fluids weight.

HX 711 module is loacted at toolbox. HX 71 has working principle of converting and filtering measured changes receives weight data from Load Cell sensors. Node MCU is placed at toolbox beside other units. Node MCU has functions as data processor and sends data to database.

3.4. Smart Infusion of Software

The software used by the Node MCU program is Arduino IDE using C language, while for developing website use html programming and php programming language, as follows:

Basic views of database is called db_infus to show main page of database.

r r " "
r "r "
rd ""'
d "d "
r r r rd d rr r
d "D d " . rr r
r
r r
" D r ". r ." R
d 1 "
r R " "
" rr r " . . " r " . rr r

The table name of database is db_infus and databases save data. In status table, there are ID of volume number and weight of fluids infusion. There are three ID's stores. Total weight of fluid infusion is 570,00grams.

Figure 6. The initial views on the Website

Figure 6. Basic views of web, displaying data of detected patient data at isolation room at clinic or community health center. Fluid infusion condition is full.

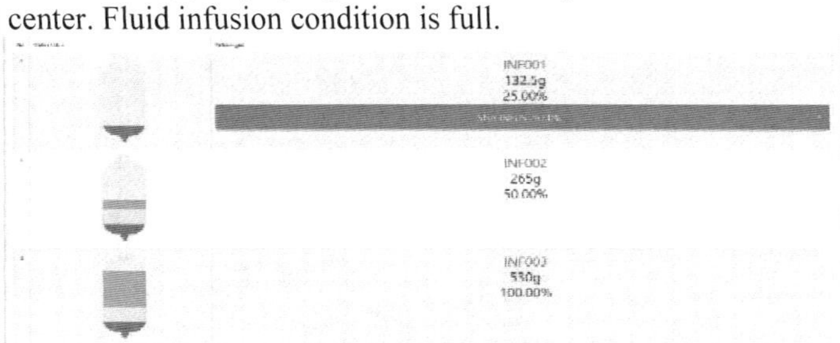

Figure 7. One of alert displays on Website

Figure 7 displays id INF-001 of 25,00%, id INF-002 of 50,00 %, and the id INF-003 of 100%. So, id of INF-001 of 25,00% will be ringing alarm and will stop alarm when volume and percent of fluids infusion is 23,00%.

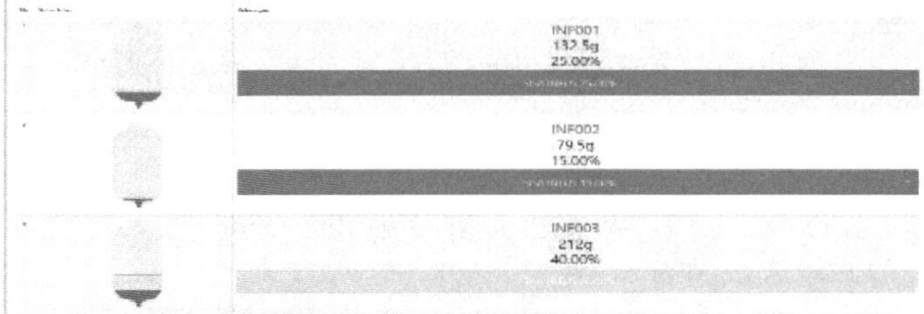

Figure 8. Two alert displays on the Website

Figure 8 displays id INF-001 of 25,00%, id INF-002 of 15,00% and id INF-003 of 40,00%. So, id INF-001 and id INF-002 will be simultaneously ringing an alarm with in a different tone.

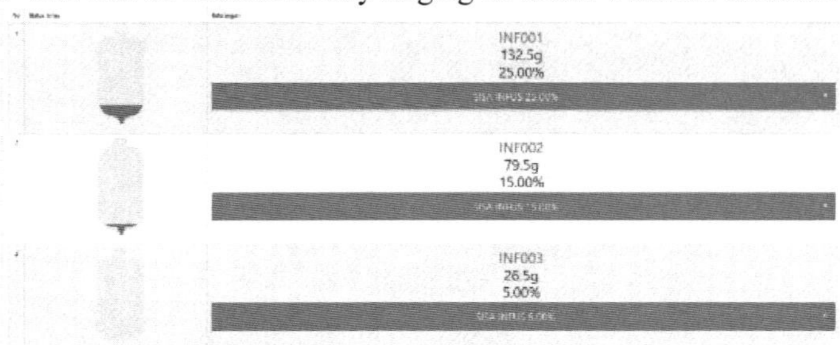

Figure 9. Three alert displays on the Website

Figure 9. displays id INF-001 of 25,00%, id INF-002 of 15,00% and id INF-003 of 5,00%. Then Id INF-001, id INF-002 and INF-003 will simultaneously ringing as an alarm with a various tone. Generally, the display basic web presents three ids such as id INF-001, id INF-002 and INF-003. When an id is selected, a full screen of that id is going to present. Thi situations display the infusion id such as INF-001 and the total weight of the infusion fluid is 530grams, with volume and 100% percentage, pointing out the infusion fluid is still full. This condition displays total weight of fluids infusion, which is 265grams, with a volume and 50,00% percentage, pointing out that the fluids infusion is in fifty-fifty full condition. When total weight of fluids infusion is 212grams, with volume and 40,00% percentage, in this condition the web display shows a yellow color. When the total weight of fluids infusion is 132,5grams, with a volume and 25,00% percentage. In this condition, web display shows a red color which is sign of warning zone and alarm will be automatically ringing until 23,00% of volume left.

When the condition of total weight of fluids infusion is 79,5grams, with volume and percentage of 15,00%. Alarm will ring until volume and percentage of 13,00% fluid infusion. This condition reveal that infusion of standby state is running out soon.

The weight on the Load Cell sensors is 570grams and weight of an empty infusion fluid bottle is 40grams. Conclusion of net weight of infusion is 530grams.

Condition of display of the infusion weight is 0grams. Volume and percentage of fluids infusion is 0%. It indicates that infusion has been run out.

Table 3. Testing of Load Cell Sensors

No	Volume Precentage (%)	Weight of Load Cell (grams)	Weight of Empty Infusion Bottle (grams)	Infusion Net Weight (grams)
1.	100	570	40	530
2.	0	40	40	40

Table 3 the weight on the Load Cell sensors is 570grams and empty weight volume of an infusion bottle size is 40grams. The conclusion of the net weight of the infusion volume view is 530grams.

Table 4. Load Cell Sensors Testing and Notifications on Website

No	Volume Precentage	Warning Colors	Warning on Website
	100 %-40%	Green	visible warning
	40 % -25%	Yellow	visible warning
1.	25 % -15%	Red	visible warning
	15 % -5%	Red	visible warning
	5 % -0%	Red	visible warning
	100 %-40%	Green	visible warning
	40 % -25%	Yellow	visible warning
2	25 % -15	Red	visible warning
	15 % -5%	Red	visible warning
	5 % -0%	Red	visible warning
	100 %-40%	Green	visible warning
	40 % -25%	Yellow	visible warning
3	25 % -15	Red	visible warning
	15 % -5%	Red	visible warning
	5 % -0%	Red	visible warning
	100 %-40%	Green	visible warning
	40 % -25%	Yellow	visible warning
4	25 % -15	Red	visible warning
	15 % -5%	Red	visible warning
	5 % -0%	Red	visible warning
	100 %-40%	Green	visible warning
	40 % -25%	Yellow	visible warning
5	25 % -15	Red	visible warning
	15 % -5%	Red	visible warning
	5 % -0%	Red	visible warning

4. Conclusion

Some conclusions can be drawn this research as follows:

1. Node MCU devices, Wi-Fi Module for Web Based ESP 8266 devices and Microcontroller base on Fuzzy Logic can monitoring also calculating the infusion fluids weight, percentage, and volume reading the data of the infusion with three colors indicators, namely colors are red, yellow, and green, with the volume and percentage of infusion fluids 100,00% - 40,00% is green color, fluids infusion volume 40,00% - 25,00% is yellow color, and 25,00% - 0% is red color.
2. The Load Cell sensors devices are successful to calculate sizes of infusion fluids weight. Scanning data from the percent and detailed volume of the Load Cell sensors is 98,00% successfully by weight infusion fluids volume and percentage.

Journal of Physics: Conference Series **2111** (2021) 012056 doi:10.1088/1742-6596/2111/1/012056

References

[1] Joni, Budi Raharjo 2011 Pemrograman C dan Implementasinya. Bandung: Informatika.

[2] Nataliana D, Taryana N, et al 2016 Alat Monitoring Infus Set Pada Pasien Rawat Inap Berbasis Mikrokontroler Atmega 8535. Elkomika: Jurnal Teknik Energi.

[3] Nuryanto, Rudi 2015 Pengukur Berat dan Tinggi Badan Ideal Berbasis *Arduino*. Jurnal Jurusan Teknik Elektro Fakultas Teknik Universitas Muhammadiyah Surakarta.

[4] Novian L W, M. Ramdlan Kirom, Reza F 2016 *Design and Application Expert Sistem for Regulatory Infuse Drip Based on Fuzzy Logic*. *e-Proceeding of Engineering,* Vol.3 No.1

[5] Prastyo Aji S 2018 Alat *Monitoring* Tetesan Infus Menggunakan *Web* Secara *Online* Berbasis ESP8266 dengan Pemrograman *Arduino IDE*.

[6] Riskitasari S, Hamida F, et al 2016 Simoci (Sistem Monitoring Cairan Intravena) Di Ruang Icu Menggunakan Zigbee. Prosiding SNST Fakultas Teknik, 1(1).

[7] Salapathy, Lait M, Samir K B 2018 *Arduino Based Home Automation Using Internet of Things (IoT)*, India. *International Journal of Pure And Applied Mathematics*.

[8] Tony Kusuma, Muhammad Tirta Mulia 2018 Perancangan Sistem Monitoring Infus Berbasis *Mikrokontroler Wemos D1 R2.*

[9] Wicaksono Y A 2017 Sistem *Monitoring* Infus Menggunakan *Load Cell* Berbasis *Mikrokontroler Atmega8535* Dan *Web* (Studi Kasus di Rumah Sakit Bhakti Wira Tamtama Semarang), Vol 10, No.1.

[10] Wadianto, W Fihayah Z 2016 Simulasi Sensor Tetesan Cairan, Pada Infus Konvensional. Jurnal Kesehatan, 7(3), pp.394-401.

[11] Yanuar Arif Wicaksono 2017 Sistem *Monitoring* Infus Menggunakan *Load Cell* Berbasis Mikrokontroler Atmega8535 Dan *Web* (Studi Kasus di Rumah Sakit Bhakti Wira Tamtama Semarang), Jurnal Elektronika dan Komputer.

[12] Zainuri A, Santoso D R, et al 2012 Monitoring dan Identifikasi Gangguan Infus Menggunakan Mikrokontroler AVR. Jurnal EECCIS, 6(1), pp.49-54.

AUTHOR INDEX

Agustianto, T ... 413
Alam, W O S Nur ... 447
Alfatih, M F ... 1
Ali, M ... 262, 329
Ali, Muhamad ... 20, 303
Anggara, N ... 188
Anggraeni, A A ... 196
Ardiansyah, M N ... 221
Arifidin, Muhammad Afif Amalul ... 403
Arifin, F ... 374, 383, 391
Arifin, Zaenal ... 303
Asnawi, Rustam ... 398
Atmoko, N T ... 94
Azis, S M ... 76
Aziz, F N ... 213
Badarudin, R ... 311, 329
Bahry, Noer Aden ... 117
Barahama, A D ... 239, 255
Budiyanto, A M ... 221
Damarwan, E S ... 286, 339, 413, 426
Darmawan, I A ... 432
Deny, P S ... 20
Dewanto, Satriyo Agung ... 133
Djaohar, M ... 420
Djatmiko, I W ... 84
Efrizon ... 188, 204
Ernawan, Ferda ... 279
Fardiansyah ... 279
Fatkhurrokhman, M ... 432
Fiducia ... 247
Gunawan, W ... 358
Gusti, W R ... 383
Habir, S A ... 447
Hadi, S ... 150, 231, 262
Hadinata, Alan ... 109
Hakim, M L ... 339, 413, 442
Hakim, Ml ... 426
Hamid, M A ... 432
Hancoyo, W ... 1
Hariyanto, D ... 329, 348, 364
Hartoyo ... 319, 403
Haryanto ... 157, 262
Hatmojo, Y I ... 76, 311
Hertanto, Deny Budi ... 398
Hidayatulloh, Indra ... 279
Husna, A F ... 262
Indrawati, B ... 311
Indrihapsari, Y ... 221
Irmansyah, M ... 188
Irmansyah, Muhammad ... 204
Jamaldi, A ... 94
Jati, H ... 221
Jati, Handaru ... 101, 133
Jayadi, F ... 144
Kassymova, G K ... 348
Khairudin, M ... 442
Khairunnisa ... 403
Kholis, N ... 339
Kusuma, Wahyu Mustafa ... 157
Laras, Bt Djoko ... 20
Lempas, J F ... 255
Lubis, A ... 420
Madona, Era ... 188, 204
Mahali, Muhammad Izzuddin ... 133
Mahir, I ... 358, 420
Malik, C A J ... 150
Manggalasari, L C ... 61
Marausna, G ... 144
Maruanaya, G J T ... 364
Maruanaya, R F ... 364
Maryadi, T H T ... 286, 311
Masek, Alias ... 51
Mashoedah ... 109, 127, 213, 279
Mataram, Nalendro ... 117
Meunmany, S ... 329
Muhfizaturrahmah ... 20
Muslikhin ... 391
Muslim, S ... 175
Mustaqim, I ... 166
Mustopa, Acep Hidayatul ... 294
Mutohhari, Farid ... 51, 157
Nasuha, A ... 374, 383, 391
Nasution, Anggara ... 204
Neng-Shu, Y ... 262
Noperi, S ... 413
Noviansyah, W ... 61
Nugraha, A C ... 84, 413, 442
Nugraha, Nikko Aji Bayu ... 127
Nurkhamid ... 221
Nurrohkayati, A S ... 94, 117, 179
Nursita, Y M ... 231
Nurtanto, M ... 432
Nurtanto, Muhammad ... 51, 157
Nuryanto, Apri ... 279
Parjiman ... 420
Pramono, H S ... 166

Pramono, Herlambang Sigit 9
Pranoto, Sigiet Haryo 117
Prasetiyo, Erwan Eko 36
Prasetya, T A 61
Pratama, G N P 28, 43
Pratiwi, H 413
Priambodo, A S 374, 383, 391
Prianto, Eko 9
Purwaningsih, S 196
Rahardjo, I A 175, 358, 420
Rahman, S A 432
Renaldo, R Y 268
Reza, F S 144
Rifan, M 175
Rizqon, Muhammad Fatih 101
Rochayati, Umi 279
Rozaq, Fadli 51
Sa'Adilah 319
Saifullizam, B P 319
Sakirah 175, 358, 420
Saputra, Bagas Woro 70
Saputro, Nuruddin Wahyu Eko 117
Sarjito 94
Setialana, P 221
Setiawan, Doni 294
Setiawan, Henri 294
Setiawan, N 43
Setiawan, Nurman 398
Setiyawan, A 61
Soenarto, S 255
Sony, Arya 279
Subekti, M 175, 358, 420
Suhanda 294
Sukir 319
Sukisno, T 286
Sunaryo 319
Sunawar, A 175
Suprapto 247
Supriyadi, Edy 268
Surimi, L 447
Surwi, Faranita 398
Suyitno, Suyitno 51, 157
Suyitno 358
Syach, Sabaruddin 117
Towip, Towip 61
Triasih, T 348
Vanany, Iwan 179
Wadjdi, F 175
Wardani, R 221, 239
Wardhana, A S J 262, 286, 339, 413, 426
Wardhana, Alex Sandria Jaya 70
Wati, M S 348
Wijayanto, R P 358
Winursito, A 28, 374, 383, 391
Yamin, M 447
Yaqin, Rizqi Ilmal 117
Yatmono, S 84, 413, 442
Yuchofif 9
Yultrisna 188
Yuniarti, Nurhening 70
Zakir, I 420
Zamtinah 166, 348